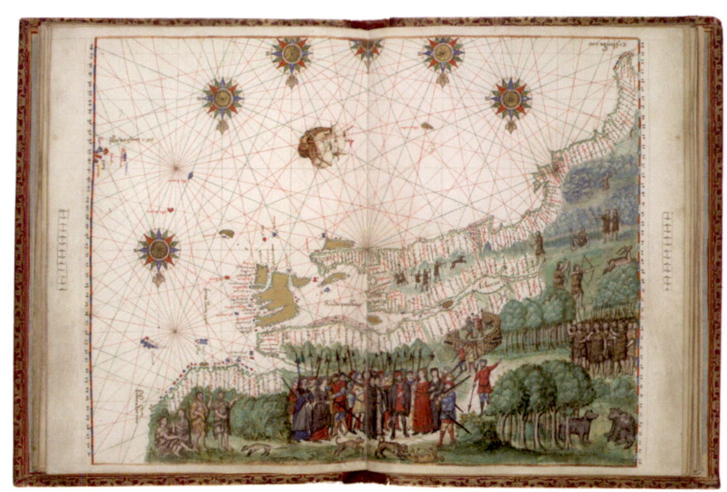

(Front endpaper)
Map from Dieppe attributed to Nicolas Vallard, 1547
Taken from an atlas by Nicolas Vallard, who is unknown to researchers, this map is a good example of the talents of Norman cartographers of the mid-sixteenth century. The contours, colours, and fine lines combine to make a complete work of art. The illustrations scattered across the St. Lawrence Valley portray Roberval's voyage to Canada and his aborted attempt at colonization. Shown are two groups of soldiers, European and Indian, both heavily armed. In the background is a fort well protected by cannons, testifying to the importance of armaments in all colonization initiatives. A number of place names, French and Portuguese, appear on the coasts, indicating that numerous fishermen from the kingdoms of France and Portugal visited North American waters in the sixteenth century. Other names along the St. Lawrence River are evidence of the presence of Indians, with whom the Europeans traded not only material objects but also geographic information.

(Back endpaper)
Map of New Belgium by Nicolaes Visscher, Amsterdam, 1655
This map by Nicolaes Visscher portraying New Belgium in the mid-seventeenth century testifies to the Dutch presence in North America at that time. The colony, also known as New Holland, covered parts of a number of today's American states, including New York, New Jersey, Pennsylvania, Connecticut, Rhode Island, and Vermont. The territory mapped is sprinkled with place names that have survived in some form to the present day, including LANGE EYLANDT (Long Island), MANHATTANS, STATEN EYL., BREUKELEN (Brooklyn), VLISSINGEN (Flushing), BLOCK ISLAND, KATS KILL (Catskill), and ROODE EYLANDT (Rhode Island). At the mouth of the Hudson River (GROOTE RIVIER), on the southern tip of Manhattan, is the capital, Nieuw Amsterdam, which became New York when it fell into English hands in 1664. At the bottom of the map is one of the oldest engravings of the city – quite a contrast with today's views of towering skyscrapers. Farther north on the Hudson River is Fort Orange (Albany), which provided the Dutch West Indian Company with a supply of furs from the Indians, especially the Mohawks and Mohicans (on the left is a portrayal of two of their fortified villages).

Publisher's note: In the legends, names in small capitals are taken from the corresponding map.

MAPPING A CONTINENT

Raymonde Litalien • Jean-François Palomino • Denis Vaugeois

MAPPING A CONTINENT

Historical Atlas of North America 1492–1814

Translated by Käthe Roth

This book was produced in collaboration with Bibliothèque et Archives nationales du Québec

McGILL–QUEEN'S UNIVERSITY PRESS

The director of Éditions du Septentrion, Gilles Herman, joins the authors in thanking the president and CEO of Bibliothèque et Archives nationales du Québec (BAnQ), Lise Bissonnette, for her enthusiastic endorsement of this project and for having provided unlimited and free access to the various collections in her institution, including those from the former Bibliothèque nationale du Québec and Bibliothèque centrale de Montréal and those resulting from the merger with the Archives nationales du Québec. The authors would like to emphasize the great availability of the general director of conservation, Claude Fournier, and thank him for his unfailing support, and they appreciate the great efficiency and personal involvement of Sophie Montreuil, director of research and publishing. She skilfully coordinated the work of many experts both outside and inside the BAnQ who were associated with production of this atlas, the authors, and the Septentrion team. Acknowledgment is also due to the following people at the BAnQ: Pierre Perrault, photographer; Michel Brisebois, expert in antique books; Marie-Claude Rioux, restorer; Monique Lord, archivist; and Michèle Lefebvre and Carole Melançon, research clerks. Finally, the support provided by Isabelle Crevier in the many interactions with institutions to which the authors made requests and the discreet assistance of Éric Bouchard were very much appreciated.

Illustrations research was helped greatly by the valuable contribution of Louis Cardinal, archivist at Library and Archives Canada; Ann Marie Holland, librarian at the Rare Books Department of McGill University; and Normand Trudel, conservator at the Stewart Museum.

Éditions du Septentrion thanks the Canada Council for the Arts and the Société de développement des entreprises culturelles du Québec (SODEC) for the support provided to its publishing program, and the Government of Québec for its Programme de crédit d'impôt pour l'édition de livres. We also acknowledge the financial assistance of the Government of Canada through its Canadian Book Publishing Development Program.

On the cover: centre, *Carte de l'Amérique septentrionale pour servir à l'histoire de la Nouvelle-France*, by Jacques-Nicolas Bellin, 1743 (BAnQ, Gagnon 971.03C478hi2); top, *De l'usaige de la presente arbaleste*, by Jacques de Vaulx, 1583 (BNF, Manuscrits occidentaux, Français 150); bottom: *Mar del Sur. Mar Pacifico*, by Hessel Gerritsz, 1622 (BNF, Cartes et Plans, Ge SH Arch 30 Rés).

Back cover: *[River of St. Lawrence, from Cock Cove near Point au Paire, up to River Chaudière past Quebec]*, by Joseph Frederick Wallet DesBarres (London: 1781), engraved map (BAnQ, G 3312 S5 1781 D41).

Publisher: Gilles Herman

Concept and general editor: Denis Vaugeois

Coordination of cartographic research: Jean-François Palomino (BAnQ), also principal author of the legends

Scientific advisors: Claude Boudreau, Louis Cardinal (Library and Archives Canada), Catherine Hofmann (Bibliothèque nationale de France), Jacques Mathieu (Université Laval), Hélène Richard (Bibliothèque nationale de France)

Selection and processing of illustrations: Jean-François Palomino (BAnQ), Denis Vaugeois

Design and layout: Folio infographie

Translation into English: Käthe Roth

Editing: Joan Irving and Jane Broderick

Indexing: Roch Côté

Permissions: Sophie Imbeault

Editorial contributors: Roch Côté, Julien Del Busso, and Sergio Paccioni

Bibliothèque et Archives nationales du Québec and Library and Archives Canada cataloguing in publication

Litalien, Raymonde

Mapping a continent [cartographic material]: historical atlas of North America, 1492–1814

Scales differ.

Translation of: La mesure d'un continent.

"This book was produced in collaboration with Bibliothèque et Archives nationales du Québec."

Includes bibliographical references and index.

ISBN 978-2-89448-527-9

1. North America - Historical geography - Maps. 2. North America - Discovery and exploration - Maps. 3. North America - History - Sources. I. Palomino, Jean-François, 1976- . II. Vaugeois, Denis, 1935- . III. Bibliothèque et Archives nationales du Québec. IV. Title.

G1106.S1L5713 2007 911'.7 C2007-941540-7

1300, av. Maguire
Sillery (Québec)
G1T 1Z3
www.septentrion.qc.ca

Distribution in Canada:
McGill-Queen's University Press
c/o Georgetown Terminal Warehouses
34 Armstrong Avenue
Georgetown, Ontario
L7G 4R9

Bibliothèque et Archives nationales du Québec
3rd quarter 2007
ISBN 978-2-89448-527-9

Contents

Preface

Between the milestone of 1814, when this historical atlas of North America concludes, and the moment when readers will open the covers of this book is a span of two centuries. What is offered here is therefore not the ever-fascinating observation of progress in cartography, one of the most established sciences of our times; rather, readers are invited to immerse themselves in a way of learning that we now find totally foreign, even in the era of space exploration.

In books on the history of New France, we have often encountered the people, both famous and unknown, who mapped North America, from *coureurs des bois* to intendants, from traders to missionaries. Very few of them were geographers, and we do not know how each contributed to the drawings of North America – from the most imprecise to the most accurate – the contours and relief features of which were finally sketched out completely in the early nineteenth century. The division of labour in the search for knowledge, today an unbreachable wall between amateurs and professionals, was nonexistent. Science was not a thing apart, but a product of life and reality.

Similarly, we find in these pages bygone modes of research, conducted in the open air. The progress of science was fuelled by the risks that hundreds of individuals assumed and the perils that they faced as they advanced, in disorganized waves, into territories that often seemed unlivable. Today, when Earth is delivering its most deeply buried secrets to observation satellites, it is touching to discover in humble, thin lines, committed to now-yellowed paper as their makers rode the watercourses of North America, a river that was navigated by human beings, a delta ten times explored before being drawn accurately, and even traces of a camp that survived the worst conditions our climate has to offer.

We are reminded of the slow, steady pace of discovery, in sharp contrast to today's impatience with the few mysteries remaining unsolved. This atlas traces a magnificent centuries-long itinerary in the development of the depiction of North America, the drawing back of the curtain from east to west to reveal a sort of eternity ahead. Along the coasts, where, at the beginning, there were many trials and errors of perspective, the lack of precision gives an indication of the immensity of the cartographers' task. It took generations for Europeans to reach the Pacific. And this was far from the end of the story, because at the same time knowledge of the geography of the centre of the continent was growing, and an infinite number of relief features were being measured. Today's children, who can find their way anywhere using GPS devices, who will never have to learn the points of the compass, should at least be taught this long history.

Although it ends at the dawn of the Romantic era, this book provides a superb illustration of one of the period's greatest inspirations. Most of those who created nineteenth-century works of literature, music, and art did not travel, but their imagination was fired by accounts of voyages. In their libraries appeared works whose descriptions of a new world fed their vision of new peoples. And if this world could not be attained, it could be experienced as a form of art. Maps, with their pen lines, colours, calligraphy, and illustrations of people and animals in fabulous lands, became aesthetic objects. As Samuel de Champlain wrote in 1632 in his dissertation on the duties of the good seaman, "... a little skill in drawing is very necessary, and the art should be practised." The accomplished drawings presented here will affect the reader much the way they did the Romantics. The North America of the early cartographers, before it was fixed and known from every angle, was a world partly of the imagination – a space in which fiction was still possible. It is not by chance that, in the twentieth century, the most powerful works of abstract art often resembled invented continents. Reality does not always hold dominion.

Our institution wanted to be part of the adventure of this book from the very beginning. Over the last decade, mergers have made the Bibliothèque et Archives nationales du Québec a unique site of convergence as guardian of the collective memory of Quebec in all its forms and media, and a major disseminator of historical, scientific, and literary history, now also through its blossoming virtual library. In this book, we found a complement to the diversity of our work. Both erudite and lucid, both elegant and accessible, *Mapping a Continent* was brought into being by a group of researchers and writers who are not only vastly experienced but also inhabited by a mission similar to ours. At a time when 400 years of the French presence in North America is being celebrated on both sides of the Atlantic – an ocean that was crossed and crossed again by the individuals described in this book – our collections and those of our partners will have given true meaning to the expression "living memory." All of those involved in the making of this book deserve our gratitude. Thanks to them, seeing and reading about North America is still a journey of our times.

Lise Bissonnette
President and CEO
Bibliothèque et Archives nationales du Québec

TRI
ONAL
TERA FRIGIDA
OCCEANVS OCCIDENTALI
RRA FIRME
NOVVS
PEL

Introduction

Before the sixteenth century, the inhabitants of Europe, Asia, and Africa had no idea that the Americas existed. Fifteen thousand years earlier, however, Siberian populations had begun to cross the narrow strait separating Asia from North America. As the landmass, not very habitable compared to other parts of the planet, gradually warmed, the Siberian hunters followed the game southward. Some groups turned back northward and settled on land that was suitable for farming, such as that in the Great Lakes region, or took up a nomadic life, roaming the entire vast, game-filled continent. Great civilizations were built over the two millennia preceding the Christian era. Only through archaeology can we gain access to a few of the secrets of this remote prehistoric period.

Western Europe was unaware, until the late fifteenth century, of these emigrants who had slipped away from the eastern tip of Asia and the territories through which they travelled. The exception was the Vikings, who settled temporarily on the Atlantic coast of North America at the turn of the second millennium. The small colonies that clung briefly to nameless shores did not pique sufficient curiosity in Europe to trigger a wave of exploration, and so they faded into oblivion. Neither a written account nor a first-hand map provides evidence of this episode in history.

In the second half of the fifteenth century, unaware of the wanderings of their predecessors several centuries earlier, northern Europeans landed on the same shores in search of food for their growing urban populations. French and English fishermen followed the ocean currents along the coast of Iceland and Greenland toward great shoals of fish, and they came upon "newfound landes." At the same time, the Spanish and Portuguese began to dream about a western route to China and India – the "old countries" that purveyed silk, spices, and other valuable goods – and looked to the landmasses emerging from the Atlantic Ocean, which appeared on early maps as a few islands with passages between them that should have led to Asia. But these lands were soon revealed to be a vast continental barrier whose size exceeded the sailors' wildest dreams. With perceptions of the world in upheaval, those who made world maps drew on their prodigious imaginations to make room for America and to trace its contours according to the presumptions of their times.

Unlike South America, North America, the subject of this book, did not offer the fascinating prospect of coveted riches to be mined. The trade in cod and beaver pelts, however, became sufficiently profitable to justify the establishment of colonies, and the settlers also turned to agriculture. From these initial settlements, for the next three hundred years explorers from all over Europe pushed across North America, hoping to find a western sea by which they could cross to Asia. Their voyage accounts revealed the geography of the continent, the contours of which European cartographers were constantly redrawing and correcting. The major zones of European influence took shape. The Portuguese shared South America with the Spanish; the latter also conquered Central America, through which they reached the shores of the Pacific Ocean. The English, who had initially been attracted to the fishing resources on the Grand Banks off Newfoundland, claimed this region and established colonies along the Atlantic coast. They also became tenacious explorers of the Northwest Passage, venturing from their Hudson Bay trading posts across the Arctic and finally reaching the Pacific Ocean. The French, convinced that they had made the best choice by colonizing the Gulf of St. Lawrence and the St. Lawrence River, criss-crossed the interior of the continent, heading not only northward but also west-

Opposite
Detail of a map from the Miller Atlas (see complete map, pp. 34–35).

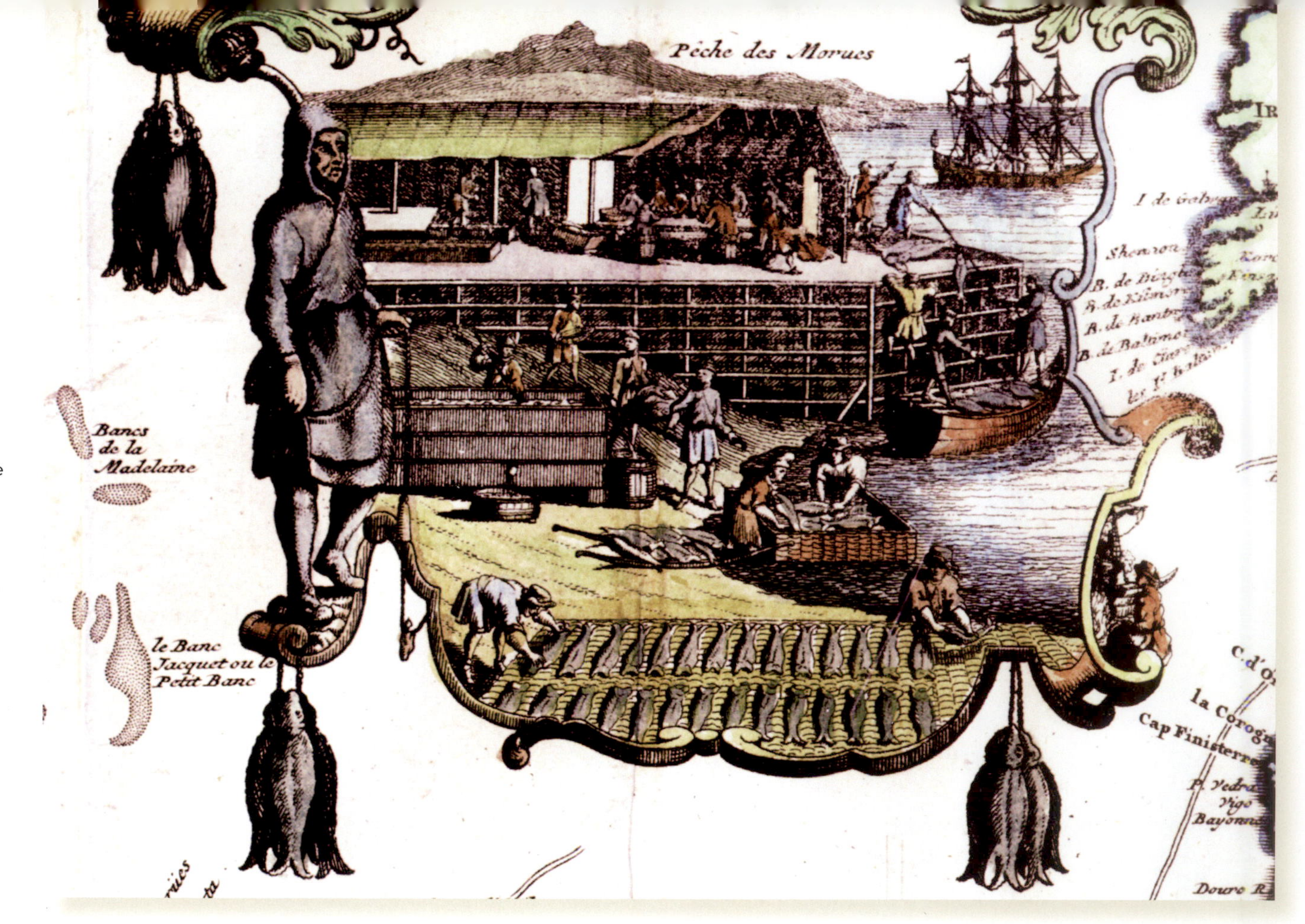

The fishing industry
This scene portraying the processing of cod was first engraved, by Nicolas Guérard, for a wall map of North America by Nicolas de Fer that was widely copied in the eighteenth century. It comes from a map by Henri Abraham Chatelain (seen in its entirety on pp. 142–43). At the time when this map was published, the fishing industry was booming on the Grand Banks off Newfoundland. Codfish brought to the shores of the island were scaled, the heads were removed, and they were gutted, filleted, washed, salted, and dried before being transported to Europe.

ward, as far as the foothills of the Rocky Mountains, and southward, to the mouth of the Mississippi River, establishing trading posts and forming lasting alliances with Indian nations. Because New France existed for 250 years, the French are a constant presence in this book. They marked their passage by providing places with names, many of which had an Indian origin, and they formulated and drew maps of the entire territory. People from other countries, in smaller groups and as individuals, also explored North America, including the Dutch of Nieuw Amsterdam (New York) in the seventeenth century and the Russians who explored and described the Bering Strait and the Pacific coast in the eighteenth century.

This world, which Europeans called "new," had been inhabited by the descendants of emigrants from Siberia for several millennia. Thus, siblings met without recognizing each other, though it was inevitable that fate would bring them together. The Europeans' impressions of the "Indians" have been extensively recorded. What the Indians thought of the Europeans is more difficult to discern, for no primary sources exist. Nevertheless, by following events and the testimonials of explorers and missionaries, historians have observed that the indigenous peoples did what they could to accommodate the newcomers – their goals being to trade with them and to solicit their military assistance. They quickly understood that the colonizers' main objective was to cross the continent to the "western sea" and the "southern sea," financing their explorations through the fur trade and mining. But who was to pass information to the explorers, guide them, feed them, and provide logistical support if not the "Savages," who were so well adapted to their environment? With their valuable assistance, explorers gained access to distant territories and pushed back the mapped borders of *terra incognita*.

For Europeans, *terra incognita* was a space with no claims on it. They could therefore name it, appropriate it, and use it as they wished. They imported their place names and their administrative and social structures, and the Indians found themselves driven off their ancestral lands. The English and French also imported their conflicts from their homelands to North America, where they then defined colonial boundaries.

More than any other source, cartography is generous in providing a record of the advance of knowledge about North America, the mobility of the borders drawn, and the economic, political, and military stakes faced by the would-be masters of the continent. Often produced not before but after exploration, a war, or a treaty, maps formed an archive that pooled the known data. They were constituted from a wealth of written and oral documentation, thus forming a foundation for historians by offering reports, in words and images, of a situation at a specific moment in time.

The authors of *Mapping a Continent: A Historical Atlas of North America, 1492–1804* wanted to write history through maps by extracting their quintessence. Following this cartographic approach, each part of the book deals with knowledge and facts that accumulated over time. Although the four parts reconstruct, broadly, the chronological history of different geographic regions of North America, there is some overlap in the description of contemporaneous events to express their inter-

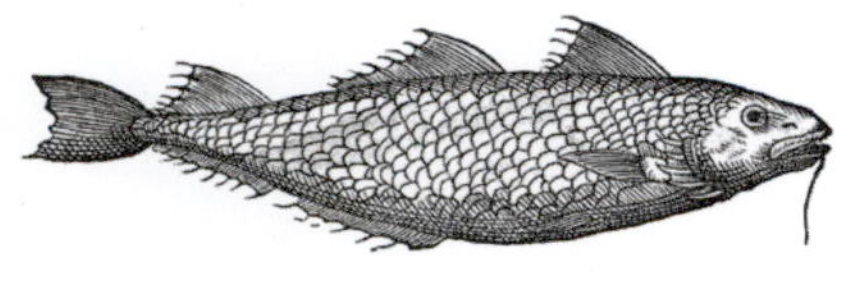

The codfish
For several centuries, cod fishing was the most lucrative commercial activity in North America. These two illustrations of codfish are taken from the *Codex canadiensis*, attributed to Louis Nicolas (top), and *Delle navigationi* by G. B. Ramusio (1556) (bottom).

dependence. As long as the borders were not set in stone, explorers and cartographers roamed freely from one end of the continent to the other. For them, "borders" were neither political nor administrative, but physical barriers to be overcome and crossed as they advanced toward the unknown. And advance they would, until they encountered an ocean, the ultimate frontier acceptable to an inquisitive mind. By bringing to light the movements of people throughout the space of North America over four centuries, we gain a broader perspective on the continent's history – that of a "new" world as dubbed by that of an "old" world that emigrated from eastern Europe and Asia.

The authors chose 1814 as the year to end *Mapping a Continent*. Although North America was not completely mapped by then, it is generally accepted that, following the Louisiana Purchase (1803) and with the synthesis resulting from Lewis and Clark's expedition, maps had fixed the contours of the continent, as well as the mountain ranges, watercourses, lake networks, and other major geographic features. Also in 1814, the Treaty of Ghent put an end to the conflict between the British and the Americans, a decisive step in finalizing the Canada–United States border. For the nineteenth century, during which remaining gaps in the maps would be filled, we provide an overview of some major stages in exploration of the Arctic and the Pacific coasts.

Several hundred old maps are reproduced in this book. The general bibliography is complemented by short lists of sources at the end of each chapter. These are only some of the references that the authors used; they also consulted various archives and their own previous work. It should be noted that the writing of the texts and legends, although assumed totally by each of the respective authors, was the fruit of detailed planning during meetings, almost-daily exchanges of e-mails, and often passionate discussions that resulted in a product that is both original and homogeneous.

The authors gratefully acknowledge the involvement of the executive level at the Bibliothèque et Archives nationales du Québec and especially its president and CEO, Lise Bissonnette, thereby confirming the institution's active participation in this intellectual production. The huge collection of old maps on which *Mapping a Continent* is based was complemented by a judicious selection of pieces from a number of other institutions in North America and Europe, prominent among them the Bibliothèque nationale de France, which generously agreed to contribute to the creation of this monumental work. Sophie Montreuil coordinated the project with consummate professionalism. Along with her colleagues in the research and publishing department of the Bibliothèque et Archives nationales du Québec, she kept a close eye on the texts as they were written, and during the sometimes heated debate among the authors, she was an effective moderator who never lost sight of the overall concept of the project. Denis Vaugeois, an experienced and indefatigable historian; Jean-François Palomino, a young yet knowledgeable map librarian; and myself, the archivist, thus pooled our knowledge and experience to create this book on the history of knowledge of North America and its occupation by humans.

Raymonde Litalien

The beaver
Illustrations of beavers taken from *Carte tres curieuse de la mer du sud* by Chatelain (top) and the *Codex canadiensis* (middle), and two "beaver" hats from *Castorologia* (bottom). Along with cod, beaver was the main economic attraction of North America. Contrary to popular belief, beaver fur was used not for protection from the cold but to produce a felt that was used to make various types of fashionable hats.

I

LANDING IN NORTH AMERICA

SIXTEENTH CENTURY

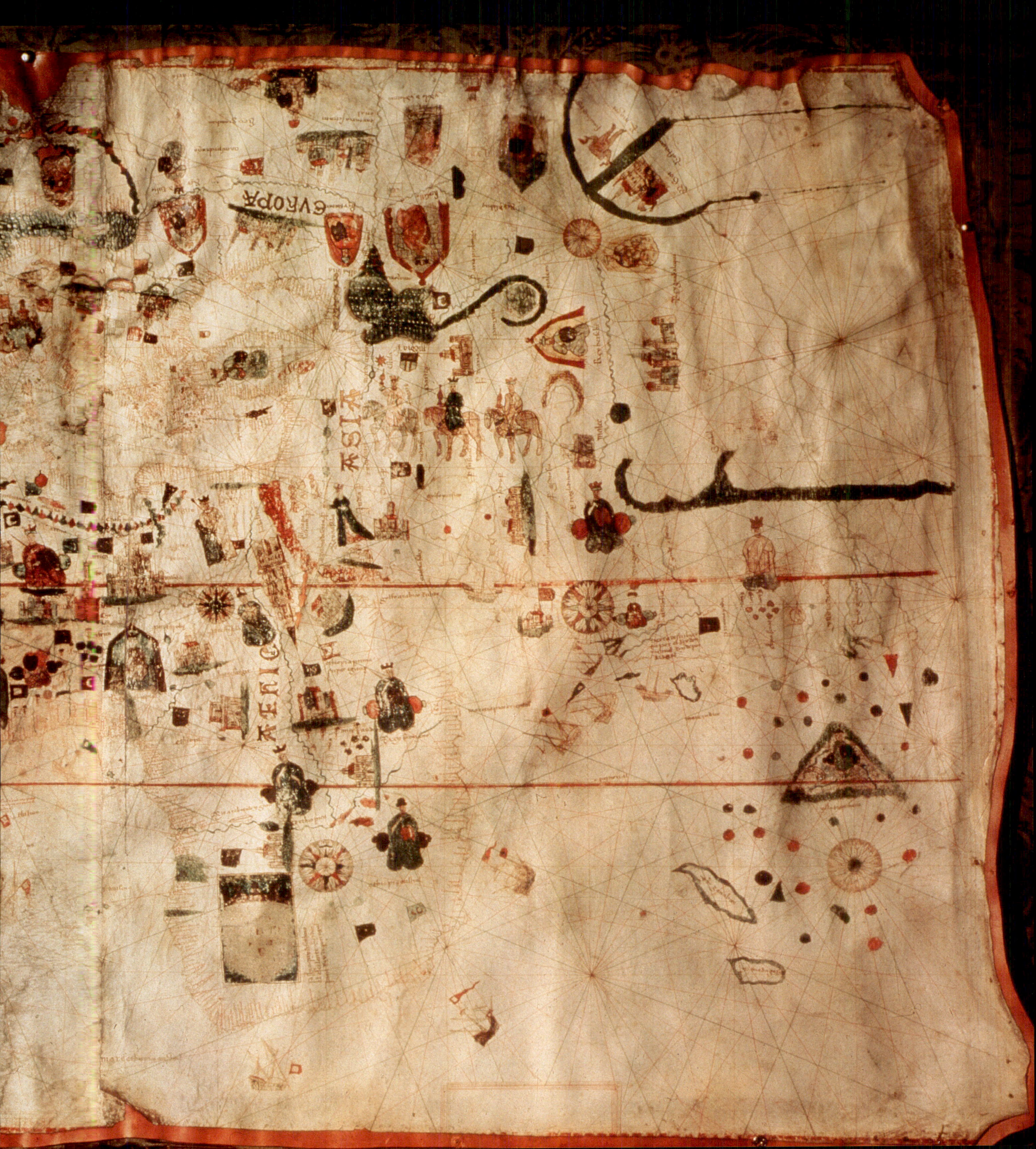
EVROPA
ASIA

TARTARIA
IAPAN
MAR NEGRO
Tropicus Cancri dat is Creefts Sonnewend of Noorder Sonnestandt
Islas de las velas, o de los Ladrones
Islas
Filipinas
ARCHIPELAGO DE S. LAZARO
MA
Linea Æquinoctialis dat is de Middellijn
Nueva Guinea
Islas de Salomon
Tropicus Capricorni dat is Steenbocx Sonnewend of Zuyder Sonnestandt
MAR
By Hessel Gerrits z.
met Octroy
vande E. H. M. Heeren
de Staten Generael
der Vereenichde Nederlanden

The Mirage of the Orient

THE SPICE ROUTE

VARIOUS SPICES – more than appreciated, all but essential – have been part of the daily life of humans since antiquity. They enhance foods with their warm colours, delicious aromas, and gentle flavours. Originally from Asia, they were for centuries carried westward by Arabs, who sold them at exorbitant prices to Venetian and Genoese merchants. There was a time when a pound of ginger cost as much as a sheep, and pepper was sold by the single peppercorn. It was the Portuguese who uncovered the secret of their provenance by finding a maritime route to the Indian Ocean. On the Moluccas (once known as the Spice Islands), they took control of the cultivation of cloves and nutmeg.

Although spices were used mainly for their flavours, they also served in preserving foods, alongside salting and drying; at the least, they masked the taste and smell of foods that had begun to spoil. Today, some spices, such as cumin and ginger, are credited with medicinal properties, but they are usually found in the kitchen alongside pepper, cinnamon, nutmeg, coriander, and cloves. Traditionally, they were sold by both apothecaries and food merchants.

Christopher Columbus, the son of a Genoese weaver, was well aware of the trade with the East. The Silk Road, like the Spice Route, was long and perilous. The maritime route established by the Portuguese eliminated the Arab intermediaries, but Columbus was certain that he could find a better one. He offered his services to Portugal but found no interest. He was, however, able to convince the Spanish sovereigns that there might be a way to reach Asia by sailing west. Columbus, a disciple of Pierre d'Ailly, himself influenced by Claudius Ptolemy, had no doubt that Earth was round and Asia within sailing distance – especially because Ptolemy had stretched its shape toward Europe and erroneously reduced Earth's circumference.

Following his own logic and the information that he had at hand, Columbus knew that he could be nowhere but India when he set foot on land in October 1492, after six weeks of sailing. He did not immediately find the coveted spices, but he was sure that there were precious metals to be found. He also noted the existence of new plants, which were destined to revolutionize life for the rest of the inhabitants of the planet. But the early explorers, Columbus included, had no idea of this, and they continued to search desperately for a passage to Asia, taking little time to note the riches that the unknown continent offered.

Giovanni da Verrazzano was one of the first exceptions. Upon his return from a voyage in 1525, he reported matter-of-factly, "What I know for certain is that America is a continent distinct from Asia and that we should perhaps show some interest in it and not be focused solely on a maritime route to China. The countryside is, in fact, full of promise and deserves to be developed for itself. The flora is rich and the fauna abundant." In his wake, and that of the dozens more explorers who were stopped by this obstacle on the route to India, the Europeans built dreams, sought to extend their empire, and set out to convert the indigenous peoples to their faith. At first the labour of sailors, soldiers, missionaries, and merchants, the colonial undertaking quickly became an affair of state.

The French made some exploratory expeditions up the St. Lawrence, then tried to establish colonies in Brazil and Florida, where they confronted, in turn, the Portuguese and the Spanish. Quickly, two old worlds, American and European, came into contact, each proposing a clearly different model, with the result that old and great civilizations clashed. The accounts of the first explorers aroused curiosity in Europe, and they were quickly translated into a number of languages and constantly reprinted. In

Preceding pages (16–17)
Map by Juan de la Cosa, 1500
Before perishing in South America, the Basque pilot Juan de la Cosa made one of the earliest maps of the Americas. Signed in his handwriting and dated 1500, this document depicts the world as it was known to Europeans in the late fifteenth century. La Cosa, owner of the caravel the *Santa Maria*, not only accompanied Columbus on the voyages of 1492 and 1493 but also took part in expeditions led by Ojeda, Pinzon, and Vespucci. The map shows the coasts reconnoitred by all of these explorers, including the first West Indian islands observed by Europeans, GUANAHANI, HISPANIOLA, and CUBA, above which floats the flag of the kingdom of Castile. To the north are five English flags and the inscription "mar descubierta por Inglese"; these have been interpreted by historians to mean that Cabot or other navigators from Bristol were present in America. Was la Cosa presenting a new continent or an extension of Asia? He prudently chose to leave the edges of the map truncated. On the bottom edge of the Gulf of Mexico, he placed the image of Columbus's patron saint, St. Christopher, bearing Baby Jesus on his shoulder.

Opposite
***Mar del Sur* by H. Gerritsz, 1622**
When he made this map, in 1622, Hessel Gerritsz was chief cartographer for the Dutch West Indian Company. The Pacific Ocean, also called the Southern Sea, was a destination of choice for European merchants, who grew spices in those regions. At the bottom of the map, Gerritsz evoked the difficult sailing conditions, perhaps referring to the Strait of Magellan.

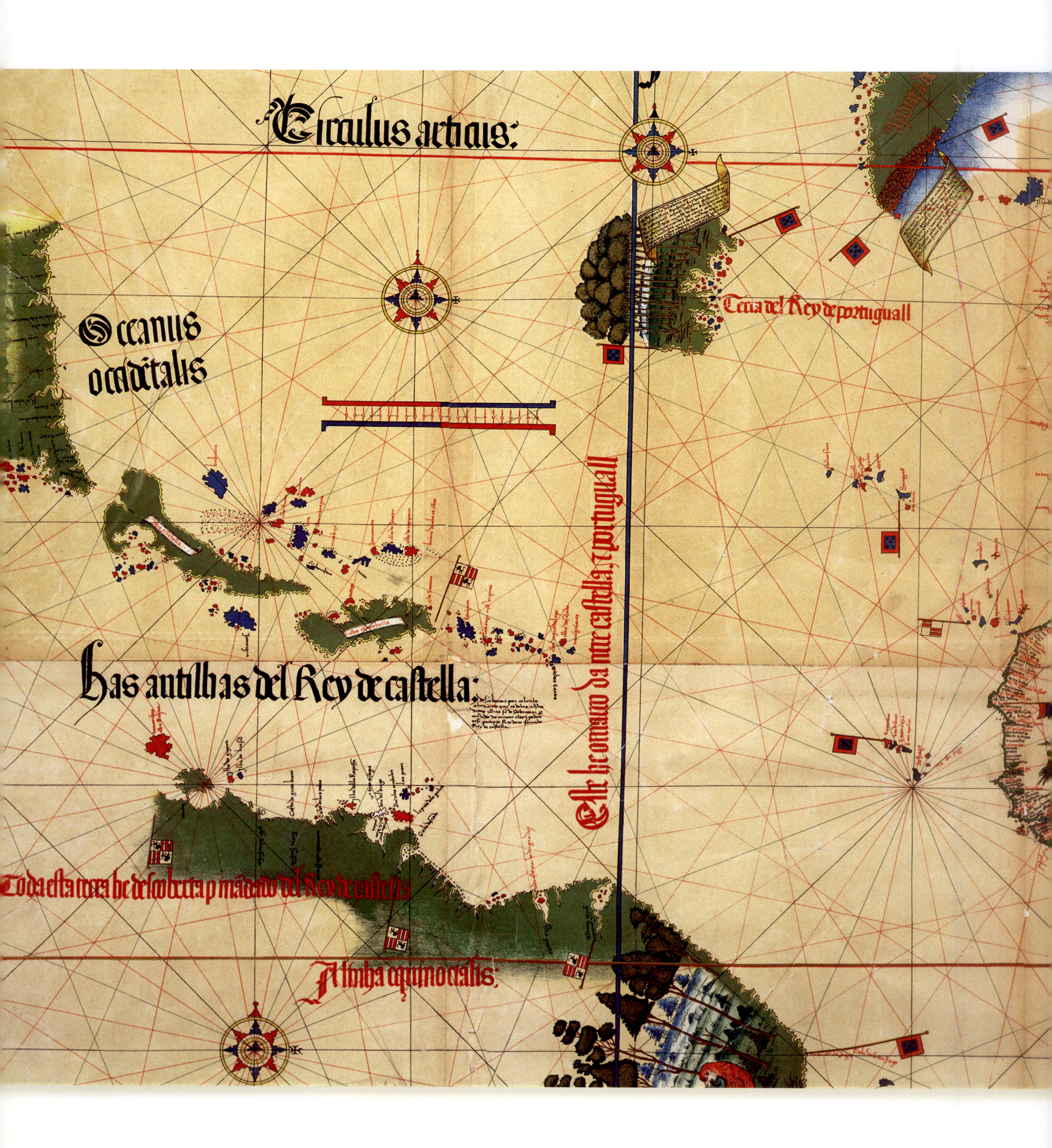

Germany, the de Bry family launched a brilliant collection called *America*, comprising the voyage accounts of Thomas Harriot, René de Laudonnière, and many others. John White's illustrations adorned maps of the Americas for decades, bringing a touch of authenticity and exoticism to cartographers' works.

In the century following Columbus's voyages, the Europeans began to define the territories that they would divide among themselves. The Spanish spread throughout the West Indies, Central America, and South America, except for the tip of Brazil, which the Treaty of Tordesillas granted to the Portuguese; the French, after much trial and error and great hesitation, undertook their conquest of North America, with the Dutch and English hot on their heels.

In less than a century, the global balance changed profoundly. After 1492, the Americas and their populations fell under the thrall of the covetous European powers, which seemed to be as interested in mining precious metals as they were in converting souls.

The original inhabitants of the Americas barely had time to profit from the trade system that had naturally formed with the Europeans. They had no resistance to the new diseases to which they were exposed, and the result was tragic. Had the Indians not been decimated, it is entirely possible that their values and ways of life might have prevailed. After all, they had the numbers, and the knowledge of the terrain. But there was no time before they were wiped out by horrible epidemics.

In North America, the Indians nevertheless had time to guide the newcomers, mainly the French. They provided them with the means to adapt and taught them ways to survive. They helped them advance through, explore, and map the continent before they found themselves consigned to a sad mode of survival or mixing their blood with the Europeans.

Main sources in order of importance

Viola, Herman J., and Carolyn Margolis (eds.). *Seeds of Change: A Quincentennial Commemoration.* Washington, D.C.: Smithsonian Institution Press, 1991. The authors adopt an original and stimulating approach. — Toussaint-Samat, Maguelonne. *A History of Food.* Translated by Anthea Bell. Cambridge, MA: Blackwell Reference, 1993. A classic on the subject. — Roth, Käthe, Louis Tardivel, and Denis Vaugeois. *America's Gift: What the World owes to the Americas and Their First Inhabitants.* Sillery: Septentrion, 2007.

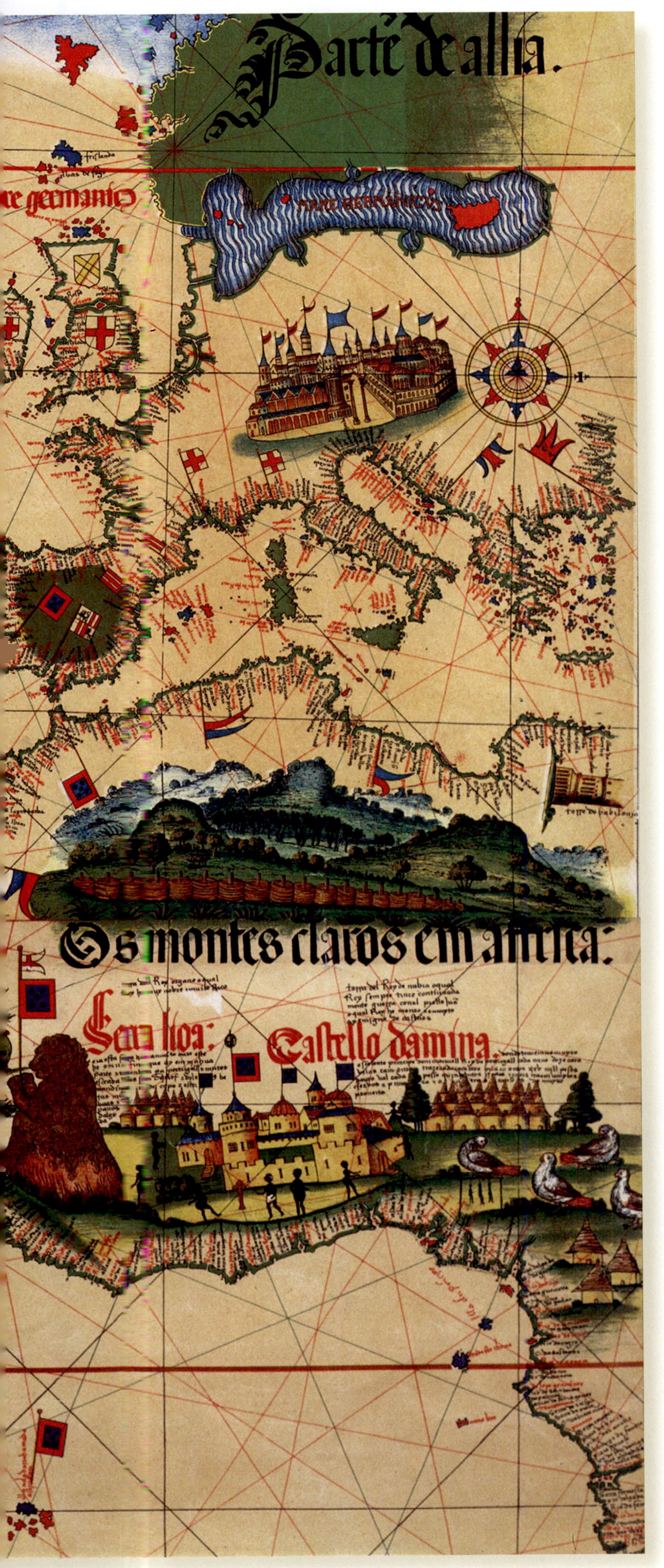

The division of the Americas between Spain and Portugal
Unfortunately, the maker of this cartographic masterpiece, dated 1502, is not known. On the back, an inscription indicates that this "nautical chart of the islands recently discovered in parts of the Indies has been offered to the duc de Ferrare, Ercole d'Este, by Alberto Cantino." Described as a secret agent by Mireille Pastoureau, Cantino was an Italian diplomat who first went to Portugal on the pretext of wanting to purchase horses. He then infiltrated the Court of Lisbon, managing to bribe a cartographer, who turned over to him this large planisphere (101 x 220 cm) for 12 gold ducats. At the beginning of the sixteenth century, the Spanish and Portuguese were in a fierce struggle for control of the seas, and maps were state secrets not to be revealed under any circumstances. Thus, it is not surprising that the identity of the cartographer was kept hidden. The prominent, thick north-south line is a reminder of one of the most stunning treaties in history: the Treaty of Tordesillas, which divided the "new world" between Spain and Portugal. To the north, the line passes west of a landmass called "Terra del Rey de portuguall," which might be Newfoundland or Acadia. Drawn much too far east, this landmass seems to have been deliberately placed on the Portuguese side. A Latin inscription indicates that it was discovered by the Portuguese noble Gaspard Corte Real, on the order of Prince Dom Manuel; the text notes that some members of the expedition returned but not the discoverer, who had disappeared. To the south, the line runs beside a large point of land occupied by parrots – the future Brazil. An inscription describes how Pedro Alvares Cabral, on his way to Calicut, was diverted from his route and found "this terra firma, which he took for a continent." Flags clearly indicate the Portuguese and Spanish possessions on either side of the dividing line. The cartographer saw these new lands as an extension of Asia, as evidenced by the following inscription: "This land discovered on the order of His Great Excellency Prince Dom Emmanuel, king of Portugal, is found to be the edge of Asia."

Le monde selon ysidre ou
xv[e] livre des ethimologies
le monde est divise en trois
parties d lune est apelee aise lautre
europe et la tierce auffrique. Ces trois
parties ne furent pas divisees egau

On the Route to India

THE WORLD WITHOUT THE AMERICAS

How did Europeans portray the world before their explorers "discovered" the Americas? Very few maps from prehistory and antiquity have survived to tell us. Those on wood and papyrus probably fell victim to humidity. Others were destroyed in wars and fires. Among the small number of ancient world maps that have survived are some copies that differ somewhat from the originals. This is why, much of the time, we must reconstruct the relevant cartographic data through literary descriptions, many of which are written in a poetic language that is difficult to interpret.

Although maps were drawn in the prehistoric, Mesopotamian, and Egyptian civilizations, it is from Greek civilization that we have the first surviving maps of the inhabited world. A number of notions about modern cartography also appeared first among the Greeks: Earth as a sphere, projections, latitude, and longitude. Great names are associated with this heritage from the Classical Age: Thales of Miletus, Pythagoras, Eratosthenes, Hipparchus, Strabo, Ptolemy, and others – all Greeks well versed in the disciplines of geography, astronomy, philosophy, and mathematics. Real progress in the graphic representation of Earth began in the third century BCE, thanks in part to the political and cultural context at that time.

The growth of Alexandria, the great Hellenic cultural centre in Egypt, had a profound effect on the evolution of sciences in general, and geography in particular. Its library contained an excellent documentation centre and provided a stimulating meeting place for scholars of ancient Greece. Alexandria thus became a site of prolific map production. The Asian conquests by the king of Macedonia, Alexander the Great (356–23 BCE), also stimulated map production, as geographic expansion of the empire provided cartographers with material and caused an upheaval in the description of the world. But the fire that Julius Caesar ordered set when he entered the city (about 48 BCE) destroyed much of the Hellenic cartographic heritage.

Eratosthenes (c. 275–c. 194 BCE) was one of the most brilliant scholars of ancient Greece. Born in Cyrene (today Shahhat, Libya), he was asked by the king of Egypt, Ptolemy III (235–21 BCE), to direct Alexandria's library and tutor the heir to the throne. Although Anaximandrus (c. 610–546 BCE) is considered the first cartographer of the Greek world, Eratosthenes is generally regarded as the founder of geography. He is known for two great achievements: the first map of the world based on measurements of longitude and latitude, and the first calculation of the circumference of Earth.

The conquest of Greece by the Romans did not stop cartographic progress; the Greeks remained at the pinnacle of geographic science. The work of two eminent men provides eloquent testimony to this: Hipparchus of Nicaea and Ptolemy. Hipparchus (c. 165–c. 127 BCE), the greatest astronomer of antiquity, demonstrated the importance of astronomy in geography and the relevance of observing the heavens to better describe Earth. He is also known for having transformed the irregular grid formulated by Eratosthenes into a complete reference system by which any geographic site could be located. Claudius Ptolemy (90–168 CE) was, throughout Roman antiquity, the reference for astrology and geography, thanks mainly to his two fundamental treatises: *Almagest* and *Geography*. The latter, building on works by Eratosthenes and Hipparchus, set out to portray the known world and systematically list sites according to their latitude and longitude.

After the fall of the Roman Empire, cartographic representation of the world was adapted to the culture,

Earth divided among Noah's three sons, circa 1460
This illumination from a fifteenth-century manuscript includes a number of characteristics of maps of the Middle Ages. North is on the left side, and east is on top. Three continental landmasses, encircled by the ocean sea, are separated by three arms of the sea, which form a T – whence the expression "T and O" used to designate these maps. This portrayal reflects the biblical conception according to which, after the Flood, God divided the world among Noah's three sons, Shem, Ham, and Japheth. Here, each is portrayed standing on the continent bequeathed to him – Africa, Asia, and Europe, respectively. In Asia is Mount Ararat, on which Noah's ark, the founding symbol of Judeo-Christian civilization, is perched.

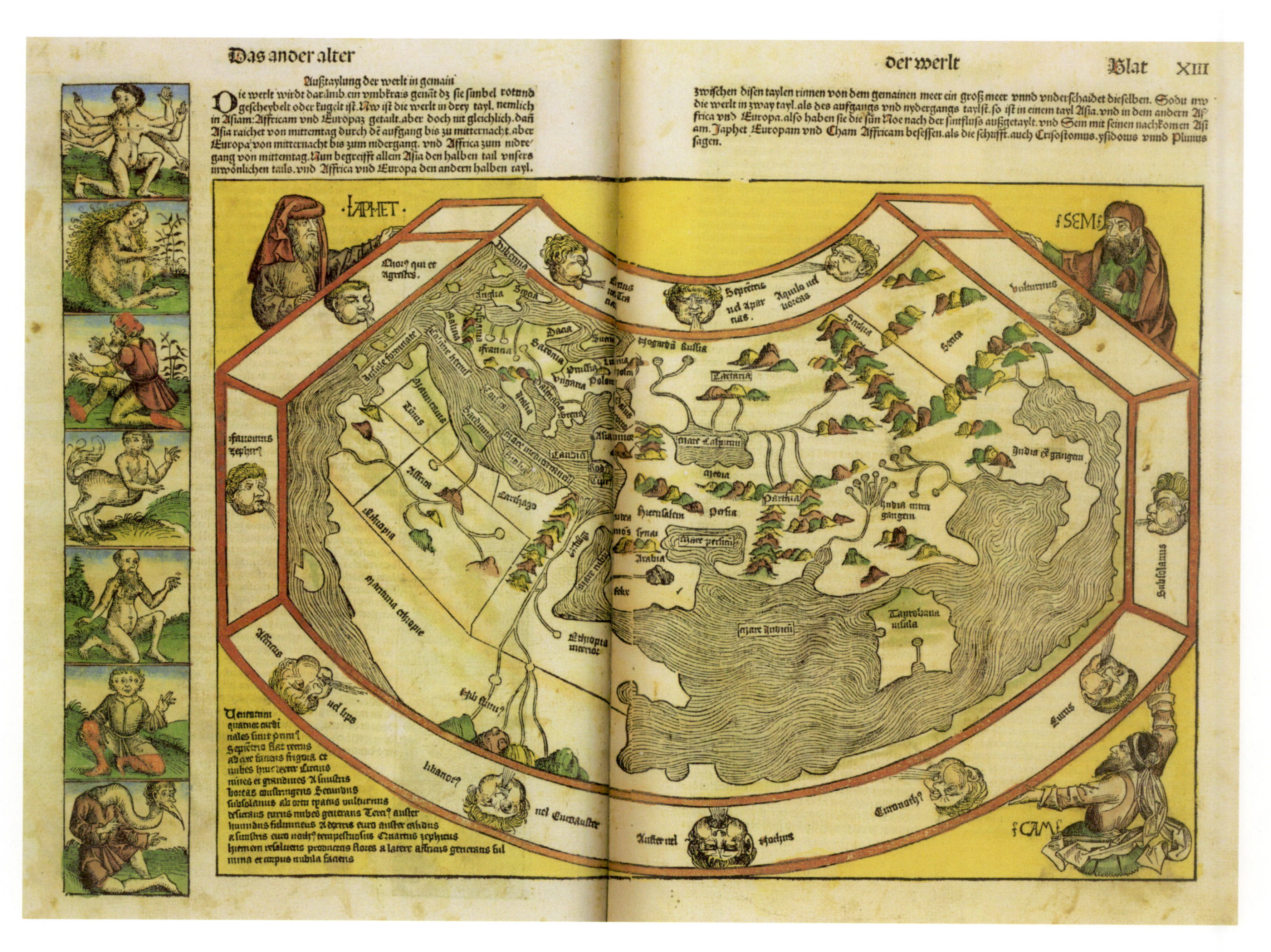

Map of the world from the Nuremberg Chronicle, 1493
Published in 1493, the same year that Columbus made his second voyage to the West Indies, this map portrays a world without America, inspired largely by Claudius Ptolemy, but also by the Old Testament – Noah's three sons are drawn in three corners of the image and watch over the fates of the continents of which they are the protectors. The Indian Ocean is a closed sea that cannot be reached by going around Africa. A procession of monsters occupies the left margin, among them a man with six arms, a satyr, a man with four eyes, and a man-crane. Inspired by medieval bestiaries, these figures are typical of portrayals of extra-European populations. The engraving is part of a famous work known as the Nuremberg Chronicle, which relates the history of the world from its origins to 1492.

beliefs, and needs of Christians. For a long time, it was believed that Earth was composed of three continents: Europe, Asia, and Africa. This tripartite division of the world originated in the Bible: after the Flood, God divided the world among Noah's three sons, Ham, Shem, and Japheth. Medieval world maps were perfect reflections of this concept. Almost all designed according to a common plan, these highly symbolic portrayals had a very recognizable form: Jerusalem in the centre; the earthly paradise to the east, where the sun rises; and three landmasses surrounded by the ocean sea (the Atlantic) – Asia, Europe, Africa – and separated by the Nile, the Mediterranean, and the Don. For ideological reasons, all of these maps were oriented with the north on the left side and the east at the top. Some maps included roads leading to Jerusalem or fictitious locations resulting from a literal interpretation of the Bible.

It was also in the Middle Ages that a Byzantine monk, Maximos Planude (c. 1260–c. 1310), rediscovered Ptolemy's *Geography,* which was circulating mainly in the Arab world. This work was translated into Latin in Florence in 1406 by Emmanuel Chrysoloras (1355–1415), and he and his students produced a number of manuscript versions. In 1475, *Geography* was printed for the first time, and it was republished almost 40 times over the following century: a bestseller, it was second only to the Bible in sales at the time. Composed of maps and tables listing up to 8,000 place names with their geographic coordinates, this work was the authoritative reference in Europe until the end of the sixteenth century and was constantly updated to take account of the most recent discoveries.

But not all discoveries were recorded in European cartography. Around the year 1000, the Vikings of Northern Europe reached North America, via Greenland, and established a short-lived settlement, Vinland. This discovery was not revealed to the rest of Europe, where the existence of America was still unknown and it was thought that Asia was on the far shore of the ocean sea.

After the Hundred Years' War (1337–1453), the European economy rebounded. The most prosperous European kingdoms began seeking new markets for their imported products and the coveted riches: silk, pepper, spices, and precious metals. The only trade route to Asia, the one taken by Marco Polo a few centuries before, was not satisfactory. To free themselves from the control of the Muslim kingdoms, but also to bypass the monopoly of the Venetians and Genoese, Europeans went in search of new maritime routes. The desire for trade, combined with a quest for intellectual knowledge, was the basis for a formidable expansion movement made possible by progress in the domains of finance, shipbuilding, navigation, and cartography. The invention of the astrolabe in the tenth century (by the Arabs), then of the cross-staff four centuries later, enabled latitude to be accurately determined. In the late thirteenth century, a Chinese invention, the compass, appeared in the West; it was more reliable than the sun and the Pole Star for orientation in foggy weather. Used as a complement to the compass, nautical guides (portolans) and nautical charts (portolan maps) enabled sailors to determine their position at sea. At about the same time, the Portuguese invented the caravel, a small and very manoeuvrable ship. This technological advance meant that navigators could envisage long voyages without having to go ashore. The caravel could also

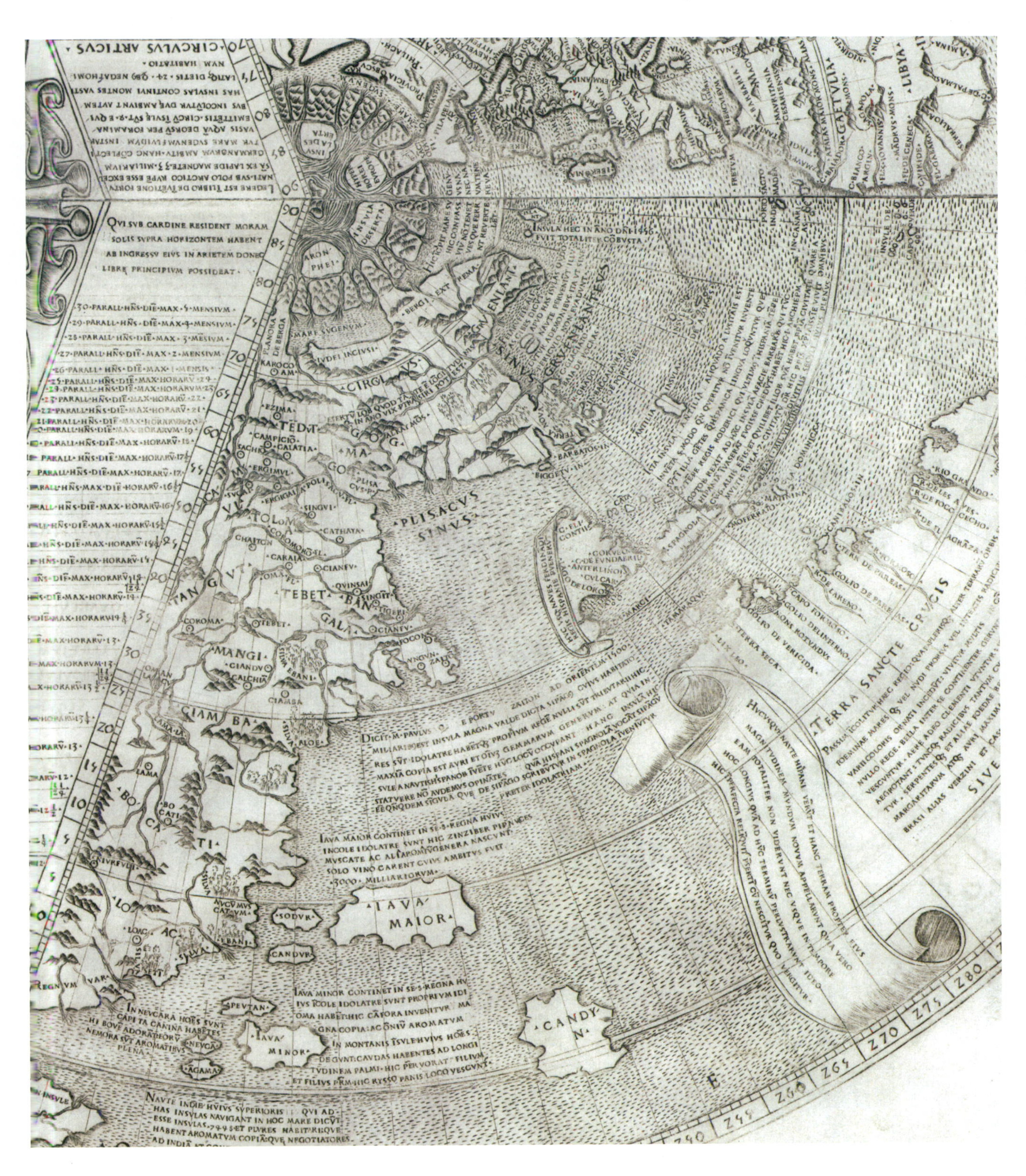

The Atlantic Ocean as drawn by Johann Ruysch, 1507
The Europeans who landed in America in the late fifteenth century thought they were in Asia – a notion conveyed in this map, which presents the islands explored by Columbus and his successors, such as Canibalos, Ispagnola, and la dominica. To the southwest, the cartographer drew an immense landmass whose western edge seems undefined (South America). To the north is the coast frequented by Basque, Breton, Norman, and Portuguese fishermen. Newfoundland (Terra Nova) is drawn not as an island, but as a peninsula attached to the Asian continent, as are Greenland (Gruenlanteus) and the Insula Baccalauras, which can be translated from the Portuguese as Island of Cod (some sources posit that this term was used by the Indians). The cartographer has a warning for fishermen: the demons that haunt the shores of this region prevent vessels from landing. West of the Newfoundland coast are the Asian landmasses of Gog and Magog (described by Marco Polo), the etymological origins of which are biblical. To the west of the Azores is the mythical island of Antilia, for which the Antilles are named, and which, according to legend, served as a refuge for a king of Spain. This map, which appeared in Ptolemy's *Geographia*, published in 1507, is the work of the Flemish astronomer and cartographer Johann Ruysch, who, it seems, visited the Newfoundland region on board an English ship.

Martin Behaim's globe, 1492
The globe made by the German Martin Behaim in 1492 is the oldest surviving spherical representation of Earth. It portrays the vision of the world held by literate Europeans on the eve of Christopher Columbus's first voyages. The influence of Cardinal d'Ailly and the Florentine cosmographer Toscanelli can be seen in this depiction of Earth, which presents a relatively small Atlantic Ocean (130 degrees of longitude). The cartographer draws the coasts of Africa, which had just been circumnavigated by the Portuguese, and sketches in, west of Europe, a sea dotted with islands seeming to offer as many ports of call for the first adventurers seeking a westward route to Asia – Antilia, Isle of Seven Cities, Brazil, Cipango, the Fortune Islands, Saint-Brandan – most of them exotic vestiges of antiquity, Celtic myth, or Marco Polo's voyage to the Orient.

tack into the wind, which meant that explorers could return to their homeports more easily (for more details on techniques and tools of sea navigation, see pp. 70–73).

In the fifteenth century, it became possible for the Europeans to venture beyond the Mediterranean. This expansion of their space was in part the doing of the Portuguese prince Henry the Navigator (1394–1460), who surrounded himself with experienced mapmakers, founded a school of navigation in Sagres, on the Algarve coast, and sent his navigators far into Atlantic waters. Sagres was the operational centre, from which explorers sailed and to which they returned with new geographic information that enabled cartographers to update maps and make new ones. Constantly being refined, these maps were thus both the prelude to and result of expeditions of exploration.

Portuguese ships reached the Madeira Islands (1418), the Azores (1432), and, in Africa, the Rio de Oro (1436), Cape Bojador (1434), Cape Blanc (1441), and Cape Verde (1443). By the time Henry died, in 1460, the Portuguese had reached Sierra Leone, but they didn't stop there. In 1487, Bartolomeu Dias sailed around the Cape of Good Hope. Ten years later, in 1497, Vasco de Gama reached Calcutta, having sailed around Africa. He returned to Europe with a cargo of pepper and the confirmation that Asia could be reached by sea. Once the islands of the Atlantic and the coasts of Africa were better known, the unthinkable became the possible: reaching the Orient by going west, by crossing the Atlantic.

In 1474, the Florentine astrologer and cosmographer Paolo Toscanelli (1397–1482) became one of the first to propose this route in a letter sent to the king of Portugal. Basing himself on Ptolemy's erroneous assessment of Earth's circumference (28,000 km instead of 40,000 km), but also on inexact descriptions by Marco Polo on the location of Japan (Cipangu), Toscanelli opined that the ocean was small enough to cross. Accompanying the letter was a map supporting his argument; this map has never been found.

Although this missive did not convince the king of Portugal, it sparked the imagination of a young Genoese navigator whose name is very familiar today: Christopher Columbus (1451–1506). Columbus obtained a map from Toscanelli, which he took on his sea voyages. But Columbus was also influenced by other sources, notably Pierre d'Ailly's cosmography treatise *Imago Mundi*. Written in 1410, this work, later meticulously annotated by Columbus, supported the idea that India could be reached by sailing westward. Estimating at 2,400 nautical miles the distance between Portugal and Japan (it is actually 10,600 nautical miles), Columbus convinced the sovereigns of Spain, Isabella of Castile and Ferdinand of Aragon, to provide backing for his projects. In August 1492, three caravels, the *Niña*, the *Pinta*, and the *Santa Maria*, left the Andalusian coast sailing toward the setting sun; some weeks later, they landed on an island that Columbus named San Salvador. Believing that he had reached Asia, he called the inhabitants Indians. Columbus made three more voyages to America (1493–96, 1498–1500, 1502–04), never once doubting that he had landed in Asia. Each time he returned, the Spanish received a bit of gold, more cases of a new form of syphilis, and new geographic data, but they never realized that Columbus and his Spanish crew had set foot in a new world.

In spite of the discoveries of Columbus and the other navigators who followed in his wake, no one imagined that there was an immense landmass between them and their destination. In 1507, one year after Columbus died, a map by Johannes Ruysch showed how the Europeans saw the world. Cuba, Hispaniola, and other islands found by Columbus are placed along the Asian coast. Farther north, Greenland and Newfoundland are shown as peninsulas of East Asia. Still in 1507, a different conception of the world first appeared in a small book that might have gone unnoticed, but it was the baptismal certificate of America.

Main sources

Boorstin, Daniel. *The Discoverers.* New York: Random House, 1983.
— *Ciel et Terre.* Virtual exhibition by the Bibliothèque nationale de France, [1998] – [www.expositions.bnf.fr/ciel/].

The Birth of America

COLUMBUS'S MISTAKE

PRUDENTLY, AND RIGHTLY, historians are generally in the habit of writing, "one of the first." Writers of dictionaries and encyclopedias, on the other hand, are not plagued with such doubts; they continue to say "the first" and use the word "discovery."

Who was the first European to reach the continent that was to bear the name America? Christopher Columbus, say the dictionaries. Yet, this is far from certain. Leaving aside legends about the settlement there of the tenth tribe of Israel, the crossing of the monk St. Brendan or Jean Cousin of Normandy, and the arrival of the Chinese around 1421, it is well established that Iceland had a colony in Greenland in the eleventh century. According to documents published on the order of Pope Leo XIII, a diocese existed there from 1124 to 1378. A few decades later, in 1410, a ship brought the last settlers back to Iceland; they had been driven out by climatic cooling.

The Vikings, who had leapfrogged from Scandinavia to Iceland (874) and Greenland (c. 985), reached Labrador and Newfoundland around 1000. This saga left no traces except, perhaps, the ancestors of the Newfoundland dog; however, it fed a tradition that Christopher Columbus likely heard when he visited Iceland in 1476–77. As he was bookish by nature, he probably paid little attention to this. Aristotle, Ptolemy, Toscanelli, and d'Ailly had more influence on him, and he was persuaded that the Atlantic could be crossed quite easily and would provide a new route to China.

The Vikings were fierce warriors and tireless traders; the Icelanders and the Basques were brave fishermen and skilful whale hunters. Starting in the tenth century, whaling became an important part of the Basque economy. Whale oil was used in lamps and as a lubricant in the manufacture of soap, whale meat fed the poorest all over Europe, and whalebone stiffened the corsets of well-born ladies. Nothing went to waste.

When whales became rarer on the Basque coast – for unexplained reasons – the Basque whalers moved to the coast of Asturias, ranging as far north as Cape Finisterre. They started using compasses in the thirteenth or fourteenth century, and then they ventured even farther.

Like all hunters who chase their prey, the Basque whalers headed forth into the unknown. Following their quarry, they sailed to Iceland, then to Labrador and Newfoundland. An Icelandic chronicle dating from 1412 mentions 20 Basque ships in hot pursuit of a pod of whales. *Balaena biscayensis,* rather solitary animals in the waters off the Basque coast, seemed to prefer living in groups farther north, whence the Basque expression *sardako balea* (herd of whales). The whales led the fishermen toward Newfoundland, where incredible shoals of cod awaited them. And so whales gained a rival: the codfish. And the news spread. The Basques were quickly surrounded by fisherman from Brittany, Normandy, and Portugal. The route to Newfoundland became the private domain of fisherman, and it remained so until the expedition by Giovanni Caboto in 1497. Then, "the Basque secret was out," noted Mark Kurlansky, the author of a small and charming book on codfish in which he joins a long tradition of scholars in expressing his conviction that the Basques reached America before Columbus. From Étienne Cleirac, in *Us et coustumes de la mer*, to Jean Le Rond d'Alembert and Jules Michelet and, more recently, Robert de Loture, in *Histoire de la grande pêche de Terre-Neuve*, this hypothesis has repeatedly been put forward. To date, concrete proof has not been found, but a tenacious oral tradition might constitute strong circumstantial evidence.

While the new territories to the north were associated with a miraculous fishery, farther south a dye made from a tree called brazilwood, or *pao brasil*, was becoming

Following pages
World map by Martin Waldseemüller, Saint-Dié, 1507
It is very likely that this map by Martin Waldseemüller was published in Saint-Dié in 1507. It was the first map to show a new continent detached from Asia, baptized AMERICA, after the Italian navigator Amerigo Vespucci. The importance accorded to Vespucci is reflected in the portrait of him at the top of the map, compass in hand and facing the great pioneer of geography, Claudius Ptolemy. The alignment of this new world, which seems deformed because of the map projection used, is in reality amazingly accurate for the time. Waldseemüller even drew a mountain range on the west coast of North America. According to known archival documents, however, neither Balboa nor Magellan – nor any other European – had yet seen and described the Pacific. Researchers, fascinated by this observation, have speculated on the possible existence of unrevealed sources. This very large 12-page map had a print run of 1,000 copies, only one of which seems to have survived; it was found in 1901 in a castle in Bade-Wurtemberg. In 2001, the Library of Congress acquired it for $10 million.

CLAVDII PTHOLOMEI ALEX
ANDRINI COSMOGRAPHI
CHORᵒ
ZEPHIRᵒ
TERRA VLTERI INCOGNITA
PARIAS
OCCEANVS OCCIDENTALIS
CIRCVLVS ARTICVS
MARE GLACIALE SIVE CONGELA
TOTA ISTA PROVINCIA INVENTA EST PER MANDATVM REGIS CASTELLE
OCCIDENTALIS OCCEANVS
EQVINOCTI
AFRICVS
AMERICA
LYBONOTHᵒ
LIBIA INTERIOR
AFFRICA
ETHIOPIA INTERIOR
SINVS MAGNVS AFFRICE
NOTVS
VNIVERSALIS
COSMOGRAPHIA
SECVNDVM PTHOLOMAEI TRA
DITIONEM

AMERICI VESPVCII
AQVILO
CECIAS
SVBSOLANVS
TIPVM ORBIS GENERALEM DESCRIBENDO: VE
TERVM INVENTA PONERE, ET EA QVE A NEO
TERICIS INTERIM REPERTA SVNT (SICVT EST
CATAIA REGIO) CONIVNGERE PLACVIT. VT
TALIVM RERVM STVDIOSI DVM VARIA COGNO
SCERE CVPIVNT VOTORVM COMPOTES LABO
RI NOSTRO SINT GRATI PLAERAQVE OMNIA
TAM PASSIM COGNITA QVAM NOVITER LVS
TRATA DILIGENTER AC DISTINCTE SVB VNO
ASPECTV COLLOCATA PROSPICIENTES.
MARE GLACIALE
ARCTICVS
CIRCVLVS
ASIA
SCITHIA EXTRA IMAV
SERICA REGIO
CHATAY
INDIA MERIDIONALIS
OCCEANVS ORIENTALIS INDICVS
SINVS GANGETICVS
INDIA
IAVA MAIOR
IAVA MINOR
SEYLAM
EQVINOCTIALIS
OCCEANVS INDICVS MERIDIONALIS
TROPICO CAPRICORNI
MADAGASCAR
AVSTER
EVRONOTVS
VVLTVRNVS EVRVS
LICET PLAERIQVE VETERVM DESCRIBENDI TERRA
RVM ORBIS STVDIOSISSIMI FVERINT: NON TAMEN
PARVM IPSIS EISDEM INCOGNITA MANSERVNT, SI
CVT EST IN OCCASV AMERICAE: AB EIVS NOMINIS
INVENTORE DICTA, QVE ORBIS QVARTA PARS PV
TANDA EST. SICVT ET VERSVS MERIDIEM APHRICE
PARS: QVAE SEPTEM PENE GRADIBVS CITRA CAPRI
CORNVM INCIPIENS/VLTRA TORRIDAM ZONAM ET
EGOCERI TROPICVM AD AVSTRVM LATISSIME PRO
TENDITVR. SICVT QVOQVE IN TRACTV ORIENTA
LI REGIO CATAIAE/ ET QVICQVID INDIAE MERIDIO
NALIS VLTRA CENTESIMVM ET OCTOGESIMVM LON
GITVDINIS GRADVM EST SITVM. QVAE NOS PRIO
RIBVS OMNIA ADIVNXIMVS: VT ISTIVSCEMODI RE
RVM AMATORES/ QVAECVNQVE SVB HANC DIEM
NOBIS PATENT OCVLIS INTVENTES / DILIGENTIAM
NOSTRAM POBENT. ID AVTEM VNVM ROGAMVS
VT RVDES ET COSMOGRAPHIAE IGNARI HAEC NON
STATIM DAMNENT ANTEAQVAM DIDICERINT CHA
RIORA IPSIS HAVD DVBIE POST CVM INTELLEXE
RINT FVTVRA.
ET AMERICI VESPVCII ALIORV QVE LVSTRATIONES

World map published in Basel, 1537
This world map is part of a collection of voyage accounts compiled by Johann Huttich and Simon Grynaeus, in which figured, notably, the writings of Christopher Columbus, Amerigo Vespucci, and Pierre Martyr d'Anghiera. Published in 1537, it illustrates the fortune of the name America, created three decades before. The New World is represented as a distinct continent that can be circumnavigated, to the north or the south, to reach Japan (ZIPANGRI). Strangely, North America is called TERRA DE CUBA, while what we now know as Cuba is called ISABELLA, in tribute to Isabelle of Castile, Columbus's patron. The iconography on the border of the map is very rich. The voyager Varthema (in the lower right-hand corner) is portrayed alongside hunters (above), a large elephant, and winged snakes. In the lower left-hand corner, the illustrator has inserted a scene of cannibals preparing a meal. He has also drawn pepper, nutmeg, and cloves, symbols of Asia. This map is also unusual because the hand positions of the angels below and above the map suggest Earth's rotational movement on its north-south axis.

increasingly popular. Fashion could be as powerful a motivator as hunger. Ships from Portugal and Dieppe had been going to Brazil since the late fifteenth century. In fact, at that time, the Atlantic – an ocean sea (or ultimate sea), as opposed to the Mediterranean, a sea in the middle of the Continent – was being crossed, north and south, by caravels, doggerboats, barges, and Biscay pinnaces. New names – Antilia, Brasil – appeared on maps of the Atlantic, including one dating from 1482 attributed to Gracioso Benincasa.

On October 12, 1492, *el dia de la raça*, the Tainos of Guanahani greeted a Genoese explorer looking for India. The men brought the crew water and fruits that were deemed "good to taste and healthy for the body." The women offering themselves to the lecherous looks of the Andalusian and Basque sailors were all as naked as Adam and Eve in their earthly paradise. Indeed, Christopher Columbus was convinced that such a paradise was not far away. On February 21, 1493, he wrote in his journal, in a burst of enthusiasm, "Wise men said well when they placed the terrestrial paradise in the Far East, because it is a most temperate region. Hence these lands that [I] had now discovered must ... be in the extreme East." Columbus was not an idiot, but he was stubborn, no doubt a bit of a crank, and he was committed to his intellectual masters' ideas.

Columbus returned to the sandbar at Saltes, Andalusia, on March 15, 1493. A few weeks later, on May 4, Pope Alexander VI proposed a north–south line to divide "the discovered world and the world to be discovered" between Spain and Portugal. The Portuguese were dissatisfied: this line, set at 100 leagues west of Cape Verde, deprived them of land that they had already partly reconnoitred to its west. At a meeting that they insisted on having, held at Tordesillas, the Spanish agreed to push the line farther west by 270 leagues, to 370 leagues west of Cape Verde.

Of course, all this brouhaha would have seemed incomprehensible if no one had known about the more or less clandestine voyages of Columbus's predecessors. But Columbus had let the cat out of the bag – to the west, there were more than just a few islands here and there – although he apparently remained convinced that he had reached the shores of Asia.

In spite of his own beliefs, Columbus was credited with discovering a new world. The terms "discovery" and "new world," of course, were erroneous: the continent in question was not waiting for Europeans to find it in order to exist; in fact, it was home to ancient civilizations. It was an old world. Could its inhabitants have taken to the sea and reached Africa or Europe? Why would they have wanted to? Their boats were designed to take advantage of the wealth and resources of their world. To the north, people had invented birchbark canoes, true masterpieces of ingenuity. To the south, people built pirogues. The Americas were rich. Their inhabitants had adapted perfectly to the land and developed their own wealth, mainly in agriculture. They were to offer this wealth to the rest of the world: their foods were ultimately responsible for spectacular population growth all over the planet. Their silver revolutionized trade. Enormous progress was made in medicine thanks to their quinine and curare. In short, the true "new world" was the result of the encounter of two old worlds, Europe and the Americas.

Europeans were slow to be convinced that there was an entire continent between them and Asia. Rather than recognizing it as such and eventually giving it a name – why not his own or that of his beloved queen – Columbus, certain that he had reached India, naturally called its inhabitants Indians; in French, corn was dubbed *blé d'inde* ("wheat of India"); a large edible bird, *poule d'inde* ("chicken of India"; the English called this bird "turkey"), later shortened to *dinde*; and a large rodent, *cochon d'inde* ("pig of India"; the English term is "Guinea pig"). In the early twentieth century, American anthropologists proposed that a suggestion by a lexicographer (J. W. Powell, in 1899) be adopted: to form, from the first letters of the words "American Indian," the word "Amerind," from which other words were coined: Amerindic, Amerindize, and Amerindian. French Canadians gradually started to use the word *Amérindien* when they spoke of indigenous peoples. Scholars retain the word "Indian" in historical texts. French dictionaries, such as the *Petit Larousse* and the *Robert 1*, recently introduced the term *amérindien* to designate the indigenous peoples of the Americas, whose descendants in North America call themselves the First Nations.

The current inhabitants of the United States have long called themselves Americans, and Europeans often mistakenly call their country America.

Main sources in order of importance

Pastoureau, Mireille. *Voies Océanes: de l'ancien aux nouveaux mondes.* Paris: Hervas, 1990. — Kurlansky, Mark. *Cod: A Biography of the Fish That Changed the World.* New York: Walker and Company, 1997. — Duviols, Jean-Paul. *L'Amérique espagnole vue et rêvée: les livres de voyages de Christophe Colomb à Bougainville.* Paris: Promodis, 1986. A marvellously scholarly work. — Nebenzahl, Kenneth. *Atlas of Columbus and the Great Discoveries.* Chicago: Rand McNally, 1990. — Harrisse, Henry. *Découverte et évolution cartographique de Terre-Neuve et des pays circonvoisins, 1497-1501-1769.* Amsterdam: N. Israël, 1968. — Ronsin, Albert (ed.). *Découverte et baptême de l'Amérique. 2nd revised and expanded ed.* Jarville-La Malgrange: Éditions de l'Est, 1992.

Sources for translation: Columbus, Christopher. "Journal of the First Voyage of Columbus." In J. E. Olson and E. G. Bourne (eds.), *The Northmen, Columbus and Cabot, 985–1503: The Voyages of the Northmen and the Voyages of Columbus and of John Cabot*, pp. 87–258. New York: Charles Scribner's Sons, 1906 [http://www.americanjourneys.org/].

With the assistance of Ms. Louisa Martin-Meras, to whom I am extremely grateful, I consulted a number of Spanish-language works at the Museo naval and Biblioteca National de España in Madrid and at Las Casas del Tratado in Tordesillas. [D. V.]

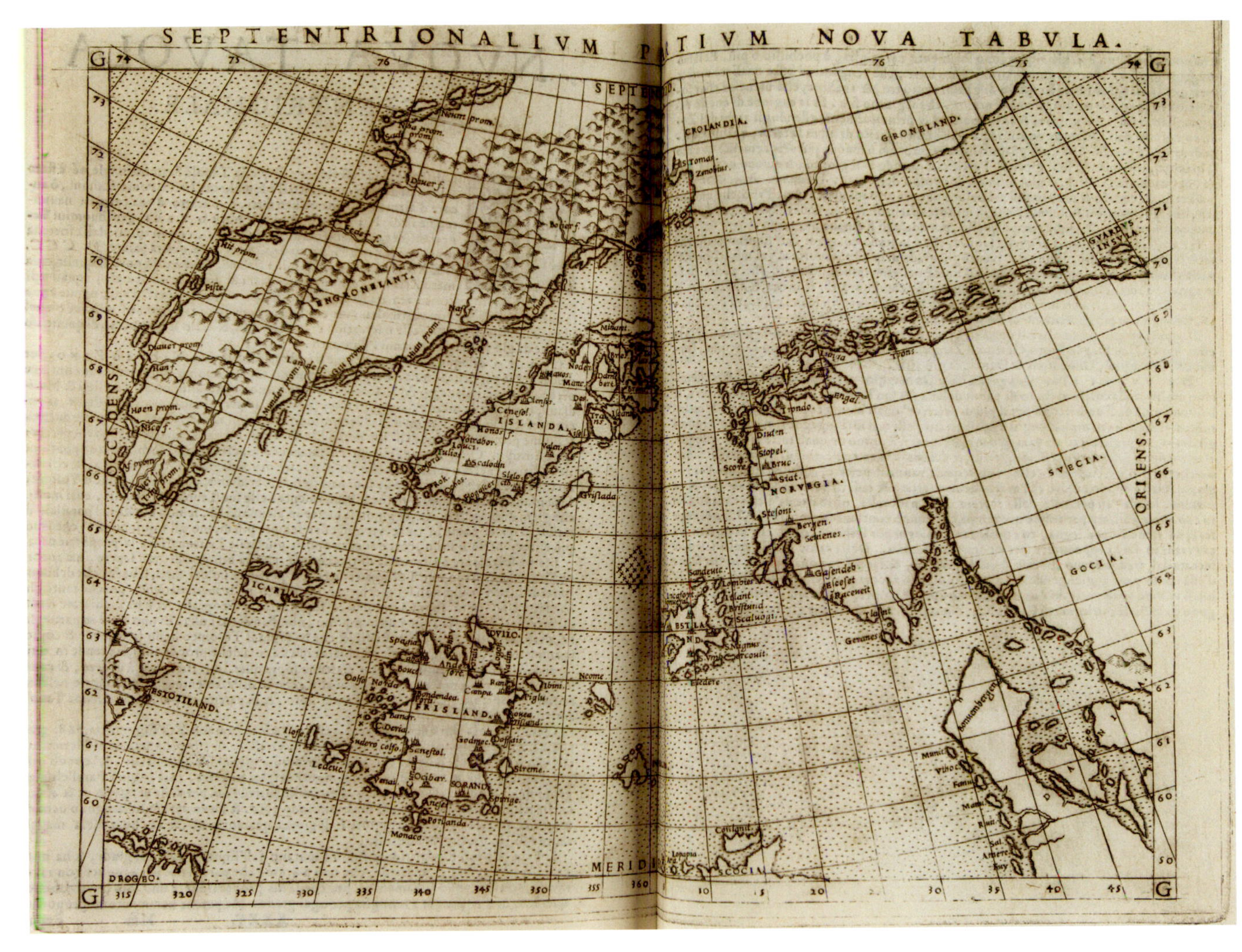

The voyage of the Zeno brothers in Ptolemy's Geographia, 1561
Many theories have circulated about the discovery of America. Phoenicians, Chinese, Irish, Scottish, Normans, and Basques have been elevated, in turn, to the rank of first discoverers. Among these were the Zeno brothers, of Venice, whose Atlantic voyage around 1380 was publicized by one of their descendants in the sixteenth century. A map of the voyage was circulating at the time in literate circles and was used in one of the editions of Ptolemy's *Geographia* published by Girolamo Ruscelli. The map shows the islands explored during this purported voyage: Frisland, Islanda, Icaria, and Estotiland, the last having been associated with the northeast point of Nova Scotia. Thus, in the sixteenth century, maps constituted a very useful tool for giving credibility to a voyage account. Today, the voyage of the Zeno brothers is considered to be apocryphal.

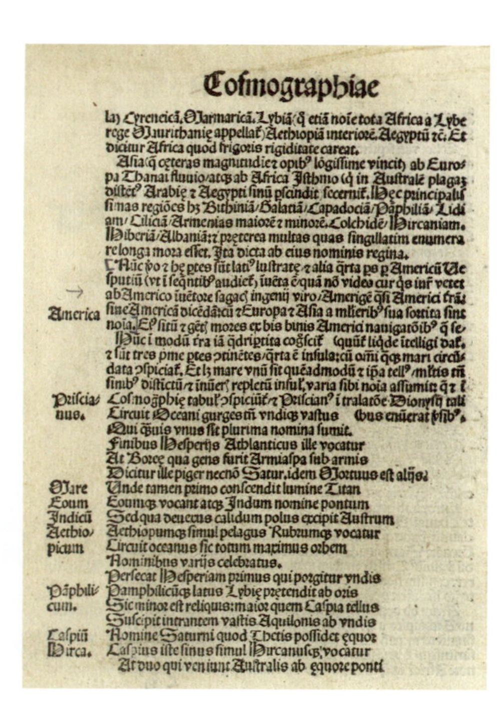

Cosmographiae

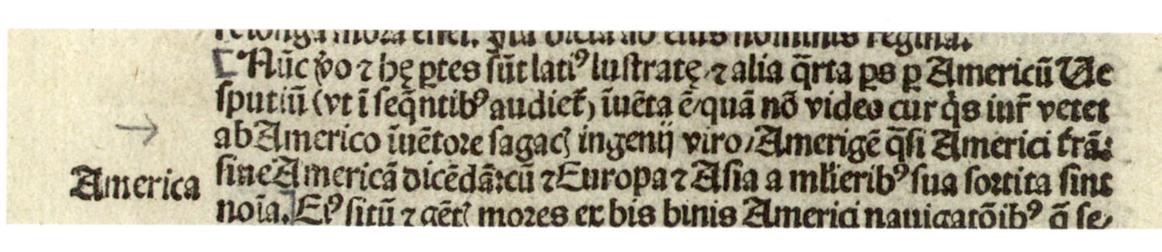

The Naming of America: *Cosmographiæ Introductio*, 1509
Where does the name America come from? The book that explains the baptism of America was published in 1507 in Saint-Dié, a small town in the duchy of Lorraine whose inhabitants were passionate about the arts and sciences. Several hundred kilometres from the sea, Saint-Dié boasted a handful of scholars devoted to the geography of new worlds; they compiled, translated, and printed any voyage account that fell into their hands. These scientists, worthy representatives of the Renaissance, were Gauthier Lud, publisher; Mathias Ringmann, Hellenist writer; Jean Basin, Latinist poet; and Martin Waldseemüller, cartographer. Around 1507, their patron, René II, duc de Lorraine, provided them with the letters of Amerigo Vespucci, an Italian navigator, who described the existence of a new world beyond the Atlantic. They printed a map of the world in Latin as well as a small explanatory booklet, *Cosmographiæ Introductio*, which contained a theoretical section on geometry and astronomy, and then the writings of Vespucci translated into Latin. In chapter 9, on the divisions of Earth, they introduced an element revolutionary for the time: the presence of a fourth continent surrounded by water and located between Europe and Asia. They attributed the discovery of this "new world" to Vespucci and, as a consequence, dubbed the landmass "Amerige, that is the land of Amerigo, or America, after the sagacious man who discovered it." Waldseemüller's map, which accompanied the book, also bore the inscription *America* on newly explored South American territories. In his later works, however, the cartographer abandoned this name, perhaps due to second thoughts about having chosen Vespucci's name over Columbus's. But it was too late: other sixteenth-century cartographers, including Sebastien Münster and Gerard Mercator, had already adopted the name.

***El Tratado* (1494)**
Tordesillas, Spain, was the site of the signing of the most stunning treaty in human history – the Treaty of Tordesillas. On June 7, 1494, Spanish and Portuguese diplomats agreed on a partition of the world, dividing "el Atlantico por medio de una raya trazada de polo a polo a 370 leguas al oeste de las islas de Cabo Verde." The Spanish sovereign approved the text on July 2, while Juan II of Portugal, who had been staying nearby (Setubal) to follow the negotiations, gave his consent only on September 5. It is likely that he was waiting for reports from his spies and wanted to be certain that his country could accept the division this time. In fact, the preceding year, Pope Alexander VI had promulgated a bull planning for the evangelization of the islands both discovered and to be discovered and had proposed "una raya o linea de polo a polo" 100 leagues west of the Cape Verde Islands and the Azores. The Portuguese wanted to protect their sea route to the Cape of Good Hope and perhaps also to ensure their possession of some islands whose existence they suspected to the west. They therefore rejected the "Alexandrine line" and demanded a new division. The Treaty of Tordesillas accorded them another 270 leagues, or about 1,350 kilometres – that is, more or less up to 46°W. Brazil would be Portuguese; the rest of the Americas would theoretically fall under Spanish authority. The museum in Tordesillas devoted to this treaty, "Casas del tratado," comprises two contiguous houses, and the Cantino map is used to illustrate the meridians proposed in 1493 and 1494.

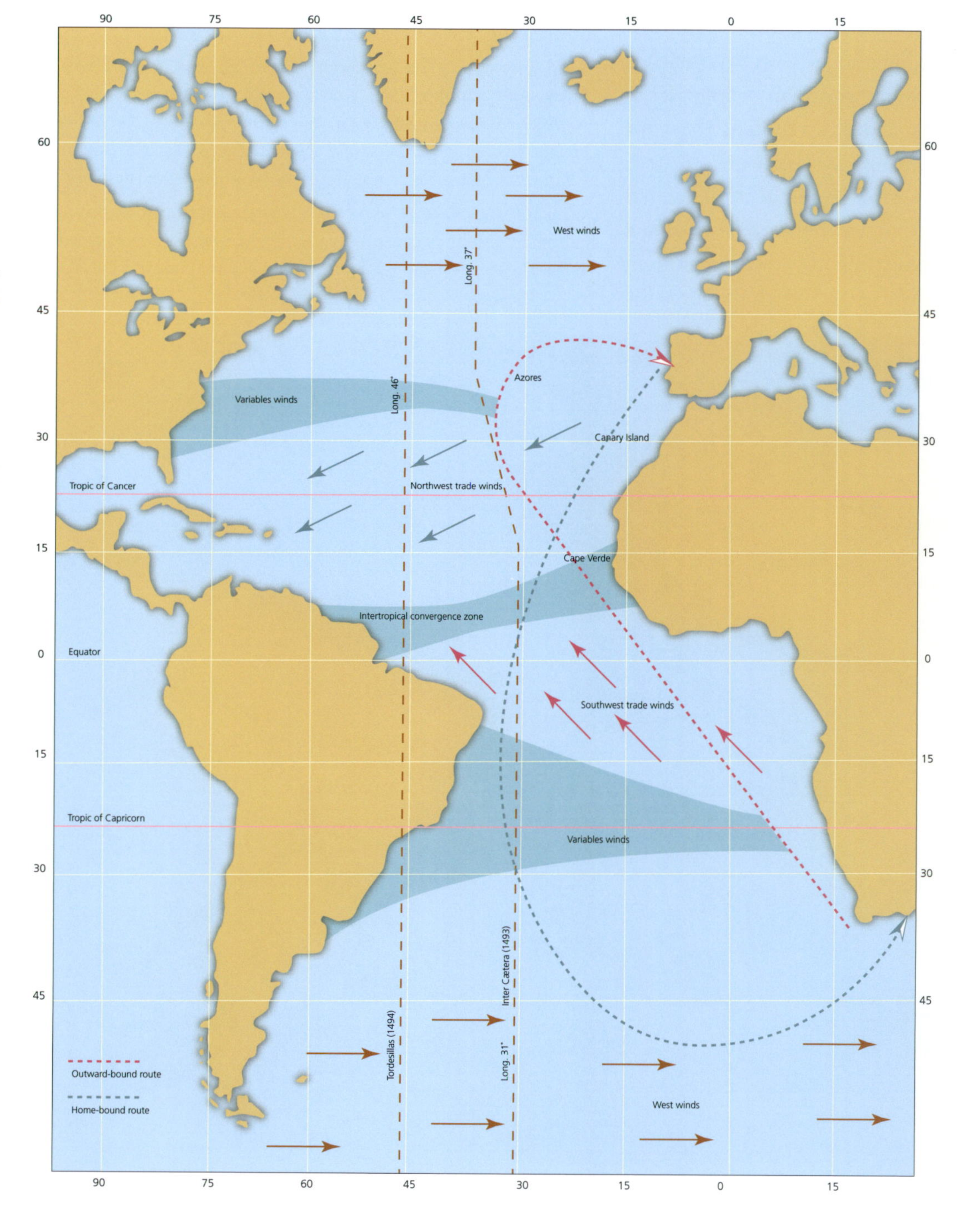

Se Chiama Francesca

FROM MAGELLAN TO THE VERRAZZANO BROTHERS

After Christopher Columbus returned to Europe in 1493, the number of voyages of exploration began to multiply. European navigators were feverishly seeking a westward passage to Asia. Even though it was very difficult to calculate longitude, cartographers sketched out various hypotheses: the Orient might be reached by sailing around the tip of Africa, or by following the coasts of the "new" landmass to the north or south.

Bartholémy Dias had reached the tip of Africa in 1488. According to the Treaty of Tordesillas, this southern route to Asia was reserved for the Portuguese. The Spanish therefore intensified their assaults on the incredibly imposing barrier thrown up by the Americas. Very quickly, the extent of the obstacle became clear; Martin Waldseemüller's map illustrated the scope of the problem. The solution had to be to explore the centre of the landmass – to search at the bottom of the Gulf of Mexico.

Chance or intuition led the Spaniard Vasco Nunez Balboa to the Panama region. He had extensively explored the coast from Cartagena to Darien. In the fall of 1513, he undertook an expedition through jungles and marshes and encountered their denizens – crocodiles, snakes, and leeches. The Indians serving as his guides understood that he wanted to reach the "great waters" and took him across the narrowest part of the continent, a distance of 150 kilometres. Balboa flattened everything in his path and conducted several massacres. After taking possession of the "sea" (Panama Bay) and the region in the name of the Spanish crown, he returned to Darien. The authorities mistrusted him but could not stop him. He cleared a passage between the two oceans and initiated several personal projects with a boldness that many found disturbing. The governor had him tried and convicted for treason and, in January 1517, he was beheaded. The conquistador had been given a lethal taste of his own medicine.

Balboa had identified a passage, but not a navigable route. The search continued. Ferdinand Magellan decided to take a bold step and go south. He left Sanlucar de Barrameda, Spain, on September 20, 1519, with five ships and 270 men recruited more or less at random – for the planned expedition struck fear into the hearts of the most experienced sailors. And the voyage proved them right. After 13 long months at sea, punctuated by two attempted mutinies, Magellan finally reached a twisting strait, which was to bear his name. On Wednesday, November 28, 1520, with his three remaining ships, he found before him a new ocean, which he later named the Pacific, the word that came to him after he had survived 580 kilometres of wind and currents in the strait north of Tierra del Fuego. Mission accomplished; now all he had to do was return home by continuing to sail "westward" – or so he thought.

After "three months and twenty days without tasting fresh food" or seeing a speck of land, Magellan grew discouraged. Where were those Molucca Islands marked on the Portuguese maps? On January 20, 1521, in a fit of anger, he threw the maps overboard. The next day, recounts Antonio Pigafetta in his diary (an invaluable source of information), they came upon a small archipelago. And then there were more islands, where the crews were able to gather provisions. Magellan finally became habituated to the length of his voyage. Time lost its importance. He was resigned, or heartened. On March 28, his Malayan slave, Enrique, acquired during an earlier voyage to India (via the Cape of Good Hope), spoke to some indigenous people in a pirogue. To general surprise, they understood each other. Enrique, after a long

Following pages
Portuguese map of the Atlantic Ocean, 1519
Taken from an atlas produced around 1519, this magnificent colour map, richly illustrated, is a tribute to the glory of Portuguese expansion in the time of King Manuel. Its maker, Lopo Homem, master of Portuguese nautical charts, expressed his art majestically. Realistic drawings of Indians, animals, and vegetation fill the unknown zones. The flags indicate Portugal's claim to the territory that was to become New France. There are numerous place names of Portuguese origin on the shores of Newfoundland, several of which have survived to the present day, such as C RAZO (Cape Race), B DA CONCEPCION (Conception Bay), and Y DAS BACALLAOS (Baccalieu Island). The map also bears traces of the first Portuguese expeditions to these northern areas, including those of the Corte-Real brothers, in honour of whom Terra Corte Regalis (today's Newfoundland and Labrador) was named.

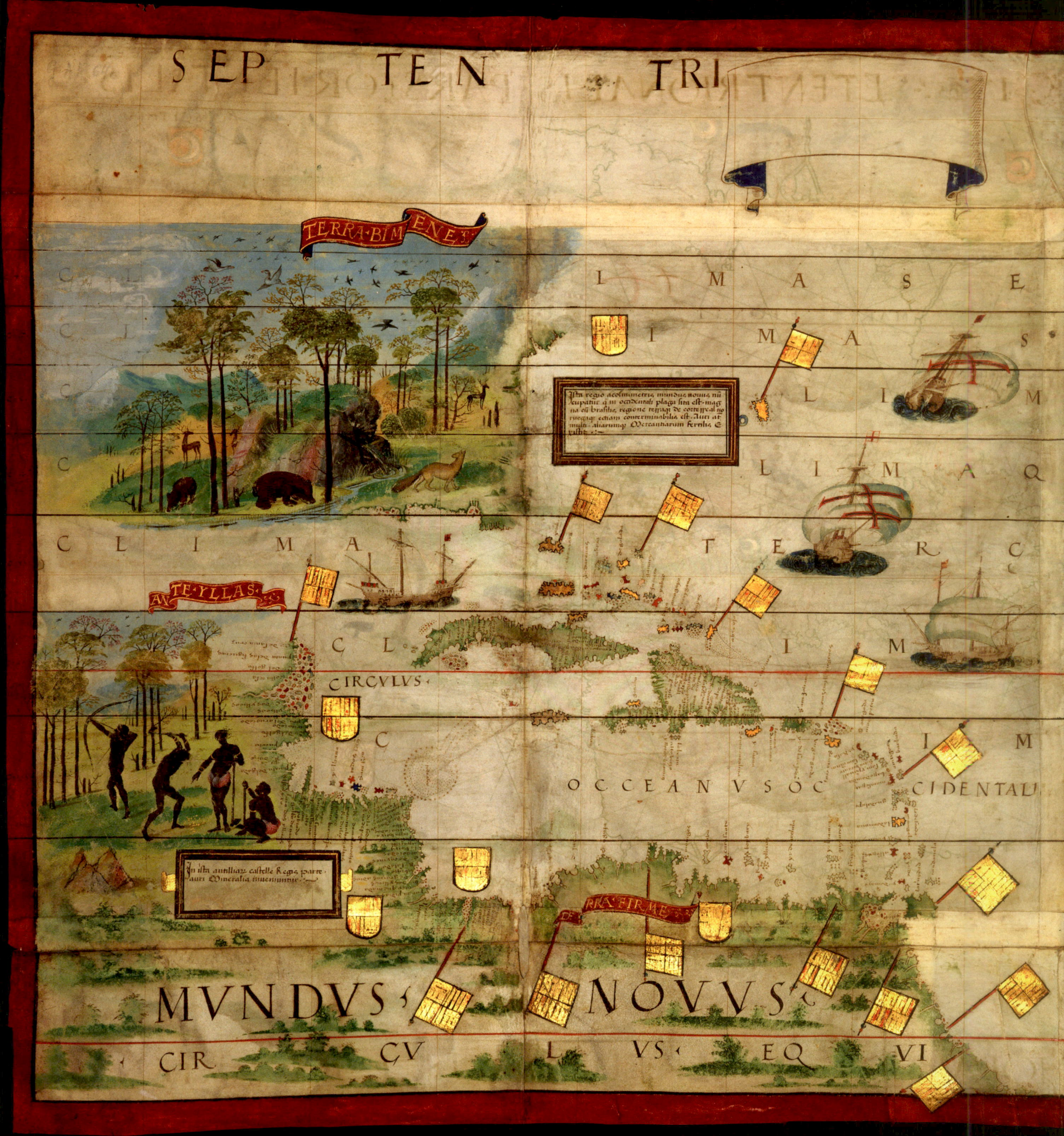

SEP TEN TRI
TERRA·BIMENES
C L I M A
C L I M A S E
I M A S
L I M
L I M A Q
T E R C
C L I M
CIRCVLVS
OCCEANVS OCCIDENTALI
ANTEYLLAS
Ista regio acolmmetris mundus novus nū
cupatur q in occidentali plaga sita est · mag
na cū brasilie regione terraqz de corte real no
rvegaqz etiam contermiabilis est · Auri at
multi · aliarumqz Mercantiarum fertilis e
xistit
In ista antilliaz castelle Regis parte
auri Mineralia inueniuntur
TERRA FIRME
MVNDVS NOVVS
CIR CV L VS EQ VI

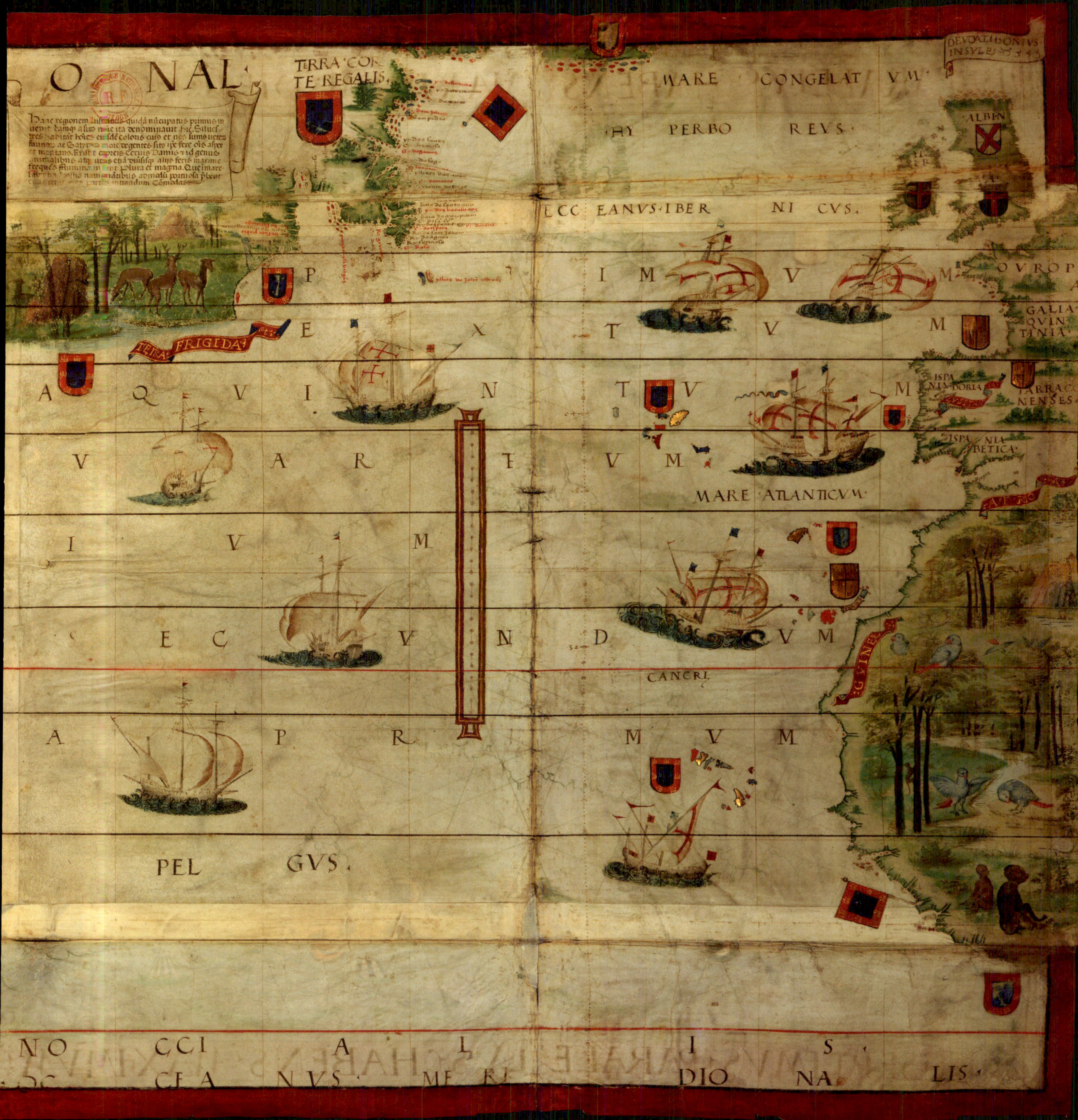

O NAL
TERRA CORTE REGALIS
MARE CONGELATVM
HYPERBOREVS
ALBIN
OCCEANVS IBERNICVS
OVROP
GALIA QVINTANIA
TERA FRIGIDA
ISPA NIA BETICA
TARRACONENSES
MARE ATLANTICVM
CANCRI
PELAGVS

America: A new island, map by Sebastien Münster, 1545
In the years following the invention of the printing press, Ptolemy's *Geographia universalis*, portraying the known world and systematically listing locations by latitude and longitude, was certainly the most popular geographic treatise. This edition, which appeared in Latin in 1545, is the work of the German Sebastien Münster, one of the best-known cartographers of the Renaissance. On this map Münster has drawn the contours of North America as a distinct continent, separated from CATHAY (China) and ZIPANGRI (Japan) by the MARE PACIFICUM (Pacific Ocean, named by Magellan). To the north, the name FRANCISCA designates the part of the continent claimed by the French following Verrazzano's voyages. CAP BRITONUM, one of the few place names in North America, is a reminder of visits by Breton fishermen.

stopover in Europe, had finally travelled around the world. His master had accomplished more or less the same feat. In fact, this was the region where Magellan had purchased Enrique. Beginning to feel that he had the soul of a conqueror, Magellan was struck with a craving for gold and spices and a mania for missionary work. After protracted negotiations, his host, the king of Cebu, agreed to have himself and his followers baptized. In return, however, Magellan had to help him vanquish a rebel vassal who had taken refuge on the island of Mattan. Leading 49 Spaniards, Magellan did battle with three successive waves of warriors, over 1,000 in total. Pigafetta relates that Magellan, though wounded, attempted to cover the retreat of his men, until he was struck dead with a "cane lance to the face" delivered by a warrior. A certain Barbasa took over command but argued with Enrique, who fomented a revolt. In the end, only 108 men remained of the original expeditionary force – none of them an officer, pilot, or astronomer.

The Basque Sebastian Del Cano, who had joined Magellan's expedition to escape a sentence of hard labour, took command; he sacrificed a ship that was too damaged and organized the return to Portugal, after lading his holds with cloves and other spices. In May 1522, having left a ship in the Moluccas, the *Victoria* sailed around the Cape of Good Hope; on September 6, 1522, almost three years after its departure, it entered the bay at Sanlucar de Barrameda. The many adventures and details of this amazing voyage would have remained unknown except for the marvellous journal kept by Pigafetta, a young, intelligent, educated Italian whose diplomatic skills landed him among the 18 survivors.

A passage to the west had been found, but it was not very practical. Another route would have to be located, perhaps north of the West Indies.

Up to this time, France had relied on its ships from the Basque, Brittany, and Normandy to occupy the seas. But then, the exploits of freebooter Jean Fleury began to cause a sensation. In 1523, this dashing captain in the employ of Jean Ango, a wealthy shipowner from Dieppe, seized a small Spanish fleet loaded with treasure stolen from Moctezuma, the Aztec emperor conquered by Cortes.

King François I had his share of ambition, and what he was hearing was encouraging him to take action. However, it was the Italian merchants and bankers who had moved to Lyon who took the initiative of financing, with royal blessing, new explorations to the northwest. They put their faith in the Florentine Giovanni Da Verrazzano.

Verrazzano was a literate man and an experienced sailor. It was sometimes said that he had accompanied the Frenchman Thomas Aubert to Newfoundland in 1508. He had no doubt met his compatriot Pigafetta in Paris. In early January 1525, he set sail and headed west, reaching the American coast at 34° latitude and turning north along the coast. He went straight ahead past the expanse of Chesapeake Bay but stopped at the future site of New York, where he gave the name Francesca to a spot sheltered behind a long island; he then sailed around Cape Cod and reached Nova Scotia. There was no "discovery" to be had here: the area was teeming with European fishermen and Basque whalers.

Throughout his six-month voyage, Verrazzano carefully observed and noted the locations that he visited. He no doubt produced maps, but they have not been found. His brother, Gerolamo, a cartographer, is credited with an interesting world map that contributes useful details on the Atlantic coast. It is included in the extensive collection at the Vatican Library, while another of his maps is kept at the National Maritime Museum in Greenwich.

From Giovanni Verrazzano, there is a long letter that he wrote upon his return. He recounts what he saw, describes the contours of the regions he visited, and above all suggests – "that which I know for certain," he emphasizes – that North America is a continent distinct from Asia and that more attention should perhaps be paid to it, rather than focusing solely on the dream of a maritime route to China. In fact, he opined, it was full of promise and was worth developing for itself. The flora was rich and the fauna abundant. The inhabitants, though sometimes bellicose, were in general beautiful, intelligent, and ingenious. Like too many voyagers, he could not resist the temptation to bring some indigenous people back to France. One day, he came upon

> ... a very old woman and a young girl of about eighteen or twenty, who had concealed themselves for the same reason; the old woman carried two infants on her shoulders and behind her neck a little boy eight years of age; when we came up to them they began to shriek and make signs to the men who had fled to the woods. We gave them a part of our provisions, which they accepted with delight, but the girl would not touch any; every thing we offered to her being thrown down in great anger. We took the little boy from the old woman to carry with us to France, and would have taken the girl also, who was very beautiful and very tall, but it was impossible because of the loud shrieks she uttered as we attempted to lead her away; having to pass some woods, and being far from the ship, we determined to leave her and take the boy only.

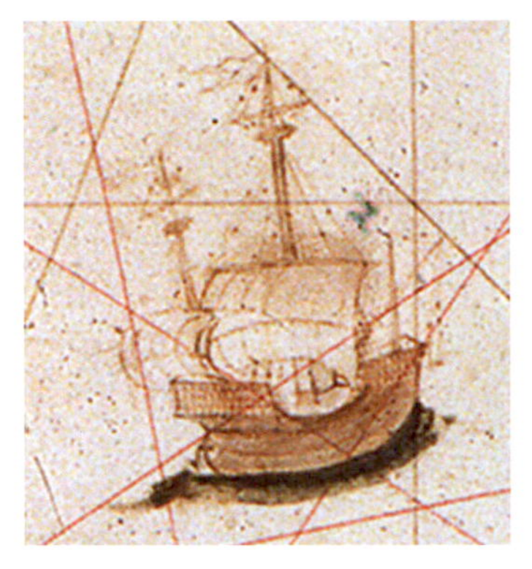

The baptism of New France by Gerolamo Verrazzano, 1529
This map, conserved at the Vatican library, could be considered the baptismal certificate of New France. It shows, for the first time, in its Latin form, the name NOVA GALLIA, replacing the name Francesca. Drawn by Gerolamo Verrazzano, it describes, among other things, the lands explored by his brother, Giovanni, in 1524; Giovanni had noted, *"Come tuta la terra trovata se chiama Francesca per il nostro Franscesco."* A little to the north of Florida, the explorer thought that he had seen the Sea of Asia hidden behind an isthmus. He was probably dreaming of an easy passage to the Orient and had no idea of the extent of North America. For the complete map, see the following pages.

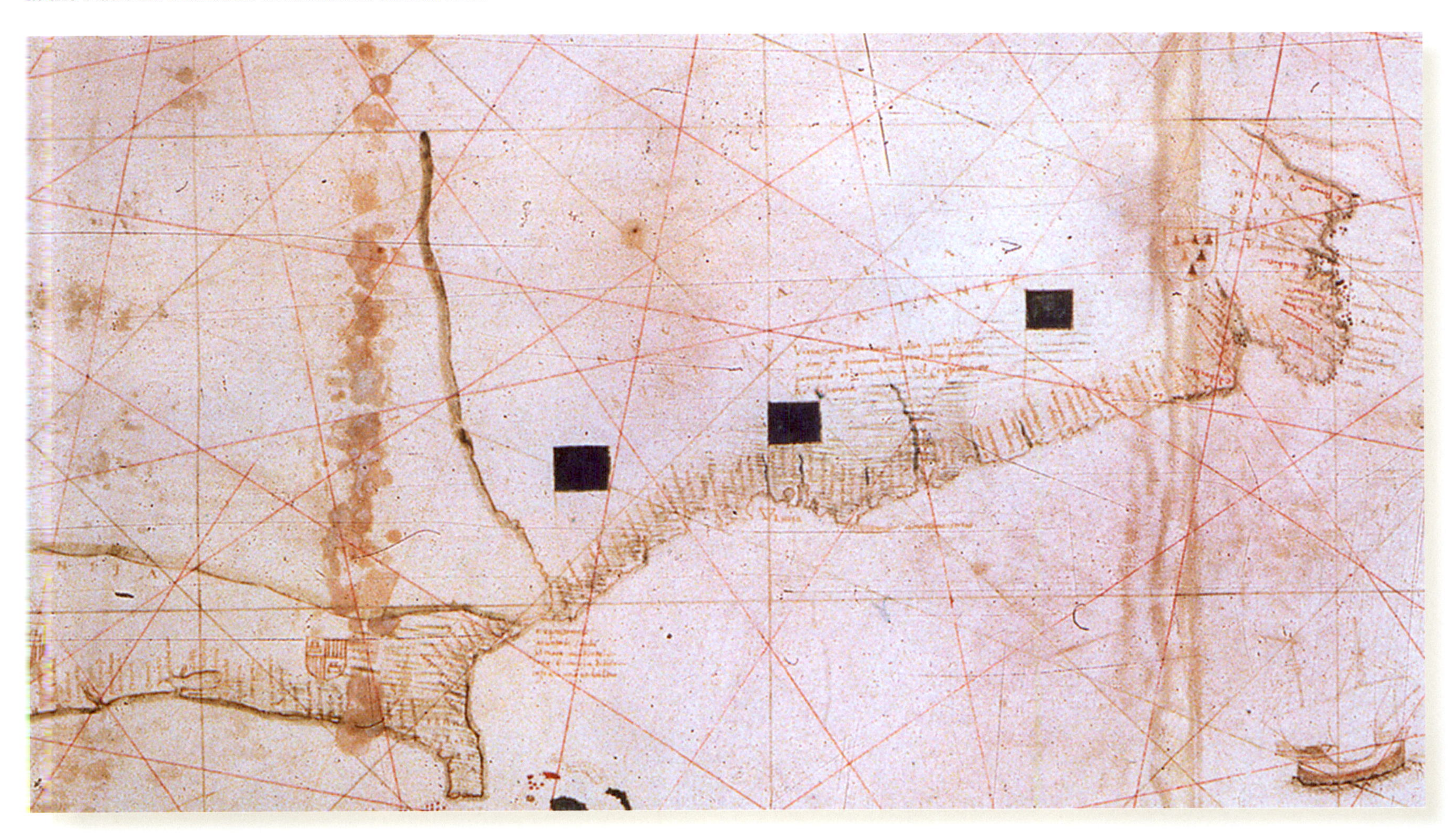

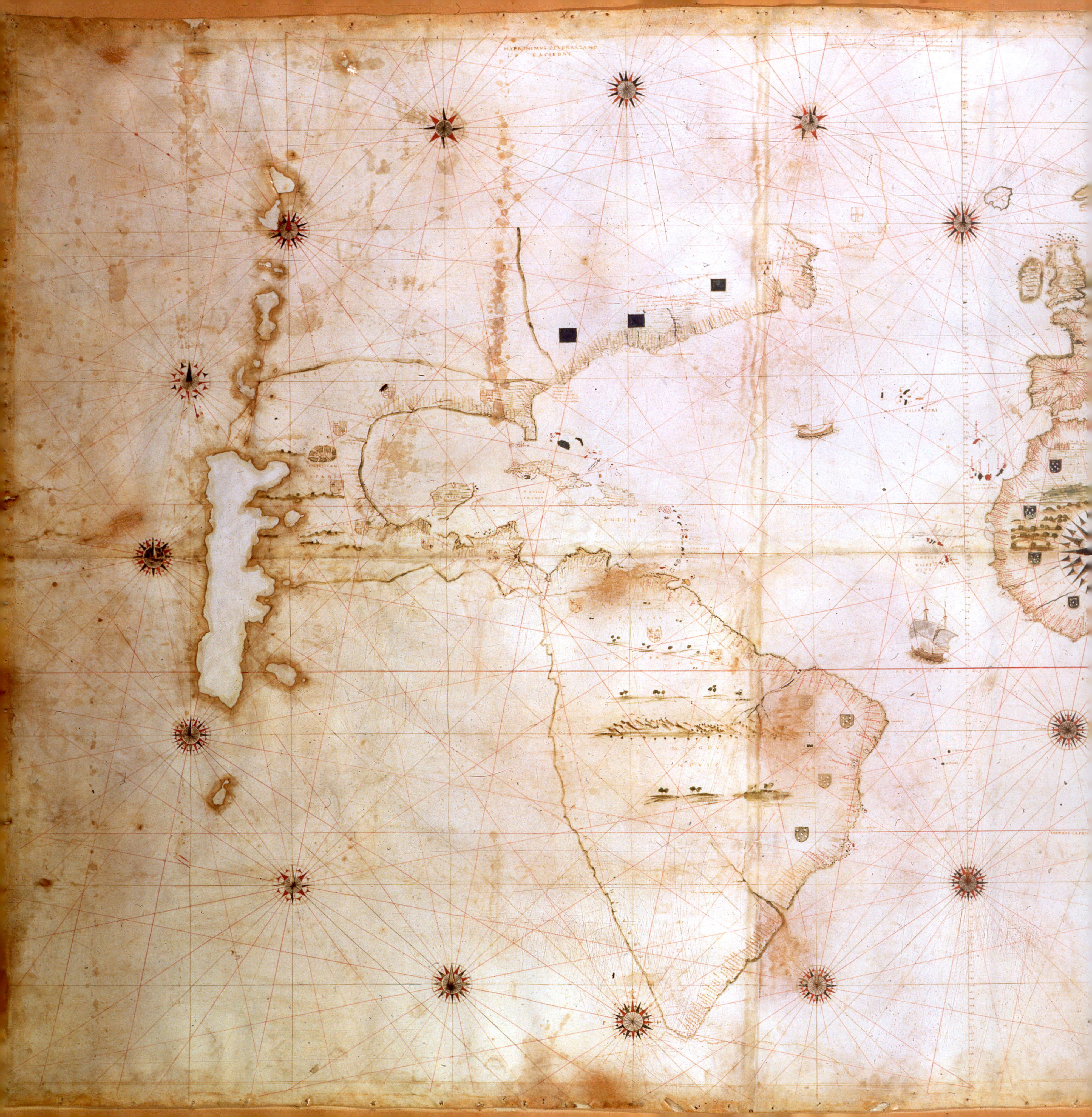

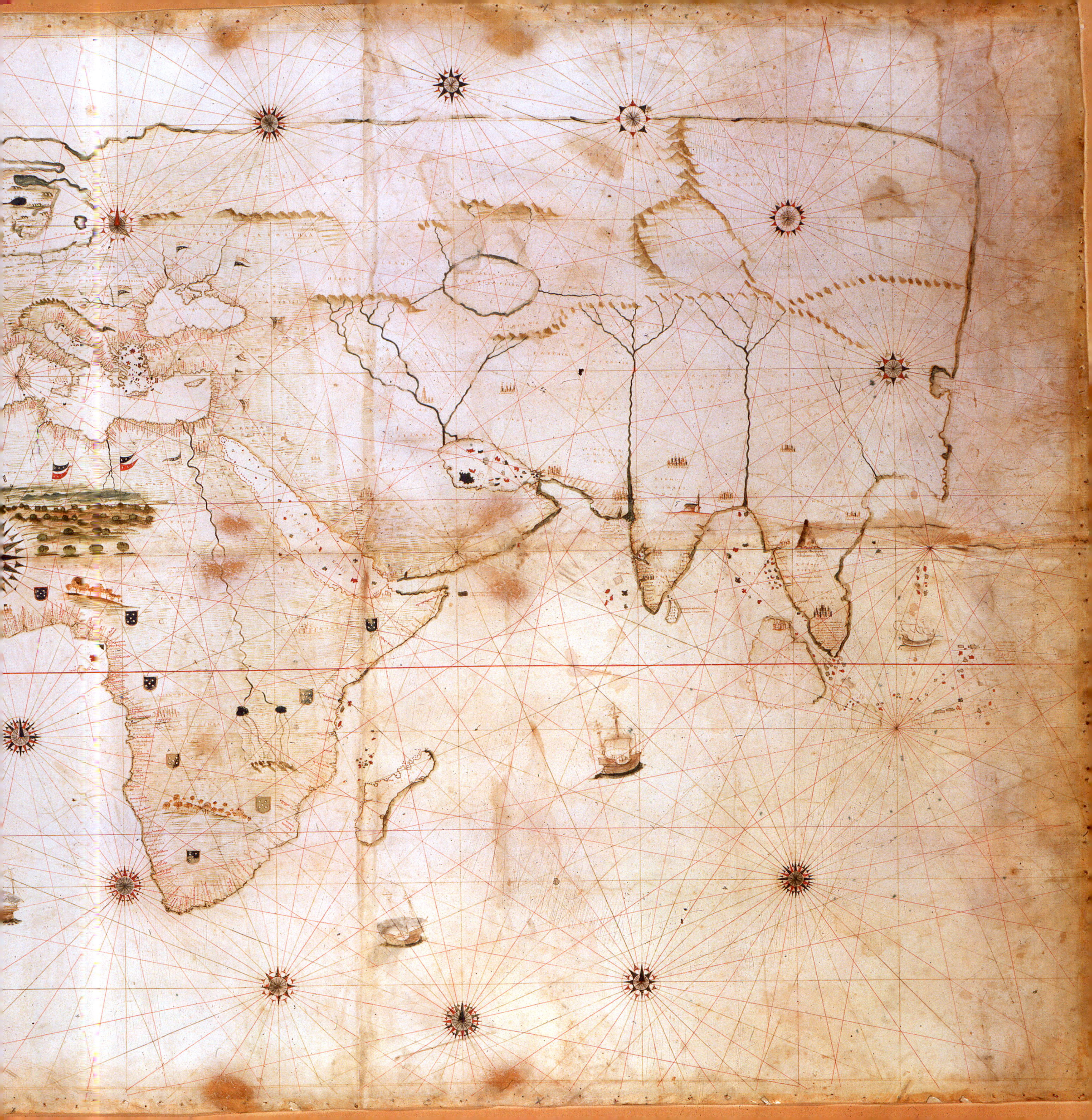

***Mondo Nuovo* by Thomaso Porcacchi, 1576**
This map of the New World is drawn from *L'isole piv famose del mondo*, an *isolario* by Thomaso Porcacchi published in Venice and dedicated to the infant Don Juan of Austria. The *isolario*, an atlas containing only maps of islands, was a prized literary genre in the sixteenth century. This book, in fact, was a success: it first appeared in 1572 and was then republished a number of times in the following years. It was common to portray North America almost in the form of an island, as is done in this map, which is drawn with simplified contours. NOVA FRANZA holds a place of honour, with capital letters, in the centre of the continent, beside LARCADIA, CANADA, FLORIDA, and TERRA DEL LABORADOR.

Later, they fraternized with other Indians "which are all in good proportion, and as such belong to well-formed men." And, he continued, "Their women are of the same form and beauty, very graceful, of fine countenances and pleasing appearance in manners and modesty."

Aside from some writings of ethnographic interest, the Verrazzano brothers produced a number of place names, but none of these were permanently adopted except for Arcadia, which became Acadia under Champlain's pen. The name "Francesca," in honour of the king, was replaced by Nova Gallia, then by Nova Franza (Zaltieri, 1566), Nova Francia (Mercator, 1569, and Wytfliet, 1597), and finally Nouvelle-France (Champlain, Sanson d'Abbeville, Franquelin, and others). The region that Gerolamo called Norembega long captured the imagination of explorers and continues to be the subject of research.

Verrazzano Bay, which it is believed that the explorer saw at the beginning of his voyage and thought was a "western sea," was named later and eventually renamed, but his admirers managed to ensure that his name graced the immense suspension bridge that crosses upper New York Bay.

Main sources in order of importance

PIGAFETTA, Antonio. *Relation du premier voyage autour du monde de Magellan: 1519-1522*. Edited by Léonce Peillard. Paris: Tallandier, 1984. — MOLLAT DU JOURDIN, Michel, and Jacques HABERT. *Giovanni et Girolamo Verrazano, navigateurs de François Ier*. Paris: Imprimerie nationale, 1982. The gold-standard reference on the Verrazzano brothers; the authors spell the name "Verrazano." — FITE, Emerson D., and Archibald FREEMAN. *A Book of Old Maps: Delineating American History from the Earliest Days Down to the Close of the Revolutionary War*. New York: Dover Publications, 1969.

Sources for translation: VERRAZZANO, Giovanni da. "Voyage of John de Verrazzano, along the Coast of North America, from Caroline to Newfoundland, A.D. 1524." In *Collections of the New-York Historical Society*, Second series, Vol. 1, pp. 37–67. New York: New York Historical Society, 1841 [http://www.americanjourneys.org].

From Cabot to Cartier

CAIN'S COUNTRY

WHO DISCOVERED CANADA, Giovanni Caboto (John Cabot) or Jacques Cartier? Who can claim to be the senior "discoverer"? Who, the English or the French, were the first? These questions, of course, are meaningless because human beings were already living there.

Cabot, like Columbus, thought that he was in the land of the Great Khan. A few brief reports by contemporaries indicate that he reached the North American coast, although he did not explore it. He would have seen signs of human presence there. Not more. But the fish were so abundant that one could fish "not only with the net, but in baskets let down with a stone," as legend was to have it.

In short, exactly where John Cabot visited in 1497 has yet to be determined. There is even less information on his second voyage (1498), as no one returned from it.

Things were different for Jacques Cartier. His first two voyages were extensively documented, even though the anonymous accounts that have survived are a translation (for the first voyage) and a transcription (for the second).

A First Official Voyage

Cartier, a captain from Saint-Malo, was chosen by King François I to go to the new territories "to discover certain islands and lands where it is said that a great quantity of gold, and other precious things, are to be found." The royal instructions ordered Cartier to go beyond Baye des Chasteaulx (Bay of Castels, today the Strait of Belle-Isle). These details, along with the rapidity of the crossing – 20 days out and 21 back – indicate that the route was well known. Beyond the Strait of Belle-Isle, Cartier had the leisure of observing, noting, and naming. But he was not alone. A little farther west, he saw a ship. Did Cartier fear that he might have to deal with the Spanish? It was very possible. The two sovereigns, François I and Charles Quint, were at war, and their armies were engaged in battle in various places. The part of the world where Cartier found himself was assigned to Spain under the terms of the Treaty of Tordesillas (1494).

Pope Clement VII, whose life was as tumultuous as that of the infamous Pope Alexander VI, was seeking an alliance with France in order to make inroads into Charles Quint's power, and it was not difficult to convince him to reinterpret the famous treaty so that other Christian nations could undertake voyages of discovery. It is said that Abbot Jean Le Veneur of Mont-Saint-Michel, an admirer and patron of Cartier's, obtained this new interpretation by arranging the marriage (1533) of François I's son to Clement VII's niece. The Spanish may have known about the new interpretation, but whether they had accepted it was definitely open to question.

West of the Strait of Belle-Isle on that day, however, there was no need for alarm; the ship encountered by Cartier was from La Rochelle. In fact, its captain didn't seem to know exactly where he was. He was looking for a harbour from which to fish. Cartier later penned a comment that became famous: "If the soil were as good as the harbours, it would be a blessing," he wrote on June 12, 1534, "but the land should not be called the New Land, being composed of stones and horrible rugged rocks … In Fine, I am rather inclined to believe that this is the land God gave to Cain."

However, Cartier observed, "There are people on this coast whose bodie-s are fairly well formed, but they are wild and savage folk … They clothe themselves with the skins of animals, both men as well as women, but the women are wrapped up more closely and snugly in their skins." They had boats "made of birch bark," from which "they catch many seals." Cartier concluded that these

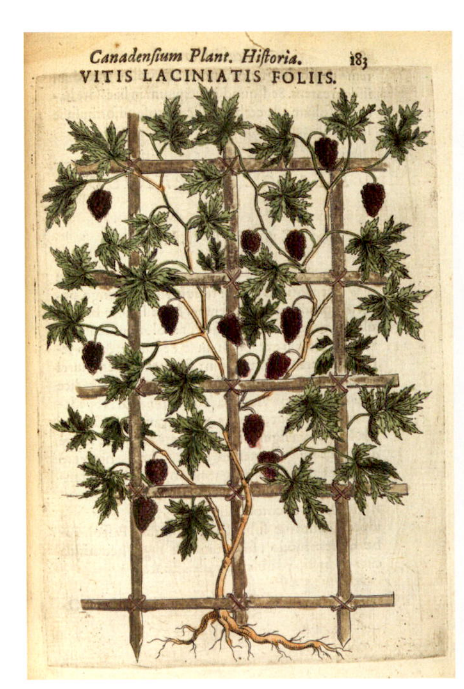

American vines, in Cornuti's history of the plants of Canada
This exceptionally beautiful colour engraving, printed in 1635, is a reminder of the fertile soils of the Americas. One century earlier, in 1535, Île d'Orléans had been dubbed Bacchus Island by Jacques Cartier, in honour of the Roman god of the vine and wine. The plant portrayed here (*Apios americana*) was grown by the Indians for its edible root.

Ilondis of the entillas
The grete occeane sey
The new fonde londe quhar men goeth a fisching
espagnolla

people came from "warmer countries" and that they were there, like the Europeans, "to catch these seals and to get other food for their sustenance."

The next day, Cartier set sail heading southwest from Newfoundland. On the island of Brion in the Magdelen Islands archipelago, he noted "fields full of wild oats ... gooseberry bushes, strawberry vines ... bears and foxes," and walruses. He continued sailing westward. In early July, he saw an opening that looked like a gulf, and he ventured into it. The Indians he found there were accustomed to seeing Europeans; the two groups talked a number of times, but Cartier was concerned above all with exploration. On July 8, he sent crews in boats to "go and explore this [Chaleur] bay." The following day, the Frenchmen had to admit the facts. At 10 o'clock in the morning, "we caught sight of the head of the bay ... At the head of this bay, beyond the low shore, were several very high mountains." They had reached the Ristigouche River. "And seeing there was no passage, we proceeded to turn back." As if to console them, Indians brought "some strips of cooked seal, which they placed on bits of wood and then withdrew, making signs to us that they were making us a present of them." Not to be outdone, the Frenchmen sent the Indians hatchets and knives. This was the signal for exchanges to begin, and there was a celebration. The Indian women danced and sang, and some "advanced freely" toward the Frenchmen, rubbing their arms with their hands then lifting their hands skyward "exhibiting many signs of joy." Cartier noted that "they are people who would be easy to convert" to Christianity. The climate was temperate and the land "the fairest that can possibly be found." Before leaving, Cartier named the bay Chaleur Bay (Bay of Heat).

Two weeks later, on July 22, 1534, the Frenchmen were in the Bay of Gaspé (a name no doubt of Micmac origin, which was later given to the bay). A new group of Indians awaited them, speaking a language that was different from that of the previous group. On the 24th, Cartier

Map by Jean Rotz of Dieppe, circa 1542
Cartographer Jean Rotz, born in Dieppe to a Scottish émigré father and a French mother, had extensive sailing experience. A pilot, merchant, and privateer, he left France in 1542 and went to the Court of England to offer his services to King Henry VIII, presenting him with a magnificent atlas known as the *Boke of Idrography*, prepared originally for François I. One of the maps in this work shows the North American coast with the south at the top. The mouth of the St. Lawrence is drawn in, but not the estuary. It is presumed that Rotz had had word of Jacques Cartier's first expedition in 1534, but didn't know about the second expedition (1535–36). Given the absence of French names on the map, it is also possible that it was made by a third person – for example, a Portuguese navigator.

had an immense cross, "thirty feet high," erected, on which was inscribed, "Long Live the King of France." The Frenchmen knelt before it while the Indians looked on, admiring. But Donnacona, the Iroquois chief who had welcomed Cartier, had second thoughts and delivered "a long harangue" to the Frenchmen, pointing at the cross and indicating "all the land around about, as if he wished to say that all this region belonged to him, and that we ought not to have set up this cross without his permission." Cartier calmly explained that the cross was simply a marker, as the French had every intention of returning to this spot. Up to then, Donnacona and the Indians with him had remained in a canoe near Cartier's ship. Suddenly, the Frenchmen caught hold of the canoe and invited its occupants to come aboard their ship, treating them with "every sign of affection." Cartier again explained that the cross would be helpful in facilitating their return. In fact, he suggested that he take several Indians back with him to talk about their world and at the same time serve as future interpreters. Domagaya and Taignoagny, who were always introduced as the chief's sons, were chosen to make the voyage to France. No one was upset, it seems. The Indians came in great numbers to make their farewells, according to the account attributed to Cartier, bringing fish and making "signs to us that they would not pull down the cross."

Cartier sailed northward, reaching the point of Anticosti and continuing along its north shore; he thought that he was in a bay, but he was in the St. Lawrence River. The weather was bad. Was summer over already? The crew was concerned; it would be better to go home. The trip back was almost as easy as the trip out had been. Cartier conversed with the two Indians. Were they anxious? Homesick? Perhaps. Certainly, they told

Cartier enough that he wanted to leave on another voyage as quickly as possible.

On October 30, 1534, a month and a half after his return, Cartier obtained a new commission from the king. He would leave in the spring with three ships, rather than two, to continue his explorations. The two Indians had given him to understand that he had missed a large river flowing from the west and leading to a large village called Hochelaga, located near torrential rapids that would impede further travel.

The Second Voyage (1535–36)

Cartier's second crossing took almost two months, after which he headed straight through the Strait of Belle-Isle and along the north shore of the St. Lawrence. On August 10, 1535, he was opposite the island of Anticosti, at about the spot where he had turned around the previous year – and at the same time of year. A contrary wind was blowing, so Cartier took shelter in a bay to which he gave the name St. Lawrence, as it was this saint's day – a widely celebrated day in Brittany. Today, this is Baie-Sainte-Geneviève; Cartier's place name was, however, passed on to the gulf and then the river, as well as to the Laurentians and Laurentia.

According to Cartier's two Indian guides, who had somewhat miraculously survived their transatlantic adventure – not so much the two crossings as the exposure to European microbes – this river led to Canada. Domagaya and Taignoagny described what lay ahead. Soon, they would cross the "kingdom of the Saguenay" and be on the "the way to the mouth of the great river of Hochelaga and the route towards Canada," a name clearly of Indian origin and adopted immediately by the Frenchmen. The water would turn from salt to fresh, the Indians told Cartier, and this river was so long that no one had gone all the way to its end. Although he was still obsessed with the search for a passage, Cartier decided, before making his leap into the unknown, to explore along the south shore of the river for a considerable distance, cross back to the north shore, return to where he had started, and then continue toward Canada.

On September 6, he was beside an island, which he named "Hazel-bush Island." This was where "the province and territory of Canada begins." The next day, his men landed on the north shore. The Indians fled upon their arrival, but when Domagaya and Taignoagny made their presence known, the rejoicing began. The Indians brought "many eels and other fish, with two or three measures of corn, which is their bread in that country." Cartier responded with gifts "of little value."

The following day, the "Lord of Canada," Donnacona, arrived with a fleet of 12 canoes and gave "a long harangue." He then spoke with Domagaya and Taignoagny, who told him how well they had been received in France. Reassured, Donnacona spent some time with Cartier, and then the Indians took their leave. Cartier chose to cast anchor at the mouth of the Saint-Charles (Sainte-Croix) River, beside the place where Donnacona was lord and where stood "his abode: it is called Stadacona, as goodly a plot of ground as possibly may be seen."

Cartier quickly took stock of the environs; he was amazed at the beauty of "Bacchus Island" (Île d'Orléans), which he named for the vines growing there. But he did not lose sight of his objective: Hochelaga. Domagaya and Taignoagny declared themselves willing to accompany him. However, in the following days things became confused. Donnacona was upset. He picked quarrels with his visitor. With every passing day, the Indians trotted out new excuses and repeated that Hochelaga was not worth the trip. Donnacona demanded that Domagaya and Taignoagny not go with the Frenchmen. To convince Cartier not to leave, the Indians heaped gifts upon him, including a 12-year-old girl – the chief's daughter, according to Taignoagny – and two younger boys. Given Cartier's determination, the Indians reassessed their offerings and then changed their minds, and Cartier felt that it would be a good idea to thank Donnacona with a magnificent cauldron. But the Indians had not yet run out of ideas. After Donnacona requested that the cannons be fired, Taignoagny took advantage of the ensuing panic to announce the death of two of his countrymen. This was quickly denied. Finally, it was time for the Great Spirit to intervene. Three Indians disguised as devils came to stop the Frenchmen, and their god Cudouagny predicted that Cartier and his men would find "so much ice and snow that all would perish."

In the end, Cartier went without any guides. He was amazed at the splendid view offered by the two shores of the St. Lawrence. Eleven days after he left Stadacona, on September 28, 1535, he arrived at a large lake ("Lake St. Peter"), beyond which there were four or five rivers and as many islands. On one of them, five Indians were hunting. They approached the Frenchmen, and one picked Cartier up in his arms and carried him to land without the slightest effort, "so strong and big was that man." The Indians confirmed that they were going in the right direction to reach Hochelaga.

On October 2, Cartier received an enthusiastic welcome from a thousand people at Hochelaga. After the feast, he explored the environs, including a mountain that he named "Mount Royal," observed the fortifications of the "citie" – as it is called in the 1906 edition of the translation – and counted about 50 longhouses, which he carefully, and quite admiringly, described. For their part, the Indians observed him with respect and naturally ascribed certain powers to him. The ill and infirm

The St. Lawrence River in the times of Cartier, circa 1547 (detail)
This map, known as the *Harleian* (in memory of the collector Robert Harley), was dedicated to the Dauphin de France before he became King Henri II. It presents the contours of the North American coast explored by Jacques Cartier. On it are Portuguese and French names that would have been very familiar to Cartier and to sixteenth-century European fishermen. Some names have lasted to the present day, including BLANC-SABLON, SEPT-ÎLES, ÎLE AUX COUDRES, and ÎLE D'ORLÉANS. Other place names were lost when the sites were renamed – for example, LAC D'ANGOULÊME (Lake Saint-Pierre), Rivière Sainte-Croix (Saint-Charles River), and Rivière SAINT-MALO. On the north coast of the St. Lawrence is a bay that Cartier called "Saint-Laurent"; the name was later given to the entire river by European cartographers. Previously, the St. Lawrence had been called Great River of Hochelaga or Great River of Canada, for the two largest Indian provinces, whose inhabitants, Hochelagans and Canadians, had disappeared by the end of the century, probably exterminated by their enemies or victims of epidemics. To the northeast, the cartographer has translated the Portuguese name Labrador to TERRE DU LABOUREUR. It is possible that he knew the origin of this name, attributed to the Azorean explorer João Fernandes. North of this territory, the map hints at the existence of a northwest passage. Farther south (near today's New York), the cartographer has drawn a narrow strip of land separating the Atlantic from the Pacific that is reminiscent of Verrazzano's famous sea. The island of Newfoundland is still in pieces, while the estuary of the St. Lawrence is too narrow. Knowledge of the interior of the continent being fragmentary, the gap is filled by various scenes, including one that some scholars believe depicts the episode of colonization at Charlesbourg-Royal in 1542; others see it as tenants paying homage to their seigneur, possibly Jacques Cartier himself.

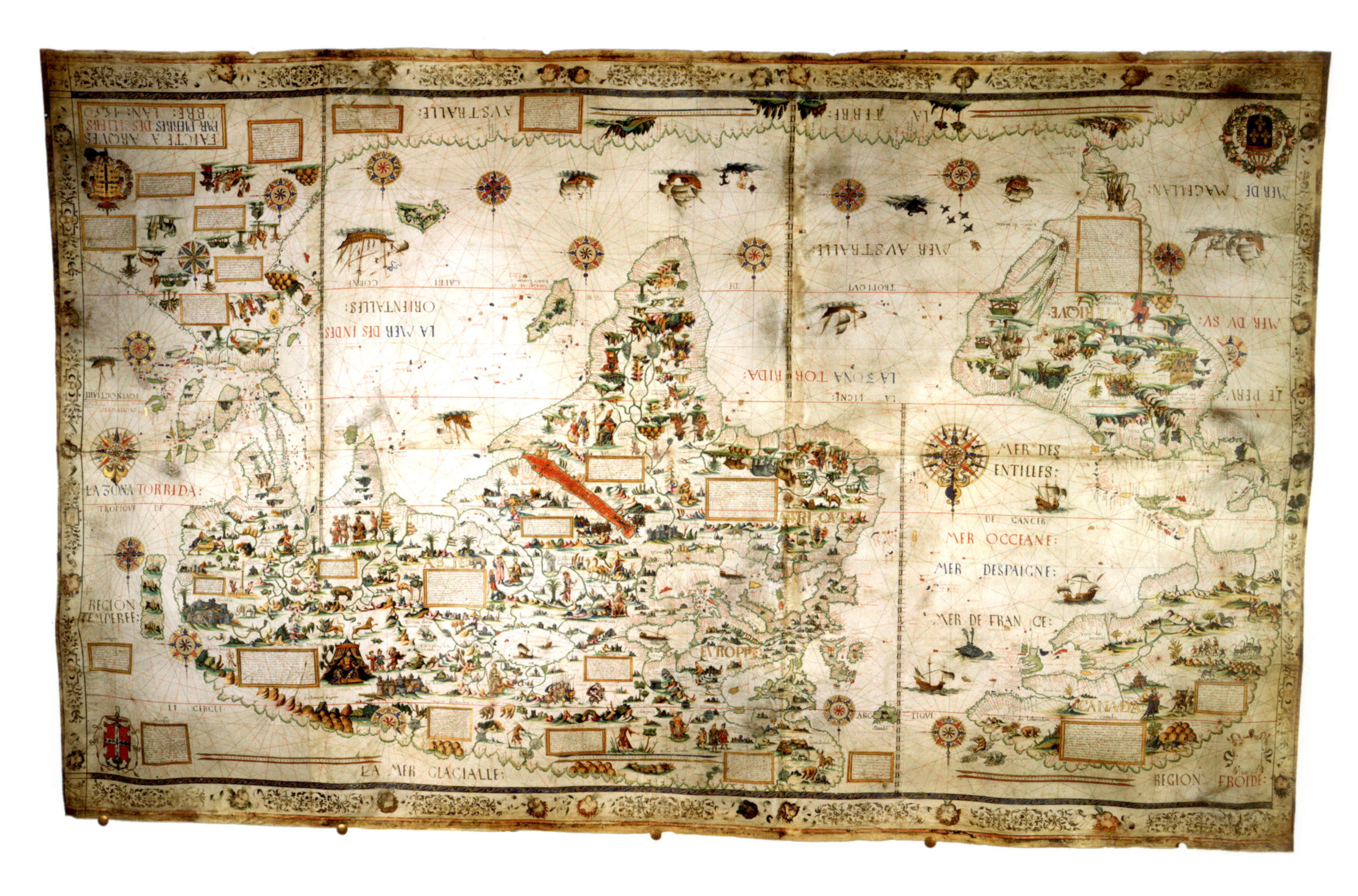

***Above* World map by Pierre Desceliers of Dieppe, Arques, 1550; *left*, detail**

Pierre Desceliers is certainly one of the best-known Norman cartographers. Considered by some to be the father of French hydrography, this priest from Arques (a village 7 kilometres away from Dieppe) was a mathematician, teacher of hydrography, and official examiner of pilots. Although Desceliers never left Europe, the positions that he held enabled him to extract from sailors the information he needed to update world maps. We know that Desceliers was prolific, yet only two of his original maps have survived to the present day; they are now conserved in England (at the British Library and the John Rylands Library). On this one, dated 1550, the North American coast, from Florida to Labrador, is generally well portrayed. The place names are Spanish, Portuguese, and especially French. The cartographer drew the St. Lawrence up to the Lachine Rapids. There is also the marvellous kingdom of the Saguenay, along the river of the same name. Some have hypothesized that the man standing at the centre, below the St. Lawrence, is Jacques Cartier. Others believe that it may be an Indian chief clothed and coiffed in Asian style. A descriptive note explains why the French abandoned the settlement of FRANCE-ROY (Cap-Rouge), on the St. Lawrence, in 1543: "The people of said country were not given to trade because of their austerity, the intemperance of said country, and the small profit; returned to France hoping to go back when it pleases the king." The map is also remarkable for its many realistic and fantastic illustrations: the bear, whale, ships, compass roses, and even a battle between pygmies and cranes! The unicorn probably comes from the accounts of Jean Alfonce, to whom Indians had described this mythical animal. At the time, there was no clearly established convention of placing north at the top of a map. This map has a double orientation: north of the equator, the text and images are upside down, leading some to think that perhaps the map was laid flat so as to be seen from both sides. In spite of this characteristic and others, such as the presence of compass points and compass roses, which were usually seen on portolans, the map was not used for navigation. This cartographic masterpiece, sumptuously illuminated, was intended, rather, for a prestigious clientele: the king of France, Admiral Claude d'Annebaut, and the supreme commander of the French armies Anne de Montmorency, whose coats of arms were drawn on the map.

presented themselves to him. At a loss, Cartier read the Gospel According to John before the gathered Indians.

His visit to the rapids, which he deemed a raging torrent, convinced him that he had reached the end of his voyage. Beyond the rapids, it would take months, he was told, to cross a country populated by "bad people" armed to the tips of their fingers, who "waged war continually one person against the other."

The Frenchmen had left the *Émerillon*, their smallest ship, in which they had travelled from Stadacona, at the end of Lake Saint-Pierre. On October 5, they cast off for Sainte-Croix, where the two other ships, the *Grande Hermine* and the *Petite Hermine*, were moored. On the 7th, they passed Trois-Rivières, where they saw four small islands at the mouth of a river that they called "The river of Fouetz." On the 11th, they returned to the harbour at Sainte-Croix, a name sadly foretelling a winter of horrors.

Donnacona and his people put on a front of being happy to see the Frenchmen return, but mistrust had crept in. They displayed "the scalps of five men" killed during a recent battle against the "Toudamans" (perhaps the Etchemins) – a bad omen. This episode piqued Cartier's curiosity, and he set about studying the habits and customs of the Indians. He was quite intrigued by their habit of smoking, though he disapproved of their way of preparing young women for life. Overall, what he saw disquieted him, and he had the fort where he and his men were staying reinforced. Unfortunately, he could not protect his men against an unexpected enemy: a mysterious illness (scurvy) that broke out in December both in Stadacona and among his ranks. Prayers, masses, and processions had no effect. Twenty-five of the 112 men had succumbed, and almost all of the others were ill, when Cartier noticed that Domagaya, who had been suffering terribly a few days earlier, now seemed to be in good health. Domagaya, not without compassion, explained that he had been cured by "the juice of the leaves of a tree and the dregs of these" and agreed to send two Indian women with these "branches," who would show the Frenchmen how to medicate themselves with *annedda*, which was what he called the tree in question. It was a miracle cure.

In the spring, the tensions dissipated. Donnacona had had a good hunt, and he gave some of his bounty to the Frenchmen, who were preparing to set sail. True to form, Taignoagny began plotting on Donnacona's behalf, spreading a rumour that there was a plot against his chief and sowing anxiety among the women. Finally, the Lord of Stadacona was convinced to go to France. He had so much to tell (and, in fact, he did not disappoint). He was to meet the king face to face. On May 6, 1536, he left his country; on July 16, he arrived in France.

Third Voyage: Roberval Chosen to Lead the Mission

François I was dazzled by Donnacona's loquaciousness: he talked of pygmies and one-legged men, of cultivation of cloves, nutmeg, and pepper, and of the presence of "infinite gold, rubies, and other riches." The Indian chief had figured out what the Europeans wanted to hear! François I was perplexed. Was the Indian chief's mysterious Saguenay the gateway to Cathay?

But time was marching on. Although he was preoccupied with Emperor Charles Quint's ambitions, François I had not forgotten Canada. Nor had his rival. In 1540, the emperor sent the commander of the Order of Alcantara on a mission to the king of France to dissuade him from pursuing his exploration plans. It took no further provocation for François I to take action. "The sun shines on me as it does on others," he supposedly said to the emissary. "I would like to see the clause in Adam's will that excludes me from sharing the world." Cartier was called back into service. This time, François I was thinking of colonization.

When the king's sights shifted from exploration to colonization, he naturally sought a lieutenant from among his entourage. Well-born though penniless, Jean-François de La Rocque, sieur de Roberval, belonged to a venerable noble family. François I made him his vassal charged with governing, defending, and developing a country over which he would hold sovereignty. Canada was inhabited by indigenous people, he knew, but there were very few of them, according to what he had been told; more important, the territory had never been occupied by another Christian prince. For the king, a discovery, to have political meaning, had to be backed up by effective occupation – even a defensive capacity.

Cartier had reconnoitred and explored a territory. In January 1541, Roberval received instructions to take possession of this territory "by friendly or amicable means" or, if need be, "by force of arms." Once occupied, the country would be developed and populated by people "of each sex," of all trades and all conditions, and would be endowed with "cities and forts, temples and churches."

François I had heard enough; Donnacona had opened his eyes to the possibilities. In bringing about this meeting, Cartier had been defeated at his own game. He was no longer the master of the situation, but was to be Roberval's guide. Roberval's powers were considerable; he would have "full power and authority" over this territory and could endow concessions as "fiefs and seigneuries." The basis for the future seigneurial system was being laid.

One question remained, however: How could the king arrange to have land that was already occupied ceded to him? He had before him Chief Donnacona. He could not feign ignorance. The account of Cartier's second voyage bore numerous mentions of "people," "peoples,"

White pine, in *Histoire des arbres forestiers de l'Amérique septentrionale* by Michaux
The text accompanying this illustration notes, "This ancient and majestic inhabitant of the forests of North America is not only the tallest but also the most valuable of the trees in them, and with its pointed crown towering in the air far above the other trees, it can be glimpsed from a great distance." This magnificent tree, now rare, was used in the construction of houses and warships.

The black spruce tree: a cure for scurvy
"It is with the young branches of *abies nigra* . . . that one makes beer. . . . One boils them in water, and one then adds a certain quantity of molasses or maple sugar, one leaves it to ferment, and one thus obtains this liquor, which is healthy and very useful in long-distance voyages to prevent scurvy" (François-André Michaux). Might this have been the *annedda* recommended to Cartier by the Iroquois?

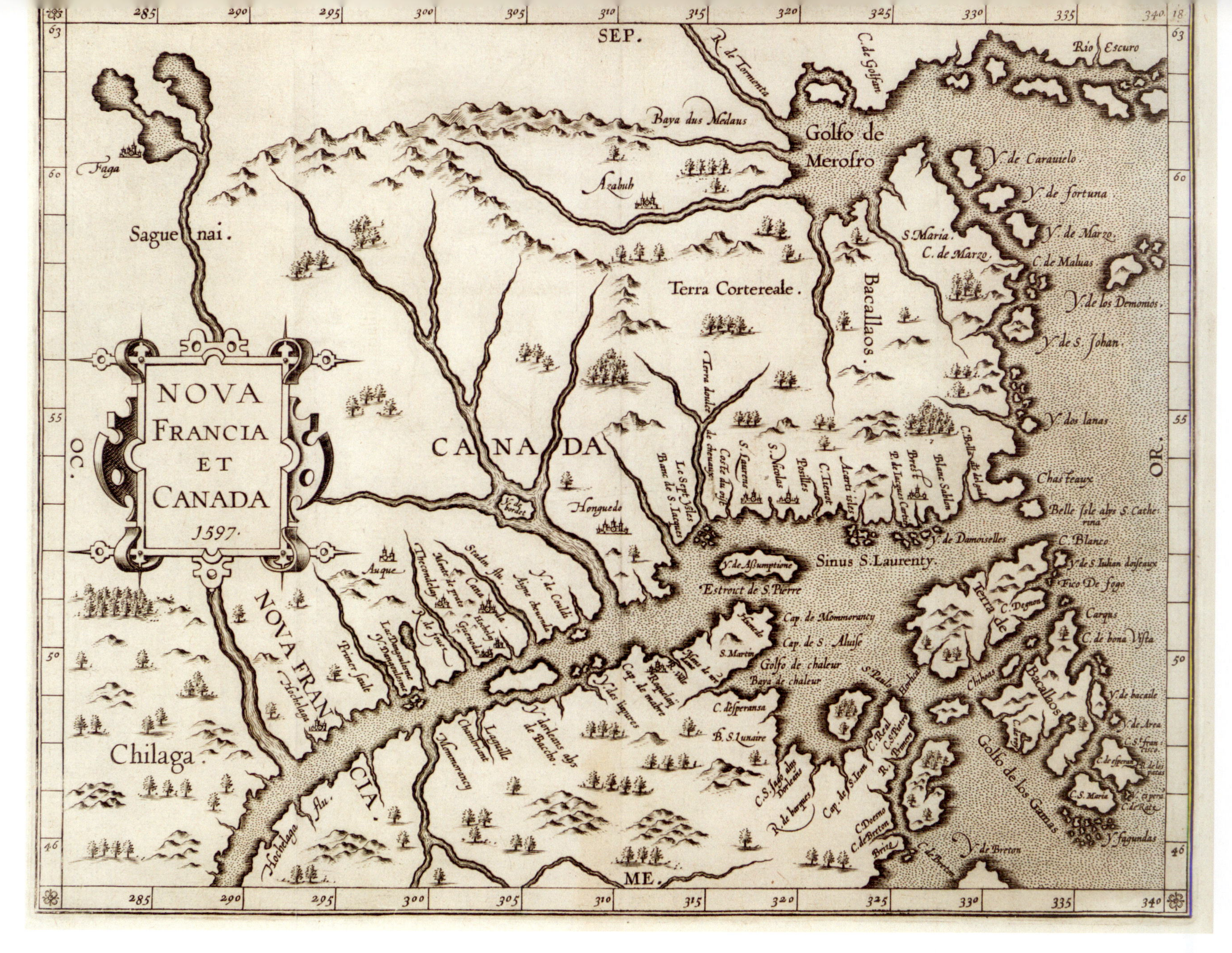

Nova Francia et Canada by Cornelis Wytfliet, 1597
This map by the Flemish Cornelis Wytfliet shows the territory explored by the French along the Gulf of St. Lawrence and St. Lawrence River in the sixteenth century. The shores of the river, named HOCHELAGA on THIS MAP, are sprinkled with pictograms of towns indicating places populated by Indians. The cartographer presents NOVA FRANCIA and CANADA as two distinct provinces. To the northwest, he offers a reminder of the Portuguese explorations in the Americas with the inscription TERRA CORTEREALE. One particular site caught the attention of the French: SAGUENAI, which Wytfliet has situated not at the source of the Saguenay River but southwest of HOCHELAGA, near a tributary that seems to correspond to the Ottawa River. The Indian chief Donnacona had held out to anyone who might reach this kingdom the shimmering hope of "infinite gold, rubies, and other riches." This map, published three times between 1597 and 1605, quickly became obsolete with the publication of the first maps by Samuel de Champlain.

"men," "persons," and "inhabitants." The word "savage," employed as both epithet and noun in the account of the first voyage, was almost completely absent from the account of the second voyage. It is not known who wrote these accounts, but their vocabulary must have resembled Cartier's. The king had surely heard these designations. In his instructions to Roberval, he himself spoke of "peoples of here well formed of body and limb and well disposed of mind and understanding." He was especially interested in them because Cartier had concluded, "They will be easy to convert." Roberval, the Huguenot, would have to see to this. He left with three ships in April 1542, and was brought home the following summer in a piteous state.

The choice of Roberval had shocked Cartier. Without waiting for his superior, but with his permission, he had taken to the sea with five ships and "1,500" people, according to a Spanish spy. He went to Stadacona but did not settle at the mouth of the Saint-Charles (Sainte-Croix) River, as he had in 1535; rather, he found a new site near the Cap-Rouge River. He had two forts built, one on the cape and the other below it. The winter was difficult as disease struck again, but he had not forgotten the benefits of *annedda*. The hostility of the Indians or the discovery of "a few leaves of gold" and some stones that resembled diamonds may have encouraged him to return to France and make a false promise to Roberval when he took him to Newfoundland in June 1542. In fact, Cartier's precious cargo had been nothing but iron pyrite and quartz. Legend seized upon his mistake: "As false as Canadian diamonds," became a common saying. Fishermen and whalers didn't care; they remained faithful to the St. Lawrence until the arrival of Samuel de Champlain.

In a context of raging religious war, the Huguenots once again assumed the lead. They headed for Brazil and Florida.

Main sources in order of importance

BIGGAR, Henry Percival (ed.). *A Collection of Documents Relating to Jacques Cartier and the Sieur de Roberval.* Ottawa: Public Archives of Canada, 1930. — POULIOT, Joseph-Camille. *La grande aventure de Jacques Cartier: épave bi-centenaire découverte au Cap des Rosiers en 1908.* Quebec City, 1934. — CARTIER, Jacques. *The Voyages of Jacques Cartier.* Translated by Henry P. Biggar, translation edited. With an introduction by Ramsay Cook. Toronto: University of Toronto Press, 1993. The Cartier quotations come from this edition. — BIDEAUX, Michel (ed.). *Jacques Cartier. Relations.* Montreal: Presses de

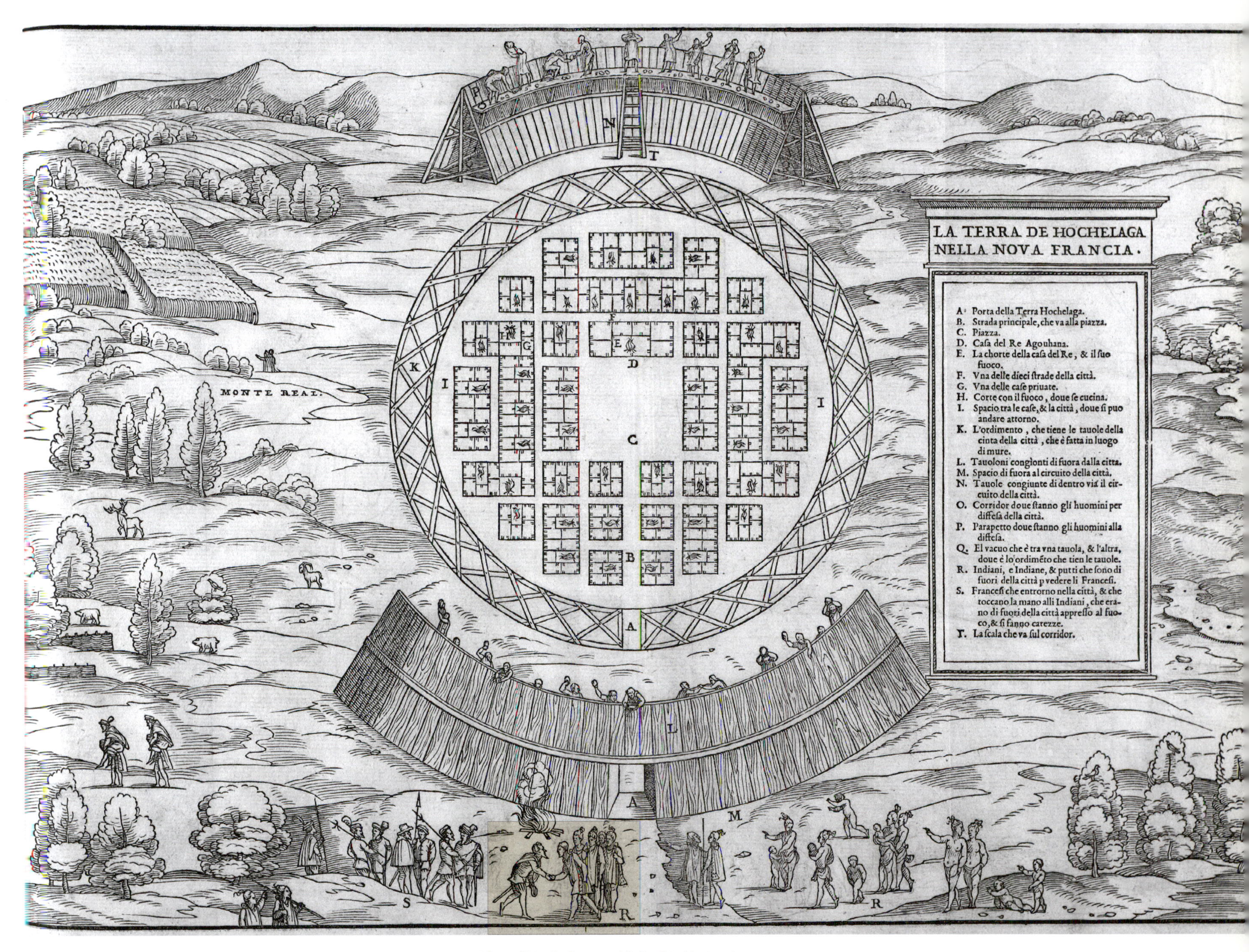

Université de Montréal, "Bibliothèque du nouveau monde" collection, 1986. — CAMPEAU, Lucien. "Jean Cabot et la découverte de l'Amérique du Nord." *Revue d'histoire de l'Amérique française*, vol. 19, no. 3 (December 1965), 384–413. — POPE, Peter Edward. *The Many Landfalls of John Cabot.* Toronto: University of Toronto Press 1997. Thorough, convincing research. — CROXTON, Derek. "The Cabot Dilemma: John Cabot's 1497 Voyage and the Limits of Historiography." *Essays in History*, vol. 33, 1990–91. — GAGNON, François-Marc, and Denise PETEL. *Hommes effarables et bestes sauvages: images du Nouveau-monde d'après les voyages de Jacques Cartier.* Montreal: Boréal, 1986. — MORISON, Samuel Eliot. *The European Discovery of America. 2 vols.* New York: Oxford University Press, 1971. — QUINN, David B. *England and the Discovery of America, 1481–1620, from the Bristol Voyages of the Fifteenth Century to the Pilgrim Settlement at Plymouth.* New York: Knopf, 1974. — WILLIAMSON, James Alexander. *The Cabot Voyages and Bristol Discovery under Henry VII.* Cambridge, UK: Cambridge University Press, 1962.

LA TERRA DE HOCHELAGA
NELLA NOVA FRANCIA.
A. Porta della Terra Hochelaga.
B. Strada principale, che va alla piazza.
C. Piazza.
D. Caſa del Re Agouhana.
E. La chorte della caſa del Re, & il ſuo fuoco.
F. Vna delle dieci ſtrade della città.
G. Vna delle caſe priuate.
H. Corte con il fuoco, doue ſe cucina.
I. Spacio tra le caſe, & la città, doue ſi puo andare attorno.
K. L'ordimento, che tiene le tauole della cinta della città, che è fatta in luogo di mure.
L. Tauoloni congionti di fuora dalla citta.
M. Spacio di fuora al circuito della città.
N. Tauole congiunte di dentro via il circuito della città.
O. Corridor doue ſtanno gli huomini per diffeſa della città.
P. Parapetto doue ſtanno gli huomini alla diffeſa.
Q. El vacuo che è tra vna tauola, & l'altra, doue è lo ordiméto che tien le tauole.
R. Indiani, e Indiane, & putti che ſono di fuori della città p vedere li Franceſi.
S. Franceſi che entrorno nella città, & che toccano la mano alli Indiani, che erano di fuori della città appreſſo al fuoco, & ſi fanno carezze.
T. La ſcala che va ſul corridor.
MONTE REAL.
A
B
C
D
E
F
G
I
K
L
M
N
P
R
S
T

Plan of Hochelaga, published in Venice in 1556

Jacques Cartier's voyages to Canada were popularized in the sixteenth century with the publication, in Italian, of a collection by Giovanni Battista Ramusio titled *Delle navigationi et viaggi*. A number of maps engraved on wood were incorporated, including this plan of HOCHELAGA, which was the first portrayal of an Indian village disseminated in Europe. The plan evokes Cartier's visit in the fall of 1535. The name MONTE REAL refers to Mount Royal, named by Cartier during his visit. The ploughed fields outside the palisades were a reminder that the Indians encountered here were sedentary and farmed wheat. At the bottom, the French and Indians seem to be on friendly terms – in contrast to the scenes on Nicolas Vallard's map. It is difficult to believe that this plan was made by Jacques Cartier or another member of his expedition. The circular palisade, the public square in the centre of the village, and the symmetrical placement of the dwellings are more reminiscent of a European fortified town than an Indian village. There is reason to believe that the artist based this engraving on the voyage account and not a survey map, as suggested by the following excerpt from Cartier's 1535 account of his voyage: "In the midst of those fields is the citie of Hochelaga, placed neere, and as it were joyned to a great mountaine that is tilled round about, very fertill, on the top of which you may see very farre, we named Mount Roiall. The citie of Hochelaga is round, compassed about with timber, with three course of Rampires, one within another framed like a sharpe Spire, but laide acrosse above. The middlemost of them is made and built, as a direct line but perpendicular. The Rampires are framed and fashioned with peeces of timber, layd along on the ground, very well and cunningly joyned togither after their fashion. This enclosure is in height about two rods. It hath but one gate or entrie thereat, which is shut with piles, stakes, and barres. Over it, and also in many places of the wall, there be places to runne along, and ladders to get up, all full of stones, for the defence of it."

Normandy and Cartography in the Sixteenth Century

In the sixteenth century, Dieppe, Rouen, Le Havre, and Honfleur were among the most active commercial ports in the kingdom of France. With their eyes turned to the ocean and to overseas territories, these Norman towns prospered through maritime trade. Norman merchants defied the Iberian monopolies by establishing lucrative trade networks in Africa, Asia, and the Americas, under the noses of the Portuguese. From India, they brought back pepper, ginger, sugar, cloves, and cinnamon. In Brazil, they exploited the brazilwood tree to make a purple dye. In the "newfound landes" of the North Atlantic, they caught a stunning quantity of fish. Fauna, flora, and indigenous people were brought ashore in Normandy, to the delight of wealthy people in search of luxury and exoticism, but also to the delight of curious minds wanting to extend the knowledge of the Ancients.

Ships from Honfleur reached North America in 1506. Thomas Aubert and his crew from Dieppe followed two years later. Aubert was also one of the first to capture a few Indians and take them back to Rouen. In 1523, Giovanni Verrazzano left from the port of Dieppe to look for a passage to Asia that went around America by the north. In 1529, Raoul and Jean Parmentier also left from Dieppe, but they sailed east to discover the Moluccas, islands in the Indian Ocean overflowing with spices. The expedition reached Sumatra (Indonesia), but the brothers died of yellow fever and the ship's crew was brought home by Pierre Crignon, the historiographer and astrologer on board.

The Norman towns were also the homeports of a few corsairs, which constantly harassed the Spanish and Portuguese galleons. The person who was responsible for much of this maritime activity was Dieppe shipowner Jean Ango (1480–1551). By financing and organizing a good number of these now-famous expeditions, Ango built a colossal fortune and became one of the most powerful men in Europe. Dieppe, with its unprecedented prosperity, was where the most skilful cartographers chose to go to seek information about the voyages of exploration. The maritime, trade, and illegal activities of the Norman towns explains, in good part, the blossoming of cartography in the province, as well as the presence of mapmakers Pierre Desceliers, Jean Rotz, Guillaume Le Testu, Nicolas Desliens, Nicolas Vallard, Jacques and Pierre de Vaulx, Guillaume Levasseur, and Nicolas Guérard. We know little about these men's lives, but they left numerous cartographic masterpieces to posterity.

The maps in this Norman corpus have a certain number of features in common. These very beautiful immense planispheres and atlases in manuscript form present the known world of the European navigators of the mid-sixteenth century. Although they have many elements found on portolan maps, such as lines corresponding to compass orientations (rhumb lines), they were not used for navigation – at least, not the ones that have survived. The cartographic layout, the illuminations, the colours, and the delicacy of the lines make them sumptuous works of art, worthy of the patrons for whom they were made, among them Henri II, Henry VIII of England, the supreme commander of the French forces Anne de Montmorency, and the admirals Gaspard de Coligny and Claude d'Annebaut.

The remarkable illustrations reflect the mental landscape of a Renaissance still imbued with the mythology of the Middle Ages. The extra-European territories evoke spices, gold, and silver, but also foreign worlds peopled with bizarre beings. Deformed creatures, exhibiting strange behaviours, occupy the margins of the known territory. Africa is overflowing with headless men (Blemmyes) and men with dog's heads (Cynocephali), bodies with a single leg (Sciapods), and faces with a single eye (Moniculi). These monsters stand beside the priest John and his mythical kingdom in Ethiopia. Maps of North America also bear a few fabulous illustrations. One of Desceliers's maps shows a unicorn and pygmies doing battle against an army of cranes, while Le Testu's atlas shows a group of men with wolf heads hunting alongside monkeys and animals resembling kangaroos.

But fabulous illustrations were relatively rare compared to realistic imagery, which conveyed in its way information on the territories mapped. For instance, on several Norman maps are portrayals of whales, caribou, bears, Indians clothed in animal skins, and Indians on snowshoes. Illustrations of people beside the St. Lawrence River on Vallard's map and another anonymous map have drawn the attention of researchers. In one of these images, we see a gathering of women and warriors led by a chief; in another, peasants are working in a field and swearing an oath of loyalty to a landowner. These scenes have been interpreted as references to the visits of Jacques Cartier and Roberval to Canada.

Norman mapmakers knew very well how to draw the contours of Europe, Africa, Asia, and the Americas – at least, the coasts bordering the Atlantic and Indian oceans, which had been described in detail. The lines clearly show capes, inlets, ports, river mouths, and islands, often highlighted in red, blue, green, and other bright colours.

Among the continental forms specific to Norman cartographers are the immense and mysterious land of Great Java, located at the eastern extremity of the Indian Ocean. Some researchers believe that this might be Australia, which may thus have been discovered by Europeans in the first half of the sixteenth century, before the Dutch arrived there. Others think that it might be Vietnam, or even a totally fictional landmass. The source of this portrayal is still controversial, but it can probably be attributed to the expedition of the Parmentiers and Pierre Crignon to Sumatra, and they would have followed a voyage by the Portuguese.

On the other side of the globe, the North American coasts, from Florida to Labrador, were relatively well defined. Two major geographic features made their appearance: the Gulf of St. Lawrence and the St. Lawrence River. The island of Newfoundland was clearly detached from the continent. Many French and Portuguese place names dotted the east and south coasts of the island, showing that European fishermen had been visiting the area for many years. A number of names on the coasts have lasted through time, including the Saint-Pierre islands, Plaisance, Baie des Trépassés (Trepassey Bay), Cape Degrat, the island of Fichot, Belle Isle, and Bonavista.

How did the sixteenth-century Norman cartographers manage to draw documents that were so exact and detailed? Although it is difficult to answer this question, there are a few clues. First, mapmaker-navigators (Le Testu, Rotz) explored some of the territories themselves. A short excerpt from Rotz's dedication in his *Boke of Idrography* (offered to King Henri VIII) confirms this: "... drawing as much from my own experiences as from certain experiences of my friends and fellow navigators." In this

Map of Newfoundland and the St. Lawrence River by Guillaume Le Testu, 1556
The maker of this map, Guillaume Le Testu, was not a conventional cosmographer. In complete contrast to the scholars in their ivory towers, this privateer from Havre-de-Grâce (Le Havre) knew the ocean sea well, as he had spent a good part of his life on it. He regularly visited the Brazilian coast, and so, in 1555, was charged with leading Durand de Villegaignon, who wanted to establish a colony known as France-Antarctique, on an expedition to Brazil. Upon his return to France the following year, Le Testu presented a superb atlas, *Cosmographie universelle selon les navigateurs, tant anciens que modernes*, to Amiral de Coligny. Le Testu knew the Atlantic well, but had he really navigated in the Gulf of St. Lawrence? It is possible, but questionable, judging from the iconographic arsenal deployed in this map from *Cosmographie*, which at first glance is unreliable; Le Testu has drawn men with wolves' heads hunting alongside monkeys and animals resembling kangaroos. The Gulf of St. Lawrence and the St. Lawrence River are relatively accurately drawn, but many of the place names are of Portuguese origin; for example, Baie des Chaleurs is translated into *baia de calleno*, which leads one to believe that the cartographer depended on sources other than his own experience. Le Testu lost his life in 1573, during an attack on a convoy of mules transporting gold and silver from Peru across the Panama isthmus. His companion at arms, Francis Drake, escaped unharmed, grabbed the booty, and went on to a very fruitful career as a privateer under the patronage of Elizabeth I of England.

dedication, the cartographer also recalls that "navigator friends" had made contributions.

North American cartography owes a large debt to the voyages of Cartier and Roberval, who preceded by a few years the apogee of Norman production. Sometimes, the reference to these explorers is explicit, as on one of Desceliers's maps: "To this country was sent by said king ... Mr. de Roberval with a large company ... to inhabit the country that was first discovered by the pilot Jacques Cartier."

Another interesting fact is that the maps show a large number of place names that are not mentioned in the published accounts. Many of these names are Portuguese. For example, we find a *b[aie] de calleno* (Baie des Chaleurs) in Le Testu's atlas, and an *Arcablanc* (Blanc-Sablon) inlet in Vallard's atlas. When we study all of the Norman maps, it is thus impossible to ignore the Portuguese influence. Did Norman mapmakers have access to Portuguese maps stolen by French corsairs? Did they receive information from Portuguese pilots tempted away by French shipowners, as the pilot Jean Alfonce had been?

In the absence of sources to verify these theories, we can only speculate on the origin of these geographic names, which are, at the very least, an invaluable trace of the presence of Europeans and indigenous peoples in North America.

Main sources in order of importance

MOLLAT DU Jourdin, Michel, and Monique DE LA RONCIÈRE. *Sea Charts of the Early Explorers: 13th to 17th Century.* Translated by L. Le R. Dethan. New York: Thames and Hudson, 1984. — MOLLAT DU JOURDIN, Michel. "Le témoignage de la cartographie." In Fernand Braudel (ed.), *Le monde de Jacques Cartier: l'aventure au XVIe siècle*, pp. 149–64. Montreal and Paris: Libre-Expression and Berger-Levrault, 1984. — MOLLAT DU JOURDIN, Michel. "Les ports normands à la in du XVe siècle". In Philippe Masson and Michel Vergé-Franceschi (eds.), *La France et la mer au siècle des grandes découvertes*, pp. 83–90. Paris: Tallandier, 1993.

SEP TEN TRI
TERRA·BIMENES
C L I M A
ANTEYLLAS
CIRCVLVS
OCCEANVS OC
RRA·FIRME
MVNDVS
NOVVS
CIR CV L VS EQ

The Black Legend

SPANISH EXPLORATIONS

In the 1520s, the search for a passage to the Orient was not the only reason for exploration. Worth as much as spices and silk was gold. The Spanish, who were searching for it desperately, found it here and there. Then, in 1521–22, Hernán Cortes hit the jackpot at Tenochtiçlan, in the heart of the Aztec empire. Ten years later, it was Francisco Pizarro's turn; he managed to lay his hands on the Inca treasures.

After the conquest of Puerto Rico, Ponce de Leon, a former companion of Columbus's, received permission to set out in search of fabulous lands to the north. He dreamed of finding the Fountain of Youth. On Palm Sunday (*Pascua florida* for the Spanish) of 1513, he landed on a coast north of Cuba. It was pleasant and deserving of the name Florida (flowery). In fact, for more than a century, this name designated a long stretch of the Atlantic coast. In 1521, Ponce de Leon attempted a colonization project, but he was repelled by Indians and died from the wounds that he sustained.

Following various missions exploring the Atlantic coast, all of them looking for a navigable route to Asia, Lucas Vasquez de Ayllon started a colony near the Savannah River, founding what is considered the first Spanish town in North America, San Miguel de Guadalupe. Once again, the Indians fought off the Europeans; in addition, the African slaves on the expedition revolted, some in the party fell ill with fever, and Ayllon himself was stricken. The 150 survivors abandoned the site in late 1526 and returned to Hispaniola (Santo Domingo).

Florida was first thought to be an island, but it was quickly found to be part of the mainland. Sailing along the coast from Florida to Mexico in 1519, Alonzo Alvaro de Pineda confirmed this beyond all doubt. Conquistadors were soon sailing in this direction seeking riches.

In 1528, Panfilo de Narvaez assembled a force of 400 men and set sail for Florida. His army began a march inland, but disaster quickly struck the expedition. The Spanish requisitioned food from the Indians, and revolt broke out soon after. The remaining men eventually reached the coast and tried to return to Spain. They cobbled together ships, but three were immediately wrecked and two others ran aground. Only four men are known to have survived, among them Alvar Nunez Cabeza de Vaca and an African slave called Estevan. From 1529 to 1534, sometimes together, sometimes apart, the four survivors wandered from tribe to tribe. Then, sometime in 1534, they joined up again, determined now not just to survive but to return to Mexico. They began to walk west, then turned southwest. In the spring of 1536, slave hunters took them in, to their mutual astonishment. By July, the four were in Mexico City. Although their account was frightening, it was also reassuring: not all Indians were enemies! Everywhere, they had found help and food.

De Vaca and Estevan wanted more. Da Vaca undertook to walk more than 1,500 kilometres to Brazil. Estevan, meanwhile, offered his services to the governor of New Galicia, Vasquez de Coronado. This time, the plan for the expedition was to reach the Pacific and sail north along the coast. Estevan was sent on a reconnaissance mission; a Franciscan priest, Marcos de Niza, who went with him told him that they were close to the legendary Seven Cities. Estevan scouted ahead toward the villages of Cibola, but he fell victim to the mistrust of the Zuni Indians and did not return alive. Marcos quickly made his way back; although he had seen little himself, he had much to tell Coronado. In 1540, Coronado set out at the head of an enormous force made up of 330 Spaniards, most of them on horseback, and some 1,000

Opposite
Detail of a map from the Miller Atlas, shown on pp. 34–35.

HAVANA Portus
St. DOMINGO
CARTAGENA
MEXICO
CUSCO
Groenlandi
Virginiani
Rex et Regina Floridæ
Novæ Albionis Rex
Mexicani
Septentrionalissimas Americæ partes, Groenlandiam puta, Islandiã et adjacentes, quod Americæ tabulæ commodè comprehendi non potuerint, peculiari hac tabella Spectatoribus exhibendas duximus.
AMERICA SEPTENTRIONALIS
NOVA FRANCIA
NOVA GRANADA
HISPANIA NOVA
LA FLORIDA
Tropicus Cancri
MAR DEL ZUR
CIRCULUS ÆQUINOCTIALIS
OCEANUS PERUVIANUS
Tropicus Capricorni
MARE PACIFICUM
AMERICÆ nova Tabula.
Auct: Guiljelmo Blaeuw.
Cum privilegio decem annorum.
TERRA AUSTRALIS INCOGNITA.

Map of the Americas by Willem Janszoon Blaeu, 1645
In 1581, the United Provinces (northern Netherlands) seceded from the Spanish empire. Amsterdam then entered a period of unprecedented economic prosperity that resulted in the emergence of local cartographers, the most notable of whom were Willem Janszoon Blaeu and his son Joan. Willem, who had been a maker of globes and scientific instruments, ventured into the publishing of maps with a large-scale project: the publication of an atlas of the world in several volumes. Upon his death, in 1638, Joan continued the work, publishing a number of volumes, first under the title *Atlas Novus*, then as *Atlas Maior*, which comprised between 9 and 12 volumes (depending on the language of publication) and almost 600 maps. At the time, Blaeu's atlases were the most expensive on the market, destined for the wealthiest aristocrats and merchants of Europe. This map, published in one of the atlases, shows the empire that Spain had built in the Americas. From their settlements on the islands of the Antilles (Hispaniola, Puerto Rico, and Cuba), the Spanish gradually conquered the territories richest in gold and silver. After subjugating first the Aztec empire (1517–22) and then the Inca empire (1533), they used Indian forced labour to mine the lodes of Potosí, Bolivia, and Zacatecas, Mexico. Other American riches were also thoroughly exploited, among them leather, tobacco, indigo, sugar, and pearls. At the top of the map, Blaeu added plans of the main cities, such as Havana, Cartagena, Santo Domingo, Mexico City, and Cuzco. Under the reigns of Charles Quint and Philippe II, the Spanish rapidly expanded their empire in all directions, southward to Chile and northward to Florida, New Grenada, and California. They met various Indian peoples, including Mexicans, Peruvians, Chileans, Floridians, and Virginians, who are illustrated on the borders of the map.

Map of New Spain, published in Venice in 1561
In the sixteenth century, Spain divided its American empire into several separate provinces, including New Spain (NUEVA HISPANIA), the first American territory conquered, which corresponds more or less to today's Mexico. The capital, Mexico City, is shown situated on a lake. Part of Cuba, occupied by the Spanish in 1511, is also visible. The north shore of the Gulf of Mexico, bordered by Florida, must have been relatively well known. The name of the "discoverer," Ponce de Leon, is shown in a bay just west of the peninsula. The SPIRITO SANTO River corresponds to today's Mississippi River. On the Pacific coast, the cartographer located CIBOLA too far south. He also drew the California peninsula and the famous MAR VERMEJO (Gulf of California), which later haunted French explorers.

Indians loaded down with baggage. To move more quickly and increase their chances of finding the Seven Cities of Cibola, the troops were divided into small groups. One group came upon the Hopis, while others reached Santa Fe and Taos. A Plains Indian, nicknamed the Turk, held out the prospect of a very prosperous city, Quivira, where the eating utensils were made of silver, the plates of gold, and "bells of the same metal lulled them to sleep." The Turk offered to serve as their guide. Unfortunately for him, paradise could not be found. Worse still, he managed to get lost. The only trails they found were those of bison, which Don Juan de Onate described as frightening creatures: "The first time we encountered bison, all the horses took flight when they caught sight of them, for they are terrible to see. They have broad, short faces and eyes two hands' breadths apart, and so prominent on the sides that they can see those chasing them." In June 1541, Coronado reached the northernmost point of his expedition, Quivira, no doubt a Wichita village somewhere in what is today Kansas. It looked pretty, and its occupants lived in straw huts.

On the way back, Coronado again crossed the Arkansas River, passing about 500 kilometres from the path of the expedition led by Hernando de Soto, one of the cruellest conquistadors of his time. Having left Florida in 1539 at the head of a small army of 200 horsemen and 400 troops accompanied by fighting dogs and, the ultimate precaution, 300 pigs, de Soto, true to form, conducted multiple raids, kidnappings for ransom, and massacres. Searching desperately for gold and treasures of all sorts, he went northward, then northeast, returned southward, then headed westward and crossed a vast river, the Mississippi, also called the Rio del Spirito Santo. The army wintered over near the river from November 1541 to March 1542. Distraught, spent, de Soto had reached the end of the line. He died in May 1542. It has been written that he died repentant, but his repentance could never measure up to his cruelty. He was one of those who fed the image of the Spaniards' inhumanity, related by Las Casas in particular in his apocalyptic *A Brief Account of the Destruction of the Indies* (1552), which, in spite of censorship and controversy, helped to create a rift in Spanish public opinion.

A young Italian voyager, Girolamo Benzoni, who had been travelling through the Spanish colonies for 14 years, added his relentless testimony in *La Historia del Mondo Nuovo,* published in 1565. This work was immediately translated into many languages – but not into Spanish.

As ruthless as the Spanish could be toward the Indians, they could be just as ruthless with other Europeans. The massacre of St. Augustine (1565), related in the next chapter, is a terrible illustration of this.

Main sources

CUMMING, William Patterson, Raleigh Ashlin SKELTON, and David Beers QUINN. *The Discovery of North America.* New York: American Heritage Press, 1971. — FRAÏSSÉ, Marie-Hélène. *Aux commencements de l'Amérique, 1497-1803.* Arles: Actes Sud, 1999. — GOETZMANN, William H., and Glyndwr WILLIAMS. *The Atlas of North American Exploration: From the Norse Voyages to the Race to the Pole.* Norman: University of Oklahoma Press, 1998. Well-constructed and extremely valuable book. — LITALIEN, Raymonde. *Les explorateurs de l'Amérique du Nord, 1492-1795.* Sillery: Septentrion, 1993. — WEBER, David J. *The Spanish Frontier in North America.* New Haven and London: Yale University Press, 1992.

The Huguenot Channel

BRAZIL AND FLORIDA

The origin of the word "Huguenot" is complex. The word "Protestant," obviously, is less problematic. A Protestant protests.

It is difficult today to imagine the state of the Catholic Church around 1500. While we are still debating whether priests should be celibate, it is disconcerting to learn that a number of popes of that time had children. Given the care that they took of their offspring, we might be led to think more kindly of them – if they had not done so in the shadow of intrigue and deceit. In short, the papacy was in the midst of an extremely serious crisis, marked by simony and corruption. A number of popes were depraved. One was strongly suspected of having bought votes for his election; another was accused of incest. The times being what they were, prelates followed the popes' lead: they purchased their positions and profited from them scandalously. The low clergy were ignorant, as were the common people. "Other times, other customs," we might say. Still, it was a bit much.

Martin Luther was among the first protesters. This Augustine monk, upset by the delinquency that he saw in the Borgias' Rome, began a vast reform movement. Under the influence of Emperor Charles Quint, himself a Catholic, who was uneasy about the schisms caused by these religious quarrels, the need for a counter-reformation became apparent. Slowly and painfully, the Council of Trent (1545–63), convoked by Paul III, set Church doctrine and imposed a bit of discipline. It was too little, too late. The objective had been to bring Catholics and Protestants together, but the Council left papists and heretics facing in opposite directions. The chasm was wider than ever.

Wars are always horrible; religious wars stand out for their fanaticism and brutality. From Germany, Luther's reform had spread to other countries in Europe, supported by other strong men of intellect, such as Jean Calvin and Ulrich Zwingli.

France was soon home to a large proportion of Protestants, called Huguenots or Reformists. The reform movement took on a political dimension, and the ensuing power struggle was marked by battles, conspiracies, and assassinations.

Gaspard de Châtillon, seigneur de Coligny, was appointed admiral of France in 1552. Attracted to Protestantism, he found himself at the centre of the troubles and thought that it would be good to create a

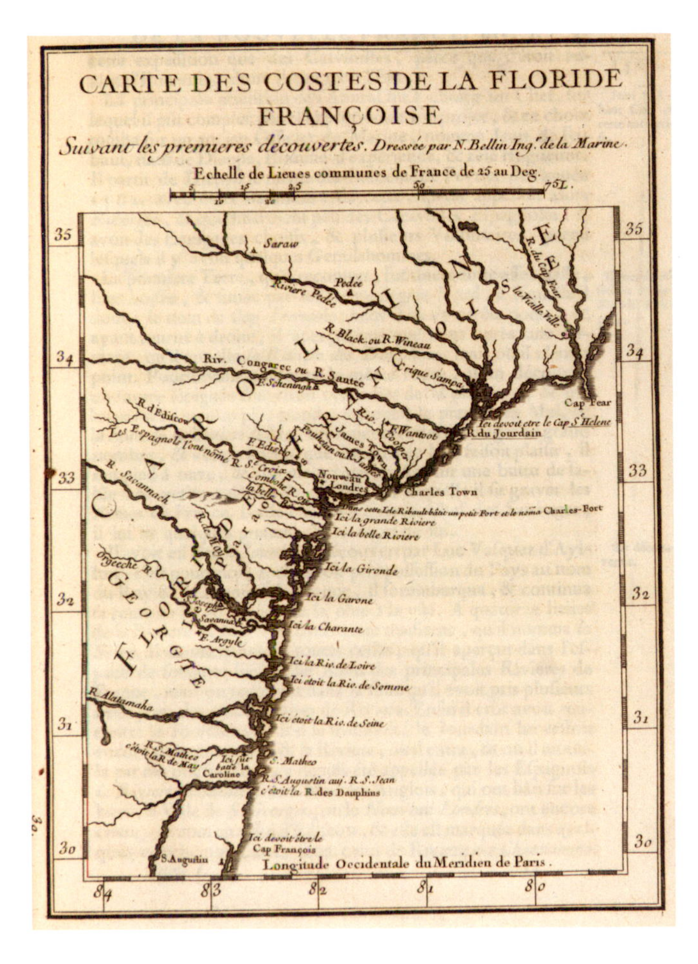

Carte de costes de la Floride françoise by Jacques-Nicolas Bellin, 1744
Some place names are amazingly resistant to passing time. In the sixteenth century, the French undertook to colonize Florida, in vain. They were violently expelled by the Spanish, who left them no hope of return. The French toponymic heritage, however, had better luck. The names created at the time – the Loire, GIRONDE, and GARONNE rivers, as well as Carolina, named in honour of Charles IX – were perpetuated in French cartography until the eighteenth century, as seen on this map by Bellin. One reason for this was quite simply that Florida and Louisiana were neighbouring territories. By conjuring up, consciously or not, the French presence in Florida, France could more easily claim Louisiana, which was coveted by both Spain and England.

Montes Apalatci, in quibus aurum argentum & æs inuenitur
Apalatci
In hoc lacu Indigenæ argenti grana inueniunt
Oustaca
Onatheaqua
Potanou
Appalou
Ehiamana
Anouala
Hicaranaou
F L O R I D A   P R O V I N C I A
AB INDIGENIS DICTA IAQUAZA
Astina
Choya
Vtina
Eloquale
Patchica
Aquouena
Cadica
Edelano
Chilili
Eclanou
Omitaqua
Calanay
Onachaquara
Mocoso
Mathiaca
Mayarca
Malica
Homoloua
Saturiova
Patica
Mairra
Marracou
Lacus aquæ dulcis
Adeo magnus est hic lacus ut ex una ripa conspici altera non possit. Distat a Charles fort 180 leucis.
Oathkaqua
Prom: Canaueral
F. Sorrochos
Sorrochos
Mocossou
Lacus & Insula Sarrope
Mexicani Sinus pars
Sinus Ioannis Ponce
F. Canoes
F. Pacis
Aquatio
C A L O S
Calos
Bimini
Æstuaria
Prom. Floridæ
Rupes
Insulæ dictæ Testudines
Scopuli dicti Martyres
Yocajouque siue maior Lucaya
Bahara
Zagareo
Hæc maris pars plena est Insulis, scopulis, breuibus et puluinis valde insidio
Sinus Magnel
Hic descendit Pamphilus Naruaez
Promontorium
Portus Ioannis Natiuitatis
F. Aethiopum
F. Florum
F. Madalena
F. Guate
F. ligeris
Charenta
Garumna
Gironda
Axona
Sequana
Hauana
Guanagmarico
Xagua
Cuspis S. Antonij
Cuba insula
Insula Pinorii
Iardines scopuli, nauigantibus formidabiles
S. Trinitatis
Albayhamo
Isabella
Portus Principis
Mons Christi
Portus Patris
Baracoa
Portus absconsus
S. Iacobi
Promont. Crucis

diversion by drawing attention away from the country. Already, he was encouraging French corsairs to "charge" the Catholic enemy. The Spanish, gorging on gold and silver, were the perfect target. More French corsairs were built. François Le Clerc, also known as Wooden Leg, and Jacques de Sores, dubbed the Exterminating Angel, became the stuff of legend.

Coligny decided that yet more had to be done. The cartographers of Dieppe spread attractive portrayals of the Americas before him. All eyes turned toward Brazil, where the French had been going for many years. In March 1554, Henri II agreed to provide a large sum of money to the "chevalier de Villegaignon" to start a French colony there. It was a failure. "That this historical episode is consigned to oblivion," writes Jean-Christophe Rufin in his wonderful novel *Rouge Brésil* (Goncourt: Gallimard, 2001), "is due to the refusal to cultivate memory and not to the absence of documents." One might say the same of the French failure in Florida. In spite of everything, including an abundance of documents, silence was the rule.

While Rufin brought to light the adventures of Villegagnon (or Villegaignon), the engravings by the de Bry family no doubt rescued from obscurity the attempts at settlement by Jean Ribaut and René Goulaine de Laudonnière. The de Brys were Lutherans who had left Liège for Frankfurt. They decided to denounce the excesses of the Spanish colonization undertakings in America by publishing travel accounts accompanied by superb engravings. (Marc Bouyer and Jean-Paul Duviols have brought them back into public view in a magnificent book called *Le Théâtre du Nouveau Monde*.)

None of the de Brys – the father, Theodor, and his sons, Johannes Theodorus and Johannes Israel – ever went to the Americas. However, they received material from remarkable illustrators: Hans Staden in Brazil, Jacques Le Moyne de Morgues in Florida, and John White in Virginia.

The few historians who study the presence of the French in Florida emphasize the difficulties they experienced – the internal quarrels, the hostility of the Indians, and a horrible massacre at St. Augustine perpetrated by the Spanish, a massacre so devastating that it seems miraculous that an account survived at all, not to mention to the present day. But Jacques Ribaut, Jean's son, managed to bring back to Europe a number of survivors in 1565, including René de Laudonnière, the carpenter Nicolas Le Challeux, and the painter Jacques Le Moyne de Morgues, who was able to save his drawings – how is a mystery. Perhaps, he reconstructed them from memory. In any case, the only surviving original (1564) portrays a Timucua chief, Athore, proudly showing Laudonnière a stela erected by Jean Ribaut two years

Map of Florida by Le Moyne de Morgues, published by Theodor de Bry in 1591
This map, published in Theodor de Bry's collection of the "great voyages," is an important testimony to the brief French presence in Florida. Its maker, Jacques Le Moyne, was a very talented artist. After managing to escape the massacre of the French colony by the Spanish, he returned to France, then moved to England, where his drawings first became known. In the seventeenth and eighteenth centuries, this portrayal was used as a model by Dutch then French cartographers who wished to establish a right of possession, even a theoretical one, for their respective kingdoms.

Coat of arms of the kingdom of Spain. It was placed on the southern part of the continent on the map, signifying Spanish occupation of the territory.

Coat of arms of the king of France, placed on the map near the Atlantic coast, where France was attempting to establish settlements from the Florida peninsula to the Gulf of St. Lawrence.

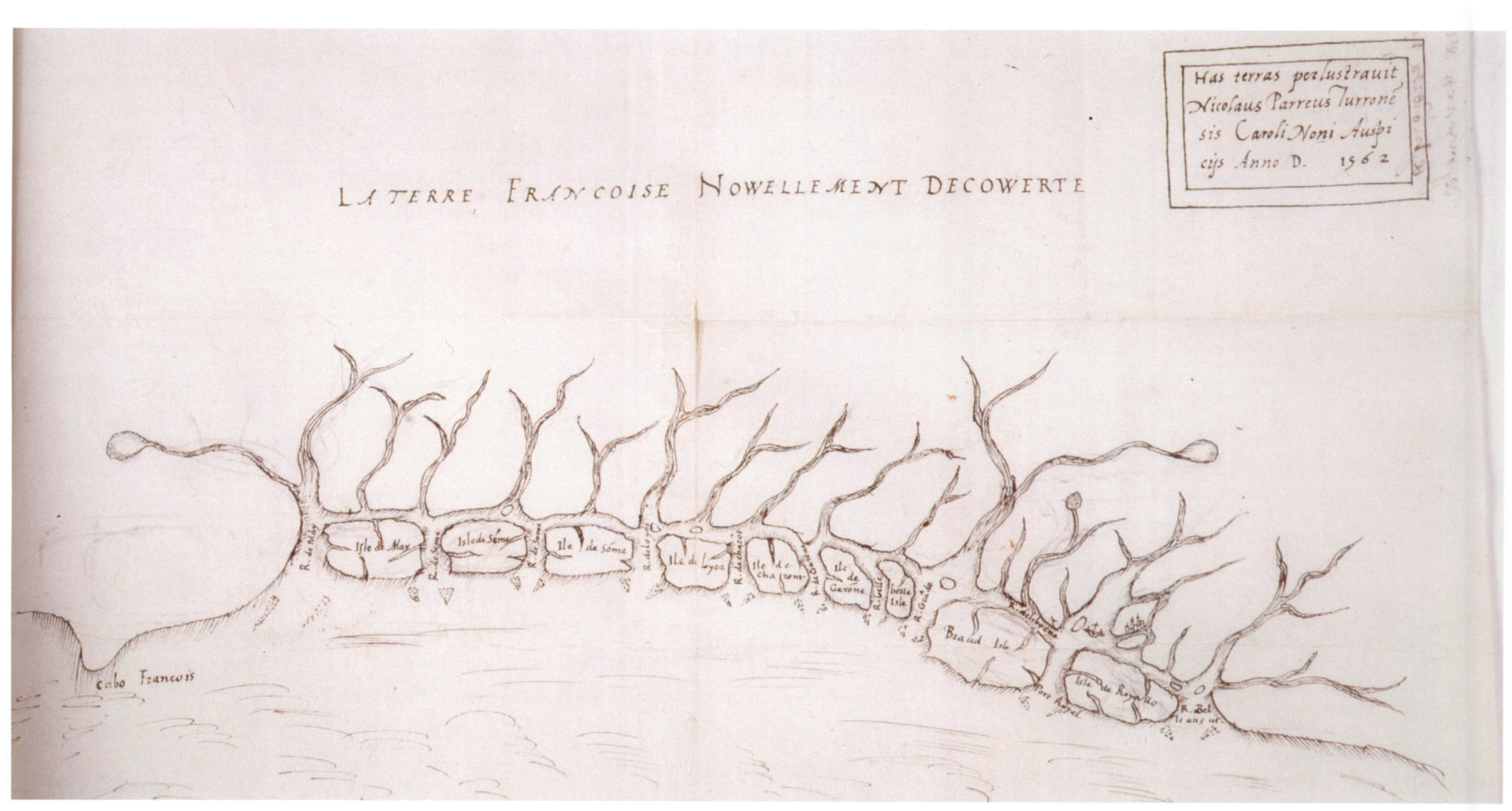
Has terras perlustrauit
Nicolaus Parreus Turroné
sis Caroli Noni Auspi
cijs Anno D. 1562
LA TERRE FRANCOISE NOWELLEMENT DECOWERTE
Cabo Francois

Map of French Florida by Nicolas Barré, 1562
In 1562, several hundred Frenchmen, financed by Amiral de Coligny, were sent to Florida to establish a Huguenot colony. This map, by pilot Nicolas Barré, is evidence of this unfortunate undertaking. It shows clearly that the French explored the coasts of the southeast United States and named places as they saw fit – Cap Français, the May and LOIRE rivers, and islands named GARONNE and SEINE – names that endured until the eighteenth century. Barré was one of the men left as reinforcements at Charlesfort. After killing their dictatorial captain, the mutineers waited for help that did not arrive. They finally decided to build a brigantine to sail themselves back to France; the Indians obtained ropes for them and showed them how to use resin. After several weeks of wandering at sea, forced to drink their urine and eat their shoes, they decided that it would be "more expedient," as Laudonnière recounts, "that one die rather than so many perish." By random draw, a man named La Chère was chosen, and his "flesh was shared equally among his companions. This is so pitiful a thing to tell that my pen even refuses to write." Eventually, the survivors were rescued by an English ship.

previously, surrounded by a group of very respectful Indians.

Le Moyne had barely escaped death. Soon after his return, he went to England, where Henry VIII had instituted Anglicanism following his break with Rome. Le Moyne took up his paintbrushes and returned to his favourite subjects: plants, fruits, and flowers. Through Protestant channels, he met John White and Theodor de Bry; the three stayed in close contact, and Le Moyne and White contributed to the de Brys' projects. When Le Moyne died, Theodor de Bry acquired a series of watercolours from which he engraved 42 plates depicting the experience of the French and many portrayals of Indians. Le Moyne also left an account, in Latin, of his adventures, in which he cast a very kind eye on Athore and his brethren.

In fact, both attempts (1562 and 1564) by the French to settle in Florida are very well documented. Laudonnière and Le Challeux made detailed reports, as did Solis de Méras, the brother-in-law of Pedro Menendez de Aviles, who commanded the small armada of ten ships sent by Philip II to rout out the French. Méras was one of the two men who put Jean Ribaut to death. In his report to the king, Menendez congratulated himself on the death of "Ribao," who "might have done more in one year than others in ten." "Indeed," he added, "we know of neither a sailor nor a corsair who was more experienced and he was very expert in this navigation to the West Indies and on the coast of Florida."

Main sources in order of importance

LUSSAGNET, Suzanne. *Les Français en Amérique pendant la deuxième moitié du XVIe siècle. Les Français en Floride.* Paris: Presses Universitaires de France, 1958. Vol. 2 contains texts by Jean Ribault, René de Laudonnière, Nicolas Le Challeux, and Dominique de Gourgues. — BOUYER, Marc, and Jean-Paul DUVIOLS (eds.). *Le Théâtre du Nouveau Monde: les grands voyages de Théodore de Bry.* Paris: Gallimard, 1992. A splendid book in every way. — GAFFAREL, Paul. *Histoire de la Floride française.* Paris: Firmin-Didot, 1875.

Discovery or Encounter?

FIRST CONTACT WITH INDIGENOUS PEOPLES

The Tainos of Guanahani, the Iroquoians of Hochelaga, and the Timucuas of Florida had several things in common, one of these being that they were the first Indians in their respective regions to "discover" the Europeans. In October 1492, sentinels in the Taino villages were astonished to see Christopher Columbus's caravels. Once the shock passed, they noticed that these floating mountains carried strange figures with beards and heavy clothing. Were they monsters? Or were they gods? The Tainos quickly came to their senses: in fact, these were human beings. They soon discerned the lust emanating from every pore of Columbus and his men, who were, of course, happy to be offered water, fish, and fruit, but whose eyes were searching for gems, precious stones, and pieces of gold. The Indian women, who "went naked" and were "very beautiful," quickly deciphered the desire that they aroused. Beyond doubt, the strangers were men like all men. And the women, like all women, were proud of the interest shown in them and attracted to the exotic foreigners.

No priests accompanied Columbus on his first voyage. They came later, and they were terribly discouraged by what they saw. The Church had recently managed to quell the habit of nudity among Europeans. A sign of poverty rather than lewdness, nudity had nevertheless been repressed by the clergy. European voyagers in the late sixteenth century subscribed to the new norm of being clothed, and so the Indians' nudity was a constant surprise.

The missionaries associated nudity with lack of virtue, even though, as Jean de Léry observed, the flirtatiousness of Frenchwomen was more of an incitement to lewdness and lechery than was the simplicity of Indian women. "I maintain," wrote this sincere Calvinist, "that finery, rouge, wigs, hair curls, large ruffed collars, farthingales, dresses over dresses, and other trifles with which women and girls over here distort themselves and of which they never have enough, are the cause of incomparably more evil than is the ordinary nudity of the Savage women: the latter, however, in terms of natural beauty, owe nothing to the former." About this, there was unanimity: the Indians were beautiful and well proportioned. Columbus dwelt upon this point, Cartier noted the Indians' "lovely corpulence," and Jean Ribaut was eloquently enthusiastic: "They are entirely naked and of good stature, powerful, beautiful, and as well built and proportioned as any other people in the world, very gentle, courteous, and naturally kind ... The women are beautiful and modest: they do not allow anyone to approach them in a dishonest manner."

The French in Florida
Athore, king of the Timucuas, proudly shows Laudonnière the stela erected two years earlier (1562) by Ribaut. Le Moyne de Morgues, who produced this painting, also wrote a very interesting text. "Athore is a handsome man, sensible, honourable, strong, and very tall, much taller than the tallest of our men," noted the artist, drawing attention to the Indian's majestic bearing.

Top: Huron woman grinding grain, probably corn, with a long pestle, likely for use in *sagamité*, a gruel composed of corn, pieces of meat or fish, and vegetables.
Centre: An Indian of the Kaskaskia tribe that lived on the Mississippi River not far from Cahokia.
Bottom: Almouchiquois woman holding an ear of corn in one hand and a squash, or *sitroule* (pumpkin), in the other.

But not everyone shared this opinion. For instance, Ignace de Loyola, a well-bred Basque, must have been rather discouraged when he read a letter from one of his missionaries: "The women are naked and refuse themselves to no one," wrote Father José de Anchieta in 1554. "They even provoke and importune men to join them in the hammocks, for it is their honour to sleep with Christians." In his writings, the valiant Jesuit never lost his Catholic and European point of view. In truth, starting in 1492, Indians and Europeans discovered each other, learned to deal with each other, and sometimes fought. There were many exchanges in both directions. Neither Europe nor the Americas would never be the same.

Balanced, Healthy Societies

The Tainos, Iroquoians, and Timucuas belonged to vast trade networks. Upon receiving "iron objects," all had the same thought: these were great trade items that would give them a clear advantage in their bartering – especially because they were exclusive to certain nations. This was a major concern, reflected in both the negotiations and the behaviours of the Indians. All of them strongly suggested that their European visitors should not venture farther, but should let them go instead. Whatever the Europeans wanted, they would bring to them. The Indians proposed themselves as intermediaries and suggested alliances. These three nations were competitors, rivals, and often enemies. The Europeans had to choose their camps – just as the Indians would later have to do.

The Tainos, Iroquois, and Timucuas were sedentary nations that had in common a highly organized community life. Their members lived in well-appointed dwellings grouped in villages that had stores of food and, a novelty to the Europeans, tobacco. All three nations grew corn, had an advanced knowledge of plants, and availed themselves of an extensive pharmacopoeia. These Indians were healthy and took such good care of themselves that they were able to conceal an almost endemic form of syphilis loose among them. They were habituated to it and did not suffer from it or quickly controlled its symptoms when necessary. The Spanish were not careful – and what if they had been? At any rate, along with samples of various plants, such as corn and tobacco, and items such as hammocks, which they adopted on the spot because they were useful aboard rat-infested ships, Columbus's men brought back syphilis, American-style. It spread with stunning swiftness in Europe, joining the French, Neapolitan, and English varieties. Even top officers, such as Martin Alonso Pinzon, did not escape it. He barely got back to Spain before taking his last breath.

The Indians knew how to treat the diseases encountered in their own land, including fevers, which they cured with decoctions of quinoa bark – in a way, the *annedda* of the south – but they were tragically unequipped to handle new European diseases such as influenza, measles, and especially smallpox.

Annihilation

The Tainos of Guanahani, the Iroquois of Hochelaga, and the Timucuas of Florida met a common fate. They were all but extinguished. Exterminated in wars? Wiped out by their enemies? No. In almost no time, the Tainos, who had numbered over one million, dropped to a population of 50,000, then to almost zero. When Cartier returned to Hochelaga in 1541, the Iroquois village was no longer there. Champlain found the St. Lawrence Valley almost completely deserted; the Indians of Cartier's time had practically disappeared. In Florida, the Tiumucuas dispersed, and the survivors joined neighbouring tribes.

The fate of these three nations was not the exception but the rule. The extent of the epidemics that mowed down Indian populations has just begun to be admitted. The historian Léo-Paul Desrosiers encountered quite a bit of skepticism when he wrote in the first volume of *Iroquoisie* (1947), "I have 10,000 deaths to explain for that year. Wars alone are not enough; they could be the cause of the death of 1,000 at most." In 1985, based on an opinion by the Catholic Committee, the Quebec ministry of education refused to approve a Canadian history textbook because the authors suggested that missionaries had helped to propagate the epidemics.

The ravages of smallpox were particularly distressing. An individual could have the disease and be contagious without showing external signs. When the symptoms appeared, people distanced themselves from the sick person; many fled to another village, where they were taken in without question, thus spreading the disease. This is how entire nations were swept from existence. It is now well known that smallpox travelled more quickly than the conquistadors' troops. All it took was for the scouts from the two sides to come into contact, sometimes just touching the same objects, for the epidemic to spread. The conquests of Cortes, Pizarro, and others are explained less by the superiority of their weapons and the fact that they had horses and fighting dogs than by the epidemics that they unleashed, combined with their schemes to divide the Indians and a shocking cruelty that spared no one – sometimes costing the lives of even the conquistadors themselves.

In spite of its errors, Spain's colonization was far more advanced than France's or England's. In 1600, while the French and English had no permanent settlements in

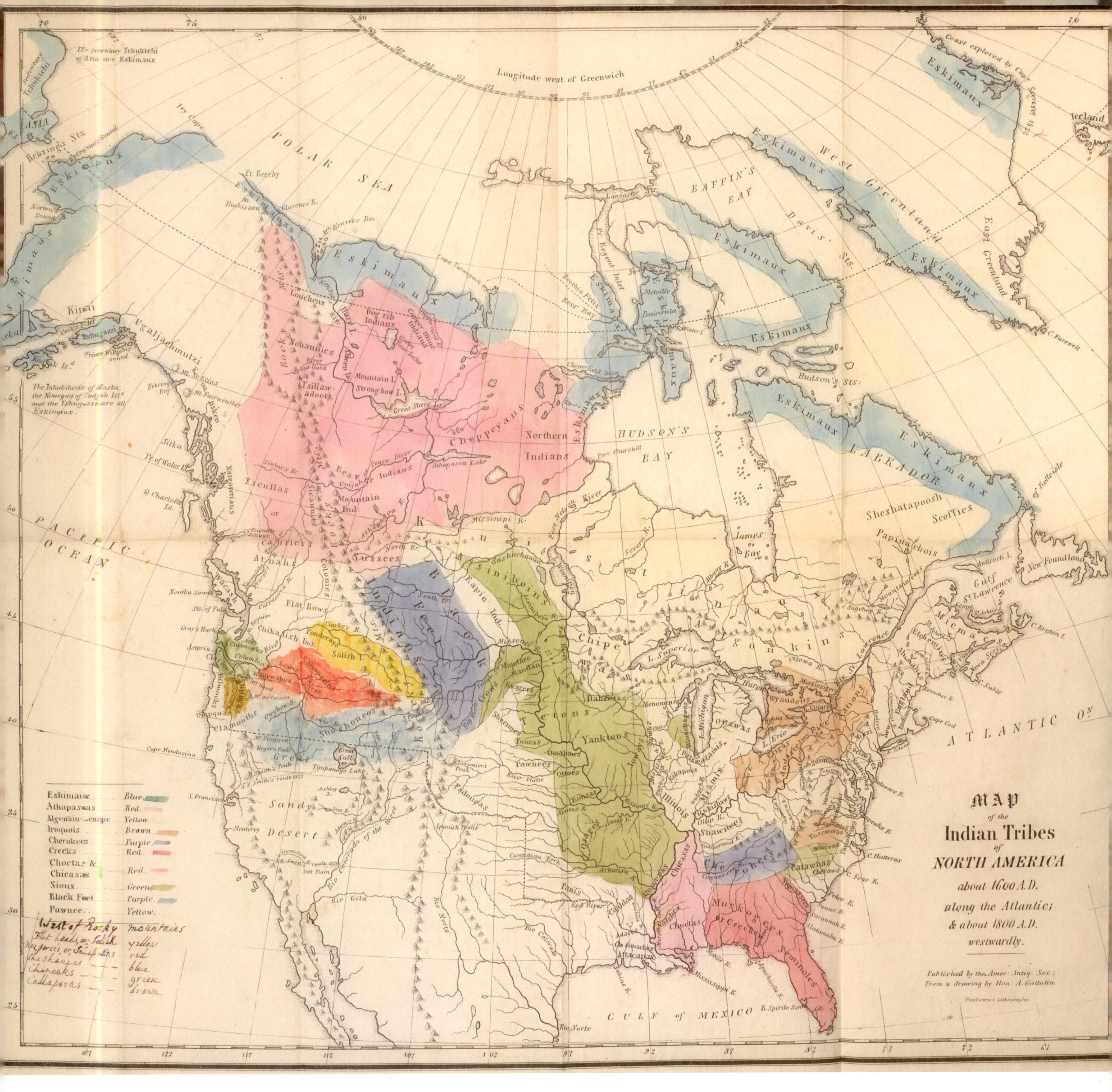

Map of the Indian Tribes of North America about 1600 A.D. along the Atlantic & about 1800 A.D. westwardly by Albert Gallatin
This attempt to classify Indian tribes was made by the patient and talented Albert Gallatin, financier, diplomat, and amateur ethnologist. He had the clever idea of compiling a survey of Indian tribes in two different time periods: in 1600 for those on the Atlantic coast and in 1800 for those in the interior. The former were wiped out or displaced; the latter were barely known to Americans before the mid-eighteenth century. Gallatin shows some 80 tribes, which he groups into eight linguistic families, identified on the map in different colours. Following Lewis and Clark's expedition to the Pacific coast, he added, by hand, new families, including the Salish, the Clatsops, the Shoshones, and the Chinooks. At the other end of the continent, he did not omit the Mountaineers, Papinachois, Micmacs, Etchemins, and Abenakis. An absolutely extraordinary document, this map was accompanied by Gallatin's *Synopsis of the Indian Tribes*, in which he described in detail the Indian languages carefully observed by Lewis and Clark, whose notes (lost today) he probably used.

Top: One person comforts another one suffering from smallpox. This well-known drawing is taken from the *Florentine Codex* by Fray Bernardino de Sahagún.
Bottom: This illustration by Theodor de Bry portrays the account of Johann von Staden, whose life was saved thanks to a fast-moving epidemic.

the Americas and were engaged mainly in privateering, the Spanish had founded about 200 towns in the Americas, in which 160,000 settlers lived.

At the time of Columbus's arrival, there were 15 major linguistic families in North America, shared among some 300 Indian nations. As they came into contact with the Europeans, these languages began to disappear. This sad phenomenon intensified over time; little by little, the Indians became aware of it.

To have an idea of the scope of the damage, we can look, for instance, at what happened when Europeans arrived in the heart of the continent, in the region around the Missouri. In their carefully kept journals, the Americans Merriwether Lewis and William Clark noted the terrible effects of the recent epidemics. On August 3, 1804, they met several Missouri Indians, the last survivors of the epidemic of 1801–02, who had taken refuge with the Otto Indians. Later, they visited the Arikaras, who were divided into three sparsely populated villages, where before they had inhabited 18. Close to where they wintered over, the Mandans had abandoned three of their five villages; the remaining two did not last long. By 1837, the end had arrived for many of these peoples.

In 1492, as many people lived in the Americas as in Europe; one population tragically collapsed, while the other flourished. The sedentary Indians were excellent farmers and better nourished than the Europeans. Corn is an almost miraculous plant; potatoes alone can keep a human being alive. The Indians had developed infinite varieties of these species, adapting them to very different soil and climatic conditions. These two foods gradually changed the nutritional habits of Europeans (and then Asians); along with them came many other fruits and vegetables – tomatoes, peppers, beans, squashes, pineapples, papaya, avocado, passion fruit, and others – and flavours without which we would have trouble imagining a dessert menu: chocolate, vanilla, and maple.

Of course, Europe was Civilization. The great Fernand Braudel, faithful to his Mediterranean perspective – for him, it was the "centre of the world" – first maintained that "the conquest of the New World was also the expansion of European civilization in all of its forms," before he finally recognized that "once it entered the life of Europe, America changed, little by little, all of its basic facts and even reoriented its activities."

The question is still open. Soon or later, we will have to re-examine the history of the Atlantic world – in fact, the history of the entire planet – and admit that 1492 marked an encounter between two worlds unaware of each other, a reciprocal discovery, the beginning of a new era of exchanges, the gradual birth of a true new world born of two old worlds.

Main sources

This article is a sort of summary of various aspects that I have discussed in previous writings, including my own *America* (Sillery: Septentrion, 2002) and L. Côté et al., *L'Indien généreux* (Montreal and Sillery: Boréal and Septentrion, 1992). [D. V.]

See also:

FENN, Elizabeth Anne. *Pox Americana: The Great Smallpox Epidemic of 1775–82*. New York: Hill and Wang, 2001. The American War of Independence against the background of a smallpox epidemic. — JONES, David Shumway. *Rationalizing Epidemics: Meanings and Uses of American Indian Mortality since 1600*. Cambridge: Harvard University Press, 2004. Fascinating, particularly on the earliest epidemics and the reactions that followed. — JULIEN, Charles-André. *Les voyages de découverte et les premiers établissements: XVe-XVIe siècles*. Paris: Presses Universitaires de France, 1948. — MANN, Charles C. *1491: New Revelations of the Americas before Columbus*. New York: Knopf, 2005. An engrossing work. — MORIN, Michel. *L'usurpation de la souveraineté autochtone: le cas des peuples de la Nouvelle-France et des colonies anglaises de l'Amérique du Nord*. Montreal: Boréal, 1997. — WALDMAN, Carl. *Atlas of the North American Indian*. New York: Facts On File, 1985.

The Thirteen Colonies

THE BEGINNING OF THE AMERICAN DREAM

WHILE THE GULF OF ST. LAWRENCE and the Gulf of Mexico were abuzz with activity, the coast from Newfoundland to today's Florida was slow to draw the attention of Europeans. Of course, the Spanish considered it their preserve, but they were unable to keep the French, English, Dutch, and even Swedish from trying to penetrate its mysteries.

Spain, at the right place at the right time, liberated from the Moorish occupation and strengthened by its recent unification, considered itself invested with a providential mission and took the lead in the Americas, since Charles Quint and his son, Philip II, had the means to impose their hegemony. Still at the beginning of her reign, Queen Elizabeth I of England considered it more prudent to keep her distance. The child of Henry VIII and Anne Boleyn, she was considered by many Catholics to be an illegitimate child and unworthy of assuming the throne. She therefore turned to Protestantism and re-established the Anglican Church from which her half-sister, Mary of Tudor, had broken during her short reign. Elizabeth also began to encourage the discontent with the Spanish that was developing in Flanders. She slowly distanced herself from her neighbours, especially Philip II, even encouraging her corsairs to harass and pillage the Spanish colonies. She revived the old dream of a route to China and was very pleased when Francis Drake succeeded with a spectacular round-the-world voyage (1577–80).

The French failure in Florida was not without its repercussions. Jean Ribaut, who had led the expedition of 1562, had to take refuge in England. He stayed longer than he would have wished, but took advantage of this sojourn to publish the account of his voyage. He was joined by Nicolas Barré – twice a survivor of voyages to the Americas, first in Brazil then in Florida – who contributed his testimony.

Over time, Queen Elizabeth became bolder, and in 1580 she sent Sir Humphrey Gilbert to explore the territories that were not yet under the authority of a Christian prince. In spite of his enthusiasm and vast knowledge, Gilbert did not get far beyond Newfoundland before he was swallowed up by the sea. His half-brother, Sir Walter Raleigh, one of Elizabeth's favourites, took up the torch and dispatched a small expedition under the command of Philip Amadas and Arthur Barlowe. On July 13, 1584, they landed on the island of Roanoke, a little north of the sites visited by the French from 1562 to 1565. They liked what they saw. The fertility of the land was amazing. "I think in all the world the like abundance is not to be found," wrote Barlowe. Better still, the Indians were "most gentle, loving, and faithfull, voide of all guile and treason, and as such live after the maner of the golden age." Barlowe related that an Indian, after visiting the English boats and accepting several gifts, returned to his boat, began fishing, and after half an hour deposited on the shore two bundles of fish, making it understood that this was his way of thanking and reimbursing them.

Raleigh could not ask for more. He obtained permission to establish a permanent settlement, and he named it Virginia, in honour of Elizabeth, who had been dubbed the Virgin Queen because she had no children. (In fact, her successor was her nephew, a Catholic.)

In Raleigh's mind, Roanoke would make not only an excellent colony, but also a base of operations for attacking Spanish convoys. Two men in particular were associated with the colony's brief history: a mathematician, Thomas Harriot, and a painter, John White. In spite of the know-how of these men (and the assistance of Manteo, a Croatan Indian who served as interpreter), the colony, dogged by bad luck, was a failure. There was drought, and then fresh supplies did not arrive as

Top: Drawing by John White titled *Regulorum Sepulcra*.
Bottom: Emperor Powhatan rules over his subjects, by John Smith. Even monarchs are mortal. Was Smith poking fun? In any case, he found the structure of the charnel house inspiring.

Detail of the map of Virginia by John White, published in Frankfurt by Theodor de Bry, 1590
The first English attempt to colonize North America took place in 1585 on the island of Roanoke (now in North Carolina). From the beginning, relations between the English and the Indians were not very cordial. Nevertheless, one member of the group of Europeans, the artist John White, was able to produce a number of drawings of Indians (from the village of Pomeiock). White also made a map of Virginia that was very accurate for the time. It was popularized by the publisher Theodor de Bry, who published it with an account by Thomas Hariot, *A briefe and true report of the new found land of Virginia*. "The map shows the island of Roanoac situated between today's Pamlico Sound and Albemarle Sound and protected by the dunes of the Hatteras Islands (HATORASK)." The cartographer also indicated the location of a number of Indian nations. Between 1587 and 1590, all trace of those who had lived on Roanoke mysteriously disappeared, and the site was later dubbed the "lost colony."

expected; the survivors had no choice but to return to England with Drake in June 1586. White arrived with a new group of settlers the following year but had trouble maintaining links with the metropolis in the context of a general mobilization against the Spanish Armada in 1588. Finally, in 1590, White managed to return to Roanoke. His settlers, as well as his daughter, Elenor, and his granddaughter, born on the island in August 1587, could not be found; several clues indicated that they had been gathered up and adopted by an Indian tribe.

It was almost 20 years before a new attempt at colonization was organized. After many twists and turns, the Virginia Company established a base in the southern part of Virginia, which became the first permanent English settlement in North America. On April 26, 1607, the English entered "Chesupioc" Bay. The beginnings of Jamestown were difficult. By December, only half of the people who had arrived with the Virginia Company remained. The instructions of the company administrators, which were to find the Roanoke survivors, to begin to search for a passage to the "East India Sea," and to discover "precious minerals," were far from the settlers' minds. Survival came first. *Primo vivere*. John Smith, designated as one of the seven members of the governing council, rapidly stepped forward as the leader. "He who does not work," Smith made it known, "will not eat." "Everyone to work" became his watchword.

Does John Smith deserve the stature of hero that was built around him? Perhaps, but it is quite difficult to sort truth from myth. Did he save the colony? Was he himself saved thanks to the intervention of Pocahontas, the daughter of Powhatan, chief of the Powhatans, an Algonquin-speaking tribe? Repatriated to heal an injury in October 1609, he returned to North America only in 1614 for a short stay. He devoted the rest of his life to recollecting and writing about his very exciting life. By the time he died, in 1631, at age 51, he had had the time to recount, and perhaps fabricate, an incredible existence. He had enlisted very young, after taking part in various wars, and found himself in Transylvania fighting the Turks, as he recounted in his many writings. He won three duels. Injured in combat, he was captured, sold as a slave, and offered as a gift to a lady in Istanbul, Charatza Tragabigzanda. She fell in love with him and entrusted him to her brother so that he could enlist in the Imperial Guard, but he killed his patron (or master) and fled across Russia and Poland to return to Transylvania, where he was rewarded for his bravery. He then travelled in Europe and North Africa before returning to England in 1604, finally ready to set out on new adventures. An encounter with Grand Chief Powhatan awaited him in North America.

Smith was a braggart, a teller of tall tales, a compulsive liar – and elusive. He certainly did not lack imagination or talent, and he did shed light on the Indians, the settlers, and his compatriots. At any rate, his accounts finally spurred European nations into action. In 1609, the Englishman Henry Hudson explored, for the Dutch, the river that would bear his name. Four years later, the Dutchman Adrien Block started a small colony at the river's mouth; in 1626, the Walloon Pierre Minuit broadened the prospects for Dutch development by purchasing the island of Manhattan from the Indians. Thus a Dutch colony took form along the Hudson River, founded, in Minuit's mind, on inter-ethnic encounter and mutual respect.

In the meantime, the English founded a second colony, this one at Plymouth. As at Jamestown, things were difficult in the early years. The winter of 1620 was terrible for the 102 pilgrims of the *Mayflower*. Half of them died. In May 1621, an Abenaki called Samoset, who had learned a few words of English in contact with the fishermen who visited the coast, gave them a bit of hope. The chief of the Massasoit Indians took things in hand and set the terms of a treaty with the help of Squanto, considered the last of the Pawtuxets. Because he had been taken to England as a prisoner, Squanto had escaped the epidemic that had wiped out his nation and, at the same time, learned English; this now came in handy. In the fall of 1621, Indians and Europeans celebrated the gathering of the harvest together. It was the first Thanksgiving holiday.

The attractions of the world on the other side of the Atlantic were stronger than the accounts of misery coming back from it. Successive crises in Europe brought to the American coast waves of refugees who founded 13 different colonies. The American dream began to take shape. It even reached Sweden. In 1638, Thomas West, baron De La Warr, founded a small colony just north of Chesapeake Bay at the bottom of a deep bay (Delaware Bay), where the Christina fort was to be erected. The Dutch, with settlements at Nieuw Amsterdam (1623) and Breukelen (1646), quickly made life difficult for the Swedes and finally annexed New Sweden, in 1655. Then, following the Peace of Breda (1667), the Dutch suffered a similar fate at the hands of the English.

Walloons, Flemish, and Dutch, along with French Huguenots, melted in among the British immigrants who were fleeing a metropolis in constant political and religious turmoil – turmoil that followed them to the new colonies, where conflicts sometimes erupted. Driven from Salem, Roger Williams instituted the Baptist religion in Rhode Island; New Hampshire separated from Maine, joined Massachusetts, then received a distinct charter; Irish Catholics settled in what became Maryland; a Puritan population dominated in New England, an area that incorporated Rhode Island, Massachusetts, New

Hampshire, and Connecticut; William Penn made the Quakers welcome in his colony, aptly named Pennsylvania; and another philanthropist, James Oglethorpe, welcomed misfits and convicts into a colony that George II conceded to him, which he named Georgia. French Huguenots revived Ribaut and Laudonnière's Carolina and took in immigrants of all origins. Germans poured in everywhere. Sometimes the initiative was under the control of the British crown, sometimes it was private owners or companies that held the reins.

The Thirteen Colonies, born one after another, were fiefs divided by rivalries, without any real links among them – nor, often, with the metropolis. Although they were refuges, their populations still thirsted for prosperity. They wanted land and so expelled the original occupants, but they remained focused on the ocean for many years. According to the rules of colonialism, they imported from England, exported to it, and compensated by developing trade with the West Indies and the rest of Europe.

The English occupied the Atlantic coast from Maine to Georgia. At first, they rarely ventured inland; to their west, the Appalachian Mountains stood in for the horizon. After the final French and Indian War (1760), they became bolder and ventured as far as Mississippi, opening up Tennessee and Kentucky. Independence (1783) put new wind in their sails. The day wasn't far off when, with the encouragement of President Thomas Jefferson, they would return to the search for a navigable route to the West. An unexpected gift from Napoleon, the Louisiana Purchase, enabled them to double their territory, opening the gates to the West. Their manifest destiny would take care of the rest.

Main sources in order of importance

APPELBAUM, Robert, and John Wood SWEET (eds.). *Envisioning an English Empire: Jamestown and the Making of the North Atlantic World.* Philadelphia: University of Pennsylvania Press, 2005. — STEELE, Ian Kenneth. *Warpaths: Invasions of North America.* New York: Oxford University Press, 1994. — TOWNSEND, Camilla. *Pocahontas and the Powhatan Dilemma.* New York: Hill and Wang, 2004. — AXTELL, James. *Natives and Newcomers: The Cultural Origins of North America.* New York: Oxford University Press, 2001.

Source for translation: BARLOWE, Arthur. "Captain Arthur Barlowe's Narrative of the First Voyage to the Coasts of America." In Henry S. Burrage (ed.), *Early English and French Voyages, Chiefly from Hakluyt, 1534–1608,* pp. 225–41. New York: Charles Scribner's Sons, 1906 [www.americanjourneys.org].

Map of Virginia by John Smith, London, 1612
The English established their first permanent settlement in Jamestown, Virginia, in 1607. The territory was mapped one year later by John Smith. To make this map, Smith explored Chesapeake Bay and its drainage basin. The English wanted to stamp the country with English place names, inspired mainly by names from the royal family (Cape Henry, Cape Charles, Jamestown), and yet the majority of place names here – a total of almost 200 – are in Indian languages. With a concern for authenticity, the mapmaker also drew two Indians, a Susqueanna warrior and Chief Powhatan, inspired largely by John White's drawings. Smith marked his map with small crosses to distinguish the territory that he had seen from the territory that he had not explored himself.

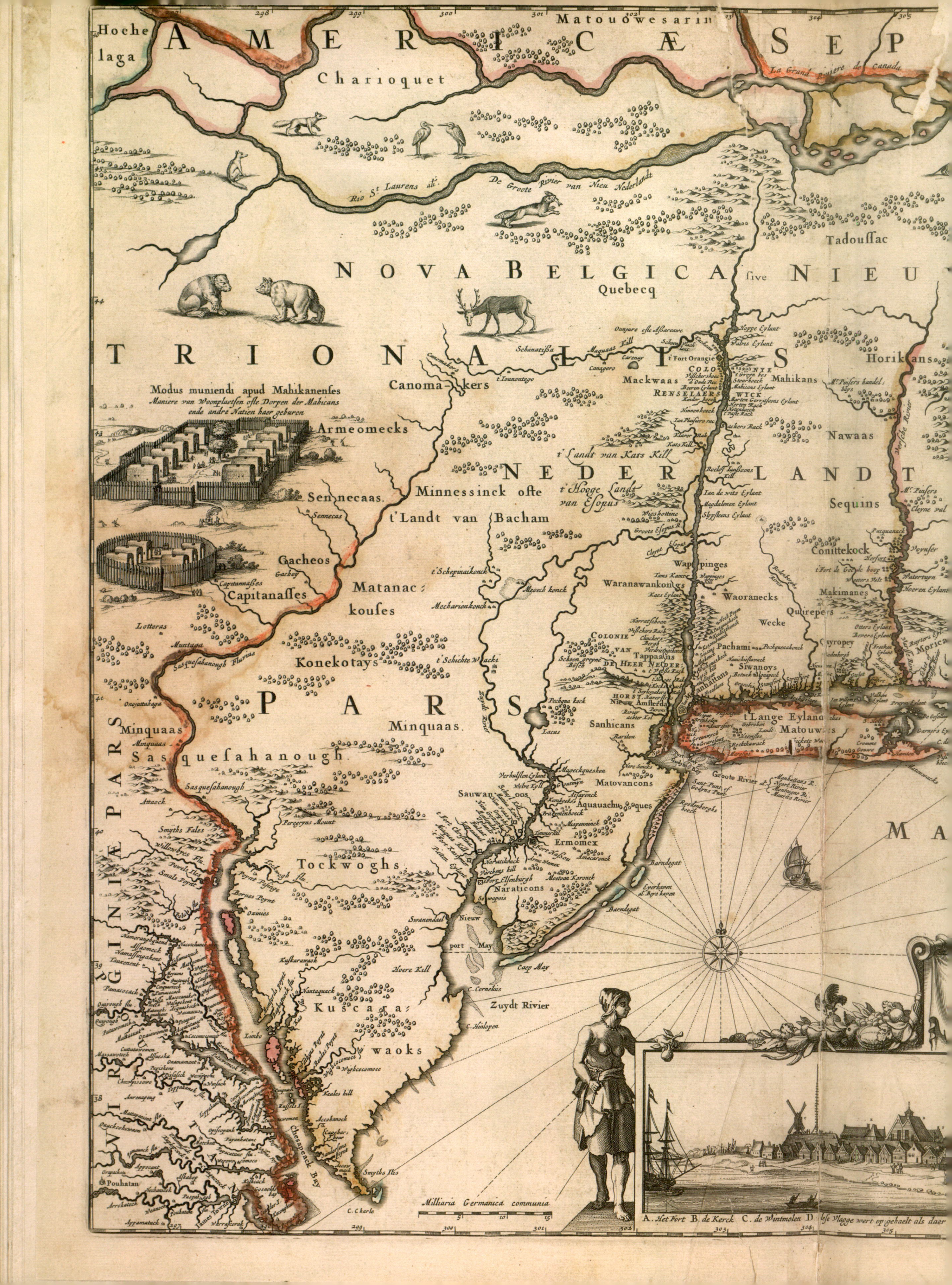

Hoche laga
AMERICÆ SEP
Matouowesarin
Charioquet
La Grand Riviere de Canada
Rio St Laurens al: De Groote Rivier van Nieu Nederlandt
Tadoussac
NOVA BELGICA sive NIEU
Quebecq
TRIONALIS
Modus muniendi apud Mahikanenses
Maniere van Woonplaetsen ofte Dorpen der Mahicans ende andre Natien haer geburen
Armeomecks
Canomakers
Mackwaas
Mahikans
Horikans
NEDER LANDT
Nawaas
Sennecaas
Minnessinck ofte t'Landt van Bacham
t'Landt van Kars Kill
t'Hooge Landt van Esopus
Sequins
Gacheos
Capitanasses
Matanackouses
Waranawankonggs
Waoranecks
Wecke
Makimanes
Quiripeys
Konekotays
PARS
Minquaas
Sanhicans
t'Lange Eylandt
Matouwas
Minquaas
Sasquesahanough
Sasquesahanough Flavius
Matovancons
Aquauachuques
Sauwanoos
Ermomex
Tockwoghs
Naraticons
Barndegat
Nieuw port May
Cape May
C. Cornelius
Zuydt Rivier
C. Hinlopen
Kuscaraawaoks
Chesapeack Bay
VIRGINIÆ PARS
Smyths Iles
C. Charle
Milliaria Germanica communia
A. Het Fort B. de Kerck C. de Wintmolen D. deze Vlagge wert opgehaelt als daer

Map of New Belgium by Nicolaes Visscher, Amsterdam, 1655
This map by Nicolaes Visscher portraying New Belgium in the mid-seventeenth century testifies to the Dutch presence in North America. This colony, also known as New Holland, covered parts of a number of today's American states, including New York, New Jersey, Pennsylvania, Connecticut, Rhode Island, and Vermont. The territory mapped is sprinkled with place names that have survived in some form down to the present day, including Lange Eylandt (Long Island), Manhattans, Staten Eyl., Breukelen (Brooklyn), Vlissingen (Flushing), Block Island, Kats Kill (Catskill), and Roode Eylandt (Rhode Island). At the mouth of the Hudson River (Groote Rivier), on the southern tip of Manhattan, is the capital, New Amsterdam, which became New York when it fell into English hands in 1664. At the bottom of the map is one of the oldest engravings of the city – quite a contrast with today's views of towering skyscrapers. Farther north on the Hudson River is Fort Orange (Albany), which provided the Dutch West Indian Company with a supply of furs from the Indians, especially the Mohawks and Mohicans (on the right is a portrayal of two fortified villages). Some Canadian coureurs des bois went to Fort Orange to trade in contraband in the 1670s, and perhaps even before. They likely did not use this map, which is very inaccurate in its description of Canada; in particular, Lac des Iroquois (Lake Champlain) is placed much too far east.

Mastering the North Atlantic: Seamen and Nautical Knowledge

IN THE TIMES OF VERRAZZANO, Cartier, Frobisher, Hudson, and Champlain, a number of conditions had to be aligned for a transatlantic crossing to be successful. Avoiding calamities at sea took solidly built and well-fitted-out ships, experienced captains and pilots, trustworthy crews, and reliable charts, navigation instruments, and sailing directions – as well as a good measure of courage and nerve. Neither the French nor the English had an organized navy; rather, traders and adventurers engaged in private commerce were obliged to serve their respective royal armed forces from time to time. Most of these rovers of the seas had not attended school but learned the rudiments of navigation on board their vessels. Thus, the best navigators received their education during cod-fishing expeditions in the cold waters off Newfoundland, a true training ground for sailors. Knowledge was passed down through families by oral tradition, since recruitment was still largely a family affair. Seamen were self-employed and enjoyed a high degree of autonomy from royal authorities; their real bosses were the financiers who fitted out and supplied their ships, hoping to make a major profit from their investment.

On board the ship, the captain was the master, second only to God. Not only did this commander-in-chief have to know the secrets of the winds and currents, he had to have the knack of making his crew obey him. He was responsible for the cargo and the men on board, from provisioning to laying up the ship. In crucial moments, he might sometimes seek the opinion of his senior officers. The account of Cartier's voyages of 1534 gives an example of this: the decision to return to France without exploring the St. Lawrence River was made after consultation with "all our Captaines, Masters, and Mariners." To bring a seagoing sailing ship home safely across the ocean, the captain needed a specialized crew of sufficient number to match the size of the vessel, each member of which had a very precise role. For example, the pilot was in charge of the course and had to have all the navigation instruments that he might need. The boatswain organized and supervised

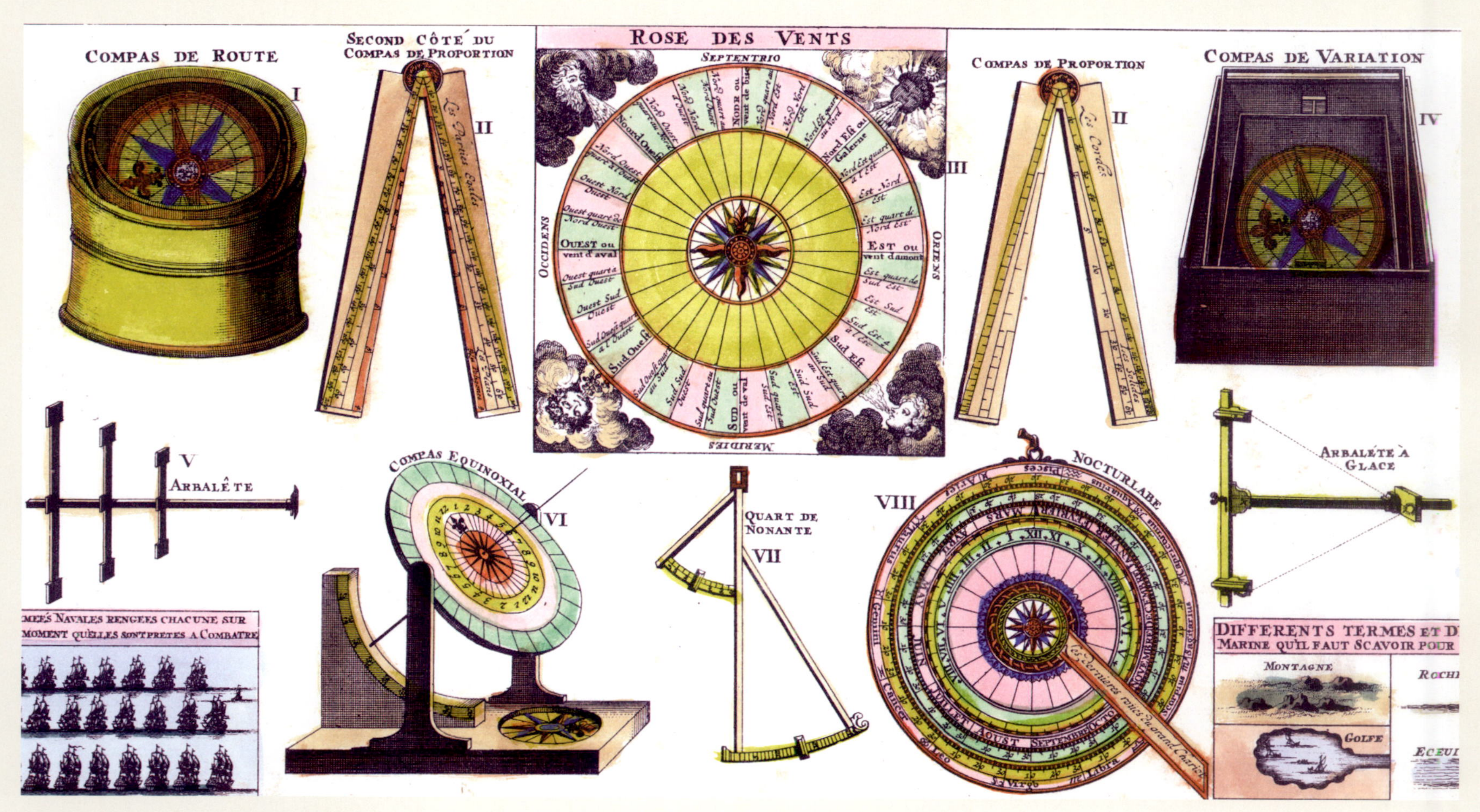

Navigation instruments by Chatelain, 1739
Europeans were able to go beyond the Mediterranean and travel the world due, in large part, to the progress in navigation after the Middle Ages. The few instruments used by pilots facilitated ocean crossings and travel on all seas for those seeking coveted riches in foreign lands, including the Americas. These instruments, illustrated in Chatelain's *Atlas Historique*, prove that by the eighteenth century navigation had become an affair of mathematics and geometry. All of these instruments helped seamen to locate their position on a nautical chart. The steering compass and the sun compass were used to find direction and measure angles; the cross-staff, the back-staff (or Davis quadrant), and the nocturnal were used to measure the altitude of the stars, both day and night, and thus to measure latitude.

the crew. The sailors specialized in maintaining the hull, the caulking, the sails, and the artillery. On a ship, it was not rare to find a cooper, a baker, a chaplain, a writer, a surgeon, and even a dog or cat to hunt down various intruders.

In the Middle Ages, the North Atlantic coasts were frequented by large Italian galleons, but sailing ships, better adapted to the ocean, quickly predominated. Technological innovations led to the development of the caravel, which enabled Europeans to venture far from the Mediterranean coasts. Invented by the Portuguese in the fourteenth century, this sailing ship, initially fitted with triangular lateen sails, could tack in zigzags into the dominant wind. In the sixteenth and seventeenth centuries, it was mainly small ships that plied the Atlantic between France and New France, compared to the 800-freight-ton (one freight ton = 800 cubic feet) and larger vessels that sailed between Lisbon and Goa. To provide better defence against the pirates that scoured the European coasts, the expeditions generally sailed in armed convoys. Even bankers (ships that went to fish the Grand Banks off Newfoundland), whose only cargo was cod, fresh or salted, might have several cannons on board.

Although there was no real state educational system before the end of the seventeenth century, there were a number of books available on nautical science. These specialized books, which became more numerous with the rise of the printing press, prefigured the progress from entirely experiential knowledge to a science that was in part in written form. The early sixteenth century saw the publication of *Le Grand Routier* by Pierre Garcie, known as Ferrande. Like all of the publications of the École du Conquet, it was interesting because it was intended for barely literate readers and used drawings and diagrams. In 1554, *Arte de navegar*, by Pedro de Medina, was translated from Spanish to French and published in Lyon. Written by a companion of Hernán Cortés, this work was considered the keystone of maritime literature. Later, Samuel de Champlain, well known for his accounts of voyages to New France, published an appendix to his 1632 account, *Treatise on Seamanship and the Duty of a Good Seaman*. As a practitioner and teacher, Champlain communicated all of the nautical knowledge that he had used during his voyages to Canada. He described various navigation techniques in great detail: taking bearings of coastlines, cartography, determination of a ship's speed, calculation of latitude with the cross-staff and of longitude by trigonometric deduction. In the area of maritime knowledge, one of the seminal works was written by the Jesuit priest Georges Fournier. This monumental 922-page work, *Hydrographie contenant la théorie et la practique de toutes les parties de la navigation*, published in 1643, was a historical, theoretical, and practical work on hydrographic science, which was then starting to be taught in the main ports of France. Indeed, in order to create a real national naval force for their kingdom, Cardinal de Richelieu and, later, Colbert created schools of hydrography to train pilots and navy officers where would-be seamen studied different subjects, both theoretical and practical, including arithmetic, geometry, navigational calculations, the use of maps and the compass, currents and tides, and course determination. Schools were set up at Dieppe, Bayonne, Brest, Rochefort, Marseille, and Toulon – and also at Quebec, where hydrography courses were given at the Jesuit college starting in 1665 and the position of king's hydrographer was created in 1686.

In the times of New France, keen judgment, a sense of observation, a good memory, and experience were the best measures of a pilot's skill. Navigating by dead reckoning in seas that were usually familiar to them, pilots guided themselves by topography and seamarks, both natural and artificial: capes, rocks, islands, crosses. In addition to these reference points, they also used various instruments that enabled them to roughly calculate their position at sea. The main tool was the compass, which

***Hæmisphærium scenographicum australe cœliti stellati et terræ* by Andreas Cellarius, 1708**
Since antiquity, people have mapped both Earth and the heavens. Geography and astronomy were long regarded as related disciplines and sister sciences. Knowing the positions of the celestial bodies made it easier to travel on the sea or in unknown lands. This plate, taken from the celestial atlas *Harmonia macrocosmica*, is a perfect illustration of the connection between sky science and earth science. With ingenuity and great artistry, Cellarius superimposed the outline of the Southern Hemisphere and that of the celestial sphere studded with constellations. On the bottom left, scholars observe the sky using scientific instruments. To the right, a professor teaches geography to an aristocrat, perhaps a prince who has claimed possession of territories overseas.

appeared in the Mediterranean in the late thirteenth century and was quickly adopted elsewhere. Affixed to a compass rose that usually had 32 directions, the magnetic needle showed the orientation of the ship or allowed the location of a seamark to be noted.

At about the same time, another valuable tool was invented: the portolan or pilot book, which contained sailing directions indicating ports and the distances from one place to another. The first pilot books for the North Atlantic were published in the sixteenth century: one by the Portuguese Jean Alfonce (1544) and one by the Basque Martin Hoyarsabal (1579). These texts included observations on courses, approaches, roadsteads, and ports in Europe and North America. For example, following his voyages on the St. Lawrence with Roberval, Alfonce gave the distance and orientation of the compass lines separating Baie des Chaleurs, Baie des Morues, Gaspé, Ognedoc, Sept-Îles, Île Raquelle, Île aux Lièvres, and so on, as far as France-Roy, the short-lived colony founded by Roberval near the site of today's Quebec City.

The best companion to the compass and the pilot book was the nautical chart, also called a portolan chart. Made starting in the late Middle Ages, first by the Genoese and Majorcans, portolan charts helped sailors recognize the islands, coasts, and ports of the Mediterranean. Up to the sixteenth century, these charts were made on a maze of criss-crossing lines, rhumbs or compass points, and seamen could situate a position at sea by following the line corresponding to the

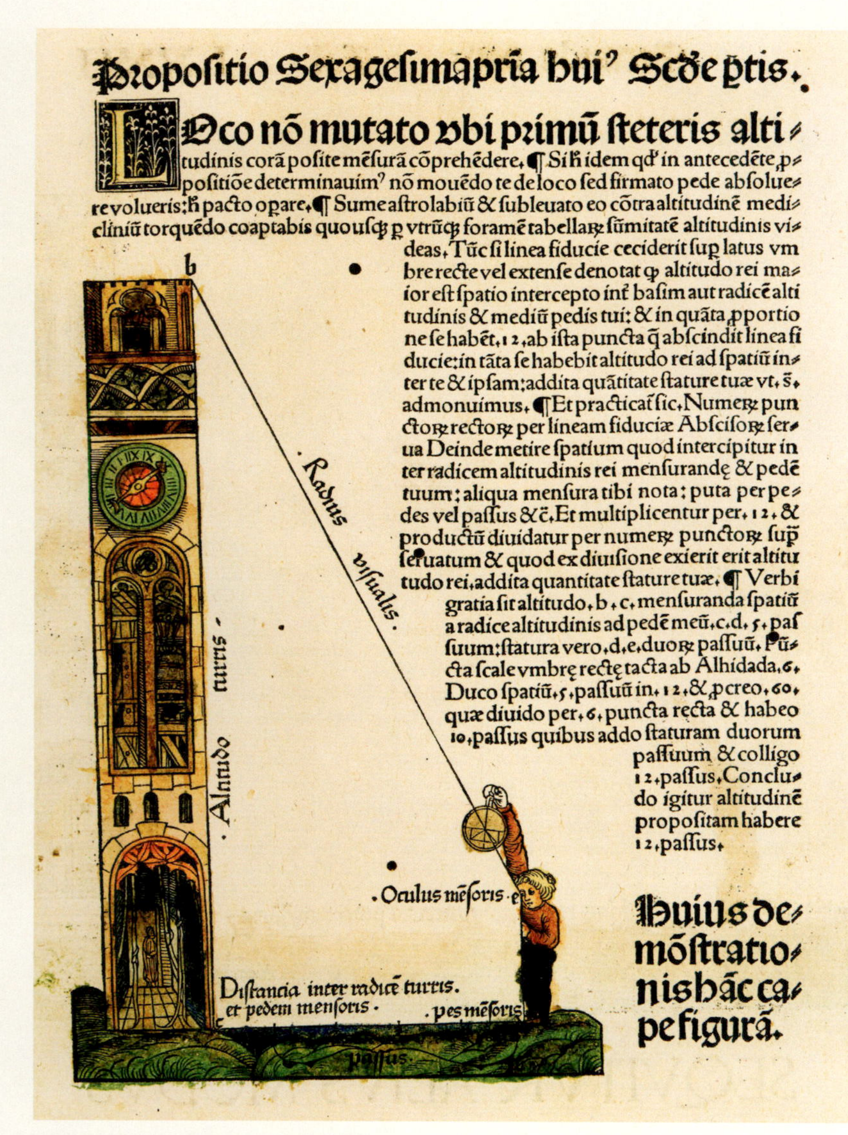

Propoſitio Sexageſimaprĩa hui⁹ Scd̃e ptis.

Loco nõ mutato vbi primũ ſteteris alti-
tudinis corã poſite mẽſurã cõprehẽdere. ¶ Si h̃ idem qđ in antecedẽte p-
poſitiõe determinauim⁹ nõ mouẽdo te de loco ſed firmato pede abſolue-
re volueris: h̃ pacto opare. ¶ Sume aſtrolabiũ & ſubleuato eo cõtra altitudinẽ medi-
cliniũ torquẽdo coaptabis quouſq; p vtrũq; foramẽ tabellarꝝ ſũmitatẽ altitudinis vi-
deas. Tũc ſi linea fiducie ceciderit ſup latus vm-
bre recte vel extenſe denotat q altitudo rei maior eſt ſpatio intercepto int̃ baſim aut radicẽ alti
tudinis & mediũ pedis tui: & in quãta ꝓportio
ne ſe habẽt. 12. ab iſta puncta q̃ abſcindit linea fi
ducie: in tãta ſe habebit altitudo rei ad ſpatiũ in-
ter te & ipſam: addita quãtitate ſtature tuæ vt. s̃.
admonuimus. ¶ Et practicat̃ ſic. Numerꝝ pun
ctorꝝ rectorꝝ per lineam fiduciæ Abſciſorꝝ ſer-
ua Deinde metire ſpatium quod intercipitur in
ter radicem altitudinis rei menſurandę & pedẽ
tuum; aliqua menſura tibi nota; puta per pe-
des vel paſſus &c̃. Et multiplicentur per. 12. &
productũ diuidatur per numerꝝ punctorꝝ ſup̃
ſeruatum & quod ex diuiſione exierit erit altitu
tudo rei. addita quantitate ſtature tuæ. ¶ Verbi
gratia ſit altitudo. b. c. menſuranda ſpatiũ
a radice altitudinis ad pedẽ meũ. c. d. 5. paſ
ſuum: ſtatura vero. d. e. duorꝝ paſſuũ. Pũ-
cta ſcale vmbrę rectę tacta ab Alhidada. 6.
Duco ſpatiũ. 5. paſſuũ in. 12. & ꝓcreo. 60.
quæ diuido per. 6. puncta recta & habeo
10. paſſus quibus addo ſtaturam duorum
paſſuum & colligo
12. paſſus. Conclu-
do igitur altitudinẽ
propoſitam habere
12. paſſus.

Huius de-
mõſtratio-
nis hãc ca-
pe figurã.

Use of the astrolabe by Jean Stöffler, 1513
Drawn from a 1513 German treatise on the manufacturing of astrolabes, this illustration explains how to take distance measurements using this instrument.

orientation of their ship. Made on parchment, the charts included diagrams of coastlines, and used alternating colours for place names. With the rise of trade on the Atlantic, their use spread beyond the Mediterranean, and cartographers used them to compile data gathered during voyages of exploration. By the end of the sixteenth century, a simple, but effective, way to measure distances travelled at sea had been devised: the log. This was a piece of wood attached to a rope with knots at equal intervals; when it was thrown into the water, the speed of the ship could be estimated by counting the number of knots that went by a fixed point over a 30-second period.

Good navigators also knew how to locate their position by observing the stars. Astronomical navigation, though impossible during cloudy periods, allowed rough estimates of position to be corrected. By the sixteenth century, nautical charts were being drawn against a canvas of latitudes and then longitudes. These increasingly popular units of measurement had been invented in ancient Greece to situate a point on Earth's surface. To calculate latitude, one simply measured the height of the sun above the horizon. Navigators used different instruments for this purpose: the cross-staff or astrolabe, and later the sextant and Hadley's octant. By the time the first Europeans arrived in America, latitude measurements were very precise, as evidenced in the accounts of Cartier's voyages and Alfonce's pilot book.

Calculation of longitude was more complex. Early measurements were based on estimates of orientation and distances travelled, both on land and at sea. But the inaccuracy of the compass and the log and the fact that ships might be pushed off course by currents and winds led to a search for new calculation methods. Very early, scholars exploited the principle of time-zone differentials, which are directly proportional to longitude differentials. Since Earth is in constant rotation, to find the distance between a given point and the prime meridian, the time difference between them is calculated. This may be done by observing predictable phenomena that occur at the same time in all parts of the world: lunar eclipses and eclipses of Jupiter's satellites. On October 27, 1633, the Jesuit Paul Le Jeune observed a lunar eclipse at 6:00 p.m. in Quebec. He had an almanac that indicated that this eclipse should occur at midnight in France, and concluded that there was a six-hour difference between the two places – thus, a distance of about 90 degrees. In 1642, Jérôme Lalemant made a similar observation at Sainte-Marie-du-Sault and obtained a difference of about five hours with Paris, France. At first imprecise, these measurements became more accurate in the latter part of the century as improvements were made to instruments and astronomical calculations. For instance, on December 11, 1685, the hydographer sent by the Académie des sciences, Jean Deshayes, observed a lunar eclipse in Quebec that enabled him to calculate a distance of 4 hours, 48 minutes, 52 seconds between Quebec and Paris – that is, 72 degrees, 13 minutes, an error of only 1.5 degrees.

However, since eclipses occur relatively rarely and very precise measurements were difficult to obtain at sea, the problem of calculating longitudes during navigation on the high seas remained unsolved. Maritime tragedies caused by insufficient knowledge of longitude became intolerable to public authorities, scholars, and ship outfitters. The stakes were so high that in 1714 the British Parliament passed the *Longitude Act,* which offered a substantial reward (up to 20,000 pounds sterling) to anyone who could precisely calculate longitude at sea. Those searching for innovative solutions and practices quickly realized that the key to the mystery was to be found not only in the hands of astronomers but also in those of clockmakers. Since longitude was calculated based on the time-zone difference between two points, it was important to have a clock that accurately kept the time of the home port. This time then had to be compared with the ship's local time, established by observing the absolute height of the sun. By the early seventeenth century, Champlain had expressed this idea in his *Treatise on Seamanship:* "When the sun is at the meridian, note forthwith by the instrument or chronometer the noon of this spot, and see the difference there is between the noon of the place whence one started and that of the place where one is, and this gives the distance traversed." Later on, in 1670, Jean Richer, at the time a mathematician of the Académie des sciences, set sail for Acadia armed with a pendulum clock designed by Christiaan Huygens to measure the longitudes of the coasts. But the pendulum broke during the journey and the expedition did not provide the anticipated results. A clock that was robust and precise enough to resist sea movement and variations in temperature and humidity had to be invented. In 1759, the English clockmaker John Harrison developed a chronometer that met all of these conditions: portable and accurate, it was slow by only five seconds at the end of a trip between England and Jamaica. The advent of the high-precision chronometer, a true technological triumph, was to change navigation practices forever by making them more precise and certain. Finally, the problem of calculating longitude at sea had been solved.

The first explorers to navigate on the Atlantic had to learn, empirically, the dominant winds and currents, and pass this knowledge on to their successors. As had happened for sailing around the tip of Africa, the

Use of the cross-staff by Jacques de Vaulx, 1583
This illustration, drawn from a magnificent treatise on navigation by Jacques de Vaulx, shows how to use a cross-staff, or Jacob's staff, which was employed in the sixteenth century to determine latitude. This instrument is composed of a long, graduated wooden rod and a visor with which to measure the altitude of the sun from the horizon. From a reading of the sun's altitude at its zenith (at noon), latitude could be calculated. Other instruments based on the same principle appeared after the cross-staff: the astrolabe, the octant, then the sextant, which was used at sea to calculate latitude until the recent invention of GPS.

major courses were rapidly marked out for the trip to the West Indies ahead of the trade winds and for Newfoundland and Canada via the north. Although the latter route was not as long and perilous as the Carreira da India, it nevertheless had its share of difficulties. After four or five weeks of navigation, heading west at 45°N, the Grand Banks of Newfoundland were the first seamark encountered. When the weather was good, the voyage might be as short as 20 days, as it was for Champlain's crossing in 1610. But the trip could also be much longer, if the skies were not clear. When Intendant Jean Talon went to Quebec in 1665, his ship took 117 days to make the crossing. In general, ships left French ports in April since, with an earlier departure, a fleet might run into a wall of icebergs. Champlain experienced this on a foggy night in the spring of 1610, when he saw "ice" passing under the bowsprit and by the side of the vessel. The next day, he saw other icebergs, "which looked like islands from a distance." One of them was so enormous that they sailed into it rather than going around it. But they could find no exit and their ship was trapped. Champlain emotionally described the crew's panic:

> More than a score of times we thought we should not come off alive. The whole night was spent amid difficulties and labours. Never was the watch better kept; for nobody had any desire to sleep, but rather to struggle to get out of the dangerous ice. The cold was so great that all the ship's running rigging was so frozen and covered with big icicles that we could not work it nor stand upon the vessel's deck … We remained four or five days in this extreme danger, until one morning as we looked about us in all directions, we [finally saw a passage] … Being outside, we praised God for our deliverance.

Other dangers also lay in wait for the travellers. Contrary winds, storms, and the lack of wind might slow a crossing, sometimes putting it in peril. The missionary Paul Le Jeune, who went to Canada for the first time in 1632, mentions the vagaries of the weather in his first published relation. This excerpt expresses well the mind of a voyager made anxious by the caprices of the weather:

> We had fine weather at first, and made about six hundred leagues in ten days; but we could hardly cover two hundred on the following thirty-three days. After this fine weather we had little but storms and contrary winds, except a few pleasant hours which were vouchsafed us from time to time. I had sometimes seen the angry sea from the windows of our little house at Dieppe; but watching the fury of the Ocean from the shore is quite different from tossing upon its waves. During three or four days we were close-reefed, as sailors say, our helm fastened down. You would have said that the winds were unchained against us. Every moment we feared lest they should snap our masts, or that the ship would spring a leak; and, in fact, there was a leak, which would, as I heard reported, have sunk us if had been lower down. It is one thing to reflect upon death in one's cell, before the image of the Crucifix; but quite another to think of it in the midst of a tempest and in the presence of death itself.

Due to the foul and precarious living conditions on a ship, the crossing to New France was anything but pleasant. One had to be very brave to confront the agonies of the sea, pirates, malnutrition, scurvy, damage to the vessel, reefs, and storms that each expedition entailed. But without these seafaring adventurers – intrepid captains, sailors, and voyagers – neither New France nor any other European colony would have been possible.

II

EXPLORING AND MAPPING NORTH AMERICA

SEVENTEENTH CENTURY

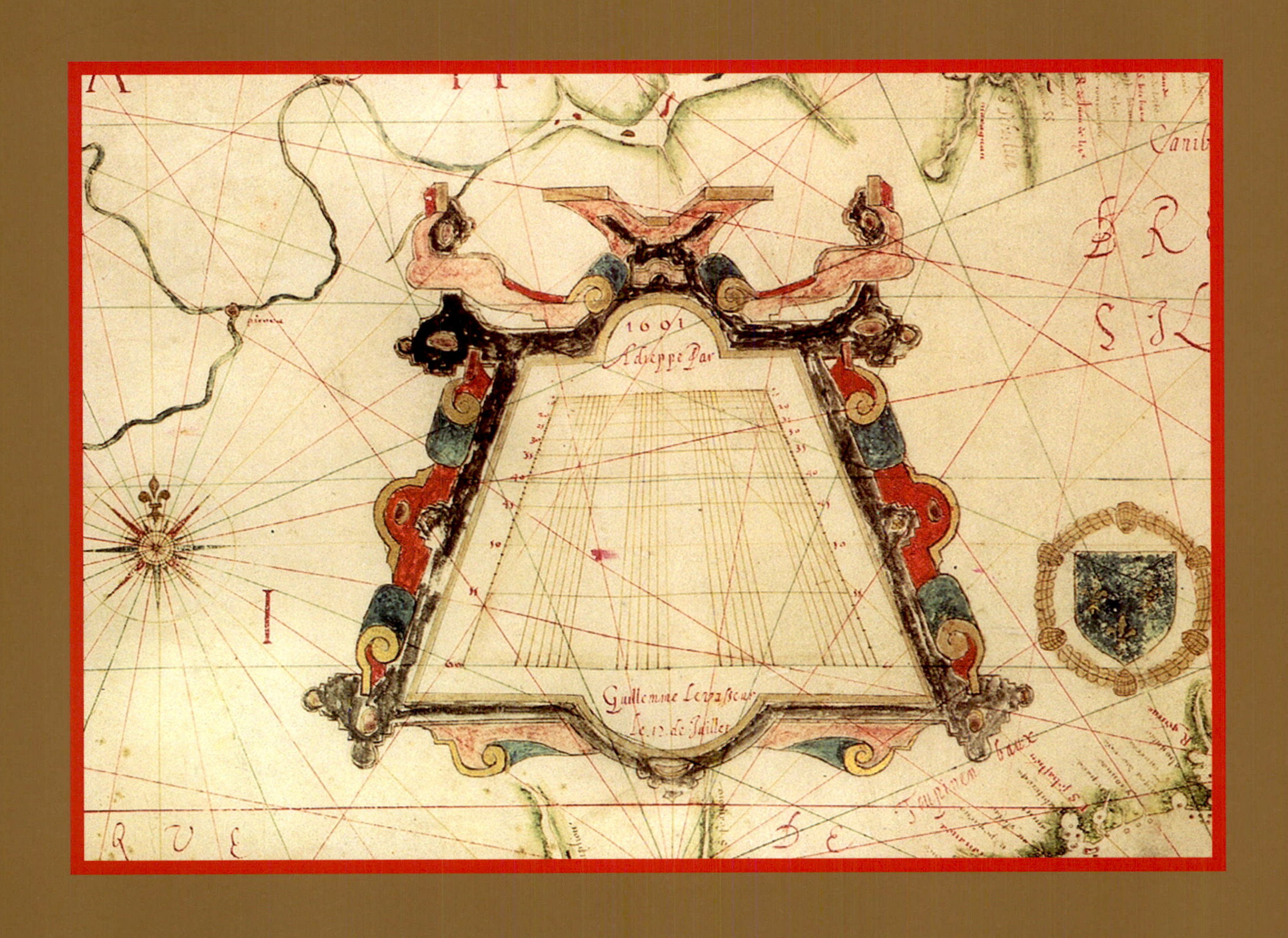

FLORIDE
TROPIQVE
A M
E R I
Q V E
BRE
SIL
1601
TROPIQVE

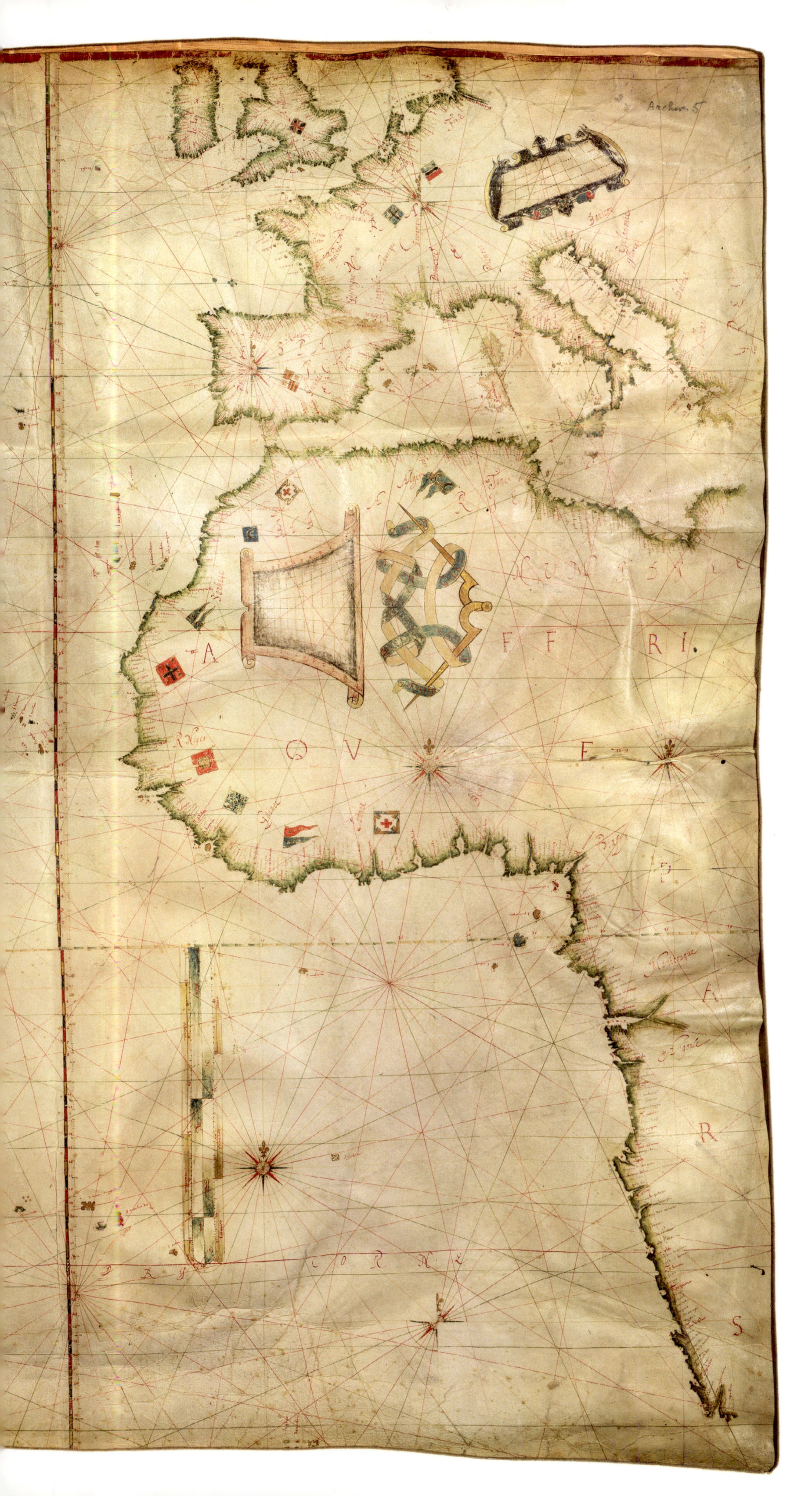

Map of the Atlantic Ocean by Guillaume Levasseur, Dieppe, 1601
Produced in 1601 in Dieppe by Guillaume Levasseur, this portolan chart of the Atlantic Ocean is a very striking document. The cartographer wrote the name QUEBECQ along the St. Lawrence River two years before Champlain made his first voyage to Canada (1603) and seven years before the city of the same name was founded (1608). Farther west is the name 3 RIVIERES, visited by the trader François Gravé Du Pont, who probably told Levasseur about it. Between these two sites, Levasseur added a number of names, including HOCHELAY, FORT DE CHARTRES, and VILLAGE DE CANADA. Still farther west, he located LAC DENGOULESME (Lake Saint-Pierre). HOCHELAGA, MONT ROYAL, and RIVIÈRE DU SAULT complete the mapping. In the Gulf of St. Lawrence, the island ST-JEAN (Prince Edward Island) also makes its appearance. All of these names testify to the presence of European fishermen and traders on the Atlantic coast and on the St. Lawrence River before the arrival of Champlain. The many place names of Indian origin are evidence of the continuing relations between these fishermen and the indigenous peoples. With its quantity and originality of place names, and also with the projection used (Mercator projection), this map shows Guillaume Levasseur to have been a remarkably skilled cartographer.

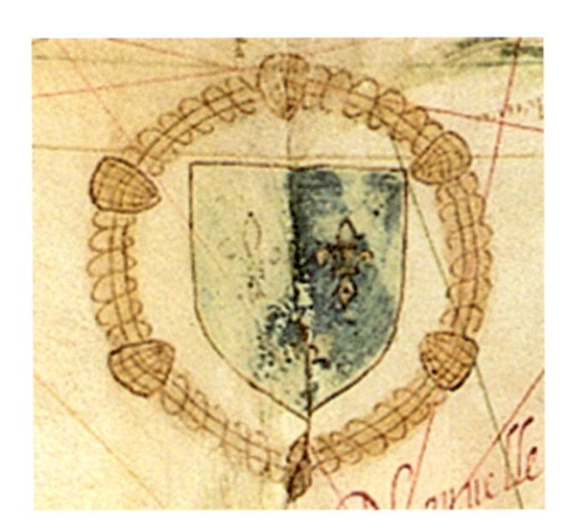

TERRES · NOEVFVES
Montaignais
baye de canada
la cadye
NOVVELLE FRANCE
les Vierges
COSTE · DE · LA · FLORIDE
TROPICQVE DE · CANCER
Isles du Perou
Noeufues Espaignes
MER · DV · SV
LE · PERV
Riuiere des Amazones
Maraonen
les cambal
La France Antartique
LE · BRESIL
TROPICQVE · DE · CAPRICORNE

Trade, Religion, and Exploration

INROADS INTO NORTH AMERICA

One century after John Cabot explored the "newfound landes," there were still no European settlements there. Although the French landed on the shores of the Gulf of St. Lawrence and the St. Lawrence River to fish cod and trade furs, they were in no hurry to make homes there; reaping the harvest from the sea and taking aboard the pelts gathered by the Indians did not require their constant presence. France was thus in a good position to found a colony on the banks of the "river of Canada," to travel up that river, and, if the information from the Indians was to be believed, to reach the "western sea."

The expeditions led by the Frenchmen Jacques Cartier and Jean-François de La Roque de Roberval between 1534 and 1543 were concentrated in the St. Lawrence Valley, where they felt certain that they would find the sought-after passage. Thus, in the early seventeenth century, much was known about the Atlantic coast and the St. Lawrence River as far as Montreal – enough to bring over colonists for settlement. The fishermen – explorers whose names are lost to us – passed their observations to their compatriots. Looking at the map by Guillaume Levasseur (1601) of Dieppe, with its place names of both French and Indian origin, it is easy to see that these coastal territories and their inhabitants had become familiar to French seamen.

François Aymar de Chaste, amiral du Ponant (admiral of the Western Fleet), got the ball rolling. In 1600, he was governor of Dieppe, centre of cartography and North Atlantic trade, when a Dieppe shipowner, Pierre Chauvin de Tonnetuit, received the mandate to found a settlement at Tadoussac. De Chaste was interested in the "new world," and in the Spanish and Portuguese conquests. He also believed in the importance to France of building an empire in North America. A report written by Champlain about his stay in the Antilles and Mexico (1599–1601), *Brief discours des choses les plus remarquables reconnues aux Indes Occidentales,* provided the final proof. After Chauvin's death, in February 1603, de Chaste took over the trade monopoly in New France and sent a ship to trade at Tadoussac; he asked Champlain to be on board. Then, when de Chaste died, he was replaced by one of his friends, Pierre Dugua de Monts, Ordinary Gentleman of the King's Chamber and previously the king's lieutenant in the town of Honfleur, who had served under Commander de Chaste during the campaign against the League.

At the turn of the seventeenth century, Henri IV disregarded the reluctance of his main advisor, the duc de Sully, and established a base for New France in Acadia, which was then extended into the St. Lawrence Valley. The king and his explorers had no idea how much time and effort would be required to reveal the northern part of America to the rest of the world. A network of highly placed civil servants in Henri IV's entourage who were closely associated with the Huguenot party found themselves at the vanguard of the French undertaking in Canada. To these nobles were added experienced seamen, such as François Gravé Du Pont, a ship's captain familiar with long voyages, and Samuel de Champlain, an explorer and cartographer who was called upon to undertake so many official functions that he soon appeared to be the cornerstone of the French edifice in America. Managing the new colony were a succession of trade companies formed of merchants from the great maritime ports, each one directed by one of the king's faithful men. The Company of New France, also known as the Company of One Hundred Associates, created in 1627 by Cardinal de Richelieu, was charged, as its predecessors had been, with combining trade, settlement, and

Map of the Atlantic Ocean by Pierre de Vaulx, 1613
Although portolan charts were initially designed for use in navigation and trade, some were so beautiful that they became prestigious documents serving to assert France's sovereignty over the territories explored. This was the case for this magnificent brightly coloured map, a tribute to French colonization in the Americas. It was the work of a pilot and geographer from Le Havre, Pierre de Vaulx, who used sumptuous flags to highlight for the king the territories that he had the right to claim overseas. In Brazil, de Vaulx recalls the disastrous adventure of France Antarticque, which did not survive religious dissent and murderous raids by the Portuguese. In North America, he notes the existence of Nouvelle France, its contours still uncertain, located between Terres Neufves and the coste de la Floride. The coste de Cadye, a variant of the name Acadia, was visited by seamen because of its abundance of fish and furs. Near to today's Bay of Fundy, the mapmaker penned in, in small letters, a second *Coste de la floride,* much farther north than the peninsula now called Florida. According to notarized documents from Normandy and La Rochelle, this was one of the best places to fish cod and an excellent place to trade with the Indians.

Three details from the map by Pierre de Vaulx (1613). Above, the Montagnais. Below, very unmonstrous "sea monsters." They seem to be mermaids, one with a mirror and the other with a comb!

the conversion of the Indians. By 1663, the company had managed to settle barely 3,000 French inhabitants in the colonies; at that point, Louis XIV decided to take over their administration himself.

Given the two objectives set for it – the fur trade and discovery of the western sea – and the fact that it was handicapped by a perpetually small European population, New France could not really exist without a system of alliances with the Indians. Evangelization, included in all colonization mandates, was a natural part of this; it elevated, in the view of other European states, the status of a project that was otherwise essentially materialistic and devoted to conquest. And so, missionaries also became explorers and ethnologists. From the time of Champlain's "discoveries" to the founding of Detroit (1701), Recollet and Jesuit missionaries, with their "lay brothers" and some interpreters, were the main reliable informants with regard to the Hudson Bay, Great Lakes, and Mississippi drainage basins. Following the missionaries, who reconnoitred the waterways and signposted the places where missions and trading posts might be built, came laymen such as Simon-François Daumont de Saint-Lusson (1671), René-Robert Cavelier de La Salle (1682–87), and Pierre Le Moyne d'Iberville (1698–99). As they searched for clues to the route to the western sea, they advanced farther and farther into the central and southern parts of the continent.

During the first century of France's presence in Canada, government, trade, and religion were tightly interwoven – but not just any religion. In 1604, under Henri IV, an effort at Catholic and Protestant cohabitation was made when a priest and a pastor embarked with settlers on the ship headed for Acadia. History has recorded only their constant disagreement during their brief stay at Sainte-Croix, where neither survived the first winter. The experiment was not repeated. After the death of Henri IV, the Counter-Reformation took over; in fact, it became compulsory to be Roman Catholic to settle and trade in New France – a rule imposed by the Company of the One Hundred Associates. François de Montmorency-Laval arrived in the settlement of Quebec in 1659 as the vicar apostolic; in 1674, he became the first bishop of New France. Also a member of the Sovereign Council, he was to increase the Church's involvement in the colony's government, notably with regard to trade practices. In spite of frequent disagreements between the bishop, the governor general, and the intendant, episcopal authority was strong, as administered by the secular clergy in 35 parishes (1688), by missionaries, and by communities of nuns devoted to teaching and nursing. The France of the Catholic Reformation, through its structures and its penetration of the territory, was to have a long-lasting effect on the new colony.

On the edges of New France, the English presence, already sizable at the time of the first settlement in Acadia, quickly made itself felt in this desirable Atlantic region. Martin Pring and George Weymouth almost crossed paths with Champlain and Dugua de Monts on the coasts of Norembega – or New Englande, as John Smith called it on his 1616 map. When the Pilgrims arrived, in 1620, settlers were already living farther south in Virginia. By the end of the century, they had absorbed even the Dutch colony of Nieuw Amsterdam (New York), for whose founding Henry Hudson had explored the river as far as Albany, so that the entire Atlantic seaboard belonged to England. More inclined to sedentary activities than to exploration, the English settlers tended to concentrate between the Appalachians and the ocean, in a space that measured only 300 kilometres at its widest point. The Mississippi constituted the central axis of the vast French empire south of the Great Lakes before the Treaty of Utrecht.

The English focused their exploration efforts in the northern part of the continent, between Newfoundland and Greenland, where they searched for a northwest passage. Over half a century, from 1577 to 1631, a dozen expeditions cleared a path through the ice. Henry Hudson's, in 1610, to the bay that bears his name, led to major advances in knowledge of the Arctic. After a number of other incursions and with the opening of trading posts, this region began to siphon the best furs of the tundra off to British trade. Once the Hudson's Bay Company was founded, in 1668, and in spite of brief periods of French occupation, the region of the Northwest Passage and Hudson Bay remained definitively beyond the purview of French exploration.

During the seventeenth century, the European spheres of influence in North America were roughly defined: England posted its markers in the game-filled regions around Hudson Bay and a vast Atlantic littoral. France set down roots in the St. Lawrence Valley, certain that an opening to the western sea had been found through the centre of the continent. For both powers, the fur trade was the main, if not the only, source of financing for exploration. France, however, had to bear the additional expenses of settlement and creating an administrative and religious infrastructure.

Main sources

Ickowicz, Pierre (ed.). *Dieppe-Canada: cinq cents ans d'histoire commune.* Paris: Magellan & Cie, 2004. On the first contacts by fishermen and cartography in the sixteenth century. — Pariset, Jean-Daniel (exhibition curator). *Henri IV et la reconstruction du royaume.* Paris: Éditions de la Réunion des musées nationaux, Archives nationales, 1990. A good explanation of the Huguenot party and its role in the foundation of the first settlements in New France.

From Arcadia to Acadia

BETWEEN UTOPIA AND REALITY

THE TERRITORY WHERE THE FANTASIES of the explorer-conquerors played out was Acadia, the part of North America closest to Europe after Newfoundland. Could this be the Vikings' fertile *Vinland*? Compared to the rocky coasts and soil of Newfoundland and Labrador, Acadia, farther south and more hospitable, was the "promised land," the land of milk and honey, where the "good Savages" lived. As in Newfoundland, there was incredible, miraculous fishing. And there was an abundance of animals whose luxurious pelts, carefully prepared by the inhabitants and exchanged for cheap beads and tools, brought wealth.

Acadia also represented the hope, if not the oft-expressed certainty, of a passage across the continent to Asia. The possibility of realizing this ancient dream justified Europeans' desire to settle there. Those who wished to possess this space – French, English, and Dutch – began to engage in armed conflict soon after the Spanish and Portuguese voluntarily withdrew to the south.

Historians generally agree that Giovanni Verrazzano gave the territory its name. In his travel account (1524), he states, "We named it Arcadia on account of the beauty of the trees." Some feel that Verrazzano, a literate man, was comparing the fertile land in the area to that in Arcadia of antiquity, the spelling of which, over time, evolved to Cadie, Accadie, and Acadie – and, in English, Acadia. Others disagree with this theory and suggest that it is an indigenous name bearing a resemblance to place names such as Passamaquody and Tracadie. Whatever the case, it was a happy coincidence that served to bring the Indians and explorers closer. In fact, the French and Indians of Acadia eventually worked together as partners both for trade and to fight off invasive neighbours and competitors.

The location of Acadia is even more uncertain than is the origin of its name. In his account, Verrazzano situated it in today's Nova Scotia, but the cartographer Giacomo Gastaldi placed it south of the Bay of Angoulesme (New York) and did not draw in its borders. In the late sixteenth century, it corresponded to approximately the area that was to be recognized during the New France colonization period – that is, the territory from Baie des Chaleurs to the Penobscot River and from Cape Breton to the St. John River. It thus incorporated large chunks of the present-day province of Quebec and state of Maine.

As soon as it existed, and although its borders were not clearly defined, Acadia was coveted by the English, who were well established in Newfoundland and in northern Virginia, which had been discovered and named by Walter Raleigh in 1585. In 1602, English ships began to go directly to New England, which was considered the northern part of Virginia. The writer John Brereton, a passenger on the 1602 passage of the *Concord*, captained by Bartholomew Gosnold, gave the first description of New England after Verrazzano's: *Briefe and true relation of discoverie of the north part of Virginia.* In 1603, Martin Pring went to southern Maine and loaded his ship with sassafras, believed to be a miracle cure for syphilis. George Weymouth explored the same territory in 1605. The Court of England then decided that the North American Atlantic coast offered enough assets to justify its officially taking possession of it, although the Court made sure to limit the border to parts of the littoral not effectively occupied by another European power. In 1606, King James I accorded a charter to the Virginia Company for a large part of the seaboard, from 34°, north of Cape Fear, to 45°, south of Bangor. In spite of protests by Spain, France, and Holland,

Wild ginger
In 1635, Jacques Philippe Cornuti published, in Paris, the first botanical treatise on North America. *Canadensium plantarum historia* describes 40 Canadian species that may have been taken back to France by Louis Hébert, the first apothecary and herbalist in New France. Canadian wild ginger, whose rhizome has the same properties as ginger, was used for its taste and to treat fever. "The root produces fairly long fibres with a very pleasant fragrance.... These fibres add a particular flavour to wine if, once ground and wrapped in a tied cloth, they are placed in an unfermented barrel of wine under a weight that keeps them at the bottom ... this same root, when chewed, sweetens the breath."

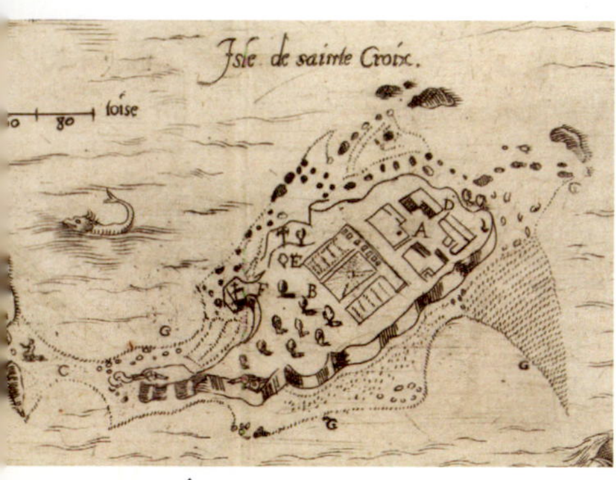

Île Sainte-Croix (detail)
In 1604, accompanied by Champlain, de Monts settled a party of men on Île Sainte-Croix, at the mouth of the river of the same name. They fortified the island and erected dwellings, a storehouse, a cemetery, a chapel, a well, a kitchen, and a bakery. But it proved to be a disastrous choice of location for a settlement. The 79 Frenchmen who stayed there quickly exhausted the available resources, were unable to find potable water and game, and became prisoners of the small, desolate island.

Port-Royal by Marc Lescarbot
This map by Lescarbot portrays one of the great achievements of French colonization in North America: Port-Royal, founded on what is now known as the Bay of Fundy, in 1605. When the king revoked the trade monopoly accorded to Dugua de Monts, Port-Royal was abandoned. Lescarbot's patron, Poutrincourt, took charge of the colony after a long negotiation with the Court. Lescarbot's toponymy tended to obey the imperatives of patronage. He took advantage of the showcase offered by this map to immortalize its main actors and backers: BIENCOURVILLE Island for his patron, Poutrincourt; MONT DE LA ROQUE in honour of a certain De la Roque Prevost de Vimeu, a friend "who wanted to have land there and send men." Other names still shrouded in mystery – FORT DUBALDIM, ILE CLAUDIANE – were probably drawn from Lescarbot's entourage.

the Virginia Company's charter was confirmed. Its mandate was to form two colonies: the Virginia Company of London was to create permanent settlements between 36° and 41°, and the Virginia Company of Plymouth was to do the same between 38° and 45°.

Colonization plans were also drawn up in the Court of France. In the late sixteenth century, France found Acadia very attractive, both for the excellent "market cod" that was so appreciated by the hundreds of sailors who went there every year to fish and for the possibilities of trading for pelts with the particularly hospitable Indians. French navigators were encouraged to undertake expeditions there. The one by Étienne Bellenger in 1583 brought back to Rouen furs from and information on Acadia, as did those by the Breton Troilus de La Roche de Mesgouez, holder of a 10-year trade monopoly that started in 1598. De La Roche's success as a trader was, however, sullied by the unfortunate incident of the colonists transported to and then abandoned on Sable Island. After the hardships endured by the settlers at Tadoussac in 1600, the French turned once again toward Acadia because of its agricultural and commercial potential.

On November 8, 1603, King Henri IV of France accorded Pierre Dugua de Monts, a Huguenot from Royan and Ordinary Gentleman of the King's Chamber, the commission of lieutenant general and an appointment as the king's personal representative "to the country, territory, coasts, and confines of Lacadie. To begin at the fortieth degree, up to the forty-sixth." De Monts, a Protestant, was to establish a colony, look for mines, and convert the "Savages" to the Christian faith. The island of Sainte-Croix was probably chosen for settlement because it was already familiar territory, visited for at least a half-century for fishing and trading, as revealed in the notarized acts preserved in the Norman ports; on the 1550 map by Pierre Desceliers, a "rivière de Lacroix" is indicated at the approximate site of the first settlement. Bellenger's and de La Roche's expeditions had also targeted this latitude. The fact that the location was an island undoubtedly made it attractive to seamen: as well as being accessible by boat, a small island such as Sainte-Croix offered excellent security conditions, as its inhabitants could see attackers coming and prepare to protect themselves; because it was close to the mainland, boats could be used to fetch provisions. Since the earliest explorations, "the Islands," the subject of fantasies, had been sailors' primary, and often ultimate, objective. But Sainte-Croix, much farther north, in winter became a lifeless prison, from which boats could not escape because of the ice floes in the river. This island thus became synonymous with disenchantment.

During the exploratory expedition of 1604, de Monts, Samuel de Champlain, and Jean de Biencourt de Poutrincourt noted the Port-Royal basin, which had appeared on a 1556 map by Ramusio. The site was an ideal location for a trading post, and for subsistence farming. The sailors finally agreed to turn their backs to the sea and settle on the mainland. Thus, in 1605, the colony was moved from Sainte-Croix to Port-Royal, which was considered the capital of Acadia for several years. On February 25, 1606, Henri IV confirmed the concession of the territory to Poutrincourt, lieutenant general of Acadia, on condition that he found a colony there. New arrivals in the summer of 1606 provided reinforcements for the 40 who had survived the previous winter at Sainte-Croix. Marc Lescarbot, a lawyer and writer from Vervins (home also to Poutrincourt), near to the properties of the Roberval family, was curious about this French undertaking in America and made an effective contribution to the settlement and organization of the colony. When he returned to France, he published *Histoire de la Nouvelle-France* (Paris, 1609), featuring an updated map of Acadia; this work provided excellent promotion of Dugua de Monts's project, which was then being threatened with elimination.

In spite of the colonists' laudable efforts to feed themselves from hunting, fishing, and gardening, Port-Royal had to finance all of its other costs, the largest of which was provisioning of the ships, through the fur trade. In 1607, the monopoly was withdrawn from Dugua de Monts, due to strong pressure by Breton and Norman competitors and the aversion to colonization of Sully, Henri IV's trusted minister. Very regretfully, the colonists left their Indian allies and friends, the village of Port-Royal, and the gardens that they had created, and returned to France. This pause slowed the development of Acadia, just as settlement in New England was picking up steam. Poutrincourt nevertheless managed to preserve his Port-Royal concession, to which he returned several years later, in 1610. Acadia was not spared new perils, but its existence was now recognized – well rooted in a soil for which France would fight its enemies fiercely.

Main sources in order of importance

BINOT, Guy. *Pierre Dugua de Mons, gentilhomme royannais, premier colonisateur du Canada, lieutenant général de la Nouvelle-France de 1603 à 1612.* Vaux-sur-Mer: Éditions Bonne Anse, 2004. A full portrait of the lieutenant general of Acadia and New France from 1603 to 1612. — THIERRY, Éric. *Marc Lescarbot (vers 1570-1641): un homme de plume au service de la Nouvelle-France.* Paris: Honoré Champion, 2001. Places the first writer in Acadia in context.

A Cartographer in North America

SAMUEL DE CHAMPLAIN

Champlain was a man of duty. His entire career, spent serving his king in the Americas, is surely proof of this. Although his first mandate was to be an observer, he soon became an explorer and began producing descriptions of unprecedented accuracy. Charged with finding a place appropriate for creating a French colony, he acted as a diplomat in establishing trade relations, if not alliances, with the riparian nations from the Gulf of St. Lawrence to the Great Lakes. As commander and administrator of the Quebec post, he founded a "habitation" there, a crossroads for the fur trade, where a colonial nucleus capable of resisting military attacks and other perils took root. Champlain's multi-volume journal is chock-full of geographic, ethnological, zoological, and botanical information, often copied onto drawings and maps that were more detailed than anything previously produced on North America.

In the account that he gave to Admiral François Aymar de Chaste of a first secret mission to the Antilles and Mexico from 1599 to 1601, Champlain showed his talent as an observer and drawer. He left for Tadoussac in 1603 on a ship captained by François Gravé du Pont, a St. Malo seaman who had made a number of fishing and trading voyages to Canada. Thanks to two Indians on the ship

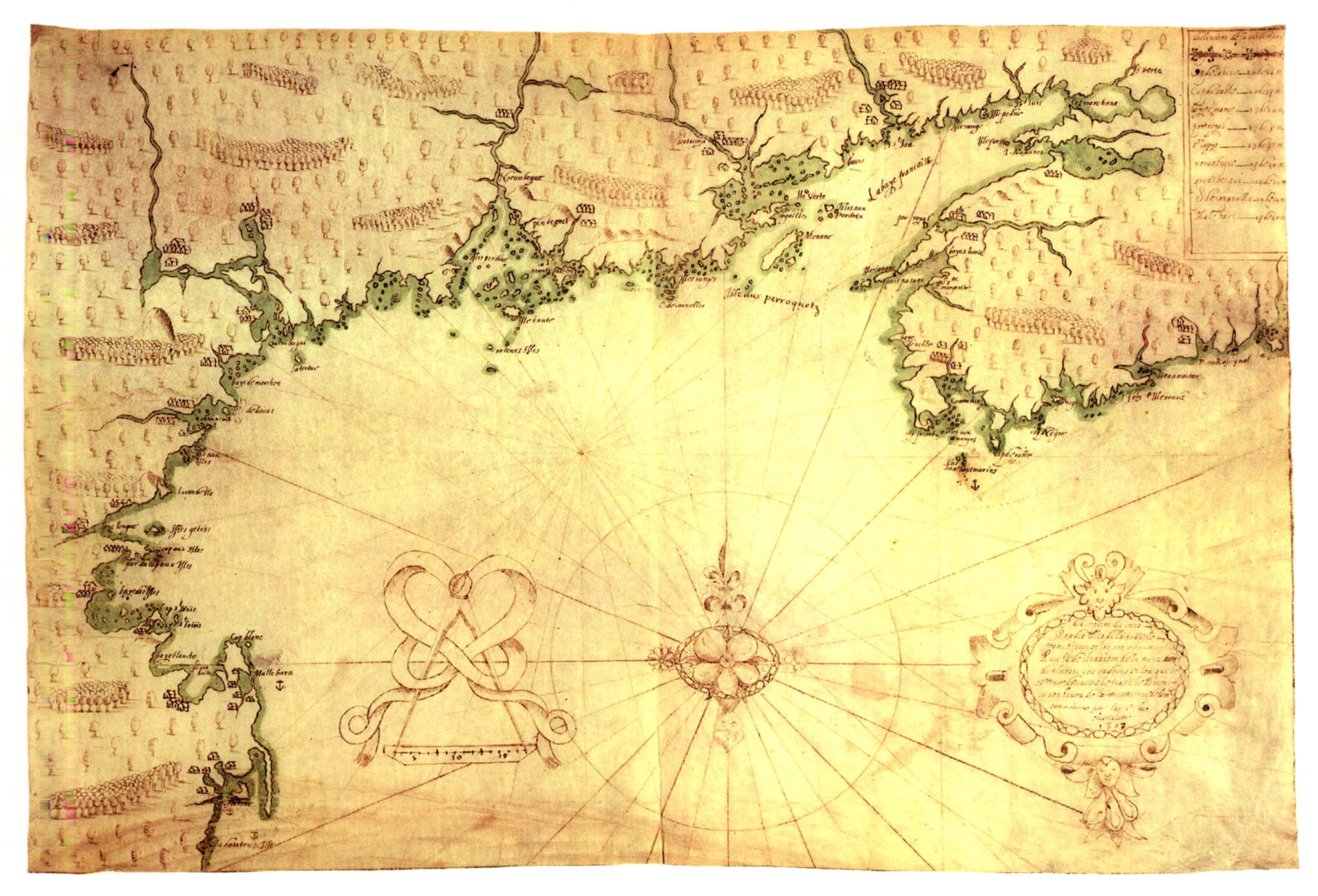

Manuscript map of Acadia by Samuel de Champlain, 1607
Now elevated to the status of father of New France, Samuel de Champlain might also bear the title of father of modern cartography in North America, as evidenced by his *Description des costs, pts., rades, illes de la Nouvele France,* dated 1607. The only manuscript map by Champlain that has survived to the present day, it portrays the colonization episode at Sainte-Croix and then Port-Royal between 1604 and 1607, as well as the main sites reconnoitred by the French in search of gold or silver mines, a passage toward Asia, and a new settlement site. Champlain carefully observed the capes, islands, rivers, bays, and ports around BAIE FRANÇOISE (Bay of Fundy) and recorded this information in map form. In all, he noted about 80 place names, all situated between CAP BLANC (Cape Cod, Massachusetts), and LA HÈVE, Nova Scotia. Most of these names were also used on Champlain's later maps, printed between 1612 and 1632. This unique document was left by the bibliophile and historian Henry Harrisse to the Library of Congress, where it is considered a priceless patrimonial jewel.

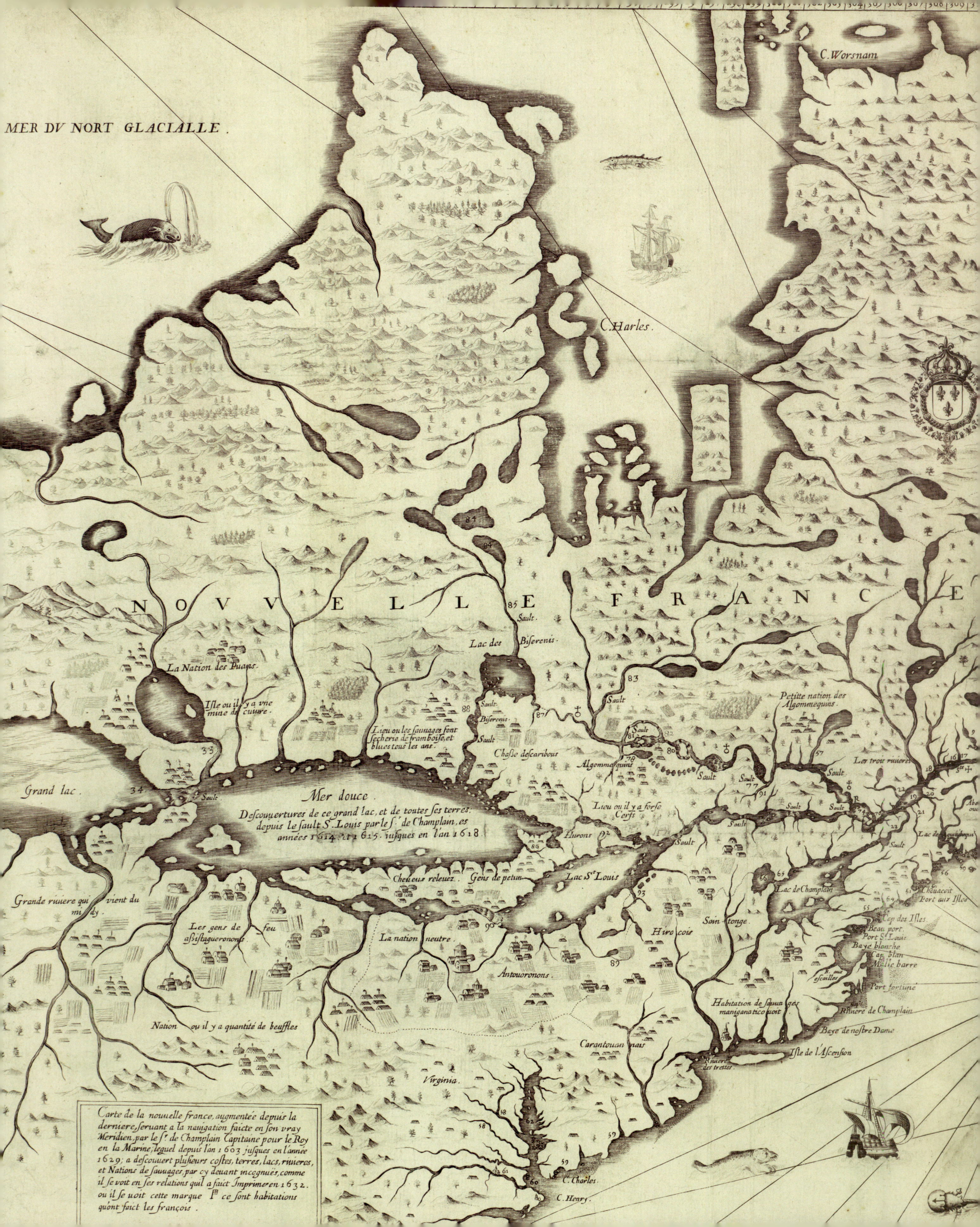
MER DV NORT GLACIALLE.
C. Worsnam
C.Harles.
NOVVELLE FRANCE
La Nation des Puans.
Isle ou il y a vne mine de cuiure.
Grand lac.
Mer douce.
Descouuertures de ce grand lac, et de toutes ses terres, depuis le sault S.t Louis par le S.r de Champlain, es années 1614. et 1615. iusques en l'an 1618.
Lac des Bisserenis.
Lieu ou les sauuages font secherie de framboise et bluës tous les ans.
Chasse des caribous
Algommequins
Petitte nation des Algommequins
Les trois riuieres
Lieu ou il y a force Cerfs
Hurons
Lac S.t Louis
Cheueux releuez.
Gens de petun
La nation neutre
Les gens de feu
Grande riuiere qui vient du midy.
Antouoronons
Hirocois
Saintonge
Lac de Champlain
Habitation de sauuages manigaticoiuoit
Carantouan nais
Virginia.
Nation ou il y a quantité de beuffles
Isle de l'Ascension
Riuiere des tretres
Baye de nostre Dame
Riuiere de Champlain
Port fortuné
C. Charles.
C. Henry.
Chouacoit
Port aux Isles
Cap des Isles
Beau port
Port S.t Louis
Baye blanche
Cap blan
Malle barre
Sault
Carte de la nouuelle france, augmentée depuis la derniere, seruant a la nauigation faicte en son vray Meridien, par le S.r de Champlain Capitaine pour le Roy en la Marine; lequel depuis l'an 1603 jusques en l'année 1629; a descouuert plusieurs costes, terres, lacs, riuieres, et Nations de sauuages, par cy deuant incognuës, comme il se voit en ses relations quil a faict Imprimer en 1632. ou il se uoit cette marque ⊞ ce sont habitations qu'ont faict les françois.

C. Elizabeth
Lomle Inlet
Groen lan dia
Terres de la Brador
Esquimaux
Brest
Croix blanche
Belle isle
Isle fichet
Cap de grat
Cap rouge
Groye
Baye dorge
Isle aux apouois
Baye blanche
Cap St Iean
Isles des fougues
Isle de moy
C. de bonne viste
Isle de bacallos
Baye Ste Claire
Baye de la Conception
Isles despoirs
Cap de raza
La grande baye
Montagnairs
Bersia
Chisdec
Baye des rochers
Port aux Ours
Le Golphe St Laurens
Terre neuue
Gaspay
Baye des molues
Isle percée
Isle bonne aduenture
Baye de Chaleu
Miscou
La Magdelene
Isle Brion
Isle aux oyseaux
Isle ramée
Cap de ray
Isle St Iean
Passage de Canceau
Cap breton
Conceau
Nouuele
France
C. des mines
C. de Poitrincourt
Souricois
Port royal
Port au mouton
Cap de sable
Isles aux loups marins
Isle de sable
Banquereaux
Rochers
Le grand banc
5 10 20 30 40 50 60 70 80 90 100
Faicte l'an 1632 par le sieur de Champlain

Preceding pages (84–85)
Map of New France by Samuel de Champlain, published in Paris in 1632
In 1629, the Kirke brothers, privateers from Dieppe serving the king of England, captured what Champlain called the "habitation" at Quebec. Forced to return to Europe, Champlain moved heaven and earth to ensure that France recovered its colony. In 1632, he published new voyage accounts that highlighted French undertakings in the "new world." Through the descriptions of his voyages of discovery, he justified French claims to the North American territories. He also included this map, which gives an overview of the territory explored and was probably produced in the context of negotiations to recover New France. The map shows, among other things, Newfoundland, Acadia, the St. Lawrence, Lake Champlain, and part of the Great Lakes, all drawn with a great amount of detail for the time, if compared, for example, to maps by Lescarbot or any other European cartographer. An experienced navigator, Champlain knew how to calculate distances and take precise latitude measurements. He also incorporated information supplied by his Indian allies, which, although not as reliable as direct observation, was always useful for sketching out areas to be explored. Although the shores of the continent are mapped precisely, the western part of the continent is less accurately portrayed. Hudson Bay, drawn according to the voyages of Henry Hudson, seems to be divided and bears the name MER DU NORD GLACIALLE. While leaving open the possibility of a passage to the Pacific Ocean in the environs, Champlain erased all trace of the English. Farther south, Lake Huron, visited in 1614, is named MER DOUCE. At the western edge of the map, Lake Superior is called simply GRAND LAC. Champlain was apparently referring to the Mississippi when he indicated the "Grande rivière qui vient du midy," 40 years before the French first explored the area. To better promote the colonization of New France and assert the primacy of his discoveries, Champlain unhesitatingly used information supplied by his predecessors (Cartier, Dugua de Monts, Lescarbot, Hudson), without crediting them.

during the Atlantic crossing, Champlain was probably able to communicate with the Algonquins at Tadoussac who were his hosts from May 26 to July 11. With them, he travelled up the Saguenay for some 50 kilometres, learned of the existence of a saltwater sea to the north, and deduced that this was not the "sea of Asia" but one of its outlets. In 1603, seven years before the English discovered it, Champlain established the location of the inland sea that was to be named Hudson Bay. He then explored the St. Lawrence River, at the time called the "river of Canada," and named several geographic features, including Lake Saint-Pierre and the Richelieu River, or "Iroquois River." Like Jacques Cartier, he went to Hochelaga, where the "Saut Saint-Louis" (Lachine Rapids) blocked his progress upriver. With the help of the Indians, he was able to sketch out the Great Lakes network, although it was impossible for him to resist the idea that the "sea of Asia" was not far away. On the voyage back to Le Havre, Champlain met another man from St. Malo, Jean Sarcel de Prévert, in the Gaspé. Prévert told him about Acadia in terms that led him to believe that there were mines and an opening to Asia there. The report that he made to the duc de Montmorency, "Of Savages," was published immediately upon his return (1603). It helped him become known both as an explorer from Saintonge and for the exactitude of his observations, and he went on to construct the scenario for future expeditions.

Champlain received authorization from Henri IV to accompany the founders of a new colony in March 1604. This time, he stayed in North America for three years. With the lieutenant general of Acadia, Pierre Dugua de Monts, he explored the coast south of Île Sainte-Croix, sometimes going as far as 80 kilometres upstream on the rivers, and skirted the bays southward to Mallebarre (Nauset Harbour) south of Cape Cod (named by the English captain Bartholomew Gosnold in 1602). Champlain faithfully reported the entire adventure, covering more than 1,500 kilometres, in the form of estimations of distances and latitudes, along with various descriptions, noted first on a map "according to its true meridian," dated 1607, and then in a report presented to the king of France and published in 1613. Never had accounts of voyages on the North American Atlantic shore contained such a wealth of information in the form of writings, drawings, and maps (16) and plans.

When he left a third time for New France, in 1608, Champlain was designated by de Monts as his lieutenant and advisor; they looked for a new site for settlement farther inland, as the Acadian coast had not met their expectations. This time, they found the ideal spot. The site of the future Quebec City was an excellent meeting place for Indian nations trading with the French; it later became the administrative capital of New France. Champlain unhesitatingly returned to the site that Jacques Cartier had occupied during his second voyage, on the bank of the Saint-Charles River (north of the present-day Quebec City), where he found the remains of a previous settlement.

Champlain then returned to exploring the Saguenay and its environs. The Indians who he met confirmed the existence of a "great salt sea" 40 or 50 days' travel north of Tadoussac. He presumed, with reason, that this was the bay partially explored by the English, but his guides refused to continue beyond Chicoutimi. Thus, he encountered the Indians' strategy of not communicating their knowledge to the French except in return for various advantages, the most important of which was military assistance. He was unable to find a way around this rule, and he thereafter became embroiled in a complex tangle of Indian alliances and military support.

The next year, 1609, in exchange for information on the copper mines and to satisfy an agreement made with his Montagnais allies, Champlain had to return to the heart of Iroquoisia – to Crown Point, south of the lake to which his name would be given. A French military strike succeeded thanks to the superiority of arquebuses over arrows, to the great satisfaction of his Indian friends. With this voyage on the Richelieu River, Champlain expanded the map of New France and inaugurated a route that was to be a strategic avenue for Europeans for the next two centuries. Lake Champlain was the "great lake" that the settlers of New England had situated in an imaginary "Laconia" province, for which they searched in vain until the 1630s. Champlain thus found himself in coveted territory, and he was not the only one to explore these watercourses. He left the Richelieu River on July 30. On September 4 of that year, a little farther south, the Englishman Henry Hudson, with a commission from Holland, left Nieuw Amsterdam (New York) and travelled north to Albany on a river that was to bear his name, thus introducing Dutch domination into a region that remained beyond French influence.

Champlain did not conceal his amazement at the Montreal archipelago, where he found himself in 1611, and he imagined that a city could be founded on the largest of the islands. He named the island close to the best harbour Sainte-Hélène, in honour of his young wife, Hélène Boullé, a recent convert to Catholicism. West of Montreal, he became the first European to cross the Saut Saint-Louis rapids by canoe, a feat that raised his prestige considerably among the Indians.

Champlain's account of his exploratory voyages was presented in a second publication, *The Voyages of Sieur de Champlain, of Saintonge, Captain in Ordinary to the King in the Marine* (1613). In this work, Champlain showcased his talents as a geographer and seaman with

his descriptions of the "coasts, rivers, ports, and harbours, with their latitudes and the various deflections of the Magnetic Needle." His observations were backed up by many maps, both detailed and general. As a consequence, his power was extended and he was appointed lieutenant to Prince de Condé, viceroy of New France. In addition to command of the colony, Champlain was given the mission to "try to find, through the said country, the easy route to the country of China ... so far as may be possible along the coasts and on the mainland" with the assistance of the indigenous peoples. Champlain's rise was also the result of skilful interventions with influential people, notably those made by his father-in-law, Nicolas Boullé, secretary to the King's Chamber.

On the basis of his new mandate, Champlain began to explore in the direction of "Huron country." Between 1613 and 1616, he travelled more than 1,000 kilometres gathering information on a region that was rarely visited by Europeans. From Saut Saint-Louis, he went up the Ottawa River, part of the major trade route toward the "upper country," crossed a series of lakes to Île aux Allumettes, near today's Petawawa, Ontario, then took the Mataouan River; finally, via Lake Nipissing and French River, he reached the "freshwater sea" – Lake Huron.

Champlain finally arrived in Huron country in summertime, on August 1, 1615. He was full of praise for the territory, the beauty and fertility of which surpassed his expectations: "Along the shores one would think the trees had been planted for ornament in most places." During his stay, which lasted through the fall and winter, he accompanied the Indians on hunts and was at his leisure "to study their country, their manners, customs, modes of life." He also went to Lake Simcoe, crossed the eastern end of Lake Ontario, and visited the Petuns south of Nottawasaga Bay, as well as the Cheveux-Relevés south of Georgian Bay. However, he received little further information on the "sea of Asia" and, in spite of his requests, they refused to take him there.

When he returned to Quebec, Champlain found a valuable emissary in Étienne Brûlé, who was already familiar with Huronia. Indeed, in 1610, at 18 years of age, Brûlé had been exchanged for a young Huron, who was sent to France and named Savignon. Brûlé went as far as Lake Superior, probably saw Lake Erie, and travelled through the Susquehanna River region, in the southern part of Iroquoisia (Pennsylvania). All of Champlain's observations were published in *Voyages and Discoveries in New France from the Year 1615 to the End of the year*

***Description de la Nouvelle France* by Jean Boisseau, 1643**
In the seventeenth century, the notion of copyright for maps was rarely acknowledged. The proof is this map, published in 1643 by the king's illuminator Jean Boisseau, who had no qualms about copying a map by Champlain published 11 years earlier (see pp. 84–85). A few very slight modifications were made, including no doubt the addition of Lake Erie and of the Algonquin people, TAVANDEKOUATE. All trace of Champlain's authorship was erased. It took a sensational trial in the early eighteenth century, between Guillaume Delisle and Jean-Baptiste Nolin, for copyright customs to change and the moral rights of cartographers to win greater respect.

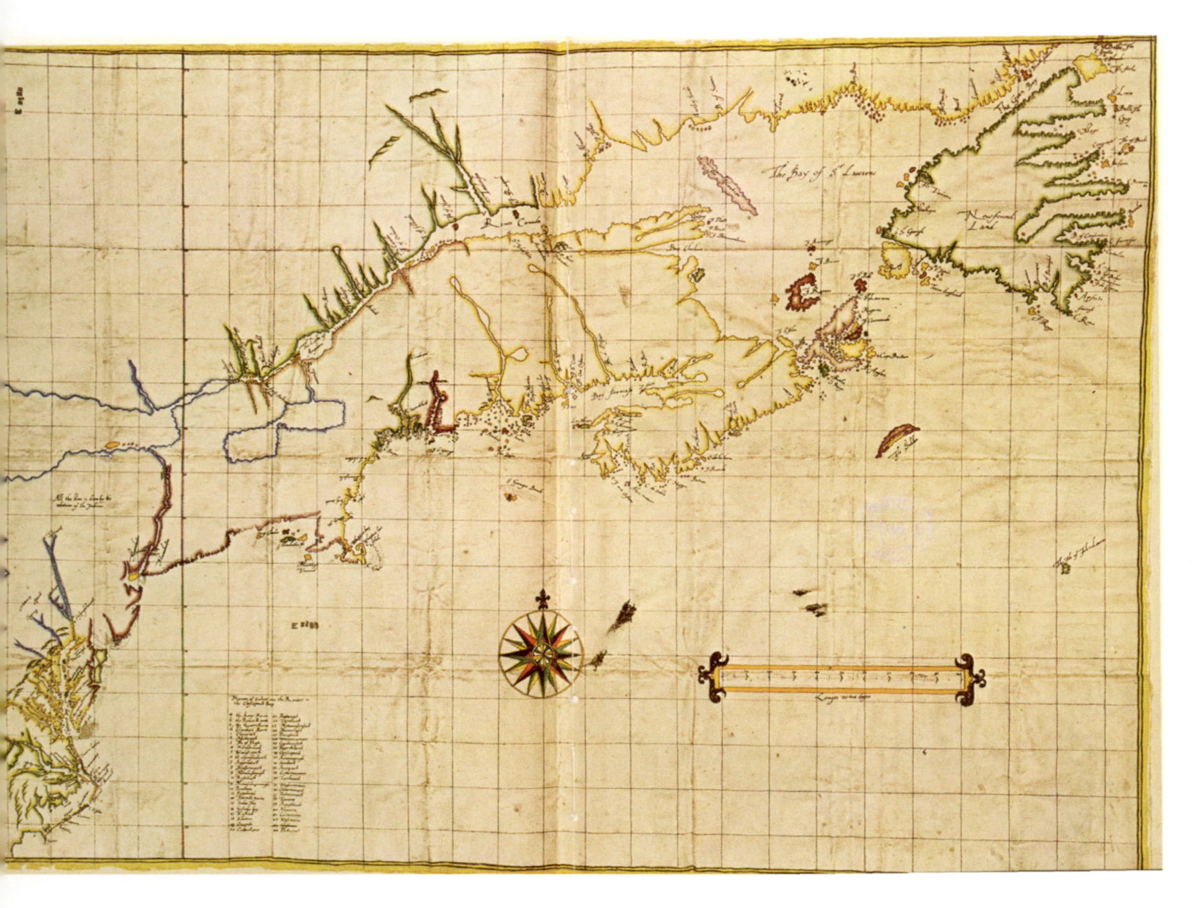

English map reproduced by a Spanish ambassador, 1610
This map, amazingly exact compared to previous representations, shows the area from the coast of Newfoundland to Virginia at the time of Champlain's explorations, as well as the St. Lawrence River from its mouth to Lake Ontario. The place names are, for the most part, creations of the sixteenth century, but there are a few more recent ones, such as Trois-Rivières, the IROCOIS River (Richelieu River), and the ALGONQUIN RIVER (Ottawa River). This last was depicted three years before Champlain sailed up it in an attempt to reach Hudson Bay. This document was known as the Velasco Map, in honour of the Spanish ambassador to London, who is reputed to have had it reproduced for King Philippe III. Its apparent anachronism and great accuracy have led certain researchers to doubt its authenticity and to date it to a later period.

1618, published in 1619. The book's ethnographic drawings were its main original contribution, reflecting Champlain's great experience with the Indian nations, but they were reviewed and modified by the engraver, who was certainly more familiar with portraits of men of the Court than of Indians of North America.

After his stay in the upper country in 1615–16, Champlain apparently stopped going on exploratory expeditions, but he continued to gather geographic information from the missionaries and certain "lay brothers," such as Jean Nicollet. From then on, he devoted himself exclusively to governing New France, hoping to provide it with the administrative and commercial structures most likely to ensure that it would become self-supporting. Up to then, Champlain had had three main objectives: to found a colony, to structure the fur trade, and to explore the territory with a view to finding a passage to the "western sea." These three challenges were partially met, but New France, having grown too fast, already suffered from gigantism! The colony lacked a framework and was too dependent on the inadequately controlled fur trade. Over the next 20 years, New France struggled for survival, not just to attain economic self-sufficiency, but also to block the expansionist ambitions of the English. In 1629, after the Kirke brothers captured the settlement of Quebec, Champlain met with the French ambassador in England to demonstrate, map in hand, the geopolitical importance of Canada and to plead on behalf of French possession of the colony. His arguments were convincing, and Quebec was returned to France in the Treaty of Saint-Germain-en-Laye (1632). Before his final return to North America, Champlain summarized his actions in a new edition of *Voyages*, published in 1632. The book was complemented by a magnificent map, the sum of knowledge acquired on New France, and by the "Treatise on Seamanship and the Duty of a Good Seaman," a methodological discourse and assessment based on "having spent thirty-eight years of my life in making many sea voyages and run many risks . . . at the conclusion of my discoveries in New France in the west, to compose for my own satisfaction a little treatise that may be understood and be of advantage to those who are willing to make use of it." One of the capacities required of a good seaman was to "know how to make charts, so as to be able to recognize accurately the lie of the coast, the entrances to ports, the harbours, roadsteads, rocks, shoals, reefs, islands, anchorages … to make pictures of the birds, animals, fishes, plants . . . and everything unusual that is seen; for this a little skill in drawing is very necessary, and the art should be practised." Champlain's accomplishments have largely proved the validity of these principles, which he applied throughout his life.

In a total of 23 maps and many drawings, Champlain depicted 4,300 kilometres of coastal territory. He left six voyage accounts, most of which were published during his lifetime, containing an inventory of the indigenous flora, fauna, and peoples. Here again, Champlain innovated. In fact, seamen and explorers were reluctant to disseminate exploration accounts; rather, they tended to keep the results of their voyages secret to avoid feeding the envy of competitors and foes. Not only did Champlain travel through and explore an immense territory, but he knew how to relate his experience. His books were technical works created by a scientific mind. When he died, on December 25, 1635, the colony of 150 settlers still faced threats from within and without, but it was now able to find the resources and defences that it needed to survive.

Main sources in order of importance

GLÉNISSON, Jean (ed.). *La France d'Amérique: voyages de Samuel Champlain, 1604-1629.* Paris: Imprimerie nationale 1994. A masterful synthesis of Champlain's activities and an erudite presentation of his *Voyages* (1632). — HEIDENREICH, Conrad E. *Explorations and Mapping of Samuel de Champlain, 1603–1632.* Toronto: B. V. Gutsell, "Cartographica" collection, 1976. — LITALIEN, Raymonde, and Denis VAUGEOIS (eds.). *Champlain: The Birth of French America.* Translated by Käthe Roth. Montreal and Sillery: McGill-Queen's University Press and Septentrion, 2004. The chapter "Samuel de Champlain's Cartography, 1603–32," by Conrad E. Heidenreich and Edward H. Dahl, is an indispensable complement to this chapter.

Source for translation: CHAMPLAIN, Samuel de. *The Works of Samuel de Champlain,* Vol. 3, translated and edited by H. H. Langton and W. F. Ganong. Vol. 6, translated by W. D. LeSueur and H. H. Langton. Toronto: Champlain Society, 1922–36.

Exploring the Great Lakes

MISSIONARY CARTOGRAPHERS

The missionaries were a natural choice to continue Champlain's explorations. Their zeal to convert new peoples took them beyond the established posts to places and along paths that they knew very little about, and their determination enabled them to overcome all obstacles. It should not be forgotten that New France was entirely a product of the Catholic Reformation, at a time when proselytizing was at its apex. In the Court of France, Henri IV's successors bound the colonial undertaking and the development of trade tightly to propagation of the Christian faith. Emigration by Huguenots was forbidden, in theory at least, by the statutes of the Company of One Hundred Associates (1627). Catholics therefore assumed complete responsibility for the religious aspect of colonization. In fact, they were to play a unique role in expanding the French presence in North America.

The Jesuits, members of the order to which Pierre Coton (scrupulous confessor of Henri IV, Louis XIII, and Queen Marie de' Medici) belonged, were the designated emissaries of the Court. After a brief stay in Acadia, the Jesuits landed in Quebec, in 1625, ten years after the Recollets. Both the Jesuits and Recollets were taken back to Europe by the Kirke brothers in 1629. Only the Jesuits would return to New France with Champlain in 1633, followed by the Sulpicians in 1657, then by the Recollets not sooner than 1670.

The Jesuits settled in "Huron country," founding a number of missions between "Lake St. Louis" (Lake Ontario) and the "freshwater sea" (Lake Huron). They made it a priority to live among the sedentary populations that seemed open to accepting evangelization. In addition, this territory opened the door westward toward the China Sea – and China itself, where Jesuits had also settled. The Sainte-Marie mission, created in 1639, was 800 kilometres from Montreal – an advance deep into the heart of the continent. During the first half of the seventeenth century, the only Europeans who had permission to be in the Great Lakes region were the Jesuits and their "lay brothers, the young, devoted, and zealous men associated with the apostolic ministry." In 1634, the lay brother Jean Nicollet, probably mandated by Champlain, explored beyond the "freshwater sea." Guided by Hurons, he was probably the first European to see Lake Michigan and Fox River, to encounter the Winnebagos (or Menominees), and to hear talk of a great nation, the Sioux, located farther west. He also heard vague reports about the "great river" (the Mississippi), a navigable route that he was led to believe would lead to the "western sea."

At the same time, the founding of Trois-Rivières at the confluence of the St. Lawrence and Saint-Maurice rivers, in 1634, and of Ville-Marie at the confluence of the St. Lawrence and Ottawa rivers, in 1642, opened routes northward and westward. Montreal, already identified by Cartier, Roberval, and Champlain as a very advantageous site, became a point of access to the "upper country" and a crossroads for the fur trade. Information on the Indian nations thus arrived quite quickly at Quebec. Jean de Brébeuf and Pierre-Joseph-Marie Chaumonot decided to visit the Neutral Indians west of "Lake Saint-Louis" (Lake Ontario). They probably saw Niagara Falls in 1640–41, at the same time as Isaac Jogues and Charles Raymbault were at Sault Sainte-Marie and discovering Lake Superior. In 1646, Father Jogues and the engineer Jean Bourdon, exploring south of the Great Lakes, made a sketch of the area that may be the oldest cartographic depiction of the territories visited by the Jesuits in this region.

Published maps, such as those drawn by Nicolas Sanson in 1650 and 1656, were based largely on the maps

Drawing attributed to Claude Dablon
Every year, the superior of the Jesuit missions of New France compiled all of the relations written by the missionaries to send them to Paris, where they were read and then published. The manuscript relation attributed to Claude Dablon, unlike the printed version, had several illustrations, including this portrait of an Indian from the back.

Anonymous map of New France, circa 1640
The indigenous populations were very diverse and widespread, as shown on this map titled *Nouvelle France*. Inspired by one of Champlain's maps, its author added an impressive number of ethnonyms that were not francized. Like many manuscript maps of the time, this one bears neither a date nor an author's name. It was drawn on an animal skin several years before the Neutrals, Petuns, and Hurons were dispersed by Iroquois attacks, between 1648 and 1651. It is presumably the work of a Jesuit missionary who was in contact with the Hurons. Father Ragueneau, who was in Huronia from 1637 to 1650, no doubt contributed in some way to its making, although some researchers do not recognize his calligraphy on it. This map was preserved at the Jesuit College until 1759, and then taken to England by the military engineer John Montresor.

and descriptions of the eastern part of the Great Lakes – including lakes Ontario, Erie, and Huron, in almost their exact dimensions and shapes – published in the Jesuit *Relations*. Two other Jesuits updated the geographic knowledge acquired by their colleagues. In 1657, Francesco Bressani, who had lived in Canada from 1642 to 1650, made a detailed map with interesting scenes illustrating the life of the Indians, while François Du Creux, without the personal experience of a stay in North America, drew on information provided by the missionaries and published a map in 1660 that proved the existence of a drainage basin used for the fur trade – the entire region from Lake Superior to Montreal.

The European presence at the Great Lakes, as elsewhere in North America, brought its share of calamities. In contact with the French, the Huron population was hit with a devastating smallpox epidemic in the 1630s. In addition, the newcomers aroused anxiety among the traditional enemies of the Hurons, even though the 20 residents of the Sainte-Marie mission, like those in other missions, did not form population settlements but simply missionary centres combined with trading posts. In fact, the Hurons and French wanted to work together to control trade routes. The Iroquois farther south knew this and were on alert; with encouragement from their Dutch partners, they tried to destroy the French enterprise by attacking fur and supply convoys, Huron villages, and Jesuit missions. They succeeded; by 1649, Huronia no longer existed. The survivors were dispersed and the missionaries tortured. However, filled with incredible ardour and taking advantage of a brief truce, Jesuits such as Chaumonot, Dablon, and Simon Le Moyne nevertheless travelled to Iroquois country, in 1654–55, and gathered geographic information on the area from south of Lake Ontario to Onondaga (central New York state). Thanks to their sustained efforts, the Onondagas and Mohawks agreed to the founding of missions. Father Le Moyne also made a number of diplomatic voyages, including a visit to the Dutch governor of Nieuw Amsterdam (New York), in an attempt to repatriate prisoners and preserve the good relations necessary to trade.

Missionary activity in the upper country certainly helped to maintain French interest in expanding the colony. However, it was not until Louis XIV took charge of the colony, in 1663, that exploration officially started up again. The objectives, now more clearly defined, were based on existing foundations: protecting the fur trade, including the control and defence of the traplines, against English competitors, and evangelization of the Indians, – and, through these activities, the maintenance of alliances. The appeal was enhanced, of course, by the attractiveness of the space and a fascination with the unknown. These inextricably linked elements breathed energy into an exploration movement centred on the Great Lakes region. Thus, when Intendant Jean Talon arrived in Quebec, growth was part of his plan, and he envisaged glorious discoveries and taking possession of vast territories. One of his fond hopes was to see the colony become self-sufficient while satisfying the king by sending back to France certain products that it lacked. Under Talon's first mandate (1665–68), the fur trade burgeoned, but new sources of wealth, such as precious metals and the fervently desired passages to the "southern sea" and the "western sea," remained elusive.

To support these prospects for expansion, the Jesuits did not hesitate to return to the Iroquois. In 1665, Father Allouez described the Great Lakes, including the northern part of Lake Superior from Lake Nipissing to Lake Nipigon; in 1669, he followed the route previously taken by Jean Nicollet from Baie des Puants (Green Bay) to today's state of Wisconsin, where the Mascoutens lived. He then learned that the "great river," the Mississippi, was no more than six days' travel away. The last piece of the exploration puzzle in the Great Lakes was filled in by the Sulpicians Dollier de Casson and Bréhant de Galinée. Intending to found a mission southwest of Lake Superior, they travelled south of the Great Lakes from the summer of 1669 to the summer of 1670. After defining the contours of the lakes, they returned northward via the "traders' route." Their main contributions were to show that the five lakes were linked together and to reveal the navigable route to Montreal.

In 1669, Talon began to search for copper ore, picking up where Champlain had left off. Jean Péré, coureur des bois, and Adrien Jolliet, the brother of the famous explorer, were ordered "to go and reconnoitre to see whether the copper mine situated to the north of Lake Ontario is rich and of extraction." After stopping to trade furs at Sault Sainte-Marie, Péré in fact found a copper mine farther west, in the Lake Superior region. With this new information, on September 3, 1670, Talon deputized Daumont de Saint-Lusson to "search for the copper mine in the country of the Ottawas, Nez-Percés, Illinois, and other nations discovered and to be discovered in northern America beside Lake Superior or the Freshwater Sea ... carefully search, whether by lakes or rivers, some communication with the Southern Sea that separates this continent from China."

Daumont de Saint-Lusson called a great assembly of all the Indian nations that he was able to contact. On June 4, 1671, at the Sainte-Marie-du-Sault mission, before the representatives of the "Fourteen Nations" and with as much pomp as possible, the explorer proclaimed Louis XIV "Captain of the Greatest Captains" and officially took possession of all the land extending from there to the "northern sea" (Hudson Bay) and the Gulf of Mexico – in effect, appropriating all of North America. The minutes of the ceremony were signed by Saint-Lusson and Nicolas Perrot; by the Jesuits Claude Dablon, Claude Allouez, and Gabriel Druillettes; and by the other Frenchmen present.

The great accomplishment of Daumont de Saint-Lusson's mission was not his reconnoitring of the copper mine, which had already been located, but his appropriation for France, in theory at least, of an immense piece of empire the extent of which was not yet known. By opening the door to the southern part of the continent, the expedition allowed the French to envisage the "Vermilion Sea" by which it was hoped that a passage to the western sea would be found. It was also the first exploration resulting from an administrative order, rather than being a by-product of the fur trade or evangelization. Now, voyages of discovery were to be planned and directed by colonial authorities, and they had to result in reports, journals, and maps, as well as taking possession of the territories visited.

The Jesuit *Relations* were, among other things, travel accounts. Many were accompanied by drawings and

Map of land west and south of Montreal, attributed to Jean Bourdon, circa 1646
This anonymous map shows the complexity of the drainage basin south and west of the St. Lawrence, into which the French were just beginning to make inroads in the 1630s and 1640s. Although the region is sketched in roughly, the map is more accurate than the one drawn by Champlain. The calligraphy and content of the text indicate that it was made by the surveyor Jean Bourdon. In 1646, Bourdon undertook a voyage to Mohawk country, in the company of Father Jogues, to finalize a French-Iroquois peace treaty that was too fragile to last. The map shows the route taken by the French diplomats to ORANGE (Albany) and MANHATTE (New York): RIVIÈRE DES IROQUOIS (Richelieu River), LAC CHAMPLAIN (Lake Champlain), LAC SAINT-SACREMENT (Lake Saint-Sacrement), then a portage to RIVIÈRE D'ORANGE (Hudson River). In the fall of the same year, Jogues went again to Iroquois country, where he was killed, with a hatchet blow to the head, by Iroquois who accused him of having spread an epidemic among them.

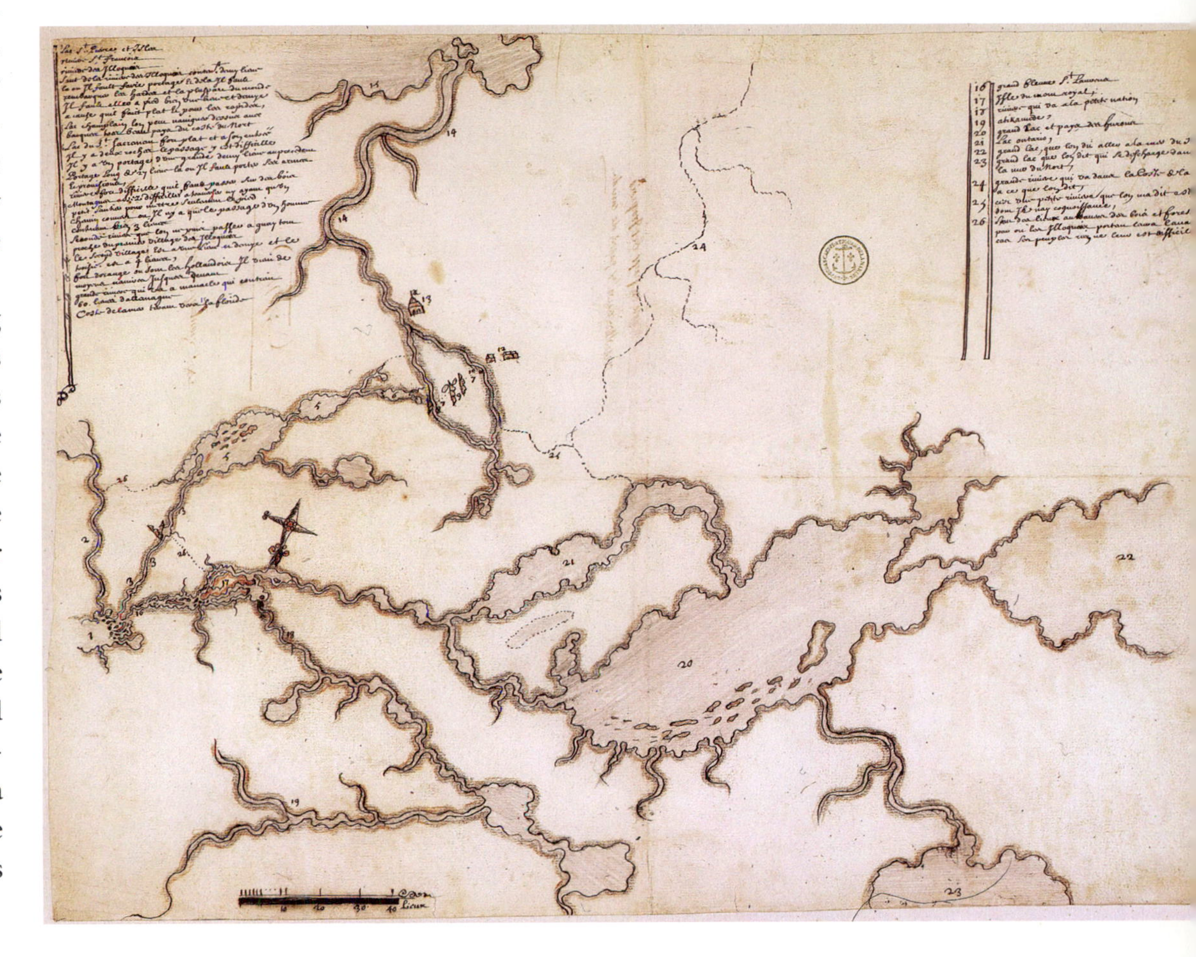

Nouveau Danemarcq.
Mer Christiane
Buttons Bay.
New Southwalles.
Golfe de Hudson, ou Hudson Bay.
New North Walles.
Hope Advanced
Port de Munck
Hope Check
Port Nelson
New Severne R.
Mansfield I.
Nottingham
Salisbury
C. Charles
Mille I.
C. Penbrock
Des troit de Hudson
Terres Arctic
Queenes Anne Forland
Gods Mercies
Iland of Good Fortune
Cap Elizabeth
Salvage I.
King James his Forland
C. Prince Henry
St. Thomas Smits Forland
Marchands Isles
Estotilande, ou
Terre de Laborador
Nouvelle Bretagne
la Grande Baye ou Delivrance
James his Bay
Cap Henriette Marie
Hope Advanced
Hoybergen
Kiristinous.
Nadouessoue ronons
Lac Superieur.
Astakouankae ronons
Skiaeronon.
Lac de Puans
Oukouararonons
Karegnon di.
Nou vel le France
Le Saguenay.
Tadoussac
Quebec
Sillery
I. d'Orleans
Nipissiriniens.
Enchek
Aossondi
Attiquamecques
Ottawa
Chiaentonati.
Chaouaeronon
Tarantou.
Ontario, ou Lac des St. Louys.
Mont Real
Saut de S. Louys
Richelieu
L. Champlain
Iroquois.
Nouvelle Angleterre
Plymouth
Malabarre
Vlielands
Nouveau Pays Bas.
t Lange Eyland.
Nouvelle Amsterdam
N. Sued
Nassau
C. de May
Riviere de Sud
Assistaeronons, ou du Feu
Ariatoeronon
Couaeronon.
Lac des Eaux de Mer
Aictaeronon.
Neutre
L. Erié, ou du Chat.
Squenquioronon.
Ontarraronon.
Attiouandarons
Virginie
Powhatan
Smyths Isles
C. Charles
C. Henry
Dematmotique
Secotan
Paquiwock
Croatoan
Wokokon
Floride
Floride Françoise
Apalache
Malique
La Floride Espagnole.
Chuca
Xuala
Canaraguay
Guaxule
Chalaque
Cofachiqui
Vlibahaly
Tali
Cosle
Ychiaba
Aminoja
Anilco
Cofaqui
Cofa
Acoste
Guacacoya
Guancane
Naguatex
Coza
Capaha
Chasquin
Vtiaunque
Colima
Ossachile
Houstaqua
France
Occident
Septentrion

***Le Canada ou Nouvelle France* by Nicolas Sanson, published in Paris in 1656**
The founder of a major dynasty of cartographers and the author of this map, Nicolas Sanson was known to his contemporaries and is now recognized by historians as one of the great geographers of his time. This former military engineer, educated at the Jesuits' school, was the geography teacher of kings Louis XIII and Louis XIV. Very prolific, he was the first Frenchman to compete with the Dutch mapmakers whose production was dominating the European market. Sanson's success must be attributed not only to his proximity to royal power, but to his association with the very energetic publisher Pierre Mariette. Sanson was a *géographe de cabinet*, taking advantage of his position at the centre of the French kingdom to gather and compile information from a great variety of sources. To produce this map of Canada, he used the accounts of foreign navigators – English, Danish, and Dutch – as well as those of Champlain and the Jesuits. The missionary priests had published many relations highlighting their evangelization efforts, including descriptions of the inhabitants and their customs, while giving a good written and cartographic portrayal of the country. Translating the texts into maps, Sanson revealed the interior of the continent – including the five Great Lakes, which are quite accurately positioned – to a broader public. Lake Superior and Lake Michigan (LAC DES PUANTS) were left unfinished to the west, as no European had yet explored their shores. Among the 50 Indian nations portrayed, most had been visited by the handful of priests sent on apostolic missions: NEUTRES (Neutrals), PÉTUNS, HURONS, IROQUOIS, NADOUSSOUE-RONONS (Sioux), NIPISSIRINIENS (Nipissings), KIRISTINOUS (Crees), ATTIKAMEKS, QUOUATOUATA and QUIOQUHIAC (Montagnais), and Ottawa ALGONQUINS.

Novæ Franciæ Accurata Delineatio by Bressani, 1657
Father Bressani landed in the settlement of Quebec in 1642 and set out for Huronia, but he was captured by the Iroquois, who tortured him. When he wrote to his superior, he apologized: "The letter is poorly written and quite dirty, because … he who writes it has only one entire finger on his right hand and he cannot keep the blood … from soiling the paper." Upon his return to Italy, he published a relation and a map that depicts quite well the territory of the St. Lawrence Valley and the Great Lakes. Bressani signposted the route between Montreal and Georgian Bay, listing the rapids and portages. Because the map was abundantly illustrated, it became an education tool for later missionaries.

geographic outlines to complement and add accuracy to descriptions of ever-more remote territories. These publications, widely distributed in France, showed not only the perseverance of the missionaries in propagating the Christian faith but also their scholarly contribution to exploration. The *Relations,* as edifying works, became excellent tools for promoting the French undertaking in America. In the mid-1670s, the advance to the centre of the continent was consolidated. The numbers of converts remained low, but the territories described and mapped were immense and the trade resources seemed unlimited. The way was now open "to the discovery of the southern sea."

Main sources in order of importance

Deslandres, Dominique. *Croire et faire croire: les missions françaises au xviie siècle.* Paris: Fayard, 2003. — Trigger, Bruce Graham. *The Children of Aataentsic: A History of the Huron People to 1660.* Montreal: McGill-Queen's University Press, 1976. — Cardinal, Louis. "Record of an Ideal: Father Francesco Giuseppe Bressani's 1657 Map of New France." *The Portolan,* No. 61 (Winter 2004–05), 13–28.

The Mapmakers' Source

THE COUREURS DES BOIS

During the early French expeditions in Canada, the fur trade was conducted essentially in the St. Lawrence Valley, which was under the control of the trade companies. Starting in the 1650s, especially after the destruction of Huronia, the Iroquois prevented the competing Algonquin nations from trading with the French and taking their pelts to Montreal. Then, without the intendant's permission, the "*coureurs des bois*" began travelling to the "upper country" – that is, the territories west and north of Montreal, so named because to reach them one travelled up the rivers from Montreal. As the expeditions continued, this indeterminate space, home to tens of thousands of Indians speaking Iroquois, Algonquin, and Sioux, constantly expanded. French explorers and traders saw the outer edges of the upper country as something that had to be pushed back. In 1658, the Jesuits spoke of "les pays superieurs," from which came the name for Lake Superior, the westernmost of the Great Lakes. The imprecision of the plural, "les pays," hinted at the possibility of unknown spaces, including the utopian "western sea."

The upper country was also the ideal space for the coureurs des bois, who met up and traded with Indian trappers or ventured into the forest to trap animals themselves. They travelled using the same means as the Indians: birchbark canoes in summer and snowshoes in winter. The young men who lived in the woods, some of them former lay brothers, became intoxicated by the "savage" life – both the charms of living in the wild and the absence of certain social constraints. For instance, they generally had the companionship of Indian women, to the consternation of the missionaries and administrators who saw them "becoming savages."

The *course des bois* was given new impetus in 1652 and 1654, when a few Huron and Ottawa Indians went to Trois-Rivières with furs – a small part, they said, of what had accumulated due to a drop in sales; they also spoke of a river, beyond their territories, "very spacious, which leads to a great sea." This information reawakened the lust for adventure among young, curious, ambitious men. Médard Chouart Des Groseilliers and an unidentified companion left Trois-Rivières in the autumn of 1654 and accompanied the Indians to their territories. For two years, they travelled through the upper country, reconnoitring and naming places along the route usually taken by the Indian traders. Des Groseilliers navigated upstream on the Ottawa River as far as Lake Nipissing, then up the French River to Lake Huron, and on Lake St. Clair to the future site of Detroit, between Lakes Huron and Erie. What happened afterward is less clear; it is assumed that the explorers reached Lake Michigan and followed the western shore, returning to the settlement of Quebec by the same route.

Des Groseillers and his companion each brought back 14,000 to 15,000 pounds of furs, as well as new geographic information. Des Groseilliers was interested in finding new trade routes in the upper country, in order to shorten costly portages and detours. He thought that there might be other good navigable routes providing access to the territory north of the Great Lakes and to Hudson Bay. In August 1659, he left Quebec on another trade mission, this time with his young brother-in-law, Pierre Radisson, in spite of having been forbidden to do so by the governor general of New France, Pierre de Voyer d'Argenson. The explorers followed the main trade route up the Ottawa River to Lake Huron and along its north shore to Sault Sainte-Marie, then the south shore of Lake Superior to its western end, where Des Groseilliers lent his name to a river that is still called Gooseberry River. When they returned to the St. Lawrence Valley, in

Beavers of Canada
These well-known insets come from a map of North America by Nicolas de Fer (top) and a map by Moll published in 1715 (bottom). This scene of beavers was originally engraved by Nicolas Guérard, who was inspired by an illustration of Niagara Falls in Louis Hennepin's account. Guérard's composition shows anthropomorphized beavers with very specific jobs: lumberjack, carpenter, wood bearer, mason, architect, caregiver for the injured, and so on.

Above, *Carte générale du Canada* by Lahontan, 1703
Here, Lahontan presents, among other things, the areas where Indians hunted beavers. Notably, the Iroquois have appropriated the Huron hunting grounds. In his account, Lahontan gave a detailed description of Indian hunting techniques: they took great care not to destroy the area and to leave a few males and females to ensure that the population would recover.

Right, *Carte du Canada* by de Couagne, 1711
Thanks to the size of this document, a great number of place names are written in on the route taken by the coureurs des bois from the St. Lawrence to the main trading posts in the interior of the continent. The routes toward Lake Michigan, Lake Erie, and Hudson Bay are particularly well marked with the names of rivers and portages that the coureurs des bois had to take.

the summer of 1660, the two coureurs des bois made a strong impression with their 60 canoes heaped with bundles of furs and an escort of 300 Indian traders. The furs were seized; Des Groseilliers was charged with having traded for furs without a permit and imprisoned on the order of the governor general.

In spite of its anticlimactic end, Des Groseillier's expedition reopened the fur trade and probably saved New France from economic disaster. However, Des Groseilliers and Radisson developed a lasting rancour toward the administrators of New France and turned instead to the English. Sharing their knowledge of Iroquoisia and the upper country, they facilitated the English conquest of New Holland (New York) in 1664, and provided assets that proved valuable in creating the Hudson's Bay Company.

For the French administration, it was an embarrassing situation. In order to exert tighter control over the fur trade, an essential activity, the colonial authorities created "trade holidays" during which individuals were authorized to trade for pelts. But the constraints of the "holidays" were so stringent that the clandestine and contraband trade with Albany continued to attract more than 100 coureurs des bois in the 1660s. Minister Colbert reminded Governor Frontenac a number of times of the need to contain the population in the St. Lawrence Valley, except to maintain a profitable level of trade or if the territories "discovered may be brought closer to France by communication with some sea that is farther south than the entrance to the St. Lawrence River." Colbert was ambivalent, to say the least: he wanted to keep the settlers on their land, but the colony could not survive without the fur trade. He was therefore forced to let the coureurs des bois go their own way. The path to trade, and therefore to exploration of distant territories, thus remained open.

Intendant Talon jumped into the breach to accelerate his discovery program: "I had resolute men go who promised to penetrate farther ahead than has ever been done before . . . In every location these adventurers must make journals and respond upon their return to the instructions that I gave them in writing. In every location

RIVIERE DANOISE
BAYE D'HUDSON
LA DOUZAINE DU BOULANGER
PORT NELSON
RIVIERE BOURBON
NIEUSAUANE
CAP HENRIETTE MARIE
ISLES AUX OURS
TERRE
ISLE DE CHARLESTON
LAC PISCOUTAGAMY
LE GRAND LAC DES TABITIBIS
R. DES ORIGNEAUX
R. DES TABITIBIS
R. RUPERT
LAC NEMISCO
RIVIERE MICHIPICOTON
R. BATCHIANON
ESCHELLE DE 80 LIEUES
LAC TEMISCAMING
MATAOUAN OU FOURCHE
GRANDE RIVIERE DES OUTAOUACS
NOUVELLE
R. NECOUBA
LAC S.T IEAN
MANITOUALIN
LAC HURON
LA MANESTIQUE
R. MARQUET
R. BLANCHE
GRANDE RIVIERE
R. NOIRE
FORT PONCHARTRAIN
CATARACOUY
LAC ONTARIO
LAC CHAMPLAIN
LAC DU S.T SACREMENT
NOUVELLE ANGLETERRE
LAC ERIÉ
LES IROQUOIS
SONONTOUAN
GOIOGUEN
ONONTAGUE
ANEIOUT
CORLAR
ORANGE
ESOPE
BASTON
BAYE DE BASTON
DORCESTER
SUFFOLC
N. PLIMOUT
BRISTOL
C. COD
ELIZABETH
NOUVELLE
NIEU JERSEY
MANATTE
OHIO OU BELLE RIVIERE

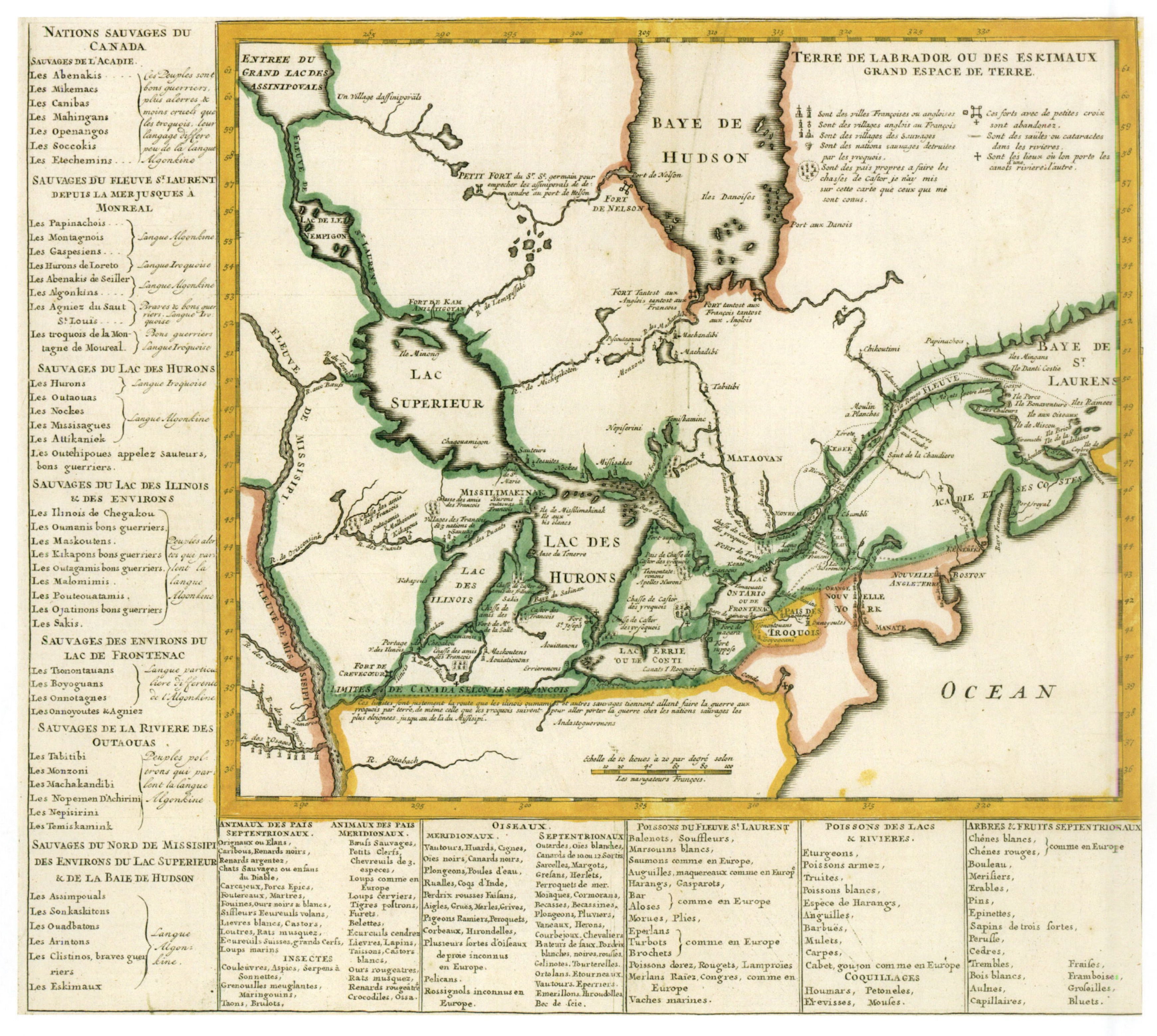

Map of the Great Lakes by Chatelain, 1719
Like most of the illustrations in Henri-Abraham Chatelain's *Atlas historique*, this map shows its maker's attention to order and classification. The map, clearly inspired by Lahontan (see p. 96), is surrounded by tables showing the merchandise traded to the Indians, the value of the furs, and the trees, fish, animals, and insects of Canada. On the left, the publisher lists almost 50 Indian nations, noting which are "good warriors," "brave warriors," and "cowards."

they must take possession, fly the arms of the King, and keep minutes to serve as title." With this strong encouragement, Jean Péré, Louis Jolliet, and others brought back to Quebec their geographic observations and travel accounts. Transposed by hydrographer Jean-Baptiste Franquelin into magnificent maps, the information circulated freely in France and elsewhere in Europe, painting a rosy picture of North America.

Main sources in order of importance

VACHON, André, Victorin CHABOT, and André DESROSIERS. *Dreams of Empire: Canada before 1700.* Translated by John F. Finn. Ottawa: Public Archives of Canada, "Les Documents de notre histoire," collection, 1982. This book is useful for its short historical syntheses and reproductions of original documents. — DELÂGE, Denys. *Bitter Feast: Amerindians and Europeans in Northeastern North America, 1600–64. Translated by Jane Brierley.* Vancouver: UBC Press, 1993. For greater comprehension of the relationships between the coureurs des bois and the Indians.

Discovery of the Mississippi

JOLLIET AND MARQUETTE

IN MAY 1673, some Illinois Indians saw half a dozen strange men, with pale faces and beards, approaching on the MICHISIPPI (Mississippi) River. One of them was wearing a long black robe and could speak several words in their language. The arrival of these newcomers caused great commotion in the village. Four Illinois elders greeted them and took them to the chief, who awaited them, naked, with a peace pipe. They were then taken to the neighbouring village of PEORIA, home to the great leader of all the Illinois. There, a feast fit for a king was prepared for them: bison, *sagamité*, fish, and dog meat, served in bison skulls. These strange-looking men were Frenchmen, among the first Europeans whom the Illinois had met. They were the coureurs des bois Jacques Largilliers, Jacques Maugras, Pierre Moreau (known as La Taupine), and Zacharie and Louis Jolliet, as well as Father Jacques Marquette, a Jesuit missionary.

The expedition was led by Louis Jolliet, who was of the first generation of French descent born in Canada. Educated by the Jesuits in the settlement of Quebec, under the patronage of François de Laval, Louis had been destined for a brilliant ecclesiastic career. But in his early twenties, he turned to a life in the woods and the fur trade, no doubt alongside his older brother, Adrien. Receiving a different kind of training from Pierre-Esprit Radisson and also captured by the Iroquois in 1658, Adrien was mandated in 1669 to locate a copper mine at Lake Superior. On his way back, he took the perilous route through lakes Erie and Ontario, a territory well guarded by the Iroquois. In 1671, as Daumont de Saint-Lusson was raising the French coat of arms to take official possession of Ottawa country and gaining the allegiance of 14 Indian nations, Louis was in Sault Sainte-Marie. There, he met the Jesuit fathers Dablon and Marquette, who had been travelling around the Great Lakes region for a number of years and were familiar with the territory. They knew several Indian languages, which both helped them with their missionary work and enabled them to obtain fairly accurate geographic information from the Indians. When they met Jolliet, they already knew about the existence of the Mississippi River and wanted to find out which sea the great river flowed into: the Gulf of Mexico or the "southern sea" (the Pacific).

The following year, Jolliet was officially mandated by Intendant Talon and Governor Frontenac to "discover the southern sea by the country of the Maskouteins, and the great river that is called the Michisippi, which, it is believed, flows into the sea of California." As was the custom among coureurs des bois, Jolliet formed a notarized partnership with six other settlers to "make together the voyage to the Ottawas, and to trade with the Savages as advantageously as possible." Thus, the expedition was not simply a voyage of discovery but also a trade mission. Jolliet and his party left Quebec in the fall of 1672 and arrived at the post of MICHILLIMACKINAC in December. There, along with Marquette, they planned their voyage of discovery, questioning the Indians who knew the Mississippi and sketching an approximate map of the route that they would take.

In the spring, they set off, their two birchbark canoes loaded with corn and smoked meat, "resolved to do and suffer everything for so glorious an undertaking." They paddled across Lake Illinois (Lake Michigan) then through BAIE DES PUANTS (Green Bay), and up a river that they named RIVIÈRE AUX RENARDS (Fox River). During this trip, Marquette made a very detailed record of the geographic features and the fauna and flora, tasted mineral waters, and gathered a medicinal herb the root of which was an antidote for snakebite. The voyagers reached the village of the Mascoutens, where the Jesuits

Capitaine de la nation des Illinois
Jacques Marquette and Louis Jolliet were among the first ethnographers of North America. Their voyage account contains a detailed description of the customs and practices of the Illinois, the main Indian group that they encountered: "When one speaks the word 'Ilinois,' it is as if one said in their language 'the men' … They are of a gentle and tractable disposition … They have several wives, of whom they are extremely jealous; they watch them very closely, and cut off their noses or ears when they misbehave … Their bodies are shapely; they are active and very skilful with bows and arrows … They are warlike, and make themselves dreaded by the distant tribes to the south and west, whither they go to procure slaves; these they barter …"

Map of the Mississippi and Lake Michigan, attributed to Father Marquette, circa 1674

This document is considered the oldest source relating Jolliet and Marquette's voyage to the Southern Sea. The presumed author, Jacques Marquette, calls the Mississippi Conception. The Missouri is called Pekitean8i; the Ohio, 8ab8skig8. In spite of its rudimentary contours, this map is a good rendering of the general orientation of the Mississippi. Marked are the Illinois villages visited by the explorers – kachkaska (Kaskaskia), Pe8area (Peoaria), Mons8pelea (Monsopelea) – then the Sioux villages of Metchigamea (Michigamea) and finally Akansea (Quapaw), the farthest point attained by the expedition. In addition, the cartographer notes a number of villages in the interior that do not appear in the written account, such as Maroa (Tamaroa), 8chage (Osage), Papikaha (Quapaw), Akoroa (Koroa), Otontanta (Oto), Maha (Omaha), and Pah8tet. If the attribution and date (1674) are correct, this map would be the first document to mention the peoples of the American Midwest. For this reason, archaeologists, ethnologists, and historians consider this artifact, preserved in the archives of the Compagnie de Jésus in Saint-Jérôme, Quebec, to be a very important historical source.

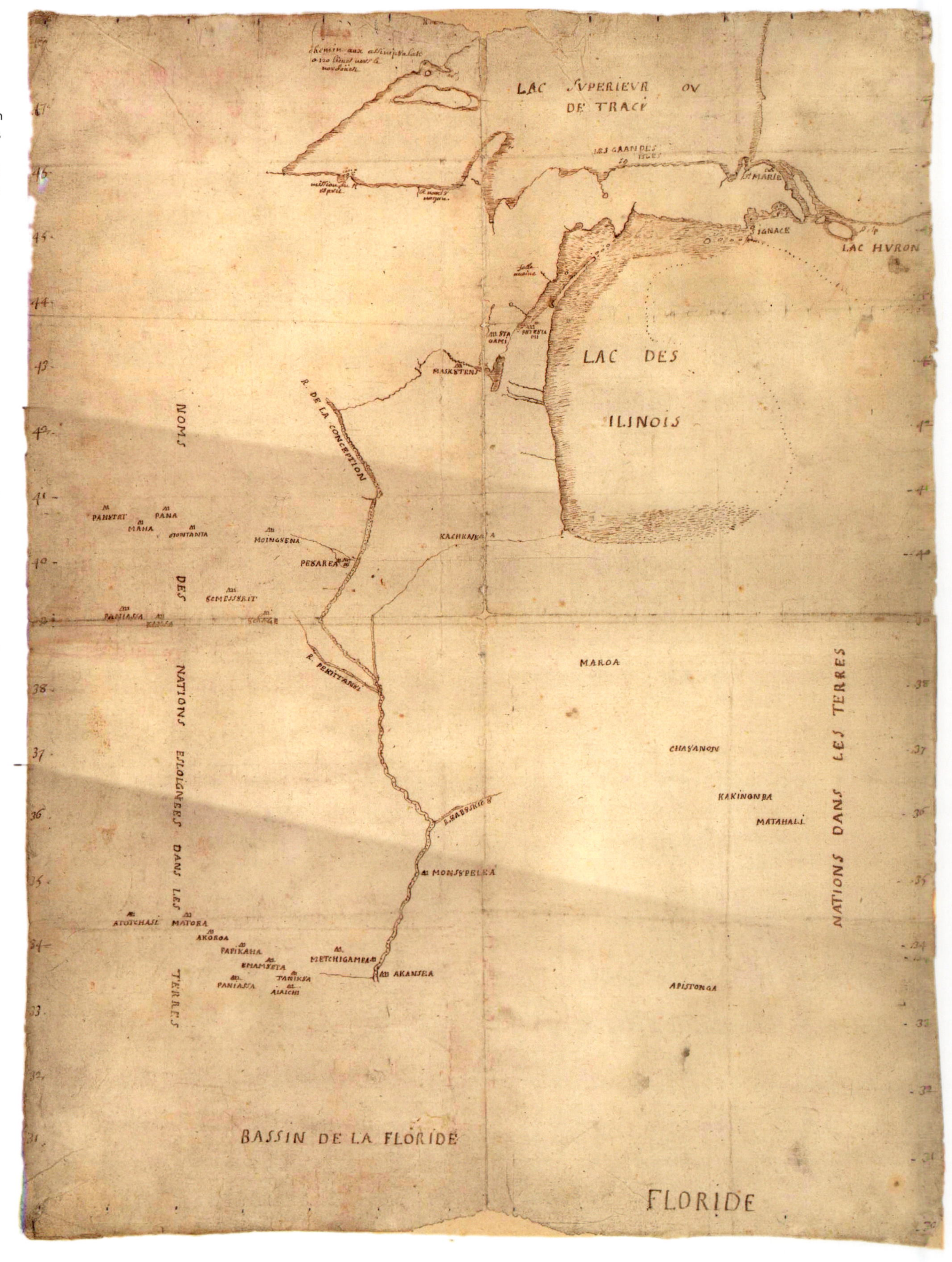

had started a mission. This location, at the extremity of the territories known to the French, was a sort of multi-ethnic centre, since, in addition to the Mascoutens (or Nation of Fire), there were Miamis, Kickapoos, Illinois – all nations that had settled there after fleeing the Iroquois. In the centre of the village, the Frenchmen saw a stunning example of religious syncretism: a cross festooned with offerings to the Great Manitou – pelts, belts, arrows, and bows.

The next part of their journey promised to be more hazardous: there were so many marshes and small lakes to cross that they absolutely needed Indian guides. Two Miamis agreed to take them to the portage that led to the MESKOUSING (Wisconsin) River, a presumed tributary of the Mississippi. In mid-June, after a few days of canoeing on this river, the Frenchmen shouted with joy when they caught sight of the coveted Mississippi.

They followed the river southward. Around 42°, Marquette noted major changes in the fauna and flora:

> There are hardly any woods or mountains; the islands are more beautiful, and are covered with finer trees. We saw only deer and cattle, bustards, and swans without wings, because they drop their plumage in this country. From time to time, we came upon monstrous fish, one of which struck our canoe with such violence that I thought that it was a great tree, about to break the canoe to pieces. On another occasion, we saw on the water a monster with the head of a tiger, a sharp nose like that of a wildcat, with whiskers and straight erect ears.

On the Mississippi, the Frenchmen encountered a number of Illinois tribes, each as friendly as the next. One of the chiefs proclaimed to his visitors, "'How Beautiful the sun is, O Frenchman, when thou comest to visit us! All our village awaits thee, and thou shalt enter all our cabins in piece.'" Navigation presented few problems. The Frenchmen had received from the chief of the Illinois two very valuable gifts: a young Indian slave to serve as a guide, and a feather-adorned calumet, a sacred object that served as a passport with the other peoples of the Mississippi.

As they paddled down the Mississippi, the group passed the PEKITAN8I (Missouri) and 8AB8SKIG8 (Ohio) rivers. South of the Ohio, the climate changed: the heat became oppressive and the mosquitoes unbearable. The explorers had left Illinois territory and entered Sioux country, where the welcome was more reserved – in fact, almost hostile. In the village of the Akanseas, not far from the mouth of the Arkansas River, the elders held a secret council during which some proposed to "break our heads and rob us," reported Marquette. In order not to jeopardize their expedition, the explorers wisely decided to continue on their way. From information obtained from the Indians and their own calculations of latitude, Jolliet and Marquette quickly deduced that the Mississippi flowed into the Gulf of Mexico.

On the way back, the travellers retraced their steps up the Mississippi then took a different tributary, the Illinois River, the main artery through Illinois country. The area impressed Marquette and Jolliet; the former promised to return to evangelize the inhabitants, while the latter intended to ask the authorities for a land concession so that he could establish a colony there. The site was strategically placed, as it connected Canada to the Mississippi

***Carte de la découverte faite l'an 1673 dans l'Amérique septentrionale*, published by Thévenot, Paris, 1681**
This account of the expedition to the Mississippi was published for the first time in 1681, in one of the voyage collections compiled by Melchisedech Thévenot. A member of the Académie des sciences, among others, and the king's librarian, Thévenot was a polyglot with an insatiable curiosity and one of the most active men of letters and sciences of his time. He invented the spirit level and penned a treatise on the art of swimming, and he introduced the French people to a number of distant countries, such as China, Persia, Japan, the Bengal, and Turkestan. Thévenot's collections were among the first of their type published in French. When possible, he enhanced the accounts with drawings of plants and animals and with maps. He published the first maps of Australia and Iraq, as well as this map of the Mississippi River, apparently in response to the appetite of the French people for information on the potentially enriching undertakings of the kingdom of France. It shows many mineral lodes that could eventually be exploited: bloodstone, copper mines, coal, saltpetre, slate, marble, iron mines. In the centre is a statue of Manitou – the Indian word for spirit or genie – who was worshipped as a divinity.

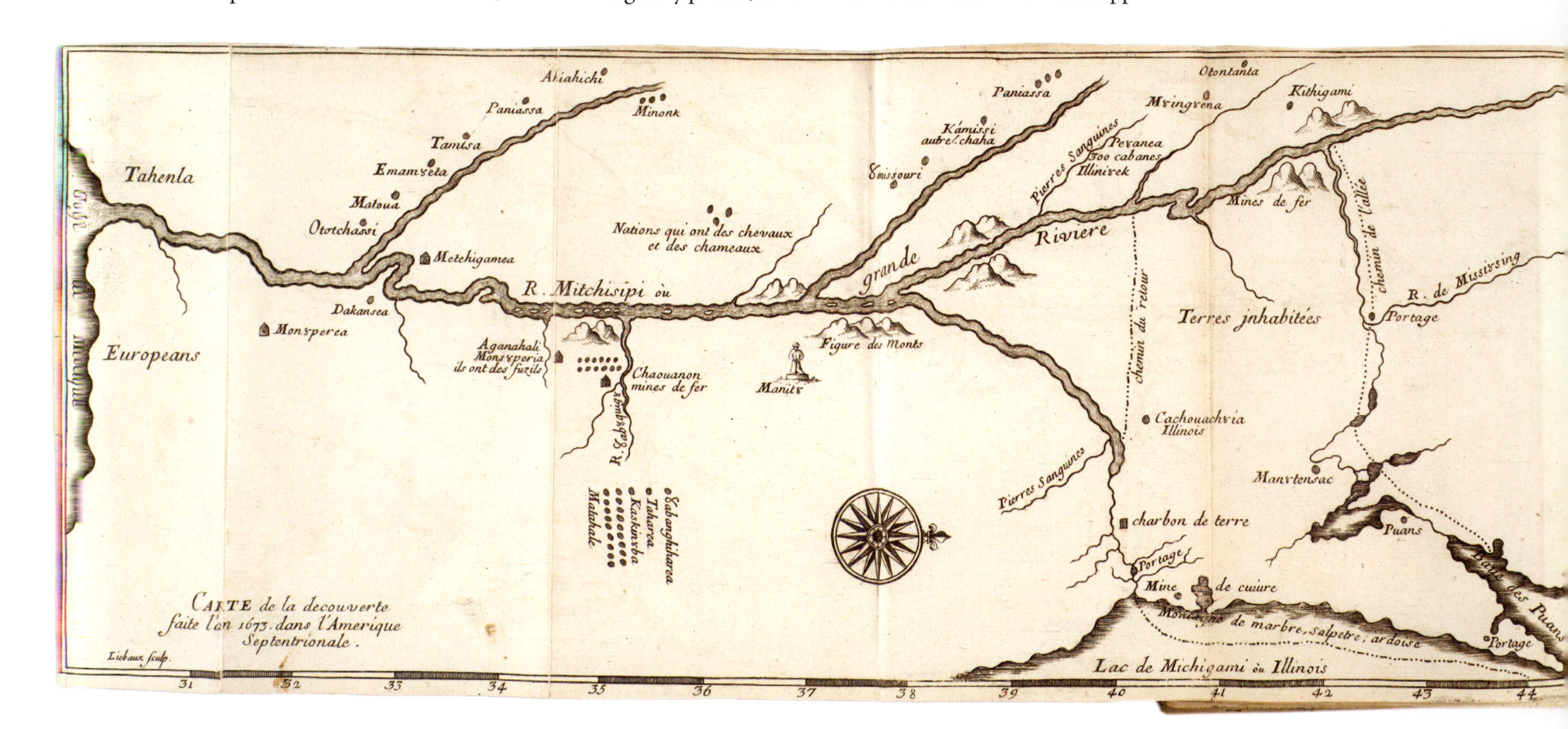

Basin. Jolliet counted more than 40 villages, the largest of which had up to 300 dwellings and 8,000 inhabitants. The explorers then returned to the Great Lakes basin via the CHECAGOU (Chicago) portage. Father Marquette wintered over at BAIE DES PUANTS, while Jolliet spent the winter at Sault Sainte-Marie, where he returned to trading and transcribed his map and journal. He went home in May 1674, with new pelts and new geographic information. Unfortunately, almost within sight of Montreal, Jolliet's canoe capsized in the last of the 42 rapids, SAUT SAINT-LOUIS (today the Lachine Rapids). Two men drowned, including the slave obtained from the Illinois, while Jolliet lost the chest containing the map and the detailed account of his voyage. It was a tragic ending for such a wonderful adventure!

The explorers should have received decent rewards, since Minister Colbert had assured Intendant Talon of this in June 1672: "Since after increasing the colony of Canada, there is nothing more important for that country, and for the service of His Majesty than the discovery of the passage to the southern sea, His Majesty wishes you to ensure a good reward to those who make this discovery." When he returned to Quebec, however, Jolliet found that Talon had returned to France, and there was no one to support his application for colonization of Illinois country. The king's policy, stated repeatedly, was very clear: to consolidate lands and colonists in the St. Lawrence Valley. Although his health had been compromised by the voyage, Father Marquette decided to return to the Illinois to found a mission. But he fell ill and died on the way to Michillimackinac, on the shore of Lake Michigan, at the mouth of a river that now bears his name (Pere Marquette River).

In spite of these disappointments, the voyage of Jolliet, Marquette, and their five acolytes remains one of the most important in the history of New France. It opened the door to an immense territory: the Mississippi basin and the future Louisiana. In his account of their travels, Jolliet stated that it was the most beautiful landscape on Earth. Growing there were corn, watermelons, plums,

Louis Jolliet's voyage map of the Mississippi River, circa 1675 (detail)
On his way back from the Mississippi, Louis Jolliet's canoe capsized, probably in the Lachine Rapids. Two men drowned in this tragic accident, including a young Indian slave who had been given to Jolliet by the Illinois. Jolliet also lost the trunk that contained his travel account and the map of the territories that he had explored. Fortunately, he was able to reconstruct his itinerary from memory, as shown on this map, which he submitted to Minister Colbert. It portrays his route via the BAYE DES PUANTS (Green Bay) and the MISKONSING River (Wisconsin), as well as the return route by LA DIVINE (Illinois River) to Lake Michigan. Farther south, the Ohio route taken by another explorer, Cavelier de La Salle, was added later. The map evokes the existence of a tributary of the Mississippi that might lead to the MER VERMEILLE (Pacific Ocean).

apples, pomegranates, lemons, blackberries, and many berries not known in Europe. Corn was harvested three times a year. The countryside was full of bison, deer, turkeys, quails, and parrots. In the ground were mineral riches: iron, hematite, saltpetre, slate, charcoal, and marble. In spite of the unfortunate loss of Jolliet's journal and map at Saut Saint-Louis, the voyage resulted in the production of a great number of maps, in itself attesting to the interest of the French authorities. These maps, which describe the drainage system linking the Great Lakes to the Mississippi River, caused a revolution in the cartography of North America: mountain ranges disappeared, fictitious rivers dried up, a majestic river made its appearance, and a multitude of indigenous peoples were revealed to Europeans. Of course, the mouth of the Mississippi was not where many hoped it would be, and no route toward the "southern sea" had yet been found. But the discovery of a number of tributaries west of the great river was to feed European optimism for more than a century.

Main sources in order of importance

Delanglez, Jean. *Life and Voyages of Louis Jolliet, 1645–1700*. Chicago: Institute of Jesuit History, 1948. — Pelletier, Monique. "Espace et temps: Mississippi et Louisiane sous le règne de Louis XIV." In Christian Huetz de Lemps (ed.), *La découverte géographique à travers le livre et la cartographie*, pp. 13-40. Bordeaux: Société des bibliophiles de Guyenne, 1997. The author, at the time director of the Département des cartes et plans at the Bibliothèque nationale de France, makes great use of the valuable collection of the Service hydrographique de la Marine, as well as other sources. — Thwaites, Reuben Gold (ed.), *The Jesuit Relations and Allied Documents: Travels and Explorations of the Jesuit Missionaries in New France, 1610–1791. Lower Canada, Illinois, Ottawas, 1673–1677,* vol. 59. Cleveland: Burrows Brothers Co., 1900.

Sources for translation: Marquette, Jacques. "The Mississippi Voyage of Jolliet and Marquette." In Louis P. Kellogg (ed.), *Early Narratives of the Northwest, 1634–1699*, pp. 223–57. New York: Charles Scribner's Sons, 1917 [www.americanjourneys.org].

Map of New France by Louis Jolliet, drawn by Franquelin, circa 1678
Although the shapes of the lakes, the contours of the rivers, and the configuration of the continent are rather schematic, the toponymic and iconographic content of this map is very rich. The drawings of many of the animals offer reference points that ease the transition between map and account, oral or written. For example, the illustration of the god Manitou marks the specific point in the expedition at which the voyagers came close to drowning. West of the Mississippi, there are exotic beasts: camels and ostriches. Because Jolliet's journal disappeared, it is difficult to explain the presence of these African animals in North America. They could have been the result of fabrication, hallucination, or incomprehension between Indians and Europeans.

Portrait of a Cartographer: Jean Baptiste Louis Franquelin, King's Geographer at Quebec

In the late seventeenth century, one man alone mapped all of French America, from the St. Lawrence to the Mississippi, as well as Hudson Bay, the Great Lakes, Acadia, and many other territories. His name was Jean-Baptiste Franquelin. His profession was king's hydrographer and geographer. With about 50 maps to his name, Franquelin was certainly the most prolific cartographer in New France. He left an exceptional testimonial to the French presence in North America; to follow his work closely is to go on a voyage of discovery of the continent. His life, however, remains largely a mystery.

Born in 1650 at Saint-Michel de Villebernin, a parish in Berry province, Franquelin left France for Canada in his early twenties. In a report submitted to the king, he noted that he had emigrated "with the intention of doing business" there. This statement is not very surprising, given his bourgeois background; his father and grandfather had been farmers-general, responsible for collecting taxes in the king's name. When he arrived at the settlement of Quebec, the young Franquelin worked as a guard for Governor Frontenac. The two had probably crossed the Atlantic together during the summer of 1672. The following year, Franquelin attended the small seminary for sacerdotal students. After three years, he left the seminary without having become a priest, and we do not know why he abandoned this path. It is certain, however, that he was welcomed into the home of Intendant Jacques Duchesneau, and it was with the urging of this new patron that he drew his first maps of America.

There is little documented information about Franquelin's education. Frontenac noted that he "draws and writes perfectly well." Indeed, in his early work, it is easy to see that Franquelin was a gifted artist. But there is no document proving that he had had training in geography. It is possible that in his youth he went to one or another of the best colleges in France. He probably studied science under the Jesuits in Quebec or under Boutet de Saint-Martin, an instructor in navigation and surveying. But these are just suppositions. In 1676, in fact, he was the only one who knew how to "make geographic maps and other plans to inform the Court and give knowledge of the environs to governors and intendants."

At the time, the need to use maps was ever more pressing. Since the Hurons, the main trading partners of the French, had been decimated by the Iroquois, New France had to find other ways to procure pelts. French voyagers were travelling to the interior of the continent in great numbers, lured by the prospect of the wealth that trading could bring and reassured by the military actions of the men of the Carignan-Salières Regiment against the Mohawks in 1665 and 1666. The routes were now more secure, protected by a series of fortified posts that defended the colony. In addition, some local leaders had dreams of grandeur and were encouraging missionaries and traders to explore all corners of the continent. They returned full of new information on watercourses, forts, Indian tribes, and mountains and other features.

Cartographers quickly became an essential resource for keeping up with the flow of data and informing the French authorities about these new discoveries. Jean-Baptiste Franquelin had his work set out for him. He first became known by mapping the Mississippi River, discovered by Louis Jolliet and Jacques Marquette and then explored by La Salle, Hennepin, and Dulhut, and he then produced plans of the lower town of Quebec, Fort Saint-Louis, and the Gouffre River, where the authorities wanted to open a silver mine.

In 1683, Franquelin went to France to present his early work to the Court. The explorer Cavelier de La Salle went with him. La Salle, who had travelled the Mississippi to its mouth, wanted to propose a project to colonize the region. The meeting of these two men was propitious, since Franquelin had the skills needed to map the region explored. However, whether through the fault of the explorer or the cartographer, on the map the mouth of the river was placed too far west, and La Salle was unable to get his bearings. After searching in vain for the river, he was assassinated by his enraged men.

When he returned to New France, Franquelin again partnered with Louis Jolliet, this time to map the St. Lawrence River. This work earned him the position of king's hydrographer. In this capacity, he taught navigation to young Canadians, a position with a fixed salary of 400 pounds per year, and his status changed from that of a precarious self-employed worker to the more secure one of government employee.

Although Franquelin was now officially the king's hydrographer, he did not hesitate to take on additional titles – geographer, master of hydrography and geography, engineer, mathematician, and professor – the variety of which matched his multifarious talents. In addition to drawing maps and plans for the colonial and metropolitan governments, he taught fortifications, geography, arithmetic, writing, drawing, mathematics, and navigation to "young men who are prone to libertinism and heading out to the woods and the traplines."

Map of the Great Lakes attributed to Jean Baptiste Louis Franquelin, circa 1678
This map testifies to Franquelin's talent as an artist and to how he was influenced by the ornamental Dutch style. The Great Lakes are presented as a passage to a more distant destination, the Mississippi River, which Jolliet and Marquette had "discovered" several years before. There is also, between lakes Ontario and Erie, the Onongiara sault, which was later named Niagara Falls.

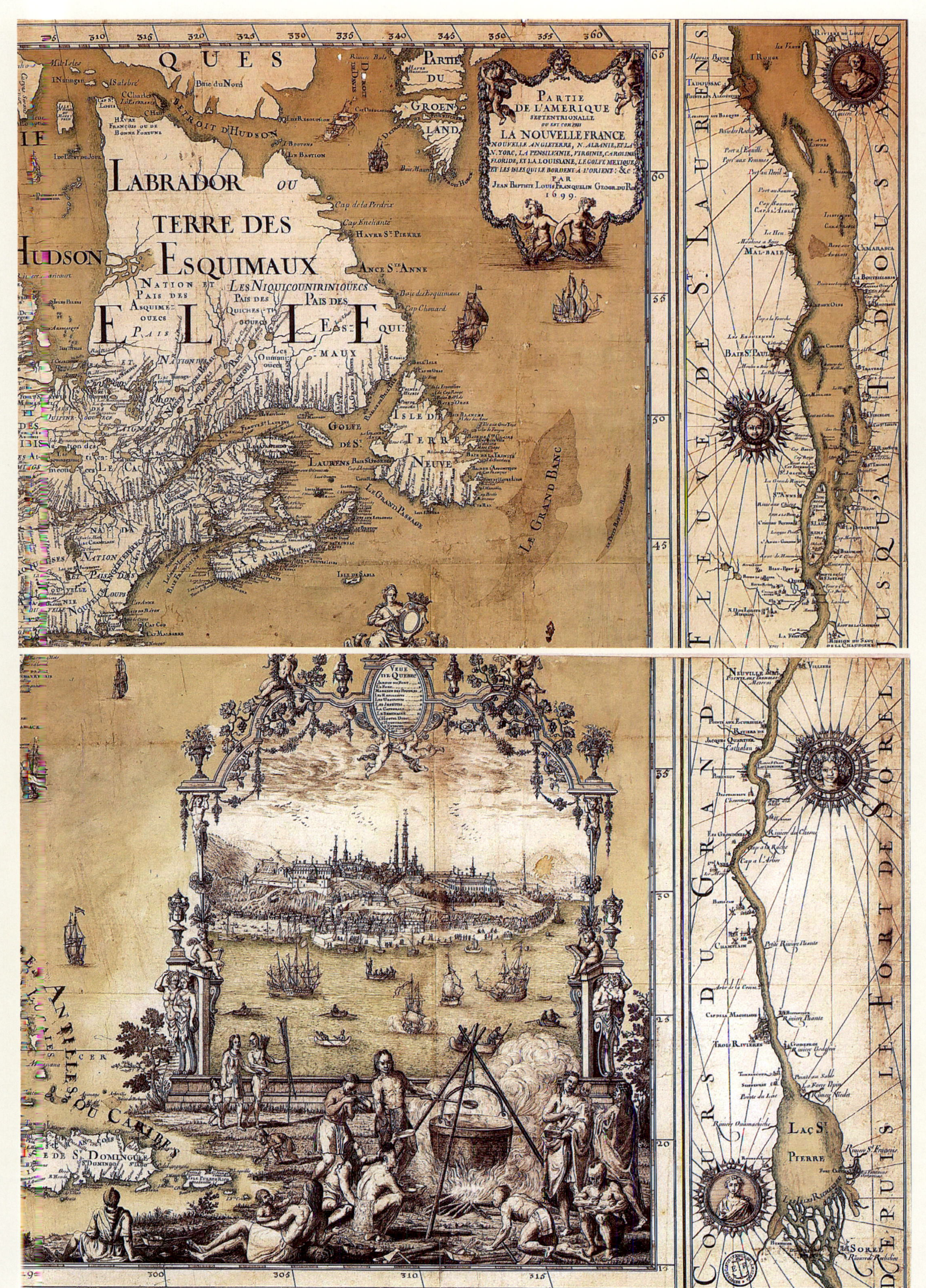

Map of North America by Franquelin, 1699
Although he left the colony in 1693, Franquelin retained the title of king's geographer and continued to be interested in New France, as evidenced by this map, made in 1699. Franquelin still had a talent for contour and drawing, and he knew how to manipulate spaces to make them stand out. In the foreground of the view of Quebec, Franquelin drew an unusual scene of indigenous life – 15 Indians assembled around a large, steaming cauldron – evoking the peaceful coexistence of the French and Indians. This exotic image no doubt pleased the aristocrats of the Court of France. On the left and right borders of the map, Franquelin traced the shores of the St. Lawrence from Lake Ontario to Tadoussac. The scale is large enough to allow the reading of the names of the earliest French settlements, even the smallest hamlets.

Map of North America by Jean Baptiste Louis Franquelin, Quebec, 1688
A concern with aesthetics was one of the highlights of Franquelin's cartography; this map shows a series of unique ornamentations. The symbolic compositions, combined with the delicacy of the lines, give it an overall impression of originality. From the king's crown are hung the necklaces of the Order of Saint-Michel and the Order of Saint-Esprit. The closed crown sits atop a globe, overlaid with three fleurs-de-lys, on which the outlines of New France are visible. Occupying the ocean space is a splendid view of Quebec, projected on a curtain held up by two cherubs, revealing to French administrators the capital of their empire in North America. On the lower left, Franquelin portrays the main tools used to draw his maps. More than 350 place names, solid milestones of French expansion in the continent, are written on the map.

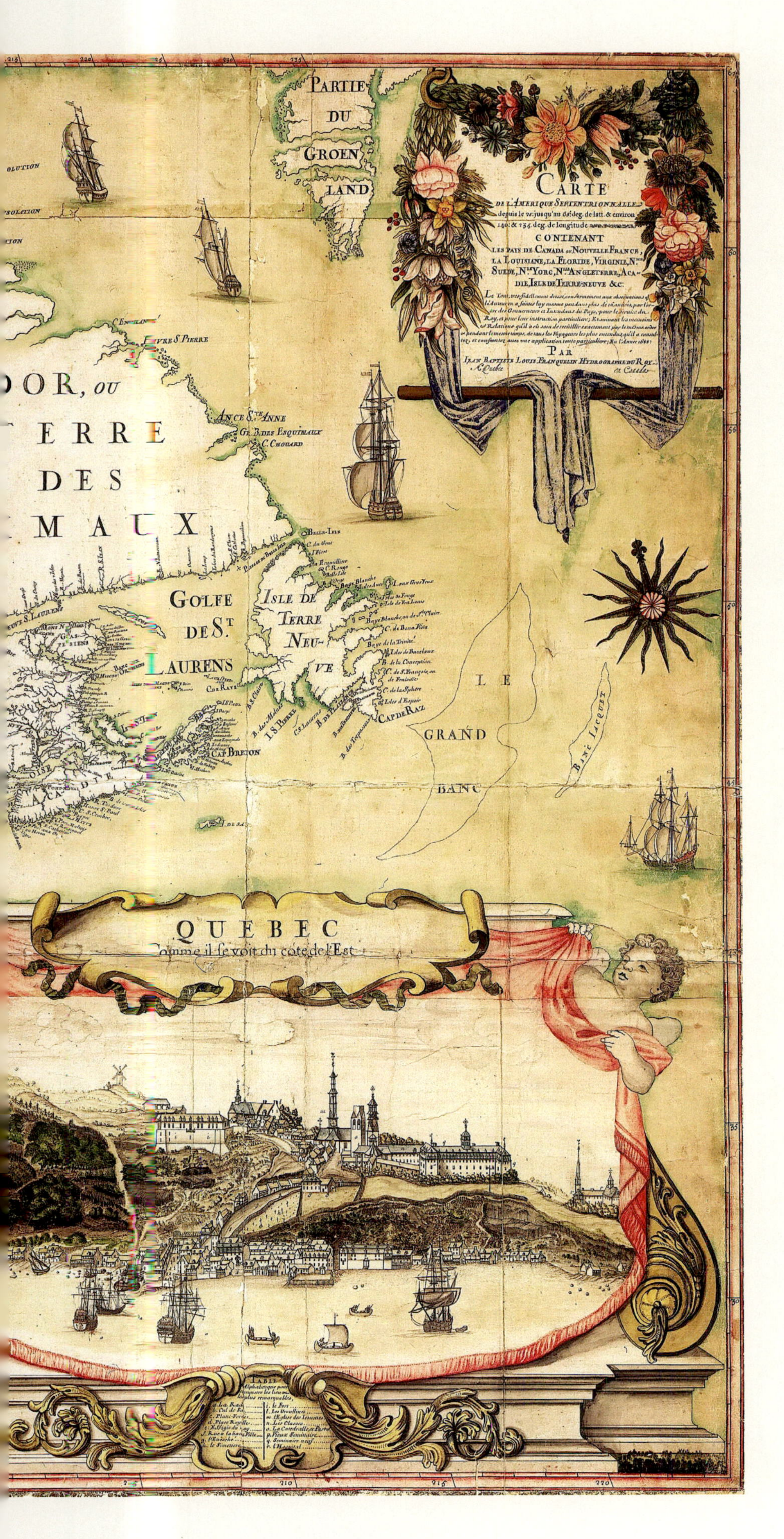

In 1686, Franquelin accompanied the new intendant, Jacques De Meulles, on a voyage to Acadia, taking advantage of this opportunity to draw more-detailed maps of the Atlantic coast. He made a general map of the trip, but also small charts of the harbours that he visited: Percé, Beaubassin, Chibouctou, La Hève, Chédabouctou, Port Rossignol (today Liverpool), and Port-Royal. Two years later, in 1688, he presented his latest work to the Court. The map that drew particular attention was the one of northern America that included a well-rendered view of Quebec and a suggested border between New France and New England. Franquelin then presented a proposal for the cartographic work that remained to be done in the colony, whereupon he was mandated by the king to "make the general map of our New France, visit all the countries where our subjects have been, and even to discover others, draw borders, erect boundary markers, parade our arms wherever needed, reconnoitre the properties of the territories, their climates, and examine the ores and minerals, and all other uses and advantages profitable for business." To finance this ambitious project, Franquelin was awarded the privilege to trade merchandise wherever he went. In New France, of course, the bonds were strong between science and economics. From Champlain to La Vérendrye, the fur trade helped to finance exploration and cartography projects.

Unfortunately, the confidence shown in Franquelin came to naught. In the spring of 1689, war broke out between France and England: Franquelin resigned himself, changed his plans, and began devoting his efforts to the defence of the colony. During the siege of Quebec, he worked as an engineer, supervising the work needed to protect the settlement. In 1692, Frontenac sent him, along with Lamothe Cadillac, to reconnoitre the New England coast to gauge the merits of launching an attack against Boston or Manhattan. The following year, having returned to France, Franquelin received terrible news: the king's ship the *Corrosol* had been wrecked in the Sept-Îles archipelago. His wife and children, on their way to join him, had drowned.

After this tragedy, Franquelin never returned to North America. In 1697, he was working for Vauban, the general commissioner for fortifications. Nevertheless, he continued to be interested in New France and, until he died (date unknown), he produced a number of maps of extraordinary quality.

Franquelin was constantly updating his maps based on information reported by travellers. When he was living in Quebec, close to the seat of colonial government, he was able to intercept voyageurs returning from the west and peruse their travel journals, when they had kept them. He could update his work much more rapidly than the European *géographes de cabinet* could, who published their maps in printed form. During his career, Franquelin transcribed the voyage accounts of a number of famous explorers, among them Louis Jolliet, Jacques Marquette, René-Robert Cavelier de La Salle, Daniel Greysolon Dulhut, Jean Péré, Pierre de Troyes, Pierre Allemand, Louis Hennepin, Pierre LeSueur, Antoine Laumet de Lamothe Cadillac, and Pierre Le Moyne d'Iberville. His work, however, never gained the recognition that it deserved. Because his maps remained in manuscript form, they did not draw the attention of historians and bibliographers. And yet, the most esteemed geographers of Paris, including Coronelli, Delisle, de Fer, and Jaillot, used his work as the inspiration for their depictions of a new America.

Septemtrionaliora
AMERICÆ
à Groenlandia, per
Freta Davidis et
Hudſon, ad Terram
Novam
Baffins Bay
New North
Wales
Buttons
Bay
New South
Wales
STRAET HUDSON
LABORADOR
NOVA BRITANNIA
The Great Bay
Cumberlants Bay
Iland of good fortuyn
De Noordelyckſte Zee kuſten van
AMERICA
van Groenland door de Straet Davis ende
Straet Hudſon tot Terra Neuf

The Northwest Passage

VIA HUDSON BAY

SINCE THEY HAD NOT FOUND A PASSAGE to China south of Newfoundland, seamen tried going north. Knowing that Earth was round, they figured that there might be a shorter northern route to the Pacific Ocean. Thus, the great wave of maritime exploration also unfurled upon the Arctic, in spite of the incredible difficulties that the ice-bound region posed to navigation: strong winds, devious currents, blinding fogs, the brevity of the navigable season, and terrifying icebergs lurking everywhere. Few seafarers had experienced these latitudes or the discomfort of living aboard ships with no heating other than the stove in the galley.

Some historians hypothesize that Gaspar Corte Real sailed as far as the entrance to Hudson Strait and Davis Strait in 1500–01. The alleged voyage of the Venetian brothers Nicolo and Antonio Zeno to the islands of the North Atlantic, or "Estotiland," around 1380, and their account, published by a descendant in 1558, made a huge impression on sixteenth-century English explorers. The imaginary locations shown on the map accompanying the account were even used by Mercator (1569) and Ortelius (1570) in their maps. This fiction, situating the north coast of North America south of 60° – thus easily accessible – explains in part the infatuation with finding a passage to Asia in this region.

At this very time, England, ruled by Elizabeth I (1558–1603), was preparing to set out on Atlantic trade voyages for the enrichment of both the Crown and individuals. The gold-filled galleons of the Spanish merchant marine had caught the imagination of English speculators, who eagerly invested their capital in exploratory expeditions. This new conquering spirit had solidly respectable intellectual foundations, such as the scientific research of John Davis and the colonial theories of John Hawkins. In addition, England had a major asset with its highly qualified and daring seamen, who had performed brilliantly in various naval battles, including the one in which they defeated the Spanish Armada (1588). In fact, as Spain's naval strength waned, England became the master of the seas.

English exploration picked up again thanks to Sir Humphrey Gilbert, whose pamphlet *A Discourse of a discoverie for a new passage to Cataia* was published in 1576, the year of Martin Frobisher's first expedition. Gilbert's map, a cordiform projection, illustrated the conviction that a short route existed from England to the Moluccas via the northwest. Frobisher, a military officer who had trafficked in slaves until 1576, was drawn to the prospect of unknown seas. Unscrupulous and colourful, neither a top-notch navigator nor an experienced explorer, he nevertheless proved to be an intrepid, imaginative, and popular captain. Under the queen's patronage, he made three voyages to the Arctic. In 1576, he sailed north along the coast of Labrador to 63°, where he entered a large bay that he took for the passage to China and named after himself. The Indians whom he met wanted to trade, but Frobisher tried to persuade them to guide him to the "western sea." Sailors from his crew left to prepare for the expedition with the "Eskimos," but they did not return to the ship quickly enough and he was forced to weigh anchor to avoid being caught in the ice. The misunderstanding between the English and the Inuit lasted a long time.

Believing that he had found gold, Frobisher brought back from his second voyage (1577) no less than 200 tons of ore, which turned out to be only marcasite. Like many other explorers, he also took Indians back to Europe. The spectacle of an "Eskimo" paddling his kayak on the Thames was an extraordinary success; Shakespeare even alluded to it in *The Tempest*. The presence of the "Indians"

Hudson Bay according to Frederick de Wit, Amsterdam, 1675 (detail)
This map is the work of the Dutchman Frederick de Wit, one of the most prolific cartographers of his time. The Dutch were masters in the arts of engraving and cartography and dominated the market for map publishing from the late sixteenth century up to the end of the seventeenth century. Mercator, Hondius, Blaeu, Ortelius, and then de Wit helped to forge a tradition of excellence in the field. Noted for their exceptionally beautiful wall maps, the Dutch also had a corner on the market for atlases. Amsterdam was the economic centre of Europe at the time; the Dutch merchants' trade network extended to Africa, Asia, and the Americas, from which the Dutch East Indian and West Indian companies brought back great riches. This map, no doubt taken from an atlas, presents Hudson Bay covered with compass points and rhumb lines, the functions of which were both aesthetic and utilitarian. A rhumb is the angle between two of the 32 divisions of the wind rose and measures 11°15'.

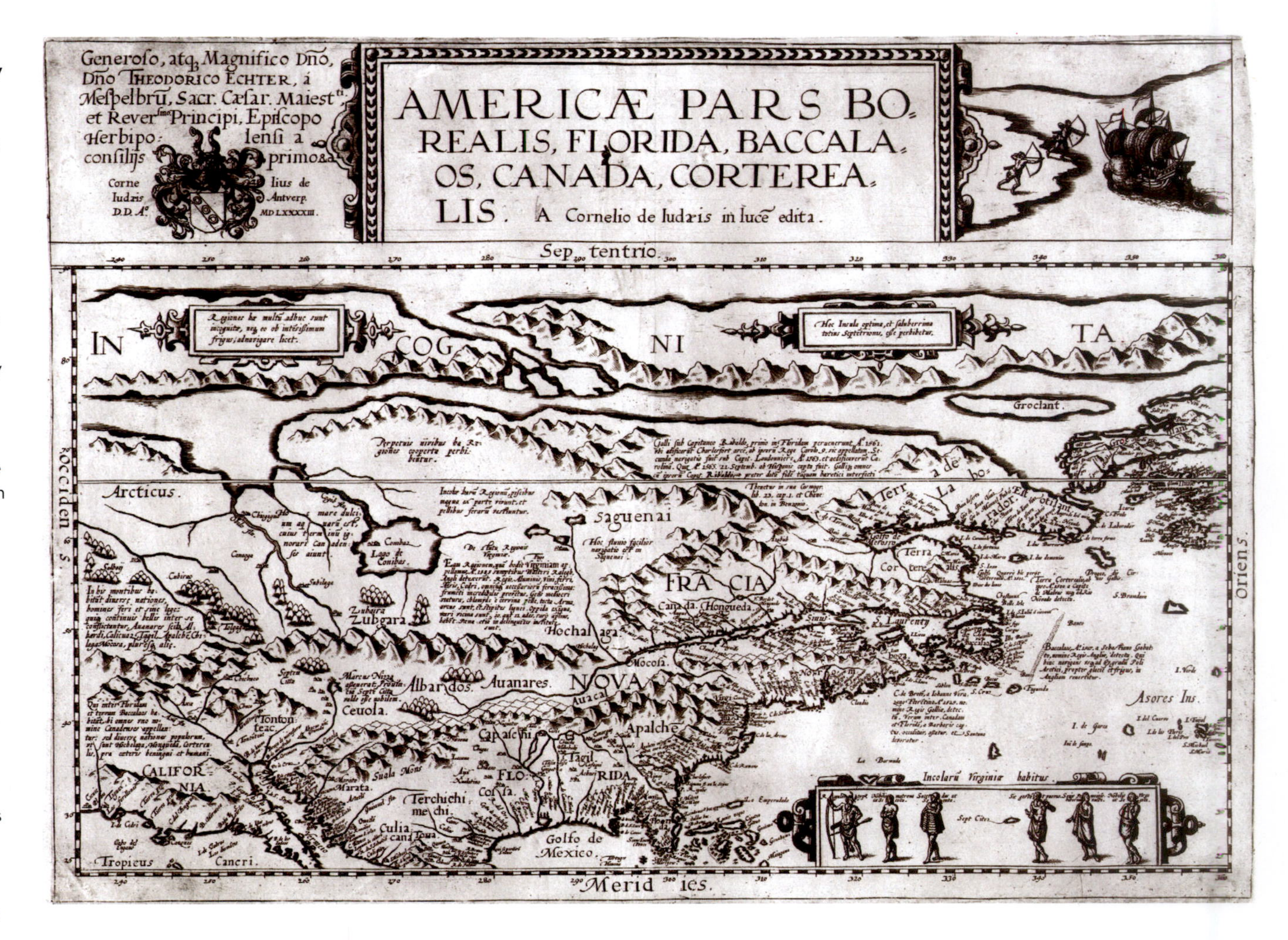

***Americæ Pars Borealis, Florida, Baccalaos, Canada, Corterealis* by Cornelis de Jode, Antwerp, 1593**
This map by the Flemish publisher and cartographer Cornelis de Jode illustrates what the Europeans desperately hoped to find north of NOVA FRANCIA: an unimpeded passage to Asia. However, it opens the possibility, in an inset in the upper right-hand corner, of the difficulties that European explorers might encounter in contact with hostile peoples: what is presumably Martin Frobisher's ship is being repulsed by Indians armed with bows and arrows. A resident of Antwerp, the largest port in Europe at the time, Jode based his maps on the most up-to-date information from multiple sources. On the east coast of North America, the influence of two maps published in Theodor de Bry's *Les Grands Voyages* – John White's *Americae Pars, nunc Virginia* (see p. 66) and Le Moyne de Morgues's *Floridae Americae Provinciae* (see p. 59) – is evident. In the lower right-hand inset, Jode copied Indians first drawn by White. Again on the east coast, he evokes the misadventures of Verrazzano, who was captured, killed, and devoured by Indians. In Florida, he locates the sites where Ribault and Laudonnière landed, as well as the main areas explored by the French: FORT CAROLINA, P. REGALIS (Port-Royal), R MAYO (May River), and CHARLEFORT. In the southwest part of the continent, he portrays the expedition of a Franciscan from Nice, Brother Mark (better known by his Spanish name, Fray Marcos de Niza), who brought back a fabulous description of the mythical kingdom of the Seven Cities (SEPTE CITTA) of Cibola (CEVOLA).

in London was fleeting, however, as they all died less than one month after their arrival. Although Frobisher did not cross the northwest on his third voyage (1578), he made major inroads, clearing the way for mining projects and trading posts to be introduced to the North.

A few years later, in 1585, John Davis, a navigator with an education in the sciences of the sea, searched farther south than had his predecessor, on the coast of Labrador, but also farther north, along what was to be called Baffin Island. On his third voyage, in 1587, Davis sailed along Greenland to Sanderson Hope, above 72°N. Believing that he had found the passage to China through Cumberland Sound, he calculated that this corridor was nearby, about one degree of latitude away, in Lancaster Sound. His detailed descriptions of the coasts, other geographic observations, and accounts of discovery were recorded by Emery Molineux on a terrestrial globe engraved in 1592. In the pamphlet that Davis published in 1595, *The Worldes Hydrographical Description*, he tried to prove that a northwest passage existed and was obstacle-free, and that the climate was tolerable in the region. He also reported that there were rapids and unpredictable variations in compass declination that previous nautical charts had not taken into consideration. Davis thus laid the scientific foundations for navigation in polar seas. A great innovator, he invented the backstaff, also known as the Davis quadrant, which was the official instrument for assessing latitude for the next century and a half, until the invention of the quadrant of reflection, in 1731.

Confident that the North was full of riches, speculators formed new trading companies to continue financing voyages of discovery: the East India Company, in 1600, and the North West Company, which received its royal privilege in 1612. During the first third of the seventeenth century, British explorers concentrated essentially on Hudson Strait. The first two to take their chances there were George Weymouth and John Knight. In 1602, Weymouth, certain of his future success because he had obtained from the queen a letter addressed to the emperor of China, navigated to 62°30'N, entered Hudson Strait, which he took for a bay, and then returned southward. In 1606, John Knight landed in Labrador and got lost; he was never seen again.

When he left, in his turn, to find a northwest passage, Henry Hudson was an experienced navigator who had already made three voyages to North America, as well as others to Spitzberg and Greenland. In 1610, he entered the strait and then the bay named after him. Advancing southward, in October he found himself at the bottom of what would later be called James Bay, and he was forced to winter over in an "endless labyrinth," probably near the Rupert River. Knowing neither the duration nor the harshness of the winter, and naturally not knowing how to cope with it, a number of his men, already tired and weakened by the long crossing, soon succumbed to scurvy. Hudson began rationing provisions without telling the crew how much food was available, raising in their minds the possibility that he was reserving food for himself. At this point, some of them apparently thought about ousting Hudson and jettisoning the sickest men in order to keep the provisions for those who were still able-bodied.

In the spring of 1611, when the game reappeared, the sailors obtained from the Indians the bare minimum that they needed to survive, but they revolted against Hudson and refused to explore the west side of the bay. Abandoning Hudson, his son, and six other members of the crew in a boat near the coast, they returned to England under the command of Robert Bylot. When they arrived in Europe, there remained only eight survivors out of the 30 who had left, and they were so weak that "only one of them had the strength to throw himself on the rudder and steer." Neither Hudson nor any of his companions were ever found.

Nonetheless, the English were convinced that Hudson had found the route to China. To fund ensuing voyages, the Company of the Merchants Discoverers of the North West Passage was formed. In addition to looking for a navigable route to Asia, these explorers received instructions from King James I to carefully observe all hydrographic aspects, such as height, compass declination and variation, and direction and strength of the tides. Thomas Button, in 1612–13, and William Gibbons, in 1614, concluded when they returned from their expeditions that if a passage existed, it would have to be found north of Hudson Bay.

In 1616, William Baffin, who had been to Greenland a number of times, took Gibbons's recommendation and sailed directly to 77°45'N, a latitude that would not be reached again for 236 years. He noted the three main exits from Baffin Bay and the Arctic archipelago, without knowing that one of these straits, Lancaster Sound, would later be revealed as the entrance to the Northwest Passage. Paradoxically, Baffin's expedition, the farthest north made up to that time, triggered disillusionment among backers of the age-old search for a northwest passage. With no solid evidence that it existed, the financiers turned instead to a confirmed source of wealth, the furs of Hudson Bay.

Denmark was the only country apart from England to sponsor an expedition. In 1619, the Dane Jens Munk received a commission from his king to seek a northwest passage to India. Aware of the previous voyages, he headed for Hudson Bay, which he reached after several involuntary detours in Frobisher Bay and Ungava Bay. After crossing the mouth of the Churchill River, he settled in to winter over. Taken by surprise by the intense cold and the paucity of local food resources, the crew fell victim to scurvy, and by June only Munk and two other sailors were left of the 61 who had started out. This was the first and last Danish expedition in search of the fabled passage.

The British sent their vessels to this region three more times. In 1625, William Hawkeridge failed to recognize Hudson Strait, even though he had previously crossed it with Thomas Button. In 1631, two parallel expeditions, those of Thomas James and Luke Fox, visited Hudson

Hudson Bay according to the voyages of Henry Hudson, Amsterdam, 1612
In the late sixteenth century, the North Atlantic, from Newfoundland to Davis Strait, seemed to be the preserve of English navigators. Davis, Frobisher, Baffin, Weymouth, Button, and Hudson were among the most famous of those who tried, in vain, to solve the secret of a passage to Asia at the top of North America – the famous Northwest Passage. In 1610, one year after his voyage of discovery on the Hudson River (in today's New York state), Henry Hudson was mandated by the North East Company to find this passage. It was his last mission. He sailed up the strait that now bears his name, which took him to a huge bay. He thought that he had reached the Pacific and set his sails to head south. To his great disappointment, he ended up in a new bay with no exit, the future James Bay. His crew mutinied and Hudson was abandoned by several of his men, who managed to get back to England. One of them, Abacuk Pricket, gave an account of the extraordinary adventure and produced a map of the regions explored. Probably drawn by Hudson, the map was printed by Hessel Gerritsz, who distributed it throughout Europe. Champlain referred to it extensively in making his own maps.

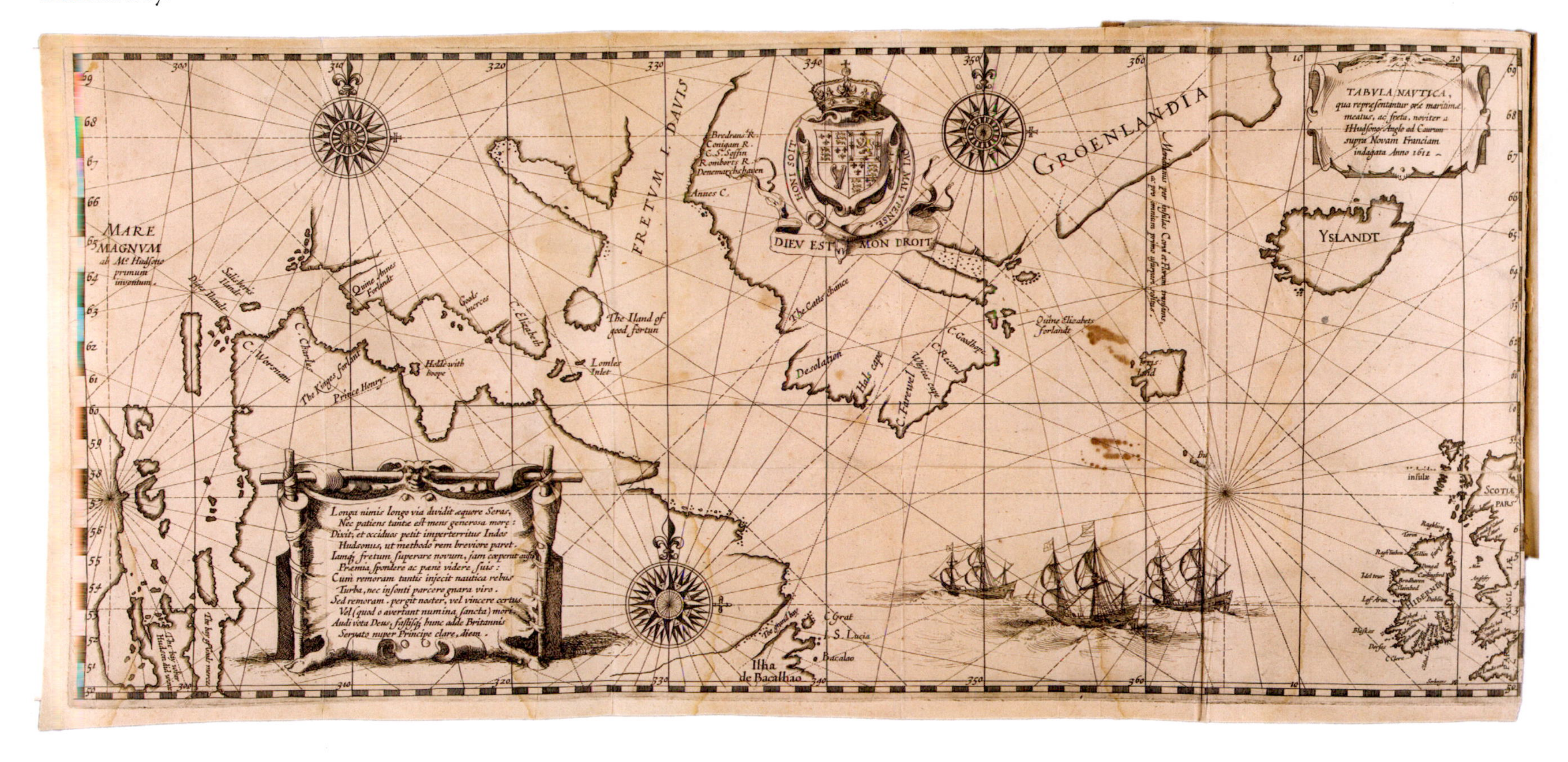

Bay. Fox entered the bay in early spring, on May 23, explored the entire west coast, sailed north to 66°47'N, and returned to England on October 31 with his entire crew – which, given previous occurrences, was extraordinary. James was not as lucky. After exploring the bay that was to bear his name, he endured a particularly rough winter, had to make repairs to his ship, and lost a number of men. However, both Fox and James demonstrated, once again, that the "western sea" could not be reached via Hudson Bay and that the surrounding land, full of marshes and rocks, was utterly inappropriate for colonization. The voyages of exploration were therefore halted, although Europeans continued to visit the North Atlantic to hunt whales and seals. On the shore of Hudson Bay, a tight network of fur-trading posts was established. The Hudson's Bay Company, which received its charter in 1670, ran a very lucrative business there, in spite of the competition from the French who had settled in Canada.

At the same time as England was sending its explorers to northwest North America, France was becoming established in the gulf and valley of the St. Lawrence and gathering information on the surrounding territory. During Samuel de Champlain's first voyage in 1603, Indians described to him a large saltwater body serving as the outlet for the rivers flowing toward it. Champlain was thinking not of the Sea of China but of "some gulf of this our sea, which overflows in the north." He probably knew about Wytfliet's 1597 map showing Lake Conibas, corresponding to what he called "the northern gulf," close to the site of Hudson Bay. Seven years before Henry Hudson's voyage, Champlain had briefly described the large expanse of saltwater that is Hudson Bay and the water routes leading to it that could be used as profitable trade circuits. However, he never saw the bay himself; instead, he relied on the notes and maps that the survivors of the Hudson expedition had brought back to England and Holland in 1610.

Absorbed by the exhausting work and struggles with establishing the settlement in the St. Lawrence Valley, people in the new colony, which was administered by trade companies, had very little time to launch explorations in search of the "northern sea." It wasn't until 1661 that the Jesuits Claude Dablon and Gabriel Druillettes, who had already travelled through most of New France and were as interested in geography as in evangelization, decided to find out whether the "northern sea" was somehow linked to the "western sea" or the "southern sea." From Tadoussac, they travelled up the Saguenay River to Chicoutimi, crossed Lake Saint-Jean, and took the rivers up to where the waters divided, where the Montagnais guides stopped and turned back for fear of encountering hostile nations.

In the late seventeenth century, the intendant of New France, Jean Talon, intrigued by the activity of the English at Hudson Bay, sent emissaries to verify whether it was really the northern sea and, if so, to take possession of the territory. The Jesuit Charles Albanel was chosen for this mission because, according to Father Dablon, "For a long time he has laboured among the Savages who have knowledge of this sea, and who alone may be the guides to these routes up to now unknown." Knowing that Minister Colbert opposed exploration and the expansion of the French territory in North America, Intendant Talon justified his decision by the exigencies of trade and security: "Three months ago I despatched Father Albanel, a Jesuit, and Monsieur de St. Simon, a young gentleman of Canada ... they are to push forward as far as Hudson Bay, present written reports on all they discover, open up the fur trade with the savages, and above all to reconnoitre to see whether there is cause to erect a few buildings for the winter there to make a storage from which ships might eventually be victualled if they should pass that way to discover the route between the two seas of the North and the South." Albanel made two expeditions to the Rupert River side of James Bay. After his second voyage, in 1674, the English, displeased that the missionary was diverting their Indian suppliers toward the French, returned him to London. Albanel later returned to the settlement at Quebec, but never to Hudson Bay.

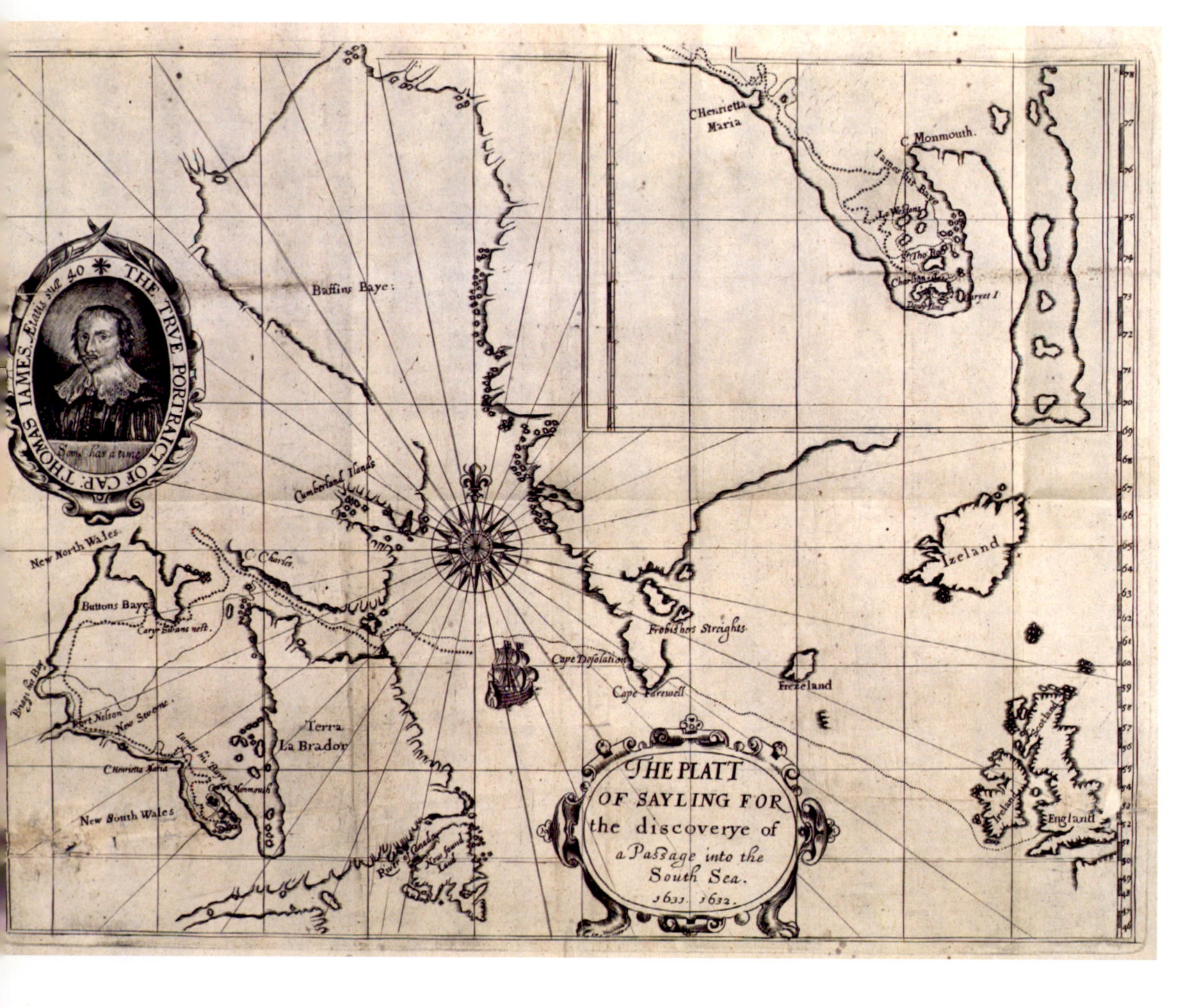

Hudson Bay according to the voyages of Thomas James, London, 1633
Hired by the Society of Merchant Venturers of the City of Bristol to lead an expedition toward the Northwest Passage, Thomas James entered Hudson Bay in July 1631, and then went on to explore the west coast of the bay. He named the New Principality of South Wales, the New Severn River, and Cape Henrietta Maria (in honour of the queen). After looking in vain for a watercourse leading to the St. Lawrence, he wintered over in a roadstead off Charlton Island. During the winter, four men died of scurvy. The rest of the crew regained their health by eating grass in the spring. In the summer, James laid claim to the site by planting a cross to which he attached a letter affirming the taking of possession and including a picture of Charles and Mary, king and queen of England and Scotland.

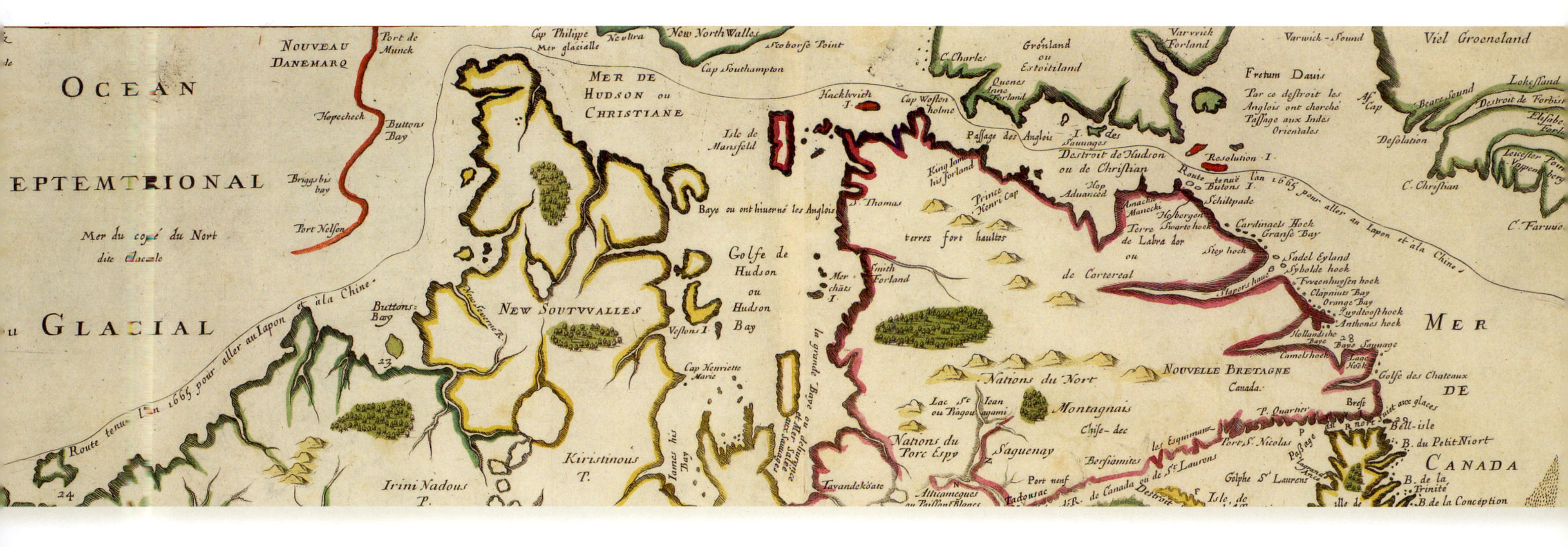

Up to the end of the seventeenth century, several other French expeditions were sent to attempt to supplant the English. The most spectacular, for both its size and its failure, was probably the one led by Laurens Van Heemskerk, a Dutch navigator who had previously gone to the Canadian Arctic in 1669 on the *Wivenhoe*, chartered by the brand-new Hudson's Bay Company, along with at least one of two French informers, Pierre Radisson or Médard Chouart Des Groseillers. A talented adventurer, spy, and storyteller, Van Heemskerk provided a very convincing description of the new territory discovered at 51°N, which he had the impertinence to name North Florida, and he subsequently received a commission from Colbert and three ships to explore this region of North America and take possession of it. He left Dunkirk on August 14 after a laborious fitting out that revealed his manifest incompetence; navigating northwest, he reached 58°N without ever seeing land. After suffering through a violent, damaging storm, the crew asked to return to Europe. The fleet turned southeast and arrived at Brest on September 30, after a month and a half at sea. In spite of his obvious failure, Van Heemskerk continued to assert that he was the only one who could find the northern route to the "southern sea." Colbert did not renew his commission, but the illusion lingered, and in 1697 Pontchartrain, the minister of the navy, asked him for clarifications on his discoveries. Van Heemskerk, then 65 years old, enthusiastically replied that he was ready to leave again, certain that he would find the passage. This plan was never taken up, and the imaginative navigator died in Brest in 1699.

In the final years of the seventeenth century, French claims to Hudson Bay and the Arctic gave rise to a number of expeditions to seize English trading posts. In 1686, Canada, which had invested considerable funds in the Compagnie du Nord, launched a first military offensive against Hudson Bay. Under the command of Chevalier Pierre de Troyes, 100 men went to seize the English Monsoni, Rupert, and Albany forts. Three of Le Moyne's brothers, including d'Iberville, participated in this extraordinary military expedition.

The approach was made by both land and river. The troops departed from Montreal and travelled up the Ottawa River, then through a series of lakes and watercourses, linked by exhausting portages, to reach the Moose River. After 85 days of constant danger, the men reached the south shore of James Bay. They launched an assault on Fort Moose, capturing it and three other forts on Hudson Bay. Commander de Troyes then designated d'Iberville to command the 40 men who were to remain on site. At the end of the summer of 1687, d'Iberville returned to the St. Lawrence Valley with a few companions; he then went to France, where he obtained from the secretary of state for the navy a cargo of trade merchandise to be used to tempt away the Indians of Port Nelson, the main English trading post. Emboldened by his previous success, d'Iberville returned to James Bay during the summer of 1688 on the *Soleil d'Afrique,* an excellent ship allocated by the Court to the Compagnie du Nord. He organized the transport of the furs that had accumulated there during the year; the English, who were trying to retake their trading posts, objected strenuously. In September 1688, with three English vessels trapped in the ice, the battle for the posts on the bay was put off to the following spring.

It was an unusual situation: the adversaries had to face winter together in an environment that was hostile in every way and in which survival was the main preoccupation. D'Iberville nevertheless plotted to preserve the advantages of his position. For instance, he refused the English permission to hunt; deprived of fresh meat, many of them developed scurvy. The 16 Canadians, as

***Le Canada faict par le Sr de Champlain* by Pierre Du Val, Paris, 1677**
Champlain's cartographic work had a major influence on his successors, as evidenced by this map published 45 years after his death. The mapmaker, Pierre Du Val, promoted his work well: he openly cited the "father of New France," whose written works were still very popular. Historians presume that for this map Du Val used a copper plate engraved for Champlain's accounts but never before printed. To the original contours, Du Val added new information, such as the names of the English colonies and indigenous peoples. He published a number of editions between 1653 and 1677. This one is particularly interesting for its tracing of a route north of Hudson Bay and along the west coast of the Arctic, which shows a passage discovered in 1665 leading toward China. Like his father-in-law, the great cartographer Nicolas Sanson, Pierre Du Val was appointed king's geographer.

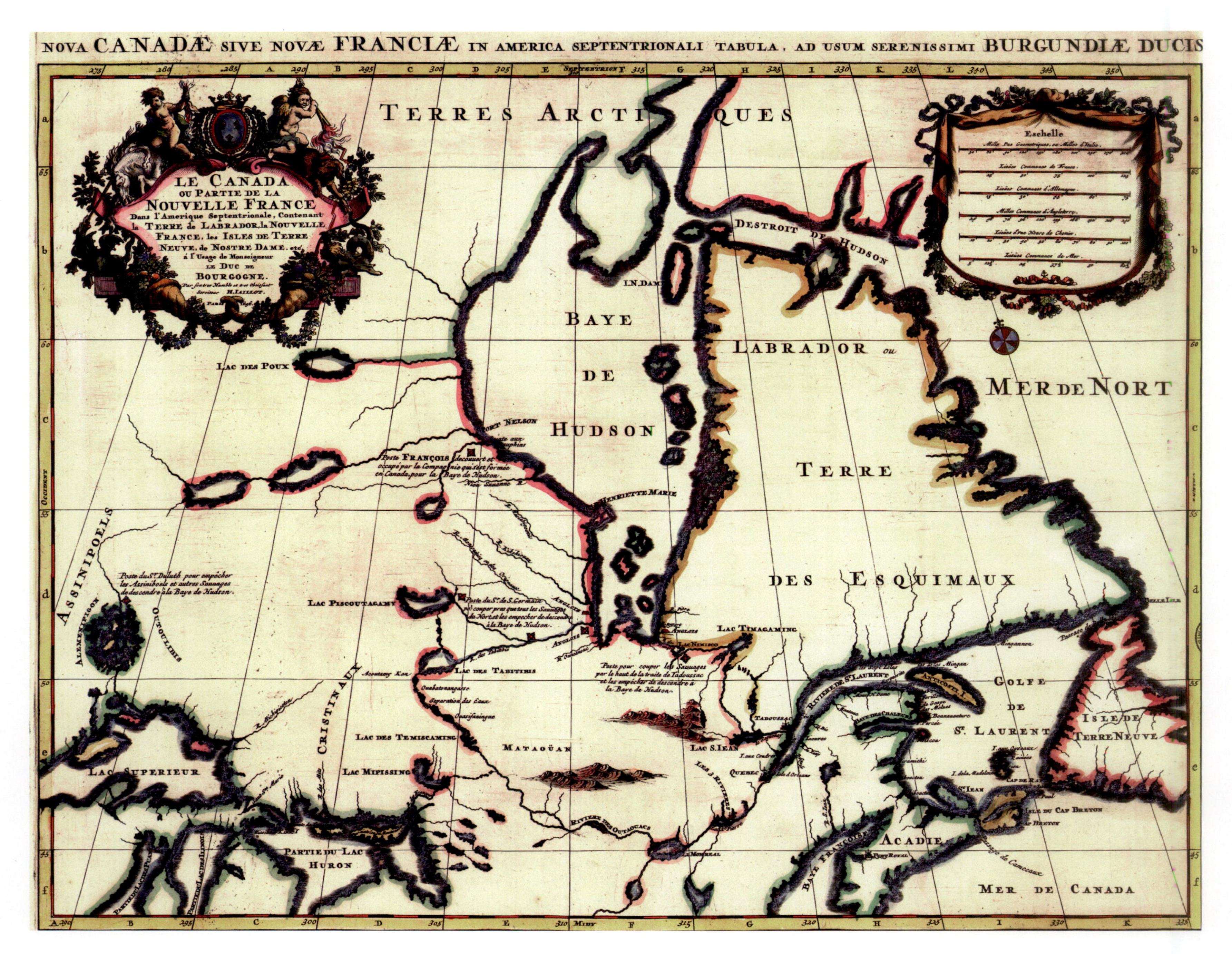

***Le Canada ou partie de la Nouvelle France* by Alexis-Hubert Jaillot, Paris, 1696**
In response to French-English rivalries at Hudson Bay, several Canadian notables founded the Compagnie du Nord in 1682. The company erected a series of fortified posts to prevent Indians from "descending to trade at Hudson Bay." As shown in this map, published in 1696, there were four of these forts, located on Lake Nemiscau (NIMISCO), Lake Abitibi (PISCOUTAGAMY), Lake Nipigon (Alemenipigon), and the Abitibi River (TABITIBIS). They were found along the intersection between the old copper route and the watershed between the St. Lawrence and Hudson Bay basins.

determined as their leader, were, by all evidence, better prepared for this type of confrontation than were the 85 employees of the Hudson's Bay Company. After 28 of their men died, the English were forced to surrender. D'Iberville returned to Quebec on October 28, 1689, with prisoners and a considerable booty of top-quality furs.

The success of this operation won d'Iberville two more commissions to command all the waters in northern New France. He took his ships to Hudson Bay in 1690–91, in 1692, and in 1694, when he received a trade monopoly. Finally, in 1697, with a single ship with 44 cannons, the *Pélican*, he sank two English warships, sent another fleeing, and obtained the surrender of the governor of Hudson Bay, Henry Baley, after only five days of battle. This last mission to Hudson Bay was the most rapid and brilliant victory in the career of Pierre Le Moyne d'Iberville. He retained his trade monopoly in the region until 1699, which made him a wealthy man. Thanks to his spectacular exploits, the French posts at Hudson Bay were not threatened in the following years and remained French property until the signing of the Treaty of Utrecht, in 1713.

Main sources

ALLAIRE, Bernard, and Donald HOGARTH. "Martin Frobisher, the Spaniards and a Sixteenth-Century Northern Spy." *Terrae Incognitae: The Journal for the History of Discoveries*, 28 (1996), 46–57. The diplomatic background for explorations in the Arctic. — LITALIEN, Raymonde. "L'exploration de l'Arctique canadien aux XVIIe et XVIIIe siècle." In France, Comité de documentation historique de la marine, *Communications 1988-1989*, pp. 203–26. Vincennes: Service historique de la Marine, 1990. One of the author's many texts on this subject. — ROYAL ONTARIO MUSEUM. "*Up North*": *The Discovery and Mapping of the Canadian Arctic, 1511–1944*. Toronto: Royal Ontario Museum, 1958. Maps essential to the search for the Northwest Passage.

King Louis's Country

THE FRENCH IN LOUISIANA

France was deeply divided on the issue of territorial expansion south of the Great Lakes. On the one hand, King Louis XIV wanted to concentrate the population in the St. Lawrence Valley in order to avoid the costs of maintaining and defending remote outposts. On the other hand, those in the fur trade, which was financing the colony, had to seek resources farther and farther from Quebec. In addition, the king coveted the ores found by Spain in the southern part of the continent, to which his Spanish wife and their descendants could become heirs. With the urgings of Intendant Jean Talon, a fervent proponent of exploration, and Governor General Frontenac, the authorities began encouraging "new discoveries."

There was no lack of volunteers willing to venture into regions still unknown to Europeans. One of them, René-Robert Cavelier de La Salle, had been in Canada in 1666, criss-crossing the Great Lakes region to conduct trade. He dreamed of discovering "the great river" (the Mississippi) and finding the way to the "southern sea and, through it, the route to China." With this objective, in 1669 La Salle joined the expedition of the Sulpician priests François Dollier de Casson and René de Bréhant de Galinée, who were setting out in search of the "beautiful river" (the Ohio) of which the Indians spoke, intending to found a mission there. La Salle was awarded the concession of Fort Cataracoui (Kingston) on Lake Ontario, along with letters of nobility that included two seigneuries, one at the entrance to Lake Erie, the other at the exit from Lake Illinois (Lake Michigan) – starting points for navigable routes to the middle of the continent. On May 12, 1678, the ambitious young man received a commission to "discover the western part of North America between New France, Florida, and Mexico." Three years passed before he embarked for the Mississippi – three long years of fundraising, false departs and restarts, forts set afire or seized by creditors, losses, thefts, discord, and desertions. Also during that period, the region intended for exploration had been devastated by war between the Illinois and the Iroquois.

Although his preparations for the expedition were problematic, La Salle and his 30 men nonetheless travelled all over the region around the Great Lakes and the sources of the Mississippi. Henri de Tonty, La Salle's trusted lieutenant, navigated the eastern shore of Lake Michigan in 1678–79. In 1679, Louis Hennepin, a Recollet missionary seeking the sources of the Upper Mississippi, went as far as 46°N, where he was stopped by a "waterfall," one of the many in the region, and taken prisoner by the Nadouesioux. In 1680, a land route between Lake Michigan and Lake Erie was discovered during an incredible march of more than 500 leagues. With five of his companions, La Salle made "the most toilsome [voyage] that any Frenchman has undertaken in America." They travelled "through forests so thickly interlaced with thorns and brambles that in two days and a half he and his men got their clothing torn to shreds, and their faces so scratched and bloody that they were not recognizable."

At the same time, a competitor was pursuing the same objectives: Daniel Greysolon Dulhut was travelling northwest of Lake Superior with seven Frenchmen and three Indian slaves. Dulhut wanted to convince the Indian nations to trade with the French and not with the English of Hudson Bay. After taking official possession of several locations in Sioux territory, in September 1678 Dulhut signed a peace treaty with all of the nations concerned. In spite of his extremely diplomatic approach, he obtained neither assistance nor permission from the Court of France to return to his explorations in the west because, unlike La Salle, he did not enjoy the support of powerful Court patrons.

Cartouche of the map on the following pages.

In early January 1682, La Salle finally left with 23 Frenchmen and 18 "Savages." He followed Marquette and Jolliet's route to the mouth of the Arkansas River, where he was amazed by the gentleness and fertility of the land. As they travelled southward, the explorers were welcomed by the Taensas, the Natchez, and the Coroas. On April 6, the expedition found itself in saltwater marshes where the river branched in three directions and explored each one to its delta at the Gulf of Mexico. On April 9, 1682, at 26°N the explorers celebrated very officially the taking possession of Louisiana. A cross was raised, bearing the arms of the king of France cut into a casting made from a cauldron. After three musket volleys and a cry of "Long live the King," La Salle advanced, dressed in a scarlet coat trimmed with gold and a great hat, and bearing his sword at his side. He read the taking of possession "in the name of His Majesty and the successors to his crown, of this country of Louisiana." The official document was registered by the notary present, Jacques de la Métairie, and countersigned by all of the Frenchmen on the expedition. La Salle had waited too long for this moment not to surround himself with pomp. The next day, April 10, 1682, he started back. Returning to the Great Lakes by the way he had come, he headed for Quebec to announce the news to the governor general, and then he continued on to France.

Seignelay, Colbert's son and his successor at the ministry of the marine, was convinced to assent to a plan to return to the Mississippi by sea, via the Gulf of Mexico. On April 10, 1684, La Salle obtained from the king a commission to command all the territory that he might discover. On July 24, 1684, his expedition left La Rochelle with 320 people, both military men and settlers for the new colony, in four ships. At the beginning of the voyage, disagreements arose between the naval officers and the commander. Once in the Gulf of Mexico, La Salle did not recognize the locations he had visited two years earlier; his ship drifted eastward and ran aground in Matagorda Bay (Texas). Relations among the men grew strained. After the store ship *L'Aimable* was wrecked, part of the crew returned to France on the *Joly*. In 1685, a number of workers died during the construction, in a particularly noxious marsh, of a small fort called Fort Saint-Louis. La Salle nevertheless persisted in exploring rivers at random with the *La Belle,* which also ran aground. Without a ship, and with his men furious or discouraged, he had to return by land. He lost his way, retraced his steps, fell ill, saw members of his expedition kill each other, then perished himself, assassinated by one of them on March 19, 1687, north of what is now Texas.

A few survivors stayed on near the Gulf of Mexico, but the majority arrived in Montreal on July 13, 1688. Much has been said and written about the controversial

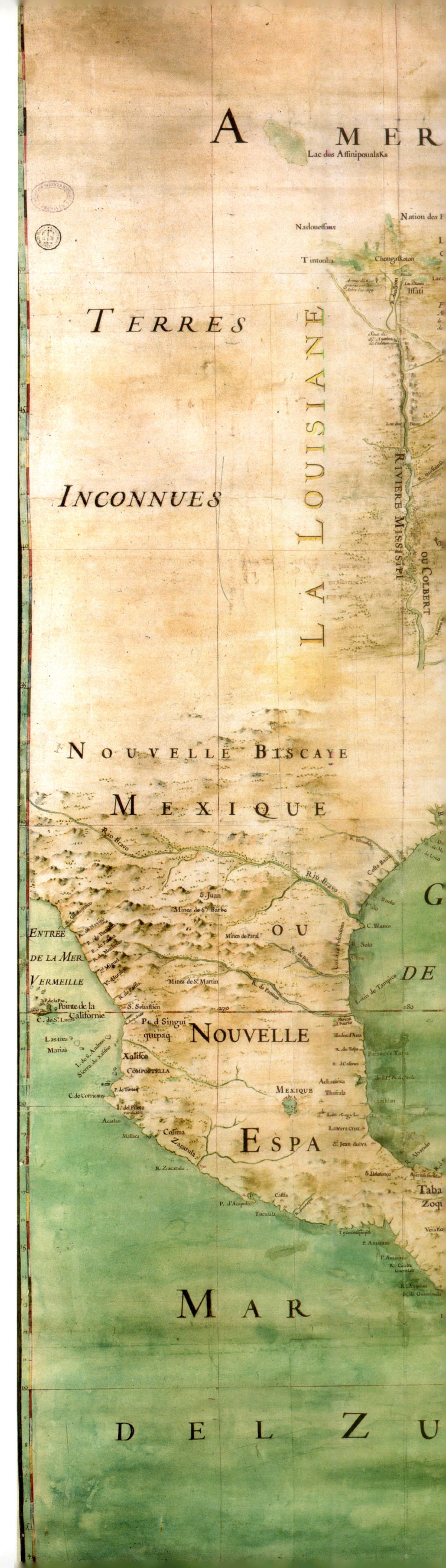

***Carte de l'Amérique septentrionale*, attributed to Claude Bernou, circa 1682**
This immense map of North America is the work of an artist who was an expert with a paintbrush and washes. The choice of colours, variations in shades simulating relief features, clear and elegant calligraphy, and spots of red and blue drapery make this a handsome work of art. The Baroque cartouche is also original, and integrates well with the ocean. Tormented people, illuminated by a cross, are perched on a globe that is emerging from the water. In spite of the obvious care taken in making this map, it is neither dated nor signed. Although he is not known to have produced any other work with this degree of refinement, Claude Bernou, a Recollet priest who was very interested in French activities in North America, has been credited with its production. In the Court of France, he supported the exploration and colonization plans of the very intrepid Cavelier de La Salle. Bernou freely admitted that he hoped to obtain an episcopal mission in New France. Whether or not he made this map, it presents territories known to the French around 1682, including the Upper Mississippi and the Sioux country explored by another Recollet, the missionary Louis Hennepin. This map strikingly portrays the border between known and unknown lands. The line of the Mississippi stops short at the confluence with the Ohio, showing that the area farther south was still unexplored. This map, which very likely preceded La Salle's voyage to the mouth of the Mississippi, bears the name Louisiana in capital letters. If the date is accurate, this would be the first document to call the region by this name.

Mer Christiane
Terre de Labrador
ou de Cortereal
Ocean Septentrional
Golfe de Hudson
Septentrionale
Nations du Nort
Nouvelle Bretagne
Esquimaux
Saguenai
Canada
ou
Nouvelle France
Riviere St. Laurens
Golfe de St. Laurens
Golfe de S.t Louis
Isle de Terre Neuve
Grand Banc de Terre Neuve
Acadie
Nouvelle Angleterre
Nouvelle York
Iroquois
Lac Erie
Mer de Canada
Mer de la Virginie
La Virginie
Floride
Carte de l'Amerique Septentrionale et partie de la Meridionale
Isle de Cuba
Espanola
Isles Antilles
Mar del Nort
Tierra Firme
Partie de l'Amerique Meridionale
Nueva Andalusia

Les costes aux environs de la riviere de Misisipi by Nicolas Àde Fer, Paris, 1701 (detail)
After descending the Mississippi River to its mouth, Cavelier de La Salle convinced the king and Minister Seignelay that Louisiana should be colonized. In 1684, he set out to reach the Mississippi again, this time by sea. While he was desperately searching for the mouth of the river, a series of misfortunes befell him: shipwrecks, Indian attacks, deaths of crew members due to drunkenness. A few survivors took exception to La Salle's authoritarian nature and assassinated him with a bullet to the head on March 19, 1687. He died without finding the Mississippi. Nicolas de Fer, famous French publisher and cartographer, published this map relating La Salle's misadventures: an inset recounts the story of his brutal death. De Fer also notes the accomplishments of Pierre Le Moyne d'Iberville, who inherited the mission of discovering the mouth of the Mississippi and founding a colony. De Fer also showed the sites of all of the Indian tribes with which the French had endeavoured to forge cordial relations, such as the Bayagoulas, Biloxis, Moctobis, Pascagoulas (Pascoboula), Houmas (Auma), and Taensas. Unlike La Salle, d'Iberville was able to find his way through the complex Mississippi delta.

character of Cavelier de La Salle, pioneer, explorer, and discoverer of 1,100 kilometres of the Lower Mississippi. His main achievement was to have accurately situated the course of the river, proving that it flowed into the Gulf of Mexico and not into the California Sea or the Pacific Ocean.

A final exploratory stage was led by Pierre Le Moyne d'Iberville, who was charged by the minister of the navy to reach the Mississippi by sea, "to discover its mouth … to choose a good site that could be defended with a few men, and … to bar entry to the river to ships of other nations." D'Iberville, the most famous of the 12 sons of Catherine Thierry and Charles Le Moyne, an interpreter and wealthy trader in Montreal, was a naval officer and thus particularly well suited to such an undertaking. Like his brothers, he had grown up in Montreal, at the time a small town devoted mainly to the fur trade and under constant attack by the Iroquois. He apprenticed as a sailor and then quickly honed his military skills so that he could help defend the young colony against both of its enemies, the English and the Iroquois.

On October 24, 1698, d'Iberville left Brest with four ships and about 400 would-be colonists. He arrived at the French colony of Santo Domingo on December 4 and then headed north, following the coast of the Gulf of Mexico to the mouth of the Mississippi River. At the entrance to Pensacola Bay, he was prevented from dropping anchor by two Spanish frigates. Warned about the French intentions, in 1693 the Spanish had taken possession of the bay found by De Soto in 1550. After crossing the "palisade," as the Spanish called the dam of tree trunks that they had built to obstruct the mouth, d'Iberville sailed upriver and found some traces of La Salle's expedition at the spot where the river split into three tributaries. The old chief of the Mongoulachas showed him a blue serge cap that had been given to him by the "iron hand," Henri de Tonty, who had left a note for La Salle, dated April 20, 1686. The note said, "The Quinipissas having smoked the calumet, I leave with them this note for you assured of my humble respect and let you know about the news that I have had of you at the fort, to wit, that you had lost a ship and had been pillaged by the Savages. On hearing this news, I went down with 25 Frenchmen, 5 Chouanons and 5 Illinois … We found the column on which you flew the King's arms, knocked down by driftwood. We raised a great pillar, and attached to it a cross and above that a coat of arms of France."

Having heard and read these accounts, d'Iberville considered the first objective of his voyage accomplished. He now began construction of Fort Maurepas in Biloxi Bay (today Ocean Springs, Mississippi), between the Mississippi and Pensacola, a Spanish fortified town. On May 3, 1699, he sailed for France, leaving on site a garrison of 81 men, including his brother, Jean-Baptiste Le Moyne de Bienville.

D'Iberville was welcomed at the Court with the recognition due a man responsible for a successful mission. He was made a Chevalier de Saint-Louis – a great honour, as he was the first Frenchman born in Canada to be admitted into the order, which had been created by Louis XIV in 1693. He soon got busy convincing the French authorities of the need to populate and colonize Louisiana in order to build an effective rampart against the Spanish and English. Otherwise, as he said in a prescient warning, he feared that "in less than one hundred years, it [the English colony] will be strong enough to seize all of America and expel all other nations." D'Iberville subsequently received the mission "to perfect and ensure a French settlement in Mississippi" and to explore the entire Gulf of Mexico region; he was instructed, however, not to offend the Spanish, who were neighbours to and allies of the French. He concentrated on the mouth of the river, while his brother Bienville, along with other Canadians, went back up the Mississippi and the Red River to Sioux country to explore a copper mine more than 600 leagues from their point of departure.

In fact, the territory that d'Iberville travelled through was not completely new. On his expeditions, he undertook to strengthen ties with the Indian tribes that he met and prepare for the creation of French posts. These expeditions went a considerable way toward confirming French ambitions for Louisiana and western Mississippi, territories also coveted by the Spanish living in Mexico and Florida and, later, by the inhabitants of the English colonies. But France had found a formidable transportation route, the Mississippi River, which crossed the entire southern half of North America. France's Laurentian settlements were now linked to the Gulf of Mexico in a monumental arc covering more than a third of the continent. From this strong position, France was able to preserve this territory for a century.

Source

Naissance de la Louisiane: tricentenaire des découvertes de Cavelier de La Salle. Paris: Délégation à l'action artistique de la Ville de Paris 1982. Catalogue for an exhibition of manuscripts, maps, iconographic documents, and navigation instruments on the history of Louisiana from 1682 to 1803.

Source for translation: La Salle, Robert Cavelier de. *Relation of the Discoveries and Voyages of Cavelier de La Salle from 1679 to 1681: The Official Narrative.* Translated by Melville B. Anderson. Chicago: The Caxton Club, 1901 [www.americanjourneys.org].

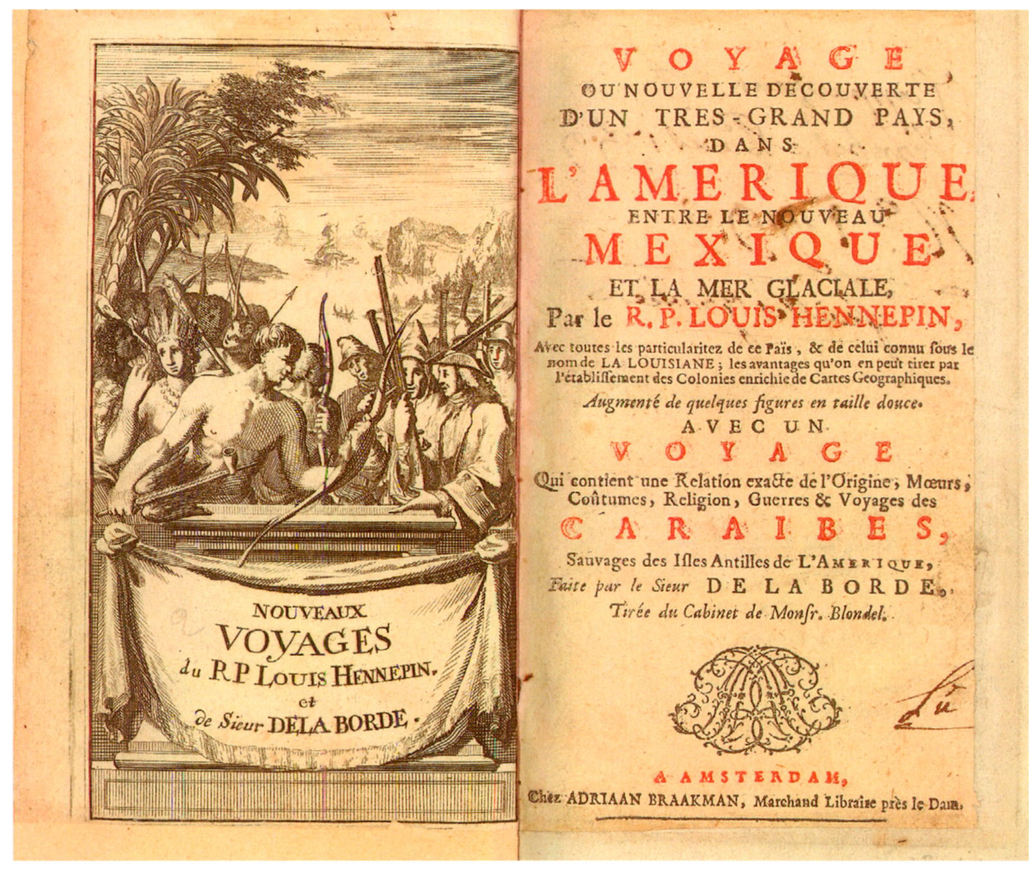

Frontispice of *Voyage ou Nouvelle découverte d'un très grand pays dans l'Amérique* by Louis Hennepin, Amsterdam, 1704

***Carte de la Nouvelle France et de la Louisiane* by Louis Hennepin, Paris, 1688**

Louis Hennepin, a Recollet priest born in Belgium, accompanied Cavelier de La Salle on his exploration voyages to the Mississippi. Sent with two other Frenchmen to scout, he was kidnapped by the Sioux and held captive for several weeks. In 1683, after he returned to France, Hennepin published *Description de la Louisiane*, a book that became very popular. The map that accompanied it showed his routes, all of which led to Recollet missions. Hennepin drew the Colbert River (the Mississippi); the last section, south to the Gulf of Mexico, is in dotted lines, implying that he had not seen its mouth. Why he claimed the contrary 14 years later, in his book *Nouvelle découverte d'un très grand pays*, remains a mystery. In the upper left-hand part of the map is an illustration relating how the king's coat of arms was carved into the bark of an oak tree to mark the taking of possession of the territory. The title cartouche illustrates the union of religion (the cross) and politics (the lily of France) and the peace and wealth that faith and the king provided to their subjected regions.

noroest passage
terre de labrador
Terre neufue
le grand banq
1674

The World of Fish

NEWFOUNDLAND AND LABRADOR

One place where history was long mythologized was the "newfound landes," where Europeans landed during the Renaissance believing that they had found an undefined space offering access to the luxuries of the Orient. A very real North American territory, the first to be shown on maps, Newfoundland had no gold or silk to offer, but the banks of its shores were found to be extraordinarily rich in fish.

Up to about 1600, European fishermen set foot on Newfoundland but did not settle there; on shore, they needed only flakes on which to dry their cod for four or five months of the year. Some cod fishermen even cast anchor on the banks to fish, and then gutted their catch on board. Fish harvested without being dried were called green cod, and they were rushed back to Europe so that they would still be fresh upon arrival. This practice developed in the late sixteenth century, once the navigation routes, the location of the fishing banks, and the best times of year for fishing were well established. Some provisioning ships were able to make two voyages per year, but this was exhausting for the crews, who were forced to suffer the discomforts of the ship for prolonged periods without time to go ashore.

Whalers – mostly Basques, but also smaller numbers of Dutch, English, Danish, Russians, and French – also came to the region. These seamen, who risked their lives to acquire the valuable oil used to light European cities, left traces of their presence in Labrador, notably at Red Bay and Saddle Island, where blubber-melting facilities have been found. *Les Voyages aventureux du capitaine Martin de Hoyarsabal, habitant de Cibiburu* (Rouen, 1532), the account of a French Basque sailor, reflects its author's profound knowledge of the maritime region of the Gulf of St. Lawrence. In 1677, the book was updated and translated into Basque by Pierre Detcheverry, and valuable maps were added.

In Newfoundland, the fishermen came into contact with Micmacs and Malecites, who were already allies in Acadia, and established equally good relations with them. On the other hand, the Beothuks, who lived in the region of Notre-Dame Bay (Boyd's Cove), took refuge in areas far from the coasts to avoid the Europeans, as they had been heavily affected by epidemics. Contacts with the "Eskimos" of Labrador, who travelled south to chase seals and gather soapstone in the Bay Verte region, were not very friendly as a result of their first encounters with whites at Hudson Bay.

Of all the countries that fished and dried their fish on the coasts of Newfoundland, England had the most lasting interest, perhaps because it was situated at the same latitude. Merchants from Bristol, among other places, had financed John Cabot's expedition in 1497, reports of which awoke Europeans to the wealth of the fishery. English explorers also found Newfoundland to be a convenient stopover during their voyages in search of the Northwest Passage. In any case, English concerns in Newfoundland soon expanded, mainly on the south shore along the eastern littoral.

The Newfoundland Company, created in 1610, received a charter allowing it to colonize what was now familiar territory, from Cape Bonavista to Cape St. Mary's. A first colony was founded at Cupids (Cuper's Cove), followed by settlements at Harbour Grace, Ferryland, and other bays in the Avalon Peninsula. The second governor of Newfoundland, John Mason, explored and described the island, recording the information on a map that was published in 1625. This was the first detailed exploration of Newfoundland by the English. Mason's successor (1638–51), David Kirke, born in Dieppe, the beacon of

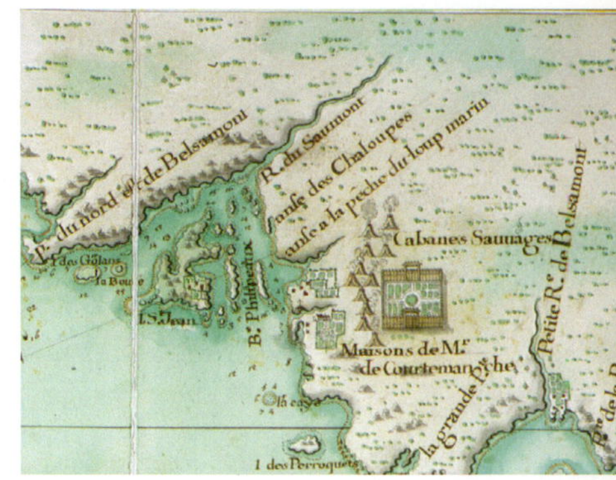

Courtemanche's house, Labrador coast, 1715 (detail)
In 1702, Le Gardeur de Courtemanche obtained exclusive rights to hunt seals and whales and fish for cod on the Labrador coast. This detail shows fortified settlements, complete with fenced gardens, in Baie de Philipeaux (Bradore Bay). Courtemanche was on good terms with the Montagnais, as some 30 families had moved to his concession and hunted for him. On the other hand, relations with the Inuit were less cordial, giving rise to the need for fortifications to protect the Montagnais fishermen and hunters from attack.

Opposite
Map of the North Atlantic Ocean by Denis de Rotis, 1674
This map, made in Saint-Jean-de-Luz by the Basque pilot Denis de Rotis, shows, to the south and east of Newfoundland, the Grand Banks and a number of small *banquereaux* with which European fishermen were very familiar. In the far north, the cartographer depicts, a bit naively, the Northwest Passage, as based on all the information brought back from this region by English explorers.

la cadie
c. sable
=les lieus de 17 & 1/2 en vn degre=
0 5 10 15 20 25 30 35 40 45 50 55 60 65 70
=les lieus de 20: en Vn de Grre=
0 5 10 15 20 25 30 35 40 45 50 55 60 65 70 75 80
I. de bonabanture
c. doest
caiscoupet
marpet
quymuybuet
port Royal
Canceu
c. Sable

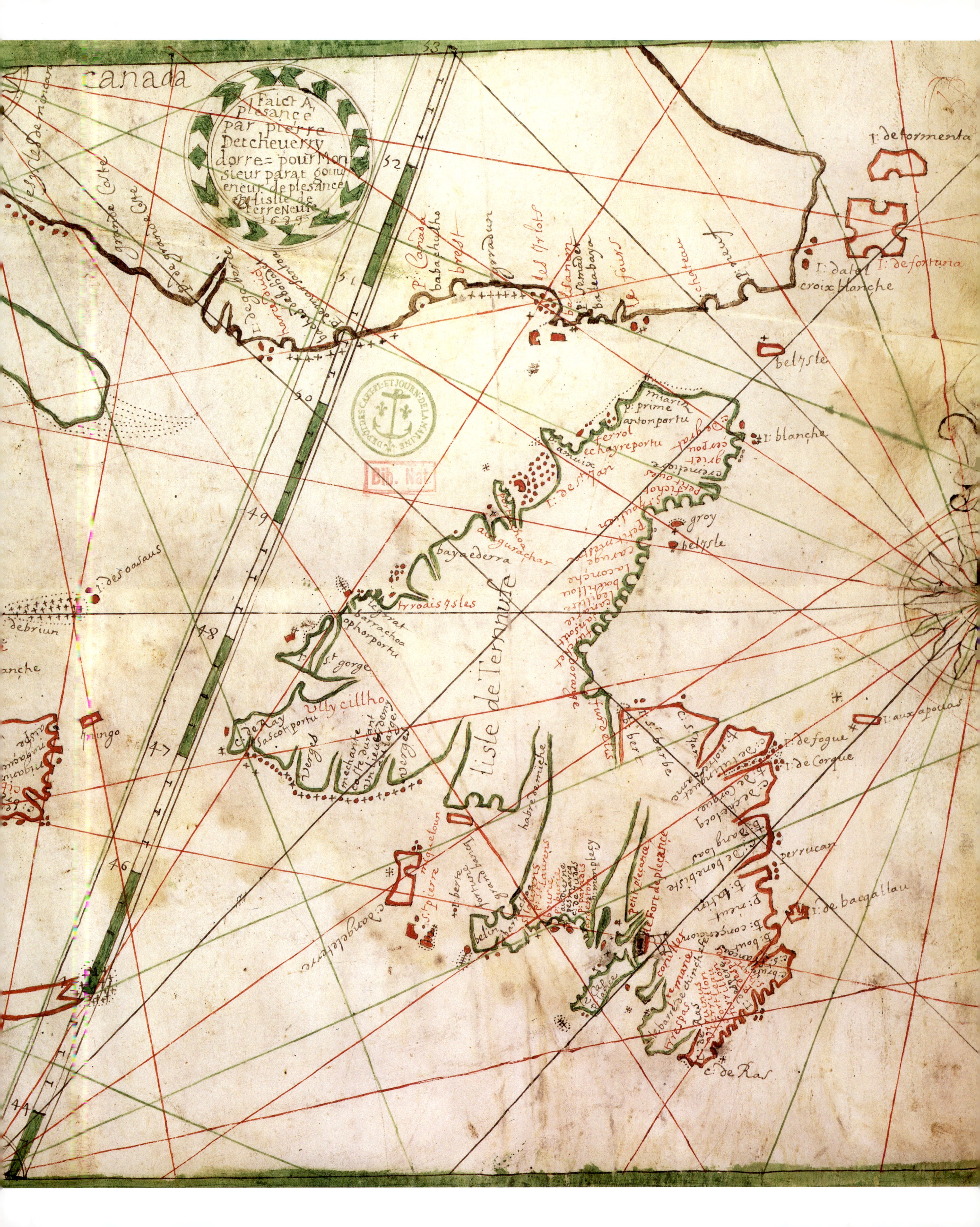

canada
Faict A
plesance
par pierre
Detcheuerry
dorre= pour Mon
sieur parat gouu
eneur de plesance
et Isle de
Terre Neufe
I: de formenta
I: de fortuna
croix blanche
bel ysle
I: blanche
groy
bel ysle
I: des oasaus
Isle de Ternnufe
St. gorge
Verges
I: de fogue
I: de Corque
perrucan
I: de bacgallau
c. de Ras

Preceding pages
Map of Newfoundland and the Gulf of St. Lawrence by Pierre Detcheverry, 1689
The fishermen who visited Newfoundland were Basques, Bretons, Normans, English, Portuguese, and Spanish. Although the Portuguese explored the littoral in the early decades of the sixteenth century, they did not visit the island as often as did the French and English. French and Spanish Basques both fished cod and hunted whales there. In the second half of the seventeenth century (1662), the French built Plaisance (today Placentia), their main settlement on the island. Minister Colbert appointed a governor and had the site fortified. Threatened by the increasingly invasive English, the French tried to reinforce their advantageous position near the Grand Banks. Plaisance, situated in a deep cove, was a perfect refuge for fishermen needing shelter from the wind. It was also a stopover port for French ships heading for Canada or Acadia and for ships returning from the Antilles to France. This manuscript map, dedicated to the governor of the day, is by a Basque, Pierre Detcheverry, who was then living in Plaisance, and bears many Basque names, such as Ullycillho, OPHOR PORTU (rest port), BARRACHOA (Barachois), and PORTUCHOA (Port au Choix). The Basques were still visiting Newfoundland at this time, despite the declining numbers of whales and the increased presence of the English. As exploitation of the fishing resource required little on-site labour, Newfoundland was under-populated compared to other French colonies. At the time when this map was made, about 250 people were living in Plaisance. In the 1713 Treaty of Utrecht, the English were given possession of the site. Some of the French therefore had to move to Île Royale (Cape Breton Island), to which the Acadians were also fleeing.

trade and French colonization in North America, knew the stakes in Canada very well, especially because he had commanded the capture of the settlement of Quebec in 1629. Appropriating a royal prerogative, he tried to impose duties on the cod fishermen, but the seasonal fishermen accused him of having thrown the industry into disarray, and he was recalled to England.

In the early seventeenth century, the master fishermen from Dorset, Devon, Somerset, and Cornwall – that is, all of western England – were annually sending to Newfoundland up to 300 fishing ships crewed with a total of about 3,000 men. The establishment of powerful trading companies did not please the resident Newfoundlanders, who had their customs, felt at home, and feared the control of a monopoly company. Representations were made to the English Parliament to stop English colonization in Newfoundland. (When settlements had been founded in Acadia in 1604, the same reaction by Breton fishermen had led to the elimination of the monopoly of Dugua de Monts in 1607.) While they were waiting for a firm position to be taken, the fishermen in the companies' settlements suffered depredations inflicted by competitors from other countries. During the 1630s, however, the situation turned to the advantage of the traders of southwest England, who finally obtained a "western charter" placing settlers and the coast under their dependency. Around 1686, 2,000 English-speaking residents were living on the shores of 35 bays or harbours on the east coast of Newfoundland. But conflicts of interest persisted, and until the end of the seventeenth century, it was impossible to enforce any regulation banning contraband fishing.

The French of Newfoundland, for their part, settled mainly on the west coast of the island, where they fortified Plaisance (Placentia) in 1662. The fort was intended to provide a safe base for fishermen, a stopover port, and, eventually, protection for the other French territories in the Gulf of St. Lawrence. In this context, Plaisance was often used as a homeport for corsairs attacking English posts. The first governor took up residence there in 1655, where he led a tiny colony: from 73 people, in 1670, the population apparently reached a peak of 640 around 1685.

A few French families settled in Labrador, on the north shore of the Gulf of St. Lawrence, and on the island of Anticosti. In 1650, François Byssot de la Rivière acquired a concession extending from Île aux Œufs to the Strait of Belle Isle, with Mingan as the main trading post. The explorer Louis Jolliet took over the seigneury when he married Byssot's daughter, Claire-Françoise. He conducted a profitable trade in furs, cod, hair seals, and whales and explored the coast of Labrador up to 56°8'N. His travel journal, illustrated with 16 sketched maps, was particularly instructive on the customs of the Eskimos. He received a 20-year trade monopoly and an appointment as "hydrography professor," teaching in wintertime at the Jesuit college in Quebec. When Jolliet died, in 1700, his sons continued to manage the territory. In some years, more than 1,000 Frenchmen and Canadians went to Labrador to fish and to trade with the Montagnais and Eskimos.

The Treaty of Utrecht, signed in 1713, brought peace to Newfoundland by awarding it to England. France preserved the right to fish on the Grand Banks and the north shore of the island, and to occupy posts necessary to its fishing activities. Possession of the Newfoundland shore to which French fishermen went each year changed according to the texts of a long series of treaties signed over the next two centuries. From the "Petit Nord" to the "French Shore," France maintained a presence in Newfoundland until 1904, when it was forced to abandon its long-time presence and restrict its landings to its possessions, the islands of Saint-Pierre and Miquelon.

Another group of Europeans, the Danish, made their mark in Labrador. In 1764, the Moravian Brothers, missionaries from the Lutheran church, which was well established in Greenland, began to found missions among the Inuit in the Strait of Belle-Isle. As interpreters for the governor of Newfoundland, negotiators between whites and Indians during conflicts over fishing, and explorers during their voyages of evangelization, the Moravian Brothers made a major contribution to European knowledge of Inuit territory, language, and civilization.

A strong wave of English and Irish emigration raised the population of Newfoundland to 7,300 inhabitants in the 1750s. Maritime traffic around Newfoundland and the Gulf of St. Lawrence was intense – the inexorable traffic of the triangular (and sometimes quadrangular) trade between Quebec and Louisbourg, the English colonies, the West Indies, and Europe. In 1748, for example, 1,200 ships and boats pulled from the water 483,000 quintals of fish from the Gulf of St. Lawrence. Because of its strong geographic position, the Atlantic region was, more than ever, a strategic zone on the east coast of the continent, but it remained exposed to all forms of greed.

Main sources

L'aventure maritime, du golfe de Gascogne à Terre-Neuve. Congress proceedings (Pau, October 1993). Paris: Éditions du CTHS, 1995. A number of articles on whale hunting and cod fishing by the Basques in Newfoundland. — LITALIEN, Raymonde. "Les Normands à l'île de Terre-Neuve sous l'Ancien Régime." In *Les Normands et l'outre-mer: actes du 35e congrès organisé par la Fédération des sociétés historiques et archéologiques de Normandie, Granville, 18-22 octobre 2000, pp. 83-92.* Caen: Annales de Normandie, 2001. — O'DEA, Fabian. *The 17th Century Cartography of Newfoundland.* Toronto: B. V. Gutsell, "Cartographica," collection, 1971.

Acadia

COVETED TERRITORY

By the spring of 1607, the approximately 100 residents of Port-Royal had overcome the main challenges encountered by European immigrants. With the assistance of the Micmacs, they had housed and fed themselves and managed to survive the winter – and they had progressed beyond simple material comforts to the luxury of taking part in games and festivals. The choice of Baie Française (Bay of Fundy) as a base for the colony was promising: the land was fertile and the fur-bearing animals were plentiful. Although in 1606 the harvest of furs had not yielded the anticipated profits, life was easy and the relations with the Indians were excellent. So there was general consternation when Pierre Dugua de Monts lost his trade monopoly in Acadia and, with the urging of Samuel de Champlain, turned his sights toward the game-filled forests of the St. Lawrence Valley.

Acadia was neither the only nor the largest settlement in New France, but it was nevertheless the port of entry. Its geographic position exposed it to many dangers against which France could provide little protection. Its institutional links with the settlement at Quebec, founded in 1608, were infrequent and strained. Acadia's development had taken a different direction, with personal initiatives taking precedence over official measures.

The effects of the elimination of Dugua de Monts's monopoly were quick to be felt. Jean de Biencourt de Poutrincourt, who had given up neither his concession nor his title of lieutenant governor of Acadia, faced great difficulty in financing his operations. Merchants likely to invest in the fur trade gravitated toward the new post at Quebec. Poutrincourt and his son, Charles, were forced to accept private financing tied to an expressly evangelistic objective: a lady-in-waiting to Queen Marie de' Medici, Antoinette de Pons, Marquise of Guercheville, underwrote a large part of the cost of the 1611 expedition, on condition that two Jesuits make the trip with the 60 settlers. Sainte-Croix and Port-Royal absorbed most of these settlers. Others moved to new posts created at Saint-Sauveur (Mount Desert Island, Maine) in 1612 and at the mouth of the Saint John River (New Brunswick).

The sites chosen were well suited to trade and food procurement. They were in the heart of the settlements of the Micmacs, Malecites, and Abenakis, the main suppliers of furs, as their English neighbours to the south were well aware. Barely were the Acadian posts established before they were attacked. Sir Samuel Argall, admiral of Virginia, had been charged with expelling the French from the territory awarded to him and establishing trade relations with the Indians there. Notwithstanding the notion that "possession is nine-tenths of the law," Argall felt justified in seizing Port-Royal in 1613 and razing it, along with Saint-Sauveur and Sainte-Croix.

In the early seventeenth century, there was what could be called a rush on occupation of the Atlantic littoral. Although Champlain had mapped and named the entire coast as far south as Cap Blanc (Cape Cod) in 1605, the French occupied the territory only north of the Penobscot River. The Dutch made incursions there, but it was the English who gradually made it their domain. In 1614–15, John Smith completed a description of this region, which he called New Englande, and portrayed it in a map published in 1616. Soon after, in 1620, the Pilgrims landed at Provincetown and then settled at Plymouth. Except for the small settlement on Cape Cod, however, New England was not colonized before 1630.

Acadia thus had a few years of respite. But the English had not forgotten the charter of the Virginia Company, which authorized them to extend their territory to 45°N – the latitude of Port-Royal. In 1621, James I, king of England, conceded to the Scotsman William Alexander the territory north of the Sainte-Croix River, up to 45'N,

"Homme Acadien," eighteenth century
Drawing by Claude-Louis Desrais, engraving by Jean-Marie Mixelle, in a work by Sylvain Maréchal, who presents, with great nostalgia, the "Mœurs et coutumes des Acadiens" (morays and customs of Acadians). His portrayal of the Acadian man is somewhat surprising. Did he intend to underline the prevalence of the mixing of blood, as his text suggests?

a territory that was dubbed "New Scotlande," or "Nova Scotia," and endowed with a coat of arms and a flag. It was a radical royal decision, but the territory was not occupied until 1629.

After the Kirke brothers attacked and captured the settlement of Quebec, in 1628–29, the Treaty of Saint-Germain-en-Laye returned to France both of its North American colonies. Port-Royal was liberated and the few colonists who had stayed in the environs were found, living rather peacefully, fishing and delivering furs to the La Rochelle merchants in exchange for annual provisions. This was when the Company of New France (or of the One Hundred Associates), created by Richelieu in 1627, began to actively exercise its monopoly over all of New France, including Acadia. In 1632, Governor Isaac de Razilly brought 300 new settlers to Acadia. He restarted the colony, pushed back the recalcitrant Scots, and settled first at La Hève then, in 1636, at Port-Royal on a point on the south shore of the basin, at the site of today's Annapolis Royal.

For 20 years, the colony blossomed. In 1635, the new governor, de Razilly's cousin Charles de Menou d'Aulnay, recruited some 15 families, for a total of about 40 settlers. Pentagouet was recaptured. The "water clearers," most of them from Poitou, brought their crop-rotation techniques to the new colony and built aboiteaux (simple sluices) to dry out the marshes. These dikes desalinized the soil, which, after a few years, became arable – well

New Scotlande by William Alexander, 1625
In 1621, the king of England, James I, awarded one of his subjects, Sir William Alexander, a concession in North America named New Scotlande; it included a large part of Acadia, which had been formed almost 20 years earlier. In order to promote colonization (which did not occur), Alexander published a map situating the borders of the territory allocated to him. To appropriate the environs, he placed on the map new names inspired by his Scottish homeland (such as Tweed River). This map, totally fictional, is a perfect example of the use of cartography as propaganda in territorial disputes.

suited to growing food crops and breeding animals. The Acadians thus became farmers, even exporting some of their surplus to New England, where the growing population presented a lucrative market.

This prosperity did not escape the notice of the English settlers. In 1654, Robert Sedgwick, general of the fleet and chief commander of the New England seaboard, besieged and captured the French posts of Pentagouet, Saint-Jean, and Port-Royal. The Lord Protector of England, Oliver Cromwell, conceded the territory to Thomas Temple, along with the title of governor of Acadia. But France refused to abandon the sites in the absence of a treaty or bilateral agreement confirming British authority. For some 15 years, therefore, parallel governments existed; finally, with the Treaty of Breda in 1667, the king of France obtained the restitution of all the land occupied by the English. In addition, Louis XIV withdrew management of the colonies from the trading companies and took control of New France himself. Acadia then became an administrative unit under the governor general residing in Quebec. After 1670, new French settlements were founded, including Beaubassin (Amherst), Grand-Pré (Wolfville), Cobequid (Truro), and Pisiquid (Windsor). The total population was about 500 Acadians distributed among 70 families, most of them living near Port-Royal.

During the half-century separating the Treaty of Breda and the Treaty of Utrecht (1713), there were, paradoxically, both an affirmation of the Acadian identity and a growing English presence. The French authority exercised its power half-heartedly and did not really protect its subjects. The English of Massachusetts, for their part, were developing trade networks with the neighbouring French colony. They supplied manufactured products in exchange for wheat, vegetables, oats, rye, barley, flax, cows, pigs, sheep, poultry, furs, feathers, and other goods. In 1686, more than 800 English ships entered Acadian ports; in 1708, 300 ships came from Boston alone. In spite of successive interdictions, Acadian fishermen continued, through convenience, to trade with their closest neighbours, eventually becoming dependent on them. Meanwhile, the Abenakis, an Indian nation living around the Penobscot River, gave priority to economic need over political considerations and traded with both French and English. On the military front, however, they were completely loyal to the Acadians. In 1703, in response to incessant attacks by Benjamin Church, commander of the armed forces of Plymouth, the Abenakis and the French laid waste to the coast occupied by England, from Canso to Wells. Thus provoked, the English responded by sacking all of Acadia.

Over time, Acadia became the main battleground for confrontations between France and England as wars in Europe waxed and waned, and also whenever there was the slightest discord between rival groups of colonists. The imbalance of power was striking. In 1710, about 1,800 inhabitants of Acadia and 16,000 in New France found themselves facing 357,000 residents of the English colonies. There were numerous confrontations, but no changes to the borders were made until the end of the War of the Spanish Succession, and thereafter sanctioned by the Treaty of Utrecht, which finally ceded Acadia to England. The Acadians, often left on their own, had developed a strongly independent spirit, as well as an enduring solidarity with the Indians. They had also learned to coexist with the English, whose authority they would now have to accept.

Plan of Port-Royal by Jean Baptiste Louis Franquelin, 1686 (detail)
As this map shows, Port-Royal was quite small in size. According to the census taken by Intendant De Meulles, it comprised 95 families and 583 individuals in 1686. Some of the inhabitants lived in the centre of the village, including Governor Perrot and Seigneur Le Borgne, whose houses are noted by the cartographer. Other settlers had cleared land along the Dauphin River. At the juncture of these two settlement groups were the parish church and the cemetery. A short distance away from the village were two houses owned by Englishmen; their presence facilitated trade between Acadia and Massachusetts. Although Port-Royal was the main site of French colonization in Acadia, it had no fortifications to defend it. The Englishman William Phips thus had no difficulty capturing the settlement in May 1690.

Main sources

Blondel-Loisel, Annie, and Raymonde Litalien (eds.). *De la Seine au Saint-Laurent avec Champlain.* Congress proceedings (Havre, April 2004). Paris: L'Harmattan, 2005. On the first permanent French settlements in Acadia. — Daigle, Jean (ed.). *Acadia of the Maritimes: Thematic Studies from the Beginning to the Present.* Moncton: Chaire d'études acadiennes, Université de Moncton, 1993.

Q. L'Ange gardien
R. S. francois
S. trou St patrice
T. Saut memorensi
V. pointe d'orleans
Y pointe de leui
Z. beau port
&. le port
Le Nord Cotte de
beau pré *
1. Silleri
2. Cap rouge
3. riuiere st charle
4. les hospitalieres
5. La brasserie
6. L'Eueschéz
7. Les jesuittes
8. La basse ville
9. Les Vrsulines
x. le chateau
xi. la haute ville
xii. La grande Allée
13. N. Dame de foy
14. La routte St jean
15. Les Recollets
16. Les islets
17. terres labourées
12. lieues de long
sur
6. de large

Occupation or Cohabitation

IN THE ST. LAWRENCE VALLEY

The joint expedition by Roberval and Cartier was a failure. As a consequence, the St. Lawrence River was forgotten by French authorities up to the end of the sixteenth century, although French fishermen did not forget that the river was teeming with fish. Henri IV sounded the wake-up call, and, after various attempts, Richelieu finally took the initiative. The Company of New France, created in 1627–28, was given a charter "in perpetuity in full ownership, justice, and seigneury," for "the entire country of New France known as Canada" from Newfoundland to Lake Huron and from Florida to the Arctic Circle. The charter provided that the descendants of French Catholic settlers, as well as "savages" who professed the Christian faith, would be presumed and recognized as naturalized French citizens with all the rights of the king's subjects. Cohabitation was thus part of the plan. There was no question of moving the Indians out, and certainly no suggestion of exterminating them. Rather, it was the familiar refrain: make the others similar to oneself to lift them out of their perceived state of inferiority.

In reality, the situation was different. In 1603, in Tadoussac, Anadabijou, the "great chief of the savages of Canada," listened to Henri IV's message, relayed by one of his men, that "they could be certain of the good intentions of his Majesty, of his desire to people their country, and either to make peace with their enemy . . . or to send troops to conquer them." Anadabijou quickly made his choice between peace and war, and he then took the time to distribute tobacco to Gravé Du Pont, Champlain, and "some other chiefs near him." Finally, speaking on behalf of his people, no doubt Montagnais, Algonquins, and Etchemins, he answered solemnly "that in truth they had good reason to be content at his Majesty's great friendship for them . . . Steadily proceeding with his speech, he said that he would be rejoiced to see his Majesty people their country and make war on their enemies, and that there was no nation in the world for whom they had more friendly feelings than for the French." Thus, the French would be allowed to settle, but they would have to share a common enemy with their allies: the Iroquois.

In granting a plot of land to Louis Hébert in 1617, Champlain became, in effect, the first surveyor in New France. Hébert's title was officially recognized in 1623, and his concession, located around what are now Hébert and Couillard streets in Quebec City, was increased by a few acres along the St. Charles River in 1626. It was not until 1634, however, that a real surveyor, Jean Bourdon, arrived. On December 4, 1635, a few weeks before Champlain died, Bourdon made his first concession certificate, for Guillaume Hubout, using the "Paris measurement," which was "eighteen feet per rod and one hundred rods per arpent." When the first concession was made at Montreal, on January 4, 1648, Maisonneuve used the same measurement system to demarcate the land awarded to Pierre Gadoys.

The immense territory granted to the Company of New France, also known as the Company of the One Hundred Associates, aroused great envy. Several members of the company had granted themselves vast properties, many of which remained untouched. Finally, the king took things in hand and repossessed the rights to land that had been abandoned. Four very large seigneuries escaped the royal chopping block: the Beaupré coast, the island of Montreal, Batiscan, and Cap-de-la-Madeleine. In addition, around Quebec, Trois-Rivières, and Montreal, there were about 50 seigneuries. No Indian villages had yet been affected. In Quebec and Montreal, the land was unencumbered. At Trois-Rivières, the Algonquin chief, Capitanal, whose tribe regularly visited the shores of Lake Saint-Pierre, asked Champlain to build "a big house" for his people and another for the

Bird's-eye view of the environs of Quebec, circa 1685
There are several types of cartographic representations, including the bird's-eye view, characterized by a displacement of the field of vision. This map of the St. Lawrence Valley, circa 1685, is a good example of an intelligent application of this view. The mapmaker elevates observers above ground level, allowing them to see the entire river from Sillery to the Beaupré coast and Île d'Orléans. In this way, he imbues the landscape with volume and life and shows the strategic position of the town of Quebec, whose name means "narrowing of the waters" in Algonquin. The region portrayed is the historic central region cleared by the pioneers of New France. The main routes to Quebec are shown, including the Grande Allée (XII) and the Saint-Jean Road (14). The mapmaker also highlighted five mills and a number of church steeples, which probably served as references when he sketched out the landscape. The archaeologist Michel Gaumond feels that this extraordinary drawing was made after 1683, when a wing was constructed adjacent to the Recollets' church (15). As well, the technique used to portray the entire grouping is similar to that used for the map on page 127; that map has been attributed to Franquelin, who arrived in New France in 1672. According to the 1667 census, there were 448 people in Quebec, 123 in Beauport (Z), 656 on the Beaupré coast, 113 on the Lauzon coast, and 529 on Île d'Orléans.

Description generalle de l'isle de Montreal Divisée par costes ou sont exactement marquées toutes les terres & les Noms de chaque habitans, la quantité de terre qu'ils possedent tant de front que de profondeur

Coste du Nord du bout de l'isle

1. gabriel Perrin 4. arpens sur 20.
2. Morisse Perrin 3 sur 20.
3. Thorel St romain 4. sur 20.

Bout de l'isle

1. la decouverte
2. milot
3. Ste genume
4. le ber 10. arpens fief
5. St Pierre
6. coulonge fief

1. St germain fief 14 arp.
2. S. andre 10. arp. fief

Coste et parroisse St Louis

1. Migeon 10 arpens sur 20.
2. ayme fonte 5. sur 20.
3. Ciboutt Courro. 4. sur 20

1. blinuille 14. fief
2. Caueliet 4. sur 20.

1. terre de l'Eglise 14. arpens
2. la londe 4. sur 20.
3. Marin 3 sur 20.
4. la plante 4 ½ sur Vinge
5. mopetit 3. id.

1. montauban 3. id.
2. coulonge 2. id.
3. la douceur 4. id.
4. Rouge lavidette 4. id.
5. a. Bourcier 4. id.
6. J. magdelaine 3.

1. Etienne Magdelaine 3 sur 20.
2. Jacques Cauchoix 2. sur 20.
3. Nello 3. arpens sur Vinge
4. pierre bouno 4. sur 20.
5. Jean quenet 10. sur 40. de profondeur

1. Brunet 4. arpens sur 20.
2. Anthoine Villeray 3. sur 20.
3. f. Lory 3. id.
4. Sabourin id
5. p. Berdue. 4. sur 20.
6. L. Lory 3. id.
7. a. Carriere 4. id.

1. Jacques J. Desnier 4. sur vingt
2. P. genume 3. id.
3. n. le Moine 4. id.
4. Jean Nouee 3. id.
5. J. charlebois 3. id.
6. Cerrier 3. id.

1. Jacques chasle 3. sur 20.
2. Brunet letang 3. id.
3. latreille 2. id.
4. J. lantier 2. id.
5. J. Bonne 3. id.
6. andre Roy 3. id
7. pierre lac 3. id.

Pointe claire

1. Jacques tenet 3 sur 20
2. Jean lecuyer 3 id
3.
4. Pointe claire 4 pr. un Moulin
5. Charles de launay 4. sur 40.
6. J. Dufresne 3. sur 20.
7.

grande ance contenant Environ cinq quarts de lieue de chemin ou sont les habitans

Lac des deux Montagnes

isle Major

Isle

Bout de l'isle

terres non concedée

Isle Concedées

De

Isle Perrot

Lac St Louis

grande ance

Rivière

beau bois

bois franc

fort Rolland

Saut St Louis

Depuis la presentation Jusques au fort Remy Sont les habitans

Depuis le fort Remy en descendant a Verdun Suit

Depuis Verdun Jusques a la Riviere St Pierre

Depuis la R. St pierre Jusques a la ville a St gabriel et Jusques au bout de la Montagne les Distances de place en place ~~auec~~ le Meilleur Ordre que lon a pue obseruer le 15 oct. 1702

forts, Eglises et moulins et le tout de suitte &

Montreal

Terres non concedées

St Michel

Laurens

La Coste de la Riuiere des Prairies ne commance a auoir des habitans que depuis Dubuisson a enuiron une lieue et demye au dessous de la Mission en descendant au bout de l'isle ainsy qu'il suiuit

1
- du Buisson .. 4 sur 20
- Boismenu .. 3 sur 20
- Jean mouner id.
- C. cardinal id.
- 1 Jean lorrain id
- Joseph lorrain id
- de Louagne 4 sur 40

2
- Rogue .. 3 sur 20
- Jaudoin .. 4 sur 20
- Brillault 9 sur 20.
- Cognon .. 3 sur vinge

3
- Lorrain ...... id.
- Thomas Charles id.
- andre Cuilloy 7 sur 20.
- Jean Dicare 3. sur 20.
- antoine Baudry id.
- Bongrain .. id.

Coste S^t^. Dominique

4
- Pierre archambault
- toussain Baudry
- Pierre Baudry
- Jean Renault
- pierre Renault
- Libersan
- p. Bazinet

3. de front sur 20.

5
- Jacques beauchamp
- p. Brouillet
- J. fortin
- les desjardins 6. sur 20.
- michel quere 3. sur vingt
- Pierre Rangé 3. id.

6
- Jiuarc ... 6 sur 20 id.
- moulin
- terre de l'Eglise 10. arp. de front
- donc 6. donnez. scauoir
- a Cadieu 3. sur vinge
- et a Boulard id.
- J. B. quesneuille 3. id.
- François Coroy .. 3. id.
- Charlie Brasso 4. id

Depuis les^r^ Montigny ou Vildaune Jusques au commencement de la Coste S^te^. anne Exclusivement.

Depuis lad. Coste S^t^. françois Jusques Jusques a la Coulée S^t^. Jean

Depuis la Coulée S^t^. Jean, au bas de lisle

Magdelaine

Preceding pages
***Description généralle de l'Isle de Montréal*, attributed to François Vachon de Belmont, 1702**
Some landowners used cartographers to define the borders and thus facilitate the administration of their seigneuries. For example, in 1702, the Compagnie des prêtres de Saint-Sulpice commissioned a land plan of the island of Montreal, which was executed by the superior of the Sulpicians in Canada, François Vachon de Belmont. It is difficult to say whether the map was intended for the Sulpicians in Paris or in Montreal. Nevertheless, on the bottom the cartographer added the names of the *censitaires*, each of whose concession was indicated on the plan. In general, the seigneur conceded narrow, deep plots of land; aligned together, they formed an ensemble called a *côte* (which in French means "slope" or "shore," though the *côtes* did not necessarily have slopes or shores). The *côtes* gave their names to the roads that ran along them. They formed the framework for rural and urban development in Montreal and eventually became major thoroughfares in the city – Côte Sainte-Catherine, Côte de Liesse, Côte Saint-Luc, and Côte des Neiges. The map also shows the forts situated on the island and on the south shore of the river, some 30 in all, which helped to defend the seigneury against Iroquois attacks. These forts, most of them made of wood, were designed not to resist European artillery, but to repulse Indian attacks. Built several kilometres from each other, they used smoke signals to communicate the arrival of enemy fighters. A number of fort names became those of cities, neighbourhoods, or boroughs: Pointe-aux-Trembles, Longue-Pointe, Rivière-des-Praires, Sainte-Marie, Saint-Gabriel, Verdun, Pointe-Claire, Senneville, Varennes, Boucherville, Longueuil, Saint-Lambert, and La Prairie. The map also shows Fort de la Montagne and Fort Lorette, near which were missions for the Iroquois, Hurons, and Algonquins who had converted to Christianity. This plan, one of the most important in the history of Montreal, is now conserved at the Bibliothèque Saint-Sulpice de Paris.

French. Champlain optimistically answered the orator, who had expressed himself with "a rhetoric so fine and nimble that one might have said he graduated from the school of Aristotle or Cicero," that "when our boys marry your girls, we will become a single people."

In addition to the rigours of winter, fear of the Iroquois was enough to dissuade potential French immigrants. Finally, in 1663, the king decided that dramatic measures were in order; he took over supervision of New France, organizing it on the model of a French province, with a governor and intendant to administer it. Two years later, the men of the Carignan-Salières Regiment marched against the Iroquois. Relative peace was established.

Once Jean Talon was appointed intendant in 1665, unprecedented efforts were initiated to encourage immigration. Minister Colbert, however, resisted; he did not want to depopulate France. Talon had an admirable response: "A new country does not make itself, if it is not helped at the beginning." Between 1665 and 1672, the population rose from 3,200 to 6,700. Some 450 demobilized soldiers accepted the government's invitation and were granted land. They became *censitaires* (individuals who had acquired the concession to a parcel of land and had to pay dues, called *cens* and *rentes*). Many had former officers from the Carignan Regiment as their seigneur: Lavaltrie, Varennes, Verchères, Contrecœur, Saint-Ours, Sorel, Chambly, Berthier, La Durantaye, and La Boutellerie.

In the year of his departure, 1672, Talon granted 46 seigneurial concessions, including 10 along the Richelieu River, a route traditionally used by the Iroquois to attack the colony. Seigneurial expansion also proceeded elsewhere: 19 concessions were granted at Quebec, 6 at Trois-Rivières, and 21 at Montreal.

Also in 1672, Talon accorded the first commissions as surveyor in New France: one to Louis-Marin Boucher and the other to Jean Le Rouge, both of whom had been taught by Martin Boutet. A musician by trade, Boutet was also a mathematician. He let the Jesuits convince him to teach mathematics, a job at which he excelled, and he made an essential contribution to surveying and the teaching of navigation. He probably taught Louis Jolliet and was the mentor of Jean-Baptiste Franquelin, with whom he worked in 1685 to produce an accurate nautical chart of the St. Lawrence River. Jolliet took the readings, and Franquelin drew the map.

To facilitate access to travel routes, the seigneuries had a narrow frontage on the river; those on the north shore had a northwest orientation, while those on the south shore had a southeast orientation. The seigneurial area was thus subdivided into long rectangles along the St. Lawrence and also along tributaries such as the Ottawa and Richelieu rivers, the latter extending to Lake Champlain.

After the royal decisions of 1663, a half-century passed before the Court began to assess the land concessions in New France. Through the Edicts of Marly of 1711, the king set out to add to the royal domain seigneuries that had not been in operation and to require *censitaires* who were not farming their land to turn it over to the seigneur. The intention was clear, but the reality was more complex. In the summer of 1712, Minister Pontchartrain observed this as he read the memoranda of Gédéon de Catalogne and examined his immense maps showing the seigneuries and settlements of the governments of Quebec, Trois-Rivières, and Montreal. These materials provided the Court with a good overview of the fiefs and seigneuries under consideration.

In several instances, the government had been particularly generous to religious communities in order to ensure them the means to oversee education and provide health care to the population. In the early eighteenth century, nearly half of the settlers in New France were concentrated in ecclesiastic seigneuries. The Sulpicians were the seigneurs of the island of Montreal, which had about 2,600 inhabitants, 1,200 of them in the town itself. The Jesuits alone held more than 10 per cent of the seigneurial zone as a whole, which included some 2,000 *censitaires*. In total, according to Catalogne, the Church controlled one fourth of the 84 seigneuries that had been distributed. In 1706, he assessed the population of European origin at 16,417.

The Great Peace of 1701

Beyond the boundaries of the land concessions was a multitude of Indian nations – nomadic ones to the north and sedentary ones to the west and southwest. In 1701, the French accomplished the feat of bringing together at Montreal, in spite of the dangers of new epidemics, some 1,300 delegates from about 40 nations that were ready to make peace. Chief Kondiaronk, one of the great crafters of this peace, was himself mown down by disease. "Our father," he declared to Governor Callière on the night before he died, "you see us here from your sleeping mat, it is not without great peril that we have undertaken such a long voyage. The waterfalls, rapids, and a thousand other obstacles have not seemed so difficult to surpass due to our desire to see you and assemble here. We have found many of our brothers dead along the river because of the sickness … But we made a bridge of all these bodies on which we marched firmly." Since the arrival of the first Europeans, epidemics had struck the Indians over and over, with tragic results. They no doubt were the reason for the rapid disappearance of the Iroquoians whom Cartier had encountered in the St. Lawrence Valley. For a long time, the population drop among Indian nations was attributed to war, but today more is known about microbe

***Carte pour servir à l'éclaircissement du papier terrier de la Nouvelle-France* by Jean Baptiste Louis Franquelin, 1678 (detail)**
Between 1676 and 1678, Intendant Duchesneau had a *terrier* (land register) made; this required that the seigneurs present a detailed description of their seigneuries, their lands, and the royalties that they charged. The resulting portrait of the colony's state of development was useful to colonial authorities. To facilitate reading of the *terrier*, Duchesneau commissioned Franquelin to draw a map depicting the lands of the seigneurs of Canada. Although its contours are somewhat simplified, the map enabled the king to situate the seigneurs to whom he had entrusted part of his territory, including Roctaillade, Godefroy, Prade, Hertel, Marsolet, Lotbinière, Bissot, Beaumont, La Durantaye, and Berthier. This cartographic information was enhanced by illustrations of animals of the Canadian forest and Indians portaging birchbark canoes.

shock and its horrible ravages. Smallpox was a particular culprit, long before inoculation and then vaccination against smallpox were envisaged. In fact, smallpox became an ally of the Europeans – involuntary for the Spanish, and deliberate, as a weapon of last resort, for the British.

Expansion of the French colony usually took place with the agreement of the Indians, who were resolutely attached to the land. Unlike the English, who regularly used territorial treaties in their relations with the Indians, especially after the Conquest of 1760, the French never concluded such agreements, which were in reality sad treaties of dispossession. In fact, because the Indians were practically absent from the seigneurial zone, the French welcomed converted Indians, often of mixed blood, called *domiciliés*. Except for the Hurons, who were settled first at Sillery and then at Lorette (1697), the Indians were settled at strategic sites on invasion routes: Iroquois at the Saint-Louis and Récollet rapids, Nipissings and Algonquins at the Isle aux Tourtes mission, Abenakis along the Saint-François and Bécancour rivers (the latter called Rivière Puante on Catalogne's map). In 1698, Catalogne had listed 1,540 Indian *domiciliés*. Over the years, these villages were moved several times, while other Indians gradually formed new villages along the Ottawa River (one is today Kanesatake) and the south shore of the St. Lawrence west of Montreal, at Saint-Régis (today Akwesasne) and Oswegatchie. These Indians remained loyal allies of the French and occasionally served as intermediaries with their nations of origin, with which the French maintained links that strengthened New France during its confrontations with the English colonies.

The Great Peace of Montreal, which brought emissaries from the far side of the Great Lakes, demonstrated the significant influence of French penetration into the west-central part of the continent. Around Detroit, between Lake St. Clair and Lake Erie, a small seigneurial zone developed that copied the model of the zone along the St. Lawrence; two others were formed along the Mississippi, one at the site of New Orleans and the other at the site of St. Genevieve.

The seigneurial mode of development was found everywhere the French settled. Associated with feudalism, mainly through its rites and vocabulary, it has often been severely criticized. However, as the historian Marcel Trudel has observed, "Under this regime, seigneurs were not feudal; nowhere were they judge and jury." Their role was to see to the settlement of people, and their responsibility was to institute a structure to receive newcomers. According to Trudel, "it was self-help built into a system." If the seigneur, like the *censitaire*, neglected his land, he risked losing it.

This land-distribution system of *cens* and *rentes* had nothing to recommend it to the conquering British who took over in 1760. James Murray quickly undertook an inventory of the territory and received instructions to switch to "free and common soccage" or "freehold tenure." In 1775, Governor Carleton was ordered to return to seigneurial tenure, which involved *rentes*, but the arrival of the Loyalists, who demanded land concessions according to the leaseholds to which they were accustomed, resulted in the reintroduction of freehold tenure, which was applied in areas surrounding the seigneurial zone. The two systems cohabited until the seigneurial regime was abolished in 1854, when 242 seigneurs gave up their *rentes* in exchange for fair compensation. The seigneurial regime that had ensured the integrity and cohesion of the French-Canadian population had outlived its usefulness, become synonymous with a closed economy, and proved an obstacle to progress.

Main sources

BOUDREAU, Claude. *La cartographie au Québec, 1760-1840.* Sainte-Foy: Presses de l'Université Laval, 1994. — CHARTRAND, René. *French Fortresses in North America 1535–1763: Québec, Montréal, Louisbourg and New Orleans.* Oxford: Osprey Publishing, 2005. — HAVARD, Gilles. *The Great Peace of Montreal of 1701: French-Native Diplomacy in the Seventeenth Century.* Translated by Phyllis Aronoff and Howard Scott. Montreal: McGill-Queen's University Press, 2001. — HAVARD, Gilles. *Montreal, 1701: Planting the Tree of Peace.* Translated by Phyllis Aronoff and Howard Scott. Montreal: Recherches amérindiennes au Québec, 2001. — LESCARBOT, Marc. *History of New France.* Vol. 2, translated by W. L. Grant. Toronto: Champlain Society, 1911. — SHORTT, Adam, and Arthur George DOUGHTY (eds.). *Documents Relating to the Constitutional History of Canada, 1759–1791.* Ottawa, 1911. — TRUDEL, Marcel. *Introduction to New France.* Montreal: Holt, Rinehart & Winston, 1968.

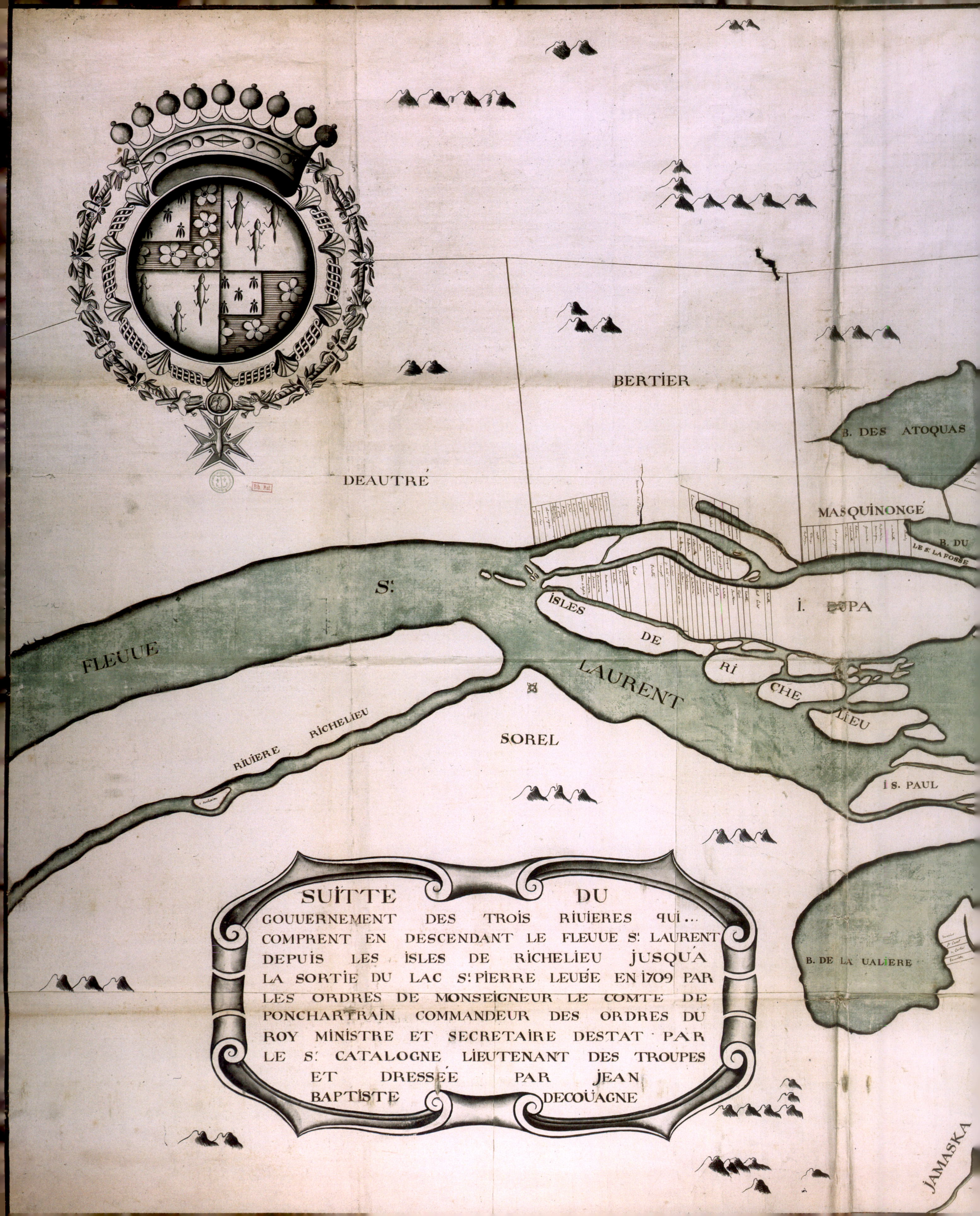
SUITTE DU
GOUUERNEMENT DES TROIS RIUIERES QUI...
COMPRENT EN DESCENDANT LE FLEUUE S.T LAURENT
DEPUIS LES ISLES DE RICHELIEU JUSQUA
LA SORTIE DU LAC S.T PIERRE LEUÉE EN 1709 PAR
LES ORDRES DE MONSEIGNEUR LE COMTE DE
PONCHARTRAIN COMMANDEUR DES ORDRES DU
ROY MINISTRE ET SECRETAIRE DESTAT PAR
LE S.T CATALOGNE LIEUTENANT DES TROUPES
ET DRESSÉE PAR JEAN
BAPTISTE DECOÜAGNE
BERTIER
DEAUTRÉ
B. DES ATOQUAS
MASQUINONGÉ
B. DU
LE S.T LA FOSSE
FLEUUE S.T LAURENT
ISLES DE RICHELIEU
I. DUPA
RIUIERE RICHELIEU
SOREL
I S. PAUL
B. DE LA UALIERE
JAMASKA

***Suitte du gouvernement des Trois Rivières* by Gédéon de Catalogne and Jean-Baptiste de Couagne, 1709**
This leaflet was part of a group of maps of the St. Lawrence Valley produced in 1709 by Jean-Baptiste de Couagne from surveys made by the military officer Gédéon de Catalogne. These maps, very valuable to historians and genealogists, illustrate how the territory was cut up into seigneuries and smaller *concessions de l'habitant* (settlers' concessions). The map mentions, exceptionally, the names of all known *censitaires*. The region portrayed here is the western part of Lake Saint-Pierre, including, at the mouth of the Richelieu River, the islands of Sorel, which are shown partially cleared. We also see a "lac des Atocas," not found today; *atoca* is the Huron term for cranberry. The geographic complexity of the area mapped shows that the division of lands, in the form of rectangles that are similar in size but varied in orientation, was well adapted to the watercourses. Although the government maps of the settlements of Quebec and Trois-Rivières are now conserved at the Bibliothèque nationale de France, those of Montreal, unfortunately, have never been found.

French *géographes de cabinet*

The best-known maps of New France were the work of *géographes de cabinet*. Unlike navigators, missionaries, engineers, and hydrographers, these cartographers did not make observations and measure the terrain themselves. They made their maps without having gone through the risks and perils of exploration voyages; instead, they used data gathered by others, which they compiled in their studios.

By the sixteenth century, European cartographers had drawn the contours of New France based on voyage accounts. The relations of Giovanni da Verrazzano and Jacques Cartier inspired Italian cartographers, who popularized the name New France in various forms (Nova Francia, Nova Franza, Nova Gallia). At the end of the century, cartography circles were dominated by mapmakers in Antwerp and Amsterdam, who produced immense wall maps and magnificent atlases. The first Frenchman to provide real competition for this Dutch production was Nicolas Sanson, who created thematic maps (of trading posts and rivers; historical maps), then published atlases for each of the four known continents. To produce his two maps of Canada (1650 and 1656), Sanson drew essentially from the Jesuit *Relations* (see pp. 92–93). The advantages enjoyed by the French cartographers became obvious: because of the increasing number of explorations into the interior of the continent, these men had access to a great number of primary sources. One talented cartographer, Jean-Baptiste Franquelin, questioned the coureurs des bois returning to the settlement of Quebec and produced many maps of North America for the Court of France. The expansion of the fur trade explains in large part the wealth of detail on French maps drawn by Sanson, Jacques-Nicolas Bellin, Vincenzo Coronelli (a Venetian who published in Paris), Alexis-Hubert Jaillot, Guillaume Delisle, and many others.

Most of these privileged geographers worked in Paris, in the heart of the political, economic, cultural, and scientific capital of the kingdom. This location provided them with a number of advantages. They were close to the seat of political power and were thus able to secure the patronage of the king and princes, and the resulting prestige gave them access to new cartographic sources. Thanks to their relationships with members of the Académie des sciences, they had access to the most recent geographic coordinates calculated from astronomical observations. In addition, a number of *géographes de cabinet* published and sold their own maps, and were thus encouraged to stay in Paris, close not only to their aristocratic and bourgeois clients but also to their printers and engravers.

French *géographes de cabinet* did not all have the same schooling. Some, like Sanson, learned cartography as military engineers. Others, like Nicolas de Fer, entered the profession by inheriting a publishing business. Jaillot and others came from artistic domains such as engraving. Guillaume Delisle was initiated into geography and history by his father, Claude, a professor at the Court of France. Trained also by Jean-Dominique Cassini, director of the Observatoire de Paris, Delisle acquired some basic knowledge of astronomy, which gave him a fresh approach to cartography.

Although *géographes de cabinet* did not travel, their work could be very taxing. Bellin noted that his "study was long, unpleasant, and hard,"

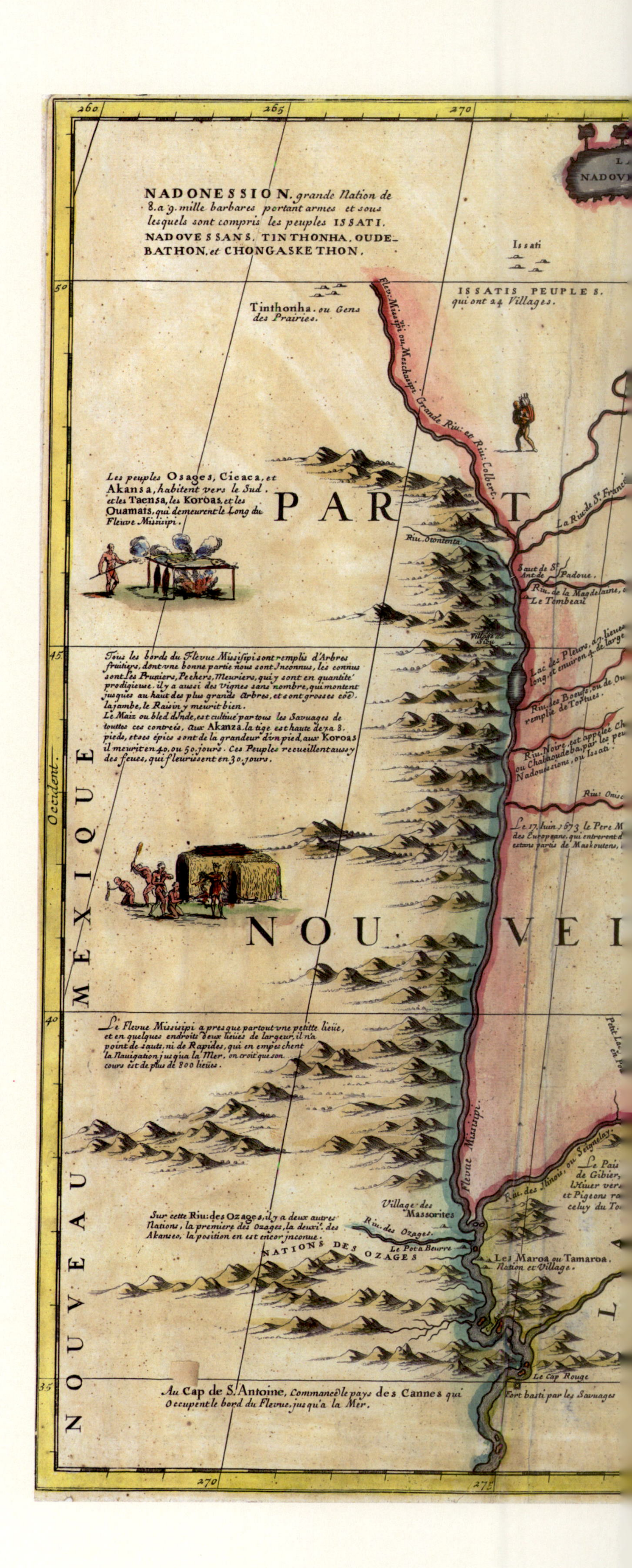

PARTIE OCCIDENTALE du CANADA ou de la NOUVELLE FRANCE
ou sont les Nations des ILINOIS, de TRACY, les IROQUOIS, et plusieurs autres Peuples;
Auec la LOUISIANE Nouuellement decouuerte etc.
Dressée sur les Memoires les plus Nouueaux.
Par le P. Coronelli Cosmographe de la Ser.me Repub. de VENISE
A PARIS
Chez J.B. Nolin Sur le Quay de l'Horloge du Palais au Coin de la Rue de Harlay a l'Enseigne de la Place des Victoires
Auec Priuilege du Roy.
1688
TERRE DE LABRADOR
Eschelle
Milles d'Italie, ou Milles Pas Geometriques.
Lieües Communes de France.
Lieües Communes d'Espagne.
Milles Communs d'Angleterre.
Lieües d'une Heure de Chemin.
Septentrion
Nadir
BAYE DE HUDSON
Baye de James
GENS DES TERRES.
Ces Nations passent leur vie a chasser dans les Forests et ne cultiuent point la Terre.
LES NATIONS
LAC DE TRACY, ou SUPERIEUR, et LAC DE CONDE. il se Decharge dans le Lac des Hurons.
On croit qu'il y a des Mines de Cuiure dans cette Isle.
Quelques vns rapportent que dans ce Lac, on trouue du Cuiure Rouge tres pur.
DU CANADA, ou
FRANCE.
LAC DES HURONS, et KAREGNONDI ALGONKINS MICHIGANGE, ou LAC D'ORLEANS.
LAC DES ILINOIS ou MICHIGAMI ou LAC DAUPHIN.
LAC FRONTENAC, ou ONTARIO, et SKANIADORIO.
LAC ERIE ou TEIOCHARONTIONG et LAC DE CONTY et du CHAT.
Montaignes du Nort.
Les Missisaghe.
Algonquins Peup.
LES CINQS NATIONS IROQUOISES.
Il y a aussi des Iroquois, qui demeurent sur la Coste Septentrionale du Lac de Frontenac.
La Nation d'Anclastogheronons, qui estoit aux enuirons du Lac Erie, a esté destruite depuis quelques annees par les Anglois, a la Solicitation des Iroquois.
Le premier Ianuier lan 1682 Mr. de la Salle entra dans la Riu: des Ilinois, et la trouua glacée, il fut obligé de trainer les Canots et Equipages enuiron 80. lieues jusqu'ou la Riu: estoit Nauigable, car le froid n'y dure que six Semaines, et il descendit en Canot jusqu'au Fleuue Missisipi, trouuant dans ce temps de chasse. La Riuiere est aussi large et profonde que la Seine a Paris.
Aux enuirons de la Riu. des Miamis, sont les Peuples Miamis, les Masconten, et les Oiatinon, autrefois les Miamis, estoient pres de la Baye des Puants.
Maskoutens, ou Nation du Feu Composée de trois Nations des Miamis, des Maskoutens, et des Kikabeux, ou Kikapous.
Fort de Creuecoeur basti par Mons.r de la Salle lan 1679.
ILINOIS.
Village des Miamis
Portage de Miamis

and that he had to "spend considerable time preparing and assembling the data needed, and often with very assiduous work, and one can hardly flatter oneself that one has mastered the difficulties that present themselves." The trade of geographer required knowledge in such areas as geometry, astronomy, and navigation. In order to produce an exact, accurate map, the *géographe de cabinet* became a relentless collector. He amassed and pored over navigation journals and voyage accounts for a specific region. It was preferable if he knew several languages so that he could consult a wider pool of sources. He transcribed the works of his predecessors, competitors, and collaborators in order to study them. He also drew coordinates calculated from observations made by the Académie des sciences, then compared all these sources, drew sketches, and hunted down errors and inconsistencies. A good *géographe de cabinet* was patient and methodical, and he had an excellent memory, an organized mind, and sound critical judgment, which allowed him to separate the wheat from the chaff.

Of course, an unscrupulous cartographer could make maps more quickly simply by copying the work of others. But, as a famous quarrel that took place in the early eighteenth century testifies, this could prove risky. In a series of open letters published in the *Journal de Trévoux* in 1700, Claude and Guillaume Delisle accused Jean-Baptiste Nolin of plagiarism. To demonstrate the originality of their work, the Delisles disclosed their sources, especially those used to locate the mouth of the Mississippi (farther east) and to draw California as a peninsula (not an island). The Delisles launched a lawsuit against Nolin, and, supported by the scientific community, won their case. In addition to opening a window on the cartography trade, this affair served to signal that the profession was developing and becoming more rigorous. Bellin facilitated this progress by publishing the sources that he used to make his maps. Cartography was taking its place on the stage of the Enlightenment.

In the eighteenth century, the best geographers established a network of correspondents in the colonies, who kept them informed of the latest discoveries. Bellin, as a hydrographer for the Dépôt de la Marine, communicated with the captains of the king's ships to obtain new information on the St. Lawrence River and the Gulf of St. Lawrence. The Delisles, father and son, obtained information from Pierre Le Moyne d'Iberville that enabled them to situate the mouth of the Mississippi River. A few years later, Guillaume Delisle was in correspondence with the missionary François Le Maire, who sent Delisle a map of Louisiana that enabled him to make his own, corrected map. The Delisles' archives

***Partie occidentale du Canada* by Vincenzo Coronelli, 1688 (preceding page) and detail of a globe by Coronelli presented to Louis XIV in 1683 (below)**
The density of place names on this map of the western part of Canada (preceding page) is exceptional, as is the newness of the names. Around the Great Lakes and along the Mississippi River, the mapmaker locates a multitude of Indian nations allied with the French, among them the Crees, Ojibways, and Sioux in the northwest, the Poutouatamis, Kikapous, Maskoutens, and Outagamis around Baie des Puants, the Miamis on the Ohio River, the Algonquins on the Ottawa River, and the Ilinois along the Illinois River. It also shows, for the first time on a printed map, the PORTAGE DE CHECAGOU and Lac de Taronto, names later associated with two great North American cities. The cartographer was not sparing in his commentaries on the regions described. Various inscriptions enhance the drawing. For instance, at Lake Superior "is found very pure red copper," and the Illinois River is "as wide and deep as the Seine." Beautiful illustrations embellish the map: scenes of hunting, harvesting crops, and cooking. How did a Franciscan monk from Venice, Vincenzo Coronelli, manage to paint such a faithful portrait of the region? How had he written in so many exotic names so accurately? In 1680, Coronelli had been asked by the French ambassador to Rome, Cardinal d'Estrées, to offer Louis XIV two gigantic globes celebrating the glory of the king, representing Earth and the celestial concave (below). This commission opened the doors to the archives of the kingdom of France, and Coronelli gained access to very rich documentation on exploration of the New World, notably the manuscript maps sent to the Court of France, as well as the most recent travel accounts, including those by Greysolon Dulhut, Cavelier de La Salle, Louis Jolliet, and Jacques Marquette. From these sources, he was able to draw an accurate picture of a hinterland full of promise for the French in Canada searching for furs and other riches vital to the colony.

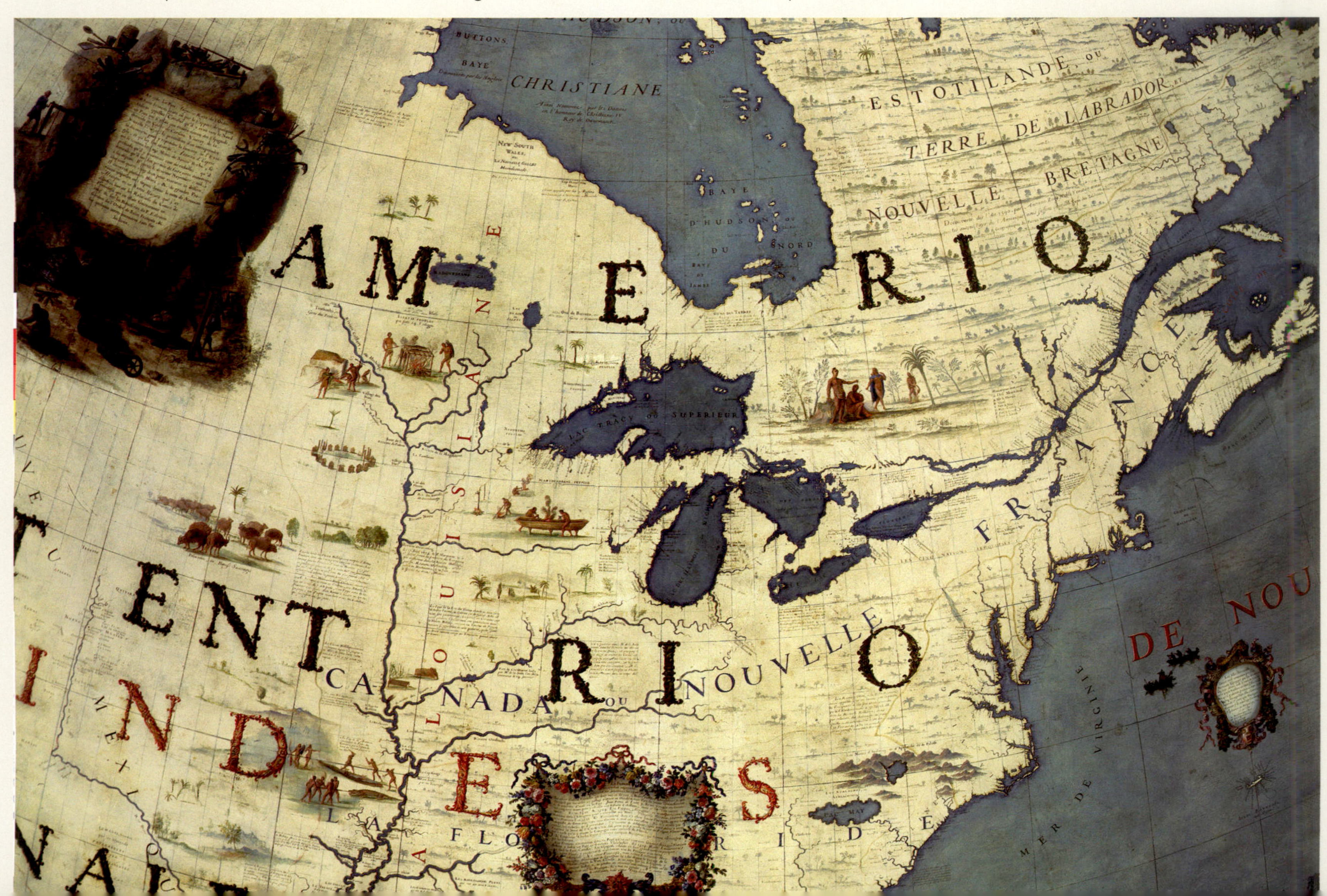

***Amérique septentrionale* by John Mitchell, translated from the English by Georges-Louis Le Rouge, Paris, 1756**
French geographers were far from the only ones to describe North America. In the eighteenth century, English geographers provided lively competition for the Dutch and French. Cartographers such as Henry Popple (see pp. 214–15), John Mitchell, and, later, Aaron Arrowsmith (see p. 249) produced works remarkable for the quality of their contours. Men of privilege, Popple and Mitchell both had access to the archives, journals, and maps of the Board of Trade and Plantations in London, the government institution that managed the American colonies. Encouraged by the authorities, they obtained recent cartographic surveys made by colonial governors. Arrowsmith benefited from access to the archives of the Hudson's Bay Company, whose explorers had gathered a huge amount of information on the western part of the continent. This map by John Mitchell is considered by American historians to be the most important in their colonial history. Surprisingly, Mitchell was not an experienced cartographer, but a physician from Virginia who was interested in mapmaking as a hobby. Mitchell considered the presence of the French on the continent more or less illegitimate. With their presence along the St. Lawrence River, the Great Lakes, and the Ohio and Mississippi rivers, the French surrounded the English colonies; in so doing, he claimed, they were impinging on territories that rightly belonged to England. Mitchell used cartography to demonstrate the foundations for his fears and recriminations to his compatriots. Basing himself on royal charters, he claimed for the English colonies a vast territory extending from the Appalachians up to the St. Lawrence in the north, beyond the Great Lakes to the northwest, and as far as New Spain to the west, thus denying Louisiana's right to exist. This very useful map was translated into French one year after it was first published, and was subsequently republished many times. The negotiators of the Treaty of Paris ending the American War of Independence (1783) used one of these editions to trace the borders between the United States and British North America (see map p. 248).

also reveal the meticulous attention that cartographers paid to the sources at their disposal. As Nelson-Martin Dawson notes in *L'Atelier Delisle*, the Delisles went through voyage accounts with a fine-toothed comb, from the accounts of Jacques Cartier to those of the Jesuits to those of Samuel de Champlain, then extracted and annotated the passages relevant to geography. By virtue of the sheer quantity of the manuscripts that they left behind, the Delisles can be considered professional cartographers.

In spite of this scientific rigour, those early maps were not devoid of errors. Bellin, for example, drew large islands in the middle of Lake Superior, an error whose origins are attributable to descriptions by Chaussegros de Léry. Under the influence of Franquelin, Delisle drew into northwest Quebec a fictional lake called Kaouinagamic. And this forerunner of French scientific cartographers drew a lake 300 leagues around – this became the famous "western sea" that preoccupied navigators throughout the eighteenth century. Most of these errors were found on the edges of maps, in zones that had been recently visited but for which cartographers lacked sources against which to check their information; they were to be expected in a context of a paucity of sources and certainly do not throw into question the colossal amount of work done in France by *géographes de cabinet*.

III

CONQUERING NORTH AMERICA

EIGHTEENTH CENTURY

CARTE TRES CURIEUSE DE LA MER DU SUD, CONTENANT DES REMARQUES NOUVE
Mais aufsy sur les principaux Pays de l'Amerique tant Septentrionale que Meridionale, Avec les Noms & la Rou
Terre de Jesso ou Eso
Terre de la Compagnie
EMPIRE OU ROYAUME DE LA CHINE
MER ORIENTALE DE LA CHINE et SEPTENTRIONALE DU JAPON
Detroit de Sangaar
Golfe de Nanquin
TARTARES
MER DU SUD
Tropique du Cancer
ISLES DES LARRONS OU DE MARIE ANNE
LES ISLES PHILIPPINES
MANILLES
ISLES MOLUQUES
MER PACIFIQUE
EQUATEUR OU LIGNE EQUINOCTIAL
Celebes
LES ISLES MOLUQUES
Nouvelle Guinee
Terre des Papous ou Nouvelle Guinee
MER DE LANTCHIDOL
Carpentarie
TERRE DE QUIR
MER DU
Isle Guadalcanal
Les Isles de Salomon
Terre de Wit ou Terre Blanche Decouverte en 1628.
NOUVELLE HOLLANDE Decouverte l'an 1644.
Terre d'Eendracht ou de Concorde Decouverte l'an 1616.
LES CALIFORNIES OU CAROLINES
NOUVELLE GRENAD
NOUVEAU M
Palmiste Epineux
Lamantin
Terre des Etats ou Nouvelle Zelande
Terre d'Antoine Diemens
MER DU NORD
MER DU SUD
Port de Acapoulco
Golfe Mexique
Port de Callao
Lac d'Eau Douce

ES ET TRES UTILES NON SEULEMENT SUR LES PORTS ET ILES DE CETTE MER,

s Voyageurs par qui la decouverte en a été faite. Le tout pour *l'intelligence Des Dissertations suivantes*

MER DU NORD

OCEAN MERIDIONAL

MER DU SUD

Tropique du Cancer

Equateur ou Ligne Equinoctial

Tropique du Capricorne

Golfe de Mexique

Amerique Meridionale

Terre Ferme

Nouvelle Grenade

Guiane ou Guaiane

Pays des Amazones

Bresil

Costes du Bresil

Paraguay

Guayra

Tucuman

Rio de la Plata

Terre Magellanique

Patagons

Isles Antilles

Isles du Cap Verd

Isles Canaries

Les Isles Acores ou Terceres

Espagne

France

Barbarie

Nigritie

Guinee

Basse Guinee ou Etats

Royaume de Congo

Cafrerie ou Costes des Cafres

Cap de Bonne Esperance

Detroit de Gibraltar

Baye de Rio Janeiro

Baye de la Conception

Premier Meridien que les François font passer par l'Isle de Fer

IGNIS
Septentrio
AMERICA
Anno Domini 1492 a Christophoro Columbo nomine Regis Castellæ primum detecta, et ab Americo Vesputio nomen sortita 1499
AMERICA
SEPTENTRIONALIS
Circul. Arctic.
Nova Albion
Groenland
Frisland
Fretum Davis
Nova Britannia
Nova Anglia
Nov. Belgium
Virginia
Florida
Granada
MAR DEL NORT
Asores Ins
Al Flandricæ
MARE VIRGINIUM
I de Cabo Verde
Tropicus Cancri
SINUS MEXICANUS
MAR DEL ZUR
Circulus Æquinoctialis
OCEANUS PERUVIENSIS
AMERICA MERIDIONALIS
Guiana
Caribana
Rio de la Plata
Fretum Magellanicum
Fretum le Maire
Tropicus Capricorni
MARE PACIFICUM
TERRA AUSTRALIS INCOGNITA
Circulus Antarcticus
L.S.
Varias a variis et antehac
editas esse, et quotidie edi tabulas toti
Mundi situm depingentes, cuiq; notum est: hac
tamen nullam accuratiorem elegantioremq; esse
libere dico; quippe quæ succinctius quam omnes
aliæ singula depingit, multosque errores aliarum
tabular. corrigit. Imprimis vero in Oceano Tartarico
et circa illum, multa aliter meliusq; delineat: nec
non in nova Guinea, et India Orientali, ut et in Ame
rica Sept. et Mar del Zur, ac versus Fretum
Magellanici plurima innovat mutatque
vale et fruere.
Meridies
AQUA
GERARDUS MERCATOR Flander.

North America

THEATRE FOR COLONIAL RIVALRIES

Exploration and development of trading posts in North America sowed the seeds for the armed conflicts that were to sweep through the continent in the seventeenth and eighteenth centuries. All of the European powers with a stake across the Atlantic rushed into previously mysterious territory, as rumours about its existence circulated more widely. Their respective spaces were quickly defined. Those who sailed from the Iberian Peninsula found coveted precious metals in South America. After several fruitless expeditions farther north, they decided to give up on the northern regions and concentrate their efforts in the south.

For more than two centuries, the French and the English refined their knowledge of the interior of North America, doing battle over each strip of territory. The ultimate goal was access to the Pacific coast, where they believed that they would find the "western sea" – which seemed inexplicably to withdraw constantly before them. And as they advanced, the colonizers had to secure territorial acquisitions that were, to various degrees, fragile and dispersed.

New France, first established on the shores of the St. Lawrence – the gulf and the river – relatively close to Europe, was a particularly accessible and coveted prey. Thus, Acadia was the site of the earliest friction between two great powers, France and England. The Abenakis, mistreated in the territories where the colonists of New England had settled, turned instead to the French, to whom they provided unstinting support against the English. After the first attack in 1613, by Sir Samuel Argall of the Company of Virginia, Acadia was captured by the English several times and then restored to France through royal charters or peace treaties; finally, starting in 1755, Acadians who remained unsubdued by the English colonial government were deported and dispersed.

Newfoundland was the site of frequent skirmishes between English and French fishermen. Both groups wanted access to the best places onshore for drying their catch, as well as access to the Grand Banks. St. John's was founded by the English in 1651, on the east coast of the island, and in response the French established a fortified post farther south, at Plaisance, in 1660.

Farther north, at Hudson Bay, where traders found the best furs, the English had built a number of fortified posts. From these bases, they pursued their search for a "northwest passage" that would take them to the "western sea".

Although the settlements in the St. Lawrence Valley were less exposed, they were nevertheless attacked by the Iroquois or by troops from New England. For English traders, the goal was to control the fur trade originating in the Great Lakes, divert it toward Albany and New York, and block convoys heading for Montreal and Quebec.

Conditions were soon ripe for conflict. The beginning of the war of the League of Augsbourg (1688–97) in Europe provided an opportunity for confrontation between English and French in North America. Forced to fight on all fronts at once, the French colony nevertheless managed to preserve all of its land by virtue of the Treaty of Ryswick, which confirmed the French victories. This first inter-colonial war highlighted the difficulties faced by the Canadians in defending such a large territory, especially as they continued to explore southward to the Mississippi River basin in search of new possibilities for fur supplies and access to the Gulf of Mexico.

During the eighteenth century, more wars between France and England gave rise to conflicts between New France and the English colonies, always with the same scenario: points of friction, latent hotbeds for war, justified violent confrontations in North America, extensions of

Preceding pages (142–143)
***Carte tres curieuse de la mer du sud* by Henri Abraham Chatelain, 1732**
This immense map of the Americas was made by the Flemish cartographer Henri Abraham Chatelain. It is remarkable for the engravings that decorate the unknown territories to the north and south. Adorned with the adjective "curious" by its maker, this map can be considered the cartographic equivalent of the curio cabinets of princes and scholars. Among the "curiosities" portrayed in three insets are "typical" scenes in New France: an Indian wedding, cod fishing, and strangely humanoid beavers building a dam. The "southern sea" evoked in the title is the old name for the Pacific Ocean, whereas the Atlantic was known as the "northern sea." The Pacific was an enticing and promising chimera for a number of voyagers seeking spices and other riches of the Orient. Chatelain refers to several of these explorers either by immortalizing them in portraits or by depicting the routes that they took against a background of oceanic cartography. Among the "discoverers" of New France, he cites Robert Cavelier de La Salle, Thomas Aubert, Giovanni Verrazzano, and Jacques Cartier.

***Opposite*: detail of a planisphere by Hondius**

TERRES ARCTIQUES
CARTE DE LA NOUVELLE FRANCE où est compris LA NOUVELLE ANGLETERRE, NOUVELLE YORC, NOUVELLE ALBANIE, NOUVELLE SUEDE, LA PENSILVANIE, LA VIRGINIE, LA FLORIDE, &c.
PAÏS DE LA NATION DES ASSINI-BOU-ELS, OU ASSINI-NIBOUELS
LAC DES ASSINIBOUELS
LAC DES CHRISTINAUX
POU-CHRISTINAUX
NATION ET PAÏS DES CHRICKS, OU
NOUVEAU DANEMARC
PORT NELSON
BAIE D'HUDSON
LABRADOR
TERRE DES ESQUIMAUX
NOUVELLE FRANCE
PAÏS DE LA NATION DES SIOUX DE L'OUEST
SIOUX DE L'EST
LAC SUPERIEUR
LAC ALIMIPIGON
MICHIGAN OU LAC DES ILINOIS
LAC DES HURONS
LAC ERIE OU DU CHAT
PAIS DE LA NATION DES ILINOIS
PAIS DE LA NATION DES PANIS
MINE DE CUIVRE
PAIS DES CICACHA
RIVIERE OHOHIO
LES BOUCHES DU FLEUVE MISSISIPI
GOLPHE DE MEXIQUE
PENINSULE DE LA FLORIDE
CANAL DE BAHAMA
LES ISLES LUCAYES
LES ISLES ANTILLES, OU CARIBES
ISLE DE CUBA
ISLE S. DOMINGUE
PENINSULE DE YUCATAN
GOLPHE DE HONDURAS
PANUCO
ECHELLE DE 200 LIEUES

the fighting underway in Europe between the colonial metropolises. In each war, France lost a bit of its land. The 1713 Treaty of Utrecht, ending the War of Spanish Succession, took away Newfoundland, Acadia, and much of the Hudson Bay drainage basin. To protect its still considerable territory, France decided to fortify Cape Breton Island by building Louisbourg. This fortress-town, "key to the St. Lawrence," became a crossroads in the trade with the West Indies.

To compensate for the loss of fur supplies from Hudson Bay, the French colonial administration embarked on a vast expansion plan. Starting from Detroit, founded in 1701, it established new trading posts in the Great Lakes, began to open settlements in Louisiana in 1717, and erected a chain of forts in the Ohio Valley. In addition, Canadians began an unprecedented wave of exploration. A first probing of the territory by the Jesuit Pierre-François-Xavier de Charlevoix strongly encouraged Governor Charles de Beauharnois de La Boische to begin searching again for the western sea. Supported by scientific circles and the minister of the navy, he authorized a major expedition in 1731.

Pierre Gaultier de Varennes et de La Vérendrye, a naval officer and fur trader, was mandated to lead the mission of discovery. It was the beginning of an adventure that would last more than two decades, taking La Vérendrye and his sons south and west of the Great Lakes. The explorers received indispensable geographic information, as well as sustenance and logistical support, from the Indian nations. The western sea was not found, and for good reason, but a wealth of geographic and ethnological data on the central plains and the Hudson Bay drainage basin were gathered. Fortified posts were built, attracting a good part of the fur trade that would otherwise have gone to the Hudson's Bay Company.

With yet another conflict, the War of Austrian Succession, the colony had to return to military concerns in the easternmost part of its territory. The colony was weakened by the fall of Louisbourg in 1745 – although it was returned under the Treaty of Aix-la-Chapelle in 1748 – and was unable to recover before the Seven Years' War began. The priority was to block the westward advance of English colonies, and so Canada chose to consolidate its positions among the Illinois in the Ohio region; the facts would bear out this strategy, since this was where the French had their most convincing victories.

The conflict that broke out in North America in 1754, and in Europe in 1756, involved two colonial empires doing battle for worldwide trade supremacy. In North America, there was an obvious imbalance: 1.5 million inhabitants in the English colonies, versus 85,000 in New France. In addition, England's Royal Navy had become the most powerful in Europe, making the military forces

***Carte de la Nouvelle-France* attributed to Jean Baptiste Louis Franquelin, circa 1708**
For 30 years, Franquelin devoted himself to collecting and organizing geographic data on North America. This map, produced during the War of Spanish Succession, was the culmination of his work. The coat of arms above the view of Quebec is that of the son of Louis XIV, Dauphin Louis de France. The sight of an immense empire that France had built in North America would certainly have pleased the heir to the throne. The profusion of place names provided, in a way, documentation of French knowledge and possession of the territory. It also shows that the colonial empire was built thanks to alliances between French and Indians. Indeed, most of the names are of Indian origin, in spite of Franquelin's suggestion that they be to translated into French. On the lower left is an Indian holding a compass and a marker showing the scale of the map – an allegory for the contribution by indigenous peoples to the scientific description of the continent.

in play unequal. In spite of its valuable Indian allies, French North America was a territory too vast to defend. After one last victory, in Carillon in 1758, Louisbourg was the first fortified place to fall. All of the others followed, including Quebec, the capital of New France, in 1759; Montreal surrendered in 1760. The Seven Years' War, the climax of French–English tensions in North America, culminated in the almost complete supremacy of England and left only a few vestiges of the colonial empires of France, Spain, and the Netherlands.

In another part of North America, the Pacific coast, a monumental struggle began over the long-sought-after western sea. Ever since the Treaty of Tordesillas (1494), Spain had laid claim to the Pacific Ocean. Firmly ensconced in Mexico, the Spanish had not had to defend their territory. New Spain was neither threatened nor occupied by other powers until the 1730s, when the Russians began making incursions to hunt sea otters to sell to China. Sailing from Mexico, a number of Spanish expeditions attempted to occupy Nootka (Vancouver). At the same time, the circumnavigation by Louis-Antoine de Bougainville (1766–69) and the expedition by Jean-François de Galaup de Lapérouse (1785–88) led these explorers to the North American Pacific coast.

The greater threats were the expeditions of the English explorers James Cook (1776–79) and George Vancouver (1792–95), who took with them new means of scientific investigation. Between 1785 and 1787, 17 foreign ships, including 14 from Great Britain, dropped anchor at Nootka, while the Russians made themselves quite at home there. In response, Spain sent eight expeditions between 1788 and 1793, with the intention of occupying the territory and imposing its authority in order to preserve its privileges in what it still considered the "Spanish Lake." The leader of the mission, Esteban José Martinez, built a post at Nootka in 1788. On July 2, 1790, the captain of an English ship sent from Macao, James Colnett, announced to Martinez that he had received orders from England to erect a permanent settlement on the site. Martinez had the ship captured and its captain arrested. When word of these events reached England, the English military forces were mobilized, and soon a powerful fleet was ready to make war against Spain. Spain asked for the assistance of France in the name of the "family pact" (1762). On August 26, 1790, in the Assemblée nationale, the deputy Honoré Gabriel Riqueti, comte de Mirabeau, denounced this pact between France and Spain as undemocratic and obsolete, condemned the arming of English vessels, and convinced the legislature not to support new colonial conquests or to use force to dominate free peoples – thus, not to support Spain against England. Without the assistance of the French navy, the Spanish Armada was not strong enough to confront the powerful Royal Navy, so Spain gave up its attack on England and withdrew from Nootka. And this is how, thousands of kilometres from the territory claimed by Spain, an intervention by a French deputy in the Assemblée nationale led some historians to say, "Mirabeau gave Spanish Columbia to England." Globalization was underway!

By supporting the English position, the Nootka Agreement opened the entire Canadian West to British imperial development. One could thus say that Canada, "from sea to sea," was born of the position taken by Mirabeau. England's maritime superiority and the abandonment by Spain of its long-time claims on the Pacific did the rest.

Detail of the map on the preceding page

Main sources

Boissonnault, Charles-Marie. "Mirabeau donne la Colombie espagnole à l'Angleterre. *Mémoires de la Société royale du Canada* (1972), 110–11. — Litalien, Raymonde. "L'expansion du territoire de commerce et des conflits armés en Amérique du Nord sous l'Ancien Régime." In *Rochefort et la mer. Guerre et commerce maritime au xviii*e *siècle*, vol. 9, pp. 45–58. Jonzac: Université francophone d'été, 1994.

Franco-Indian Relations

ALLIANCES AND RIVALRIES

The respective colonial behaviours of European countries have been compared many times. The American historian Francis Parkman summarizes the treatment of Indians by the three main colonial powers as follows: the Spanish "crushed" them, the English "scorned and neglected" them, and the French "embraced and cherished" them.

The odious conduct of the Spanish was denounced by their own people, including, vociferously, Bartolomé de Las Casas. After all, they had appropriated the land and its inhabitants in one fell swoop. The English, on the other hand, arrived in regions that were less populated, but they were numerous and they wanted all of the land. The Indians, useful at the beginning, quickly came to be seen as a nuisance. They were pushed off their land, beaten, hunted down, or deported to west of the Mississippi River. Those who misbehaved were massacred.

The French arrived a few at a time, seeking adventure and furs. Indian men presented themselves as trading partners; Indian women, as essential companions for the traders. Posts and forts were opened after agreements had been reached with the original inhabitants of the land. There were never any territorial treaties – that is, territorial dispossession. Cohabitation resulted in much mixing of blood in the areas west of Montreal.

New France lasted as long as its Franco-Indian alliances did. The texts produced at the time contain the phrases "to live together," "inseparable companions," and "close ties." Champlain went so far as to envisage a single people. The Indians, once converted to Christianity, became subjects of the French king. Anglo-Americans had difficulty comprehending what was happening in French America, which surrounded them. They considered the French and the Indians a single common enemy – evidenced by the fact that they called the succession of wars between them and the French the "French and Indian Wars."

Things started to change once the Anglo-Americans began to intermarry. Molly Brant, a Mohawk, opened the eyes of her husband, the powerful William Johnson, who became superintendent of North Indian Affairs. The British prime minister, William Pitt the Elder, proceeded cleverly to co-opt the Indian allies of the French, one by one. His Atlantic blockade stopped the flow of reinforcements and of the gifts that were essential to keeping the alliances alive. Johnson was a realist. He did not try to reverse alliances but was content with obtaining neutrality, in exchange for which he promised the Indians "all the protection that they could hope for." When the Indians gathered at Caughnawaga (Kahnewake) in September 1760, they requested "the peaceful possession of the Spot of Ground we live now upon, and in case we should be removed from it, to reserve it to us as our own."

Anadabijou's Enemies Become the Enemies of the French

At first, not all Indians were unconditional allies of the French. In traditional historiography, there is a shadow on the picture. In 1603, at Tadoussac, the chief, Anadabijou, preferred war to peace. In accepting the alliance proposed by Anadabijou, the French inherited his most ferocious enemies: the Iroquois. Champlain and the French administrators who followed in his wake had to do battle with the Iroquois. And the Iroquois were not the exception. Close examination of the history of New France in the context of North America reveals many wars between the French and the Indians. Most of these wars were accidental and brief, but at least two served to upset the colony. One involved the Fox nation; the other – an extremely violent war – involved the Natchez. The latter war has drawn more attention because the French were fascinated by the reputedly peaceful Natchez, the "People of the Sun."

Savage of Canada, eighteenth century
Drawing by Claude-Louis Desrais, engraving by Jean-Marie Mixelle. "One does not civilize these energetic barbarians by seizing their homeland and their freedom," wrote the militant republican Sylvain Maréchal. Speaking of the "savages" (Illinois, Algonquins, Hurons, Iroquois, etc.), he added, "One will exterminate them before one subjugates them, and one will not peacefully enjoy Canada until one has made it into a desert."

Map of Lake Ontario, 1688
"Ontario" means "beautiful lake" in Algonkin. This lake first appeared on Champlain's maps as Lake Saint-Louis. Later, Sanson's maps included its current name. The interpreter Étienne Brûlé was the first European to explore its shores. Although Lake Ontario formed an extension of the St. Lawrence Valley, the French did not dare to venture into Iroquois territory. Between 1673 and 1676, nevertheless, they built two forts at the western and eastern ends of the lake, Fort Frontenac (KATARAKOUY) and Fort Niagara, in an attempt to keep an eye on the Iroquois and counter English competition. This map situates the people living around the lake circa 1688, most of them Iroquois. It also shows the long portage that the Iroquois took to reach Lake Erie and attack the Illinois, a people allied with the French.

Léo-Paul Desrosiers refers to the Iroquois territory as Iroquoisia. The geographic location of this territory, south of the Great Lakes and within a triangle formed by today's New York City, Detroit, and Montreal,made it of primordial importance. Five Iroquoian nations – Mohawk, Oneida, Onondaga, Cayuga, and Seneca – had formed a federation. At about the same time, new partners emerged: to the south, the Dutch and English; to the north, the French. The Iroquois had to expand their territory to respond to new trade possibilities – that is, they had to not only eliminate their traditional rivals but also defeat their neighbours, the Hurons, Petuns, and Eries. The pressure was made even greater by the fact that their own numbers were dwindling at an alarming rate. In just 20 years, from 1630 to 1650, the total Iroquois population dropped from 21,740 to 8,734, according to the historian Dean R. Snow. These numbers may not be exact, but they give an idea of the scope of the problem. The Mohawks were also being decimated, numbering only 1,734 in 1650, compared to 7,740 in 1630. Victims of the wars that they were conducting on all fronts, they were also struck down by epidemics. They sowed terror, particularly among the French, but they lived in terror themselves.

The destruction of Huronia (1649) was an unhappy milestone. French penetration was jeopardized. New France was threatened with extinction. In 1663, the king took things in hand and replaced the company system with a royal government. In a letter dated March 18, 1664, Minister Colbert summarized the sovereign's wishes: "Entirely ruin these barbarians who are already greatly diminished following the last relations we had and the losses that they have suffered against their enemies and by a type of contagious disease that has killed a good many." The final blow was the deployment of a special military force, the Carignan-Salières regiment, comprising some 1,200 men.

Expeditions were organized against the Iroquois villages. Informed of this, and terrified, the Indians slipped away. In war, Indians operated either by ambush or by disappearing. They were no fools: they stayed alive, abandoning their houses and harvests to an enemy who, in fact, was surprised to discover how well organized their settlements were.

The Iroquois might have thrown themselves into the arms of the English, who could not have asked for more.

They did not do so, however. They were a free people and intended to remain so.

The lieutenant general, Tracy de Prouville, and his Carignan-Salières regiment resolved nothing on the military front – but there were other effects. Rather than return to the homeland, a good number of officers accepted seigneurial concessions, and almost half of the soldiers decided to stay in New France. The French authorities sent over women, the famous *Filles du Roi*, most of whom came from the region around Paris. They spoke a polished French and imposed it on those around them. In no time at all, the colony's population mushroomed; between 1663 and 1695 it quadrupled, increasing from 3,000 to about 12,500.

The wars continued. "A new generation of Canadians who knew the forest, the water, the Indian style of fighting" was born. Other expeditions – under Governors La Barre (1684) and Denonville (1687) – marched against the Iroquois. Poorly led, these expeditions were failures. The Iroquois responded in 1689 with the Lachine Massacre. The fuse was now lit. When Louis de Buade de Frontenac returned to replace Denonville, he found only devastation. He was seething. He knew what to do, and how to do it. "It would be good to see to the English in this way on their own territory," Frontenac wrote to his minister, rather than to let them "come and attack with the Iroquois from different locations, as they boasted that they could do." Small, devastating raids were conducted against posts in New England: Corlaer, Salmon Falls, Casco. The conquest of New York was envisaged. Now, the true rivals stood face to face, and a war to the finish broke out between France and England. To hold their positions in North America, the French multiplied and consolidated their alliances with the Indians. Frontenac got busy. He established the foundations for the Great Peace of 1701. Governor Louis-Hector de Callière accomplished that peace. French America had not spoken its last word. It reinforced its positions from Newfoundland to the Rockies, from Hudson Bay to the Gulf of Mexico.

The Sad Fate of the Indians of Lower Louisiana: The Natchez

As preparations for the peace of Montreal were in full swing, the Le Moyne brothers, d'Iberville and de Bienville, came into contact with the tribes of the Gulf of Mexico. The scenario of 1603 in Tadoussac was repeated. Alliances were created with the Biloxis, Bayogoulas, Houmas, Natchez, Taensas, Cénis, and Choctaws. At the same time, the French inherited the enemies of these tribes: the Creeks and the Chickasaws, who were allied with the English of Carolina. The Le Moyne brothers managed, with some success, to pacify these tribes and, especially, to stop the Chickasaws' slave raids.

The Indians' main enemy, however, was smallpox. The passage of explorers, missionaries, and merchants caused epidemic after epidemic. The Quapaws and Bayogoulas were almost wiped out; the Choctaw population dropped from 30,000 living in 50 villages (in 1699) to 8,000 dispersed in 27 villages. The Natchez, the "People of the Sun" immortalized by François-René de Chateaubriand, were reduced, in less than one generation, from some 60 villages led by 800 Suns (chiefs) to no more than six led by 11 Suns. The French encouraged "the small remnants of nations" to group together and move closer to the French posts. The Chickasaw raids and the epidemics caused great anxiety, but the French had the situation well in hand when a completely unforeseeable conflict erupted between the French and the Natchez. Early contacts with the Le Moyne brothers, who related well to the Indians, had been successful. In 1700, d'Iberville had concluded an alliance with them. In the spring of 1716, de Bienville built Fort Rosalie on the Mississippi River, in the middle of Natchez territory. The agricultural potential quickly drew numerous settlers, who came to reasonable agreements with their Indian neighbours – until the day when a greedy and arrogant new commander named Détchéparre chose a piece of land occupied by a Natchez village to develop a plantation. Though they appeared to be capitulating, the Indians were in fact planning a surprise attack. On November 29, 1729, they massacred some 250 French people and took dozens of prisoners. In accordance with the law of an eye for an eye, the French took their revenge one year later. Many Natchez were killed or deported to Santo Domingo.

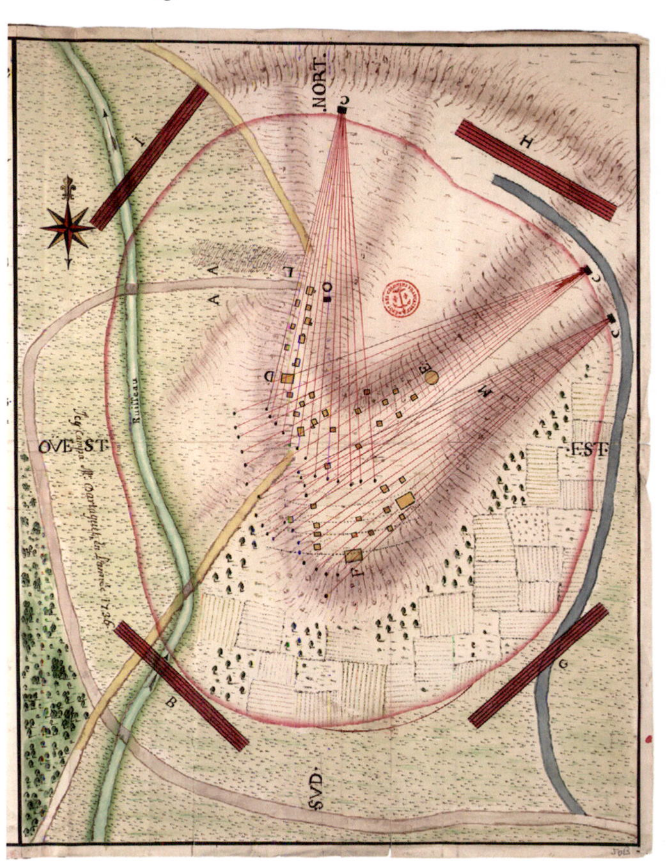

Plan of forts and attacks on the Chicacha village by Dupin de Belugard, Rochefort, 1736
In Louisiana, the Chicachas (Chickasaws) were fiercely opposed to the French. This warrior nation occupied a territory on the edges of what today are the states of Tennessee, Mississippi, and Alabama. Supported and armed by the English of Carolina, the Chicachas harassed the inhabitants of French settlements and attacked the convoys linking Canada and Louisiana. In 1736, the French decided to go to war against the enemy – a difficult prospect as this unknown territory was situated far from waterways. This plan shows how Commander Pierre d'Artaguiette's troops (indicated by the blue line and red rectangles) assaulted three Chicacha villages and suffered a devastating defeat. The reinforcements from well-hidden villages attacked the French army from the rear (yellow lines), captured d'Artaguiette and his officers, tortured them, and burned them alive.

A Ste catherine Concession du Sr de Coly, a une lieüe du fort francois et qui a eu le Meme sort; son chef y a eté Massacré.
C Natchez sauvages qui d'Amis q
etoient, ont detruits le fort et poste fra
leur Village esta une lieües de ce for
Rivierre de sainte catherine
A
C
Beausejour
La forest
Le Sr Chauvin
Le Sr mouton
Le Sr Ladouceur
Le Sr Poulains
magazin de la compagnie
chemin de sainte catherine au fort
grange
Le Sr Rousin
vieux magazin
Le Sr villeneuve
Le Sr chepart
LE FLEUVE SAINT

le marechal de belleisle a une lieue du
fort francois, elle a eu le meme sort que
luy et le Sr denoyers directeur y a-
perdu la vie ainsi que Bien dautre
elle eloignee du fort francois d'une lieue
B
LOÜIS
CARTE
du fort Rozalie des Natchez françois
ou l'on Voit la situation des Concessions
et habitations telles qu'elles etoient auant le
Massacre arrivé le 29 Nouembre 1729
et le tout par la faute de celuy que la
Compagnie des indes auoit choisy pour y
Commender
ECHELLE De Vintg Arpents
1 2 3 4 5 10 20
il en faut 80 pour une Lieue

Map of Fort Rosalie in Natchez country before the massacre of November 29, 1729, attributed to Dumont de Montigny

In 1716, the French established Fort Rosalie, named for the wife of Minister Pontchartrain, in Natchez country. This plan of the fort, while providing good documentation, refers to one of the bloodiest episodes in French colonial history. Relations between the French and Natchez were quite cordial at the start, but they deteriorated following unfortunate incidents in 1716 and 1723. Then, in 1729, the Natchez confronted the commander of Fort Rosalie (who Dumont de Montigny called the Sieur Chépare or Chépart), who had received an order "to build large storage buildings" on the land where, the Indians said, "the bones of their ancestors were at rest in their temple." Pretending to accede to the French demands, they requested "two moons" to vacate the area. Instead, they used this time to prepare for a revolt. The alert was finally given, but it did not prevent a bloody massacre from occurring. The French response was quick in coming and even more brutal.

Fox Indian
Like all Indian nations, the Foxes jealously guarded their role as intermediaries. Settled west of Lake Michigan, they controlled one of the routes leading to the Mississippi River. The Foxes made life difficult for the French. The anonymous artist who made this portrait chose to emphasize the impressive stature of the warrior Culipa. Culipa was deported to France in 1731, destined for hard labour. He died the following year in a prison in Rochefort.

The Fearsome Fox Nation and the Famous Chief Kiala

Meanwhile, the Fox Nation was suffering a similar fate. After years of resistance and combat, their chief, Kiala, and his wife were deported to Martinique. Respected by his people, feared by the French, Kiala had been preceded by his formidable reputation. The planters worried about his influence on their slaves. The correspondence exchanged among the various administrations concerned, and from the king to Maurepas, Beauharnois, Hocquart, Champigny, and d'Orgeville, bears witness to this. On October 7, 1734, Beauharnois and Hocquart informed the minister of the deportation of the Fox chief, "a man who passes for intrepid in his nation, an enemy of ours, and who we must watch closely." Kiala's wife was placed in the custody of the Hurons of Lorette but she managed to escape. Beauharnois, driven almost mad by the revolt of the Fox Nation, vowed that she would suffer the same fate as her husband. What finally transpired, apparently, is that both were sold into slavery somewhere near the mouth of the Orinoco River in Venezuela.

The Foxes had been Beauharnois's nightmare, and vice versa. Settled at the bottom of Baie des Puants (Green Bay), they controlled the route to the west. They did not want to lose this advantage. In 1701, they sent representatives to the Great Peace of Montreal; the French counted them among their allies. Nevertheless, intrigues continued and the attractions of Albany persisted. A group of Foxes decided to settle near Fort Pontchartrain (Detroit) at the invitation of Lamothe Cadillac. Cadillac's successor, Renaud DuBuisson, inherited a difficult situation. The Indians were arguing among themselves: the Foxes were taunting the Hurons and Miamis from south of Lake Michigan and were not averse to the idea of trading with the English. Governor Vaudreuil was no novice, and this situation worried him. He called the Foxes, Kickapoos, and Mascoutens to a meeting in Montreal in March 1711, at which he demanded peace and even suggested that the Foxes return to where the remains of their ancestors lay. The Foxes wanted nothing to do with this, and their provocations continued. In May 1712, the Fox village in the Detroit region came under full attack by French and Indian forces. Pemoussa, their chief, didnt'd fear anything. After all, the Foxes were immortal! In a context of false truces and trickery, they were massacred or taken prisoner. A few survivors managed to reach Baie des Puants, and a war to the finish began. In 1716, the Foxes were attacked by the troops of La Porte de Louvigny; in 1728, Le Marchand de Lignery left Montreal at the head of a company of 400 soldiers and coureurs des bois and several hundred *domicilié* Indians. Along the way, the force swelled to 1,650 men. The expedition made needless detours, and the Foxes and their allies were warned of the threat. They melted away; de Lignery replayed the unfortunate scenario of Tracy against the Mohawks in 1666 and Denonville against the Senecas in 1687, and he had to be content with contemplating the remarkable fields of crops surrounding the village. He burned down the houses and destroyed the harvests "in the name of the glory of the king and the peace of the colony."

Beauharnois had only one thing in mind: the extermination of the Foxes. De Lignery had failed. In 1730, a new expedition set out. The names of the leaders, all with the particle "de," reveal their noble blood: following in the footsteps of de Louvigny and de Lignery were Noyelles de Fleurimont, Groston de Saint-Ange, and Nicolas-Antoine Coulon de Villiers. This time, the French prevailed. Not without difficulty, they ran the Foxes to ground in a fort on the banks of the St. Joseph River in Illinois country. Villiers took the Fox chiefs to Montreal in the summer of 1731; among them was Kiala. Deemed particularly dangerous, Kiala was deported, as mentioned before. The Foxes were now ready to make peace. But peace did not come until 1738; in the meantime, in 1733. the Foxes killed Villiers.

Terribly weakened, the Foxes tried to reconstruct their nation. In 1750, they numbered almost 1,000. Among them, in a new village on the Mississippi (Rock River), Marquis de La Jonquière noted the presence of a young Fox chief who had taken the name of his father, Kiala.

The war against the Foxes was in many ways reminiscent of the war between the French and their allies and the Iroquois. In both cases, the English were in play. The Indians were not so much opposed to the French as eager to defend their own interests. In reality, the Iroquois and, no doubt, the Foxes were not for the English and against the French; they were for themselves. A number of times, they accepted a peace with the French but not necessarily with their allies. This was true of both the Iroquois and the Foxes.

These three Indian nations, which caused such problems for the French, were the exception that proved the rule: French America existed thanks to alliances with the Indians.

Sources in order of importance

DESROSIERS, Léo-Paul. *Iroquoisie.* 4 vols. Sillery: Septentrion, 1998–99. The author, faithful to the sources, delivers a breathtaking work. — HAVARD, Gilles, and Cécile VIDAL. *Histoire de l'Amérique française.* Paris: Flammarion, 2003. An excellent synthesis. — JENNINGS, Francis. *The Founders of America.* New York: Norton, 1993. A work by a great master. — JENNINGS, Francis. *The Invasion of America: Indians, Colonialism, and the Cant of Conquest.* New York: Norton, 1976. — EDMUNDS, Russell David, and Joseph L. PEYSER. *The Fox Wars: The Mesquakie Challenge to New France.* Norman: University of Oklahoma Press, 1993. — SNOW, Dean R. *The Iroquois.* Oxford: Blackwell, "The Peoples of America" collection, 1996. — AXELROD, Alan. *Chronicle of the Indian Wars: From Colonial Times to Wounded Knee.* New York: Prentice Hall, 1993.

From the Spanish Succession

TO THE TREATY OF UTRECHT

"Expansion is the most worthy and the most pleasant occupation of sovereigns," Louis XIV said to the duc de Villars, Marshal of France, on January 8, 1688, on the eve of the War of the League of Augsburg. In pursuit of this ambition, the king of France involved his country in an improbable series of wars against most of Europe. Nations formed coalitions to resist France's hegemonic designs. The conflict was extended across the Atlantic: in often spectacular raids, formidable Canadian officers retook posts from the English. In the early eighteenth century, the French empire in America reached its geographic apogee.

The truce following the signing of the Treaty of Ryswick in 1697, ending the War of the League of Augsburg, lasted barely five years, confirming France's vigorous expansionist plans for North America. In 1700, Louis XIV founded Louisiana; in 1701, he had a post erected at Detroit. Also in 1701, the Great Peace of Montreal guaranteed the neutrality of the Iroquois Five Nation Confederacy in the Great Lakes region. Through these measures, France hoped to maintain its hold on the fruitful fur trade in the "upper country" and, with its control of the Mississippi River, block the westward movement of English settlers.

The Mississippi Basin was the future Louisiana. It lay next to the Spanish colonies, whose rich silver mines were the envy of the king of France, and he laid more or less overt claimed to them. The Infanta Marie-Thérèse, Louis XIV's first wife, had renounced the succession to the Spanish throne in exchange for payment of a dowry of 500,000 gold *écus* to him. However, the dowry was never paid. When the king of Spain, Charles II, died without an heir, on November 1, 1700, Louis XIV decided to put "the rights of the queen" to good use. Quickly, on November 16, he received the Spanish ambassador at Versailles and introduced him to his grandson, Philippe, duc d'Anjou, so that "he may hail him as his king." Then, according to Saint-Simon, he addressed his courtiers in these terms: "Sirs, here is the king of Spain. Birth called him to this crown, as did the late king in his will; the entire nation wished for it and asked me for it instantly; it is the order of heaven, I granted it with pleasure." Before his death, Charles II had in fact forbidden any sharing of the Spanish heritage and designated as his successor Philippe d'Anjou, second son of the Grand Dauphin – on the express condition that Philippe renounce his rights to the throne of France. But on February 1, 1701, Louis XIV skipped over this last item and asserted his grandson's right to both thrones. Philippe V of Spain thus became the master of the rich territories of New Spain (Mexico), and Louis XIV envisaged new conquests.

On September 16, 1701, James II, dethroned king of England, died at Saint-Germain-en-Laye. Louis XIV recognized James's son as successor to the English throne under the name James III. This decision, which ran counter to the Treaty of Ryswick, seemed to be a deliberate provocation of William III and the English people, and other European powers were indignant that Louis XIV was imposing his will as the "arbiter of Europe." A coalition was formed to oppose Louis's imperialist aspirations, and on May 15, 1702, it declared war on France and Spain.

In North America, this provided an opportunity to revive the colonial animosities that had lain dormant since the signing of the Treaty of Ryswick, which had ratified the victorious raids of Pierre Le Moyne d'Iberville at Hudson Bay, Newfoundland, and the coasts of Acadia. Having firmly repelled the attack of William Phips from Boston in 1690, the government at Quebec still had its militias and its dedicated army, the Compagnies franches de la Marine. The area to be defended was immense, but

THE NADOUESIANS
OUTOLIBIS
L. SUPERIOR or NADOUESSIOUS
L. PISCOUTAGAMY
L. TABITIBIS
HUDSONS BAY
NEW SOUTH WALES
NEW BRITAIN
TERRA LA
L. TIMAGAMING
CANADA
NEW FRANCE
MATOUAN
L. MIPISSING
L. TEMISCAMING
THE ILLENOIS
THE LAKE ILLINOVECKS
THE LAKE OF HURONS
L. ONTARIO or ONTARIUS
L. ERIUS or FELIS als Cadaragua
THE FELIANS
NEW ENGLAND
NEW YORK
WEST NEW JARSEY
SEA OF NEW ENGLAND
VIRGINIA
SEA OF VIRGINIA
CAROLINA
CLARENDON COUNTY
SEA OF CAROLINA
FLORIDA
PART OF THE BAY OF MEXICO
BAHAMA
LACAIONQUE
PROVIDENCE
ATHERA
C. Florida
Augustin Bay
C. Canaveral
Fort Royall
SOUTH SEA
BAY OF MEXICO
TERRA FIRMA
R. Mitchisipi or Rio Grande
Chacouachsia Illinois
Copper Mine
Fort de Crevecoeur
Lake Erien
Quebec
Mont Royal
Kenebek Town
Boston Harbour

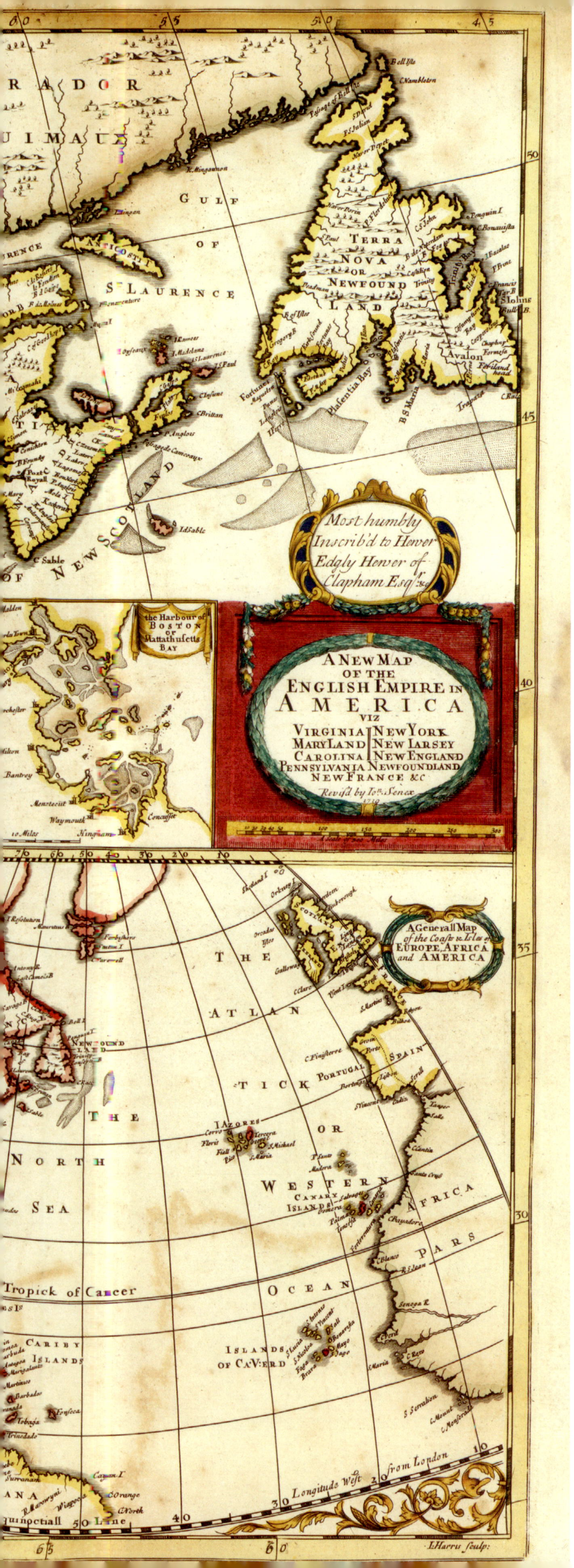

***A New Map of the English Empire in America* by John Senex, 1719**

Maps often served as propaganda tools. In 1695, the bookseller-publisher Robert Morden commissioned the engraver John Harris to produce a map of the English colonies of North America. Although inspired by the French cartographers Melchisédech Thévenot and Hubert Jaillot, this map features the English colonies. In 1719 and 1721, John Senex reproduced it with very few changes: the English colonies extend to the south shore of the St. Lawrence River and to lakes Ontario and Erie. With his territorial claims, similar to those of Herman Moll, Henry Popple, and John Mitchell, Senex paved the way for the Society of Anti-Gallicans (see p. 230). The realism of the British authorities, however, put an end to such excesses of enthusiasm. In 1763, London traced a border at the 45th parallel. The New York colony saw its appetite severely limited to the north and shared the disappointment of its neighbours, Pennsylvania and Virginia, also hemmed in to the west by an immense territory reserved for the Indians – though perhaps only "for the present," according to the text of the Royal Proclamation of October 7, 1763. The Americans were very frustrated. London had just cut short a half-century of dreams – but for how long?

Map of Canada drawn at the time of the negotiations in Utrecht, 1712
In 1713, France conceded to England a number of large territories in North America. The exact borders of these territories were not defined in the treaty, which provided instead for the appointment of English and French commissioners to set the borders within one year. This map, drawn by Jean-Baptiste de Couagne, shows the borders proposed by the French even before the treaty was signed. On the Labrador coast, a line starts at the 59th parallel and runs southwest to Lake Nemisco, then turns west to follow the 52nd parallel. During the negotiations, the English claimed the 49th parallel as the limit for Hudson Bay (a border that included the Great Lakes). On the other hand, the treaty definitively ceded Acadia, "in its entire, conformable to its ancient limits." Later, France and England would dispute the significance of this wording.

in fact the front was limited to the Atlantic region. Using the Indian tactic of surprise attacks, the Canadians ravaged the English settlements in Newfoundland, seizing posts and stores. The most brilliant successes were those at Bonavista in 1705 and St. John's in 1709. The cruellest and most devastating was at Deerfield, Massachusetts, in 1704. Even Acadia, the main stake in the war, was able to fight off English attacks until 1710.

Following these stunning Canadian military successes came a spectacular failure by the English army. In 1711, Admiral Horatio Walker, at the head of a fleet of 98 battleships bearing 12,000 men, weighed anchor at Boston and set sail for the Gulf of St. Lawrence with the intention of attacking Quebec. In darkness and fog, eight of Walker's ships broke up on the reefs of Île aux Œufs. Shocked by the accident and fearing more mishaps, 470 kilometres downstream from Quebec the admiral retreated without having fired a single shot. General Francis Nicholson, leading the English land troops who had arrived south of Lake Champlain, learned of Walker's withdrawal and did not attack Montreal. In Quebec, there was jubilation over the escape from

Treaty of Utrecht, article 15.

In the future, the inhabitants of Canada and other subjects of France will molest neither the five nations or provinces of the Indians who are subjects of Great Britain, nor the other nations of America that are friends of this Crown. Similarly, the subjects of Great Britain will behave peacefully toward the American subjects or friends of France; and both shall enjoy full freedom to travel for the purpose of trade, and with the same freedom the inhabitants of these regions will be able to visit the French and British colonies for the reciprocal advantage of trade with no molestation or difficulty by either side. In addition, the commissioners will agree, exactly and separately, on those who will be, or should be considered subjects and friends of France or Great Britain.

France. Archives du ministère des Affaires étrangères, traités multilatéraux: 17030003, pp. 65–66 (translated by K.R.).

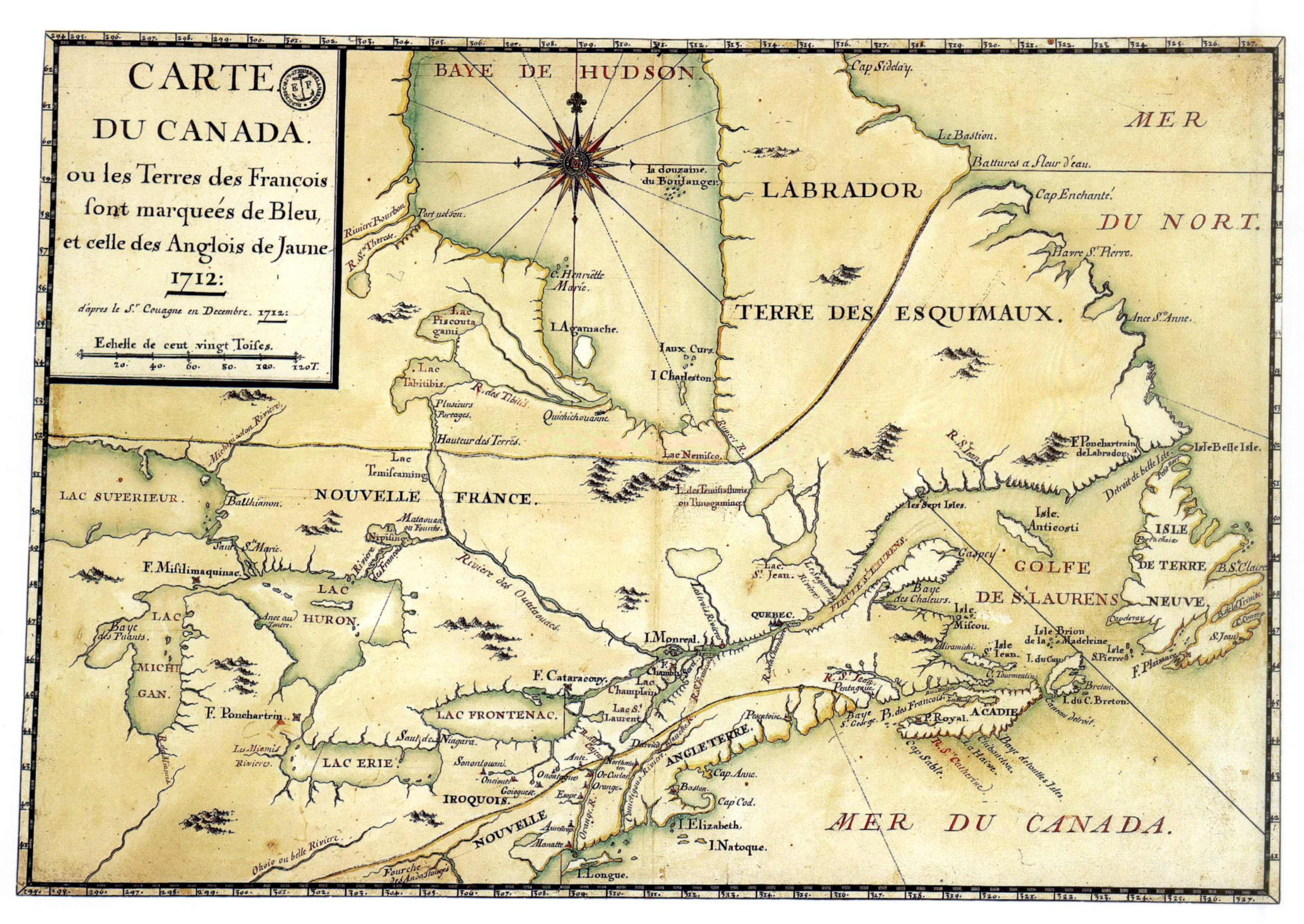

certain defeat due to a seemingly miraculous set of coincidences.

In late 1712, negotiations began at Utrecht, with a view to ending the War of Spanish Succession. On April 11, 1713, the plenipotentiaries signed a series of treaties proceeding to arbitration. For its part, France had to relinquish some of its dreams of grandeur and retrocede most of its European gains. Not only was Philippe V forced to renounce the French crown, but the other descendants of Louis XIV could no longer lay claim to the throne of Spain.

Although it was forced to give up ground, France managed to safeguard the integrity of its national borders. In North America, it lost Hudson Bay, won at such great cost, and its best posts in Newfoundland and in Acadia, whose boundaries were as yet poorly defined. English predominance thus advanced considerably on the continent, in line with the ambitious plan of the capturer of Port-Royal, Commander Samuel Vetch, who was aiming for no less than the conquest of all French possessions in North America; the Treaty of Utrecht was clearly the first step in the realization of this hegemonic project.

Nevertheless, France preserved the core of its North American empire: the St. Lawrence Valley, the Great Lakes, and Louisiana, and the continuing possibility of finding a route to the "western sea." This route was, however, strewn with obstacles, not least of which was the presence of the Iroquois, who, through the Treaty of Utrecht (article 15), went from being neutral to being British subjects authorized to trade on behalf of England. In addition, Louisiana, near the silver mines in Spanish territory, would have to be fortified; although its population was still small, it was rich in possibilities, including affording an important strategic outlet in the southern part of the continent. To the east, France no longer controlled the fishery in Newfoundland but retained its fishing rights there. It lost Acadia, but its nationals continued to live there and were to prosper on Île Saint-Jean (Prince Edward Island) and Île Royale (Cape Breton Island), where the Louisbourg fortress was built.

Through the War of Spanish Succession and the Treaty of Utrecht, France's imperial designs had been seriously impeded. However, France still had overseas territories that were much larger than all of England's colonial possessions at the time. Its 20,000 inhabitants had the means to consolidate their roots in New France, but their prosperity remained fragile, for there were 400,000 residents in the English colonies and it was only a matter of time before they would want more land.

Map of Detroit and Lake St. Clair, circa 1703 (detail)
After the erection of Fort Pontchartrain in 1701, on the shores of a narrow channel linking Lake Erie to Lake St. Clair, a number of Indians settled in the vicinity, as this map shows. It situates the Wolves, the Hurons, and the Ottawas. According to Lamothe Cadillac, about 6,000 Indians lived in the environs in wintertime. The map clearly shows the two entrances to the fort – one giving onto the watercourse, the other leading to the Huron village. Inside the fort, the buildings are laid out to create streets and alleys. In 1702, the wives of Cadillac and his lieutenant, Alphonse de Tonty, moved in; they were no doubt the first European women to settle in the Great Lakes region.

Main sources in order of importance

Mollat du Jourdin, Michel. *L'Europe et la mer.* Paris: Éditions du Seuil, 1993. European political and maritime issues in the modern era. — Miquelon, Dale. "Envisioning the French Empire: Utrecht, 1711–1713." *French Historical Studies,* vol. 24, no. 4 (2001), 653–77. The Treaty of Utrecht; very enlightening on the issue of the Spanish succession; reproduced in its entirety in the virtual exhibition available at www.archivescanadafrance.org.

A New and Exact MAP of the DOMINIONS of the KING of GREAT BRITAIN on ye Continent of NORTH AMERICA.
Containing NEWFOUNDLAND, NEW SCOTLAND, NEW ENGLAND, NEW YORK, NEW JERSEY, PENSILVANIA MARYLAND, VIRGINIA and CAROLINA.
According to the Newest and most Exact Observations By HERMAN MOLL Geographer.
JAMES BAY
PART OF CANADA
NEW SCOTLAND
NEW ENGLAND
NEW YORK
IROQUOIS
PENSILVANIA
VIRGINIA
CAROLINA
NEW FOUND LAND
ESKIMAUX
LITTLE ESKIMAUX
St. Laurens River
St. LAURENS BAY
THE FISHING BANKS OF NEWFOUND-LAND
Sholes of Acadia
Sholes of Sable Island
THE WESTERN OCEAN
The French by the Treaty of Utrecht are allowed to catch Fish, and to dry them on land, in that Part only, and no other, of the Island of Newfound-Land, which stretches from Cape Bonavista to the Northern Point of the Island, and from thence running down by the Western side, reaches as far as Point Riche: But the Island Cape Briton, as also all others both in ye mouth of the River St. Laurens and the Gulph of the same name, are given by the same Treaty to the French, with all manner of Liberty, to Fortify any Place, or Places, there.
SAPIENTIA ET VIRTUS
To the Honourable WALTER DOWGLASS Esq.r
Constituted CAPTAIN GENERAL and Chief Governor of all ye Leeward Islands in America by her late Majesty Queen Anne in ye Year 1711.
This Map is most Humbly Dedicated by your most Humble Servant Herman Moll Geog.r
1715.
The Cataract of NIAGARA, some make this Water-Fall to be half a League while others reckon it no more than a hundred Fathom.

West of Hudson Bay

FROM STUART TO HEARNE

At the same time as French explorers were advancing toward the centre of North America, the British were settling in the north. During the second half of the seventeenth century, the British concentrated almost exclusively on the fur trade, which was particularly fruitful where the climate was cold enough for the animals to grow thick pelts and beautiful fur. In the area's poor soil grew the boreal forest, in which the animals took refuge. The brutal weather did not lend itself to stable human habitation, but a small and nomadic population survived by hunting and fishing. When the English began exploring westward again, in the eighteenth century, they started from what had become their base at Hudson Bay.

The Hudson's Bay Company, founded in 1670, needed to open new territories to trade. Its employees thus became explorers, taking up their predecessors' quest for a route to the "western sea." Over time, the Europeans at Hudson Bay had learned to live decently, if not comfortably, in the region. The company had been careful to recruit its workers from an area similar in latitude and climate: the Orkney Islands, a group of small islands off the north coast of Scotland straddling 59°N, just about 3° south of the Arctic Circle. Orcadians had little difficulty adapting to the environment of Hudson Bay, which offered a way of life often better than that at home, a relatively high income, and the possibility of accumulating some savings. They were good navigators, and also mobile, frugal, solitary, and self-sufficient. In 1702, Orcadians formed the majority of the company's overseas employees; by 1799, they numbered 416 out of a total of 530.

The first European to head west by land and reach the Barren Grounds, the tundra between Hudson Bay and Great Bear Lake, was William Stuart in 1715–16. Stuart, however, was not part of the search for a northwest passage. The Hudson's Bay Company, which had employed him since 1691, sent him on a mission of reconciliation between two nations of fur suppliers, the Crees and the Chipewyans. He was also asked to gather information on the possible presence of copper mines. Accompanied by Thanadelthur, known as "the slave woman," who played an important role in the negotiations, Stuart travelled north-northwest 400 miles from Fort York to 63°N, then turned toward Great Slave Lake. No other explorer would go so far west for another half-century. In spite of his success, Stuart died of nervous exhaustion and dementia three years later, in 1719.

The ensuing expeditions were attempted by sea. In 1719, James Knight, dazzled by the mirage of a "yellow metal" evoked by the Indians, sailed from England to the Marble Islands in Rankin Inlet, where his ship ran aground. The crew tried to survive on the coast, but all died in under two years, according to an account gathered by Samuel Hearne from the Inuit 50 years later. It seems strange that the survivors were unable to reach a company post and that the company did not look for them. This remains one of the cruel and mysterious tragedies of the Arctic, and it continues to haunt and terrify visitors to Marble Island. In 1722, John Scroggs left Fort Churchill and found traces of the preceding expedition before he entered Chesterfield Strait, where promoters of the search for the passage set their hopes for several decades to come.

Christopher Middleton, in 1742, and his cousin William Moor, in 1747, sailed north along the western shore of Hudson Bay. They had received guidance and encouragement from Arthur Dobbs, an Irish member of Parliament. An armchair explorer who avidly read voyagers' accounts, Dobbs had come to the conclusion that

***A new and exact map of the Dominions of the King of Great Britain on ye continent of North America* by Herman Moll, 1715**
Herman Moll was a talented engraver, originally from Holland or Germany, who immigrated to London around 1678. A friend of the writers Daniel Defoe and Jonathan Swift, he supplied them with maps to illustrate *Robinson Crusoe* and *Gulliver's Travels*. Just after the War of Spanish Succession, Moll published this map, which shows the English colonies in North America (1715). One can see that the British interpreted the Treaty of Utrecht differently from the French, claiming all of the land up to the St. Lawrence River for their colonies of New Scotland (Acadia), New England, and New York. This map was nicknamed the Beaver Map, due to the illustration of beavers building a dam not far from Niagara Falls – an illustration that Nicolas de Fer had used in a map published 17 years earlier. Three years later, the French cartographer Guillaume Delisle published a map of Louisiana that provoked heated reactions in the English colonies, which were portrayed as squeezed between the Appalachian Mountains and the ocean (see p. 179). Delisle did not recognize the claims of Pennsylvania beyond the mountain range. He even dared to attribute a French origin to Carolina. Moll reacted with the publication of another map denouncing the French claims and putting forward the English claims. Other maps were published in the following years, including those by Henry Popple (1733) and John Mitchell (1755).

Map by a Canadian Métis drawn in a London tavern, 1744
As the title indicates, the information in this map comes from a "French Canadese Indian" named Joseph Lafrance, the son of a French trader and a Sauteux woman. Lafrance, who trafficked in contraband at Hudson Bay, was taken to London by the English in the hope that he would be able to provide them with clues to the discovery of the Northwest Passage. In London, Lafrance met Arthur Dobbs and, on the cobblestones outside a tavern, drew a sketch of the route that he had taken between Lake Superior and Hudson Bay (in today's province of Manitoba): Rainy Lake, Lake of the Woods, the large and small Ouinipique (Winnipeg) lakes, Lake Pachegoia, and the Nelson River.

a passage existed on the west side of Hudson Bay. With two powerful ships fitted out by the British Admiralty, Middleton and Moor explored, successively, Chesterfield Cove and Wager Bay, and they both stopped at Repulse Bay. But the human toll was considerable. Among the 90 men in Middleton's crew, 30 died during the winter and only seven remained more or less able-bodied for the return to England. Nevertheless, certain dreamers in London continued to pin their hopes on Chesterfield Cove, arguing that the presence of whales and strong currents indicated the existence of a passage to the Pacific Ocean and obstinately refusing to accept the denials of navigators. Although they were seen as failures, these two painstaking explorations contributed to the increasingly accurate body of knowledge on the west coast of Hudson Bay.

The Hudson's Bay Company then began once again to explore by land, for two reasons: first, the disappointing results of explorations by sea; second, the advance of the French into the western plains, notably with the voyages of La Vérendrye in the 1740s. On June 26, 1754, James Isham, manager of the Hudson Bay post, appointed Anthony Henday, one of his agents, leader of a new, large-scale expedition. Accompanied by numerous Crees, including a woman to serve as his interpreter but always introduced as his "sleeping companion," Henday completed one of the longest exploratory routes in the Canadian west: from York Factory to Red Deer, Alberta, including navigation on the Saskatchewan and Red Deer rivers – 4,167 kilometres in total, over a one-year period.

Some 15 years later, Samuel Hearne followed in Henday's footsteps. In 1769, Hearne was only 24 years old, though he already had three years of experience with the Hudson's Bay Company and was familiar with the Arctic, having hunted whales at Marble Island. He was entrusted with the mission of finding the copper mines that had been reported and looking for the passage to Asia via the Arctic. As the expedition was setting out from Fort Prince of Wales on December 7, 1770, Hearne added to his team a Chipewyan guide named Matonabbee, who would become a valued advisor on exploration. To reach the Arctic Ocean, Hearne's party walked northwest at the edge of the tundra in order to reprovision by hunting caribou and also to avoid the blizzards raging in open territory. Once they were at the copper mines, they gathered more provisions and turned north, toward the Arctic Ocean. Matonabbee's plan set out all the travel details, including the need for several dozen voyagers to supply and protect the group. The expedition involved up to 70 people, almost half of them women with their children; their presence was justified by Matonabbee, as Hearne reported in his journal:

> "Women," added [Matonabbee], "were made for labour; one of them can carry, or haul, as much as two men can do. They also pitch our tents, make and mend our clothing, keep us warm at night . . . though they do every thing, [women] are maintained at a trifling expence [*sic*]; for as they always stand cook, the very licking of their fingers in scarce times, is sufficient for their subsistence."

With such valuable advice at hand, the expedition reached its goal – Coppermine River, where there were, in fact, copper mines – on July 14, 1771. Hearne descended the river to its mouth, reaching the Arctic Ocean. The first European to see and cross Great Slave Lake, in the winter of 1771–72, he returned to Fort Prince of Wales on June 30, 1772. Encouraged by these discoveries, the company decided to open posts inland to keep the merchants of Montreal from controlling the fur market. On the basis of his 32 months exploring the West, Hearne founded the Cumberland post in Saskatchewan and took command of Fort Prince of Wales. He was still there in 1782, when Jean-François de Galaup de Lapérouse arrived, leading an expedition of 290 men and three warships. Hearne knew that his 38 men could not put up any serious resistance and turned the fort over to the French, with no discussion, on August 8, 1782; the fort returned to English hands the following year.

The results of Hearne's exploration were used widely. Other company employees gradually mapped the Canadian North: Peter Pond, in the Athabasca Basin in 1778–80; Philip Turnor and Peter Fiedler in 1790–92; David Thompson in 1795–96; and John Hodgson in 1791. Hearne's work was also studied by scholars and by other

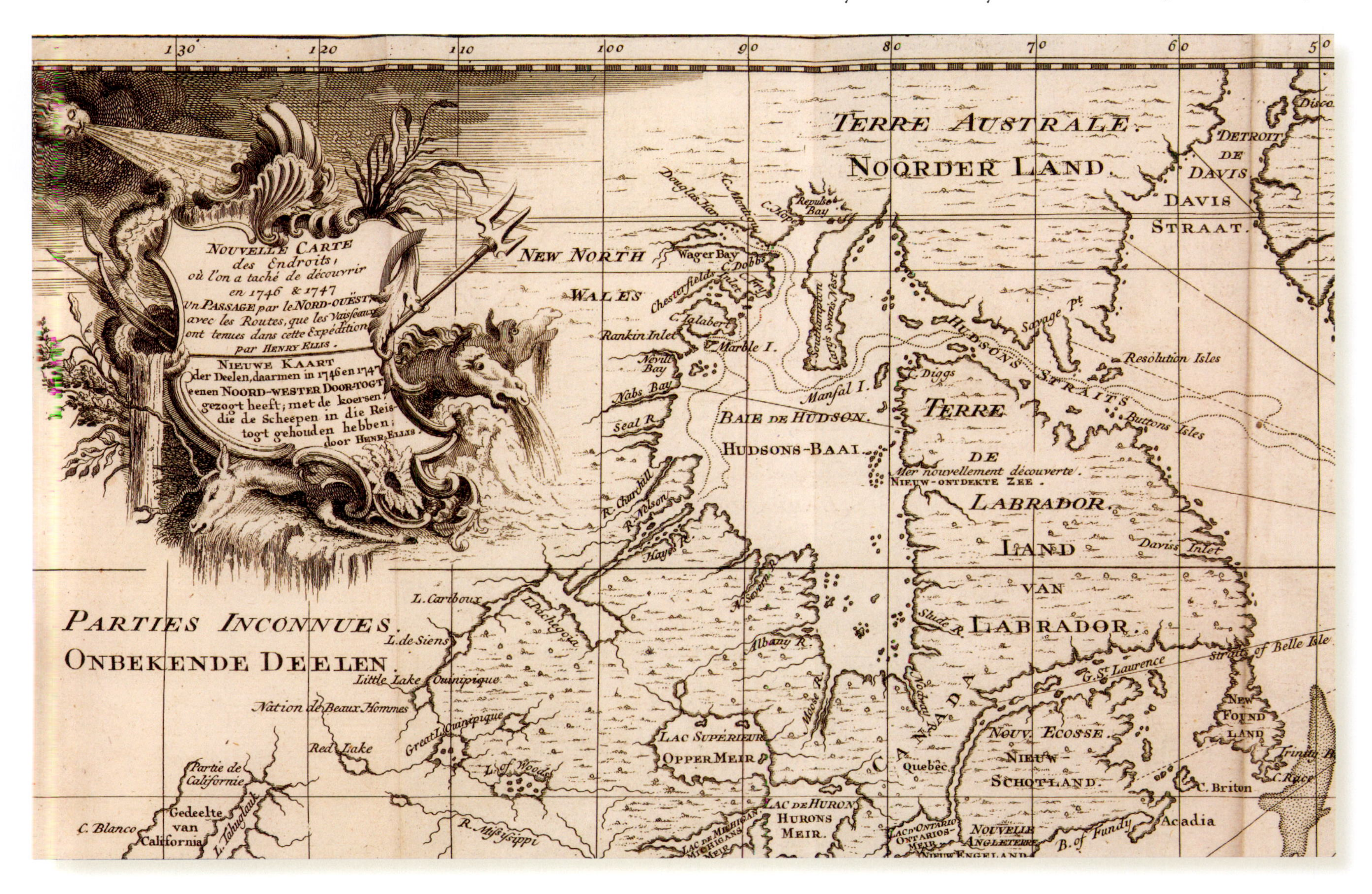

***Nouvelle carte des parties où l'on a cherché le passage du Nord-Ouest en 1746 & 1747* by Henry Ellis (detail)**
In 1746 and 1747, the maker of this map, Henry Ellis, was in Hudson Bay on a voyage of exploration commissioned by Arthur Dobbs, scourge of the Hudson's Bay Company. A dotted line shows various attempts to find the Northwest Passage, each as fruitless as the last. Farther south, the cartographer integrated information supplied by the Métis trafficker Joseph Lafrance: a succession of lakes and rivers that sparked the search for a route to the Pacific farther south. The nation of "Beautiful Men" encountered by the La Vérendrye brothers was, for Ellis, the edge of known territory.

Inuit scene
In 1746, William Moore and Henry Ellis once again set out in search of the Northwest Passage. They apparently returned empty-handed the following year. Ellis wrote an account of their voyage to Hudson Bay and provided valuable observations of the Inuit. In the background of this engraving, an Inuk paddles a kayak and brandishes a harpoon; in the foreground, two Inuit are making a fire. The artist also illustrated a practice that Ellis described: "The women have a Train to their Jackets, that reaches down to their Heels ... When they want to lay their Child out of their Arms they slip it into one of their Boots." This garment is reminiscent of the one in John White's drawing of an Inuk woman captured by Martin Frobisher during his second voyage, in 1577. This time, the woman has placed her child in her hood rather than in her ample leggings. The Inuit, Ellis wrote, "are of a middle Size, robust, and inclinable to be fat ... The Mens clothes are of Seal Skins, Deer Skins, and ... the skins of Land and Seal Fowl; each of their Coats has a Hood like that of Capuchin."

Drawing by John White

explorers, including Lapérouse and James Cook, who encouraged him to write an account of his travels. Hearne devoted the remainder of his life to this task, from the time of his retirement to London until his death in 1792.

After the explorations of Hearne and his successors, the fur trade grew substantially in the polar and sub-arctic regions. New life was breathed into cartography with the syntheses by Pond (1785) and Aaron Arrowsmith (1795). The British Admiralty continued to be interested in the Northwest Passage throughout the nineteenth century. It sent many expeditions, including one by Edward Parry (1819–20), who sailed through the straits leading to Melville Island. During his first two voyages (in 1819–22 and 1825–27), John Franklin charted large portions of the Arctic coast between Alaska and the Boothia Peninsula. At the same time, a number of other explorers surveyed the islands beside Bering Strait. Franklin's third mission was intended to fill the gaps between the discovered regions. In 1847, Franklin sailed into Lancaster Strait to Prince William Island, where his ship became ice-bound. He was never seen again, in spite of intense searches for his remains. Finally, Robert John Le Mesurier McClure, a British naval officer familiar with the Arctic, found traces of the passage in 1853–54. It was not until half a century later, however, that the entire length of the Northwest was effectively crossed in a single journey, by the Norwegian navigator Roald Amundsen, who reached Alaska via Lancaster Strait in 1906.

Main source

Hearne, Samuel. *A Journey from Prince of Wale's Fort, in Hudson's Bay, to the Northern Ocean: Undertaken by the Order of the Hudson's Bay Company for the Discovery of Copper Mines, a North West Passage, &c. in the Years 1769, 1770, 1771 & 1772*. London: A. Strahan & T. Cadell, 1795. An Arctic expedition of unprecedented scope, reported by the explorer himself in a colourful account.

Acadia in the Crossfire

BETWEEN COLONIAL EMPIRES

Officially, according to the Treaty of Utrecht (1713), Acadia no longer existed. It was renamed Nova Scotia, or New Scotlande, the name it had had from 1621 until it was retaken by the French. In 1710, Port-Royal, the oldest French post, which had become a small fortified town, was captured by the English commander Francis Nicholson. Colonel Samuel Vetch took command of the garrison in 1714 and renamed it Annapolis Royal in honour of Queen Anne of England. He then became governor of Nova Scotia.

Article 12 of the Treaty of Utrecht confirmed the cession "of Nova Scotia, otherwise called Acadia, in its entirety, in conformity with its old borders, as well as the town of Port-Royal now called Annapolis Royal and generally all that which is attached to said lands and islands in this country." In fact, Acadia was still very much on the minds of the English lawmakers; its territory overlapped with that of Nova Scotia and the names remained interchangeable for almost half a century. Furthermore, article 13, which complemented article 12, conceded that "the Island called Cape Breton, and all the other islands situated in the mouth and gulf of the St. Lawrence River, shall remain in the future with France, with the entire faculty of the Very Christian King to fortify there one or several places." To say the least, the treaty was unclear, ambiguous, and even contradictory with regard to the borders of Acadia. Thus, it is not surprising that there was recurrent disagreement between the Acadians and the English administrators regarding all of the territory at the edge of peninsular Acadia, including the Saint John River, the Memramcook River, Canso, and the Kennebec River. French Acadia, broadly defined, extended roughly from the Gaspé Peninsula to Maine.

The governor of Nova Scotia, Richard Philipps (1717–49), had the delicate task of having the Acadians take the oath of allegiance to the king that all subjects were required to take in order to enjoy civil rights, including property rights. However, the Acadians took advantage of repeated delays to postpone the decision whether to become subjects of the king of England and keep their assets or to forsake Nova Scotia, leaving all of their property to the new masters. Most wished to remain on the prosperous land that they had carefully cultivated, but they did not want to be put in the position, by pledging allegiance, of possibly having to take up arms against the French or their Indian allies. In spite of persistent efforts at integration by the British, who needed these farmers, most Acadians rejected allegiance to Great Britain. In 1730, tired of war, Governor Philipps reached a compromise in the form of an oath with reservations: the Acadians pledged loyalty to the king but agreed to remain neutral in the event of war. This verbal agreement, though momentarily satisfactory, later gave rise to endless debates and was rejected by Philipps's successor, Edward Cornwallis (1749–52).

In fact, no one wanted population migrations – neither the Canadians, for whom Acadia was a buffer zone between New France and New England, nor the English, who did not have other farmers to work the land. The Acadians were offered conditions to dissuade them from leaving, notably a proclamation issued by Governor Philipps on April 10, 1720, ensuring that "that they will enjoy civil rights and privileges as if they were English as long as they behave like good and loyal subjects of His Majesty, and that the goods that they possess will be passed down to their heirs. But it is positively forbidden to those who choose to leave the country . . . to alienate, dispose of, or take with them any of their possessions." A few dozen families moved to the French colonies of Île Saint-Jean and île Royale, while a few others settled

Acadian woman
Both the territory and the inhabitants of Acadia were a matter of dispute between the French and the English. In his *Costumes civils actuels de tous les peuples connus*, Sylvain Maréchal presented not only a portrait of an Acadian woman but also a written description of the mores and customs of this population. "Their peaceful nature, their gentle habits, their ignorance of corrupting influences gave their homeland a name similar to that of the blessed Arcadia of Antiquity." He went on, "The blood of two civilized nations ran in the presence of the Savages, the innocent cause of these political rivalries."

in the southeastern part of today's New Brunswick. But the great majority of Acadians remained on their land, and their numbers grew. According to the demographer Raymond Roy, their population increased from 2,296 in 1714 to about 13,000 in 1755, 10,000 of whom were in Nova Scotia. The English administration found itself in the uncomfortable position of governing foreigners; in reality, however, it took no better care of them than the French authorities had done. Between 1710 and 1750, English administrators of French Huguenot origin, such as Paul Mascarene, lieutenant-governor of Annapolis Royal, adhered to a policy of patient persuasion, convinced that time would turn the Acadians into loyal subjects of England. It could be that the descendants of French settlers, left to their own devices, contributed to the consolidation of their communities and the advent of what historians refer to as Acadia's golden age.

The Micmacs and Abenakis were also directly affected by the cession of Acadia to England. For almost two centuries, these Indians had welcomed the French to their country, traded goods and technologies, practised Catholicism, and been linked by mixing of blood that went back generations, and they took a dim view of the English seizing their territories, a change for which their relations with the French had not prepared them. The Abenakis, living along the coast from the Saint John River to the Saco River (Maine) and centred at Pentagouet, were the first to be affected by the ambitions of settlers in Maine and Massachusetts. Some were forced to leave their land; with the assistance of the governor of New France, they settled in the St. Lawrence Valley near the villages of Quebec and Trois-Rivières. Those who remained in Acadia or on its periphery attacked posts in New England, harassed English fishermen, and ignored any attempt to placate or negotiate with them. The Micmacs took similar actions at Île Royale and in Nova Scotia. Numbering about 1,000 in 1722, the Indians of Acadia could offer invaluable assistance to whichever side they chose to align with. In that year, they captured 18 of England's merchant ships and 18 of its fishing ships. And so the lieutenant-governor of Annapolis, John Doucett, increased the number of festivals and gifts in an attempt to win their friendship. He brought together representatives from the villages of Saint John, Cape Sable, Shubenacadie, La Hève, and Les Mines and from the Annapolis River, and a peace treaty was signed on June 4, 1726. This truce ensured the prosperity of Acadia until 1750.

A second Acadia was created by the Treaty of Utrecht: Cape Breton, renamed Île Royale. A few Acadian farming and fishing families settled there, most of them from Port-Royal. Looking to conduct free trade, they founded small settlements along the coasts, purchased schooners in New England, and began a fruitful trade in codfish with the neighbouring English colonies, as well as with the West Indies and France.

Between 1720 and 1744, France built the Louisbourg fortress, an imposing defensive port complex, on Île Royale, to provide military protection for its possessions

Plan of Port-Royal by Sieur de Labat, 1710
Located on the border between New France and New England, Acadia was always the first to take the blows of French-English rivalries in North America. In 1710, having repulsed a number of English attacks, Governor Subercase surrendered and handed over the keys to the fort to the assailant. Labat, an engineer, still hoping that Acadia would be returned to the French, drew this map of the environs of Port-Royal. While showing the extent of the destruction inflicted by the English, the map gives an excellent idea of the farmland along the Rivière du Dauphin, where the Acadians had dug weirs, called *aboiteaux*, in order to cultivate the fertile land. The defeat of Port-Royal had grave consequences for the Acadians; three years later, with the signing of the Treaty of Utrecht, France officially ceded Acadia to England.

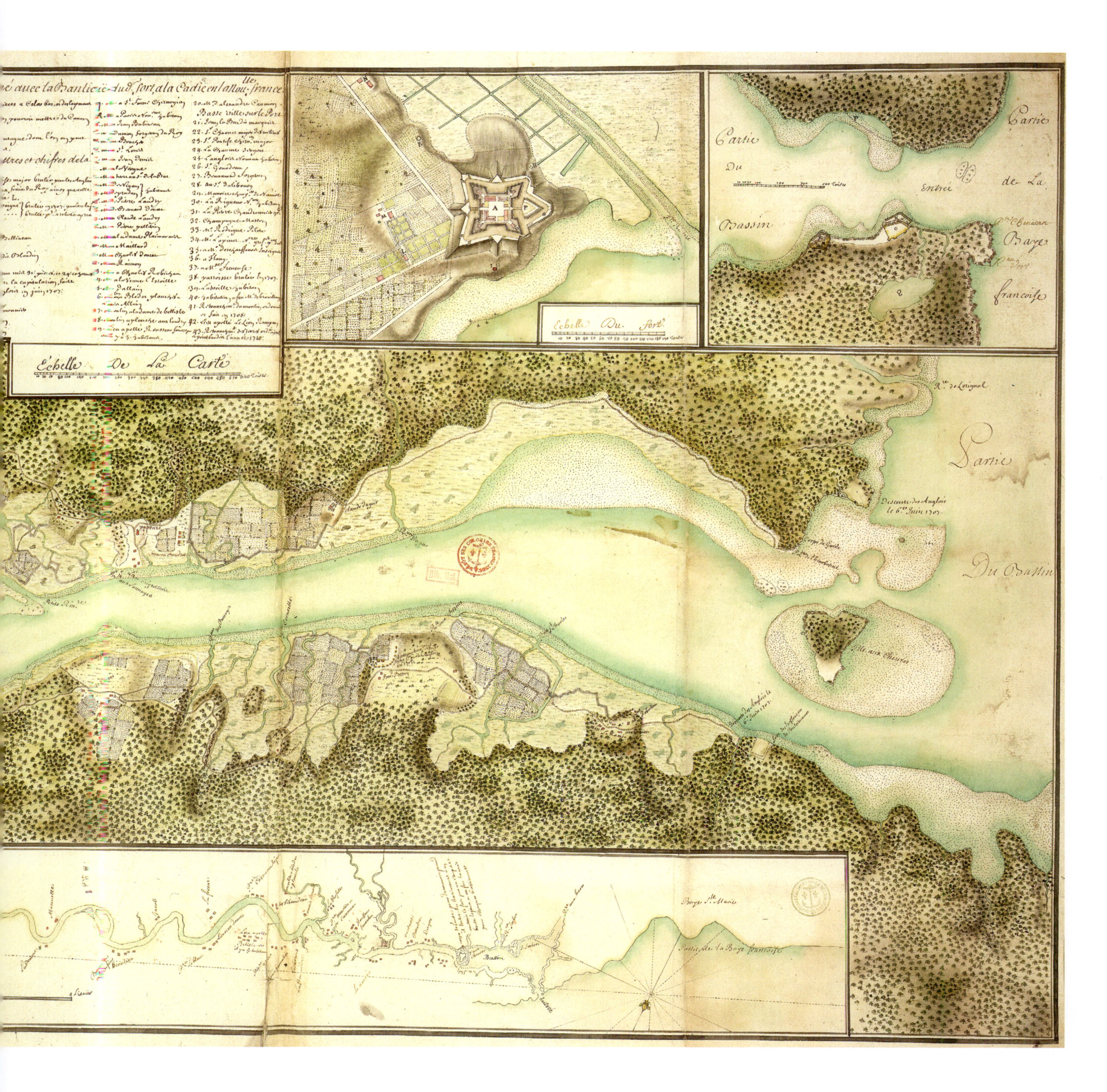

in the St. Lawrence Valley. It was a garrison town, but its privileged location at the entrance to North America made it a crossroads of international trade: French products were traded for sugar, molasses, and rum from the West Indies; cereals, meat, and wood from New England, Canada, and Acadia; and local fish. Merchants from France and New England sent their ships to Louisbourg, where a community of some 5,000 inhabitants developed during the 1750s, enjoying a level of prosperity unusual in the region. But Louisbourg's location also made it very vulnerable, as the town was accessible by sea and close to Boston, the largest city in New England. In 1745, the fortress fell a first time; it was returned to France in 1748 at the end of the War of

Nitchik-Irinioüetchs ou Nation de la Loutre
Fond de la B.e de Hudson
ou Baye
James
Partie de
dont l'Interieur est
Pays des Attik Iriniouetchs ou Gens du Caribou
Terres des Ouchestigouets
Les Ounescapi
Mistassins
Pays des Chomonchouanistes et Nekoubanistes
Attikameouecs
Pie-Koua
Isle d'Anticosti
Golphe de
St. Laurent
Fleuve de St. Laurent
Gaspesiens
Les Sept Isles
Baye des Chaleurs
I. de la Madelaine
Cap Breton
Cap Canceau
I. de Sable
Les Bancs d'Accadie Ou l'on fait la Peche
Banc de St. Georges
Cap de Sable
Port Royal ou Annapolis Royal
Baye Françoise
Halifax
Cobeguit
Havre de Baston
Cap Cod
Plimouth
Iroquois
Lac Ontario
Des Outaouais
Fort Frederic
Kayadoroses
Les Agnies
Cast Kill
Orange ou Albany
Rocher
I. des Monts Deserts

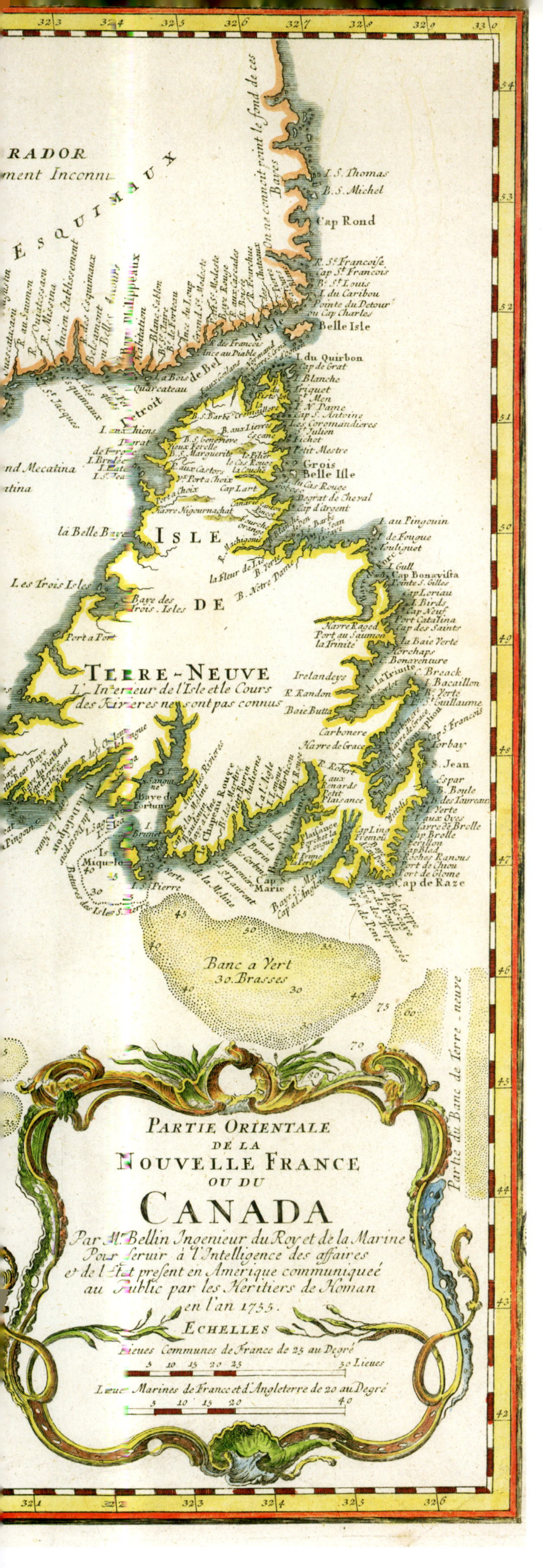

Austrian Succession; its final fall was in 1758, when it was taken in an attack by the British regular army.

The founding of Halifax, in 1749, marked a turning point in the history of Acadia: now, Nova Scotia became truly British. In response to the attack by the French duc d'Anville, in 1745, Halifax took on the role of counterweight to Louisbourg. English Protestants were encouraged to settle en masse in a territory in which Acadians continued to feel at home. By 1752, the population of Halifax had climbed from 5,000 to 6,000. As France intensified its efforts to recapture Acadia, the English of Halifax increasingly doubted the neutrality of the Acadians and, for the first time, envisaged a Nova Scotia without its population of French origin.

The recurrent problem of managing French settlers, always seen as rebels, in Acadia was solved emphatically by the military commander Charles Lawrence (1752–60). As a confrontation over supremacy in North America was building between France and England, Acadia became of paramount strategic importance. England strengthened its military presence and then moved settlers in throughout the territory. The governor of Nova Scotia, Edward Cornwallis, now required from Acadians an unequivocal oath of allegiance to the British crown that excluded all possibility of neutrality. As in the past, the Acadians counted on their negotiating power. In their petitions addressed to the Council of Nova Scotia, they once again refused to take such an oath, which would force them to take up arms against France, but confirmed their loyalty to the king of England. The members of the Council, under their chairman, Charles Lawrence, rejected any possibility of tolerance toward the Acadians, and on July 28, 1755, the Council decided to expropriate and expel them. About 7,000 Acadians were assembled and loaded onto ships headed for the English colonies along the Atlantic coast; 2,000 to 3,000 others met the same fate in the following years, up to 1762. The deportation caused many deaths due to disease, epidemics, and the punishing conditions of exile.

Starting in 1764, Acadians were allowed to return under two conditions: they had to pledge full allegiance to the king of England, and they had to disperse in small groups. They moved to the Acadian peninsula, to what is today northeast New Brunswick, as well as to Madawaska, the region around the upper Saint John River, the southeast coast of the province, the Memramcook Valley, both ends of Nova Scotia, and Prince Edward Island. The Magdalen Islands and the Gaspé Peninsula were also settlement sites, followed in the nineteenth century by a migration to the north shore of Quebec and to the Port-au-Port Peninsula in Newfoundland. Acadia ceased to exist as an administrative entity. Living in villages that were distant from each

Partie orientale de la Nouvelle France ou du Canada
by Jacques-Nicolas Bellin, 1755
Having passed into the British sphere, Acadia enjoyed an era of unprecedented prosperity. While the French and English did not agree on the exact definition of the "old borders" of Acadia cited in the Treaty of Utrecht, they nonetheless managed to coexist peacefully. The war between France and England revived the old border disputes. In 1755, when Bellin published this map, the population of Acadia was 15,000. The main population centres were around Baie Françoise (Bay of Fundy): Annapolis Royal (formerly Port-Royal), the Minas Basin, and the Chignectou (Chignitou) Basin. In that same year, the English began to deport the Acadians.

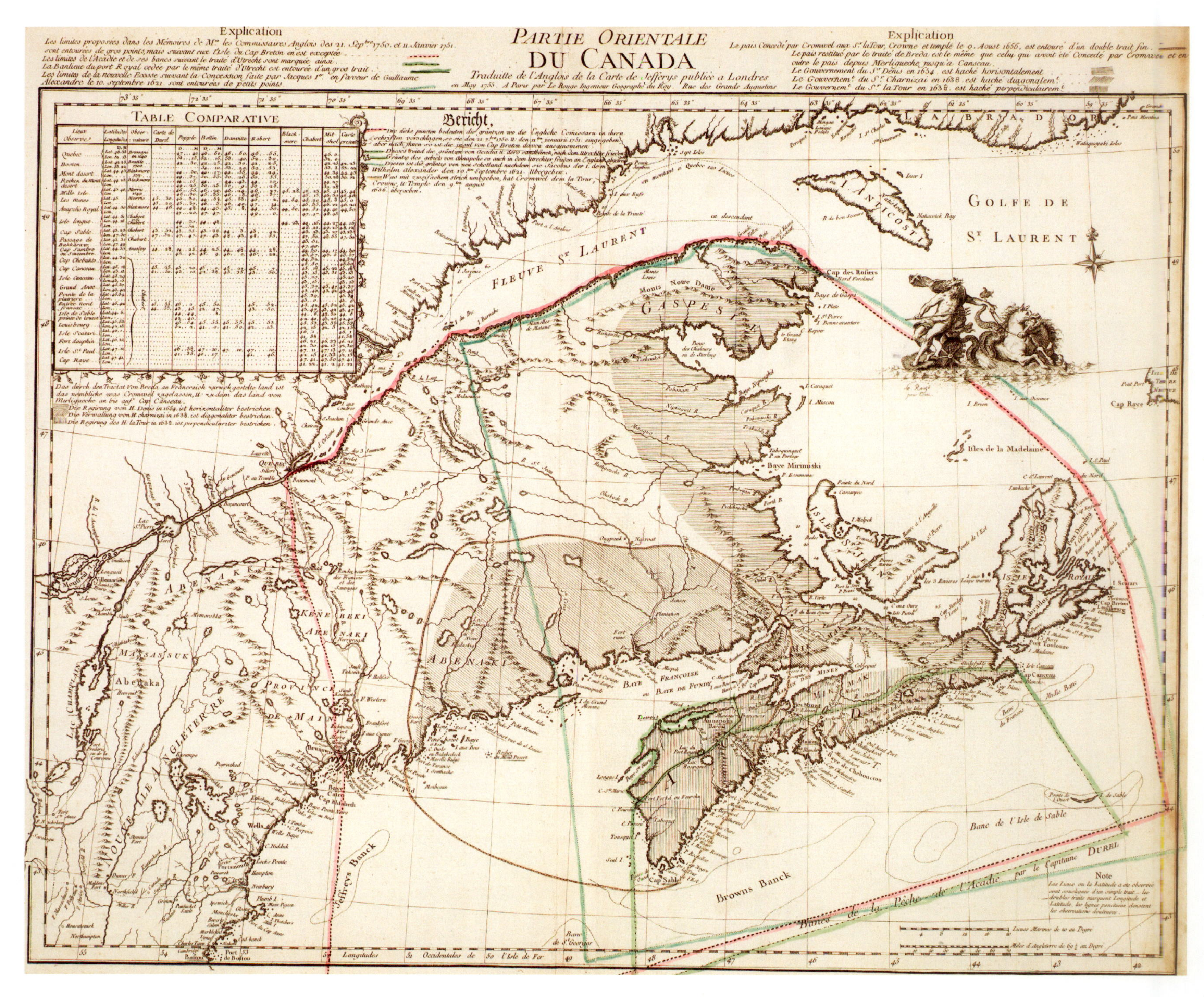

***Partie Orientale du Canada* by Le Rouge, 1755**
After the signing of the Treaty of Aix-la-Chapelle, in 1748, France and England appointed commissioners to settle the question of the borders of Acadia. The two colonial powers did not agree on the exact definition of the "old borders," which had been unclear in the Treaty of Utrecht. The English commissioners claimed that Acadia extended north to the St. Lawrence River and west to the Penobscot River. This map shows, with the dotted line, their territorial ambitions, based on various documents, including a concession by James I to Sir Alexander Sterling in 1621. There was to be no diplomatic resolution of this dispute; the Seven Years' War broke out before talks ended.

other, Acadians went through a long period of civil and economic precariousness. Even though there were many contacts between the dispersed communities, Acadia on the Atlantic was in a reconstruction stage until the mid-nineteenth century, when there emerged the first signs of what historians have called the Acadian Renaissance.

Main sources

GRIFFITHS, Naomi Elizabeth Saundaus. *The Contexts of Acadian History, 1686–1784*. Montreal and Buffalo: McGill-Queen's University Press 1992. — LANDRY, Nicolas, and Nicole LANG. *Histoire de l'Acadie*. Sillery, Quebec: Septentrion, 2001. — *Mémoires des commissionaires anglois & françois au sujet des limites de la Nouvelle Ecosse ou Acadie, &c/The Memorials of the English and French commissaries concerning the limits of Nova Scotia or Acadia*. London, 1755.

The "Western Sea"

THE GAULTIER DE LA VÉRENDRYE FAMILY

THE MIRAGE OF THE "WESTERN SEA," the product of the imagination of Renaissance sailors seeking a navigable route to the riches of Asia, persisted throughout the Enlightenment. It was revived in reports of an imaginary voyage by a fictional character, Bartholomew de Fonte, who was said to have discovered a passage between the Atlantic and Pacific oceans leading to a western sea. These reports appeared in the *Monthly, Miscellany or Memoirs for the Curious*, published in London in 1708. Although unfounded, such rumours were the stuff of dreams. They produced a wave of curiosity in the Court of France, as well as in scientific and literary circles, about the territories yet to be discovered and their still-unknown inhabitants. And the idea of a western sea raised new hopes. This infatuation was fed by the voyage to New France of the Jesuit Pierre-François-Xavier de Charlevoix, from 1720 to 1723, mandated by the regent, Philippe d'Orléans, to verify the existence of a western sea and fix its geographic location. Charlevoix spent some time in the Great Lakes region, then descended the Missouri and Mississippi rivers to the Gulf of Mexico. Although he had little to add to the body of knowledge on the western sea, he persisted in arguing for its existence.

It would take more than rumours to launch new expeditions. As it happened, when the government of New France envisaged undertaking explorations again, it did so under a number of favourable conditions. At the turn of the eighteenth century, Canada was in a serious crisis with regard to the fur trade. English competition from the Hudson's Bay Company, which was siphoning off the best furs from the North, was forcing the French to go great distances to collect and transport merchandise. To find new trade areas, the minister of the navy encouraged exploration to the south and west. In addition, a long period of peace between the French and the Indians began in 1701, and this made expansion even more attractive. And yet there remained real obstacles to reaching the "upper country." The French knew that beyond Lake Ouinipigon (Lake Winnipeg) they would be in unfamiliar territory. There would be few watercourses, and thus little opportunity for navigation, their usual mode of travel. Instead, there was a great plain crisscrossed by a few valleys and shallow rivers, with some scattered "islets of forest." The Plains Indians lived a nomadic life following herds of bison, deer, and other animals whose flesh and skin provided food, clothing, and often shelter. Cartographers found few words to describe the region that today forms the three Canadian Prairie provinces and the adjacent U.S. border states. It was this "unknown country" that was supposedly occupied by the western sea, which was portrayed as being just a short distance from the Great Lakes.

In 1726, Pierre Gaultier de Varennes et de La Vérendrye and his sons got an official foothold in the upper country by being given command of the northern post at Kaministiquia (Thunder Bay), where they prepared an exploration plan. The destination for their first expedition, in 1731, was Lake Ouinipigon, where they would build a fortified post. In Montreal, they recruited some 50 men, a number of whom were members of the La Vérendrye family, and took on trade merchandise and minimal provisions for the first days of travelling. Thereafter, the voyagers fed themselves by fishing, hunting, and trading with the Indians until they reached Michillimakinac, about 26 days away by canoe and portage.

To continue toward Lake Winnipeg, the convoy returned to Lake Huron, followed a channel leading to Lake Superior, then paddled along the northwest shore

Details of Visscher's map (see pp. 68–69)

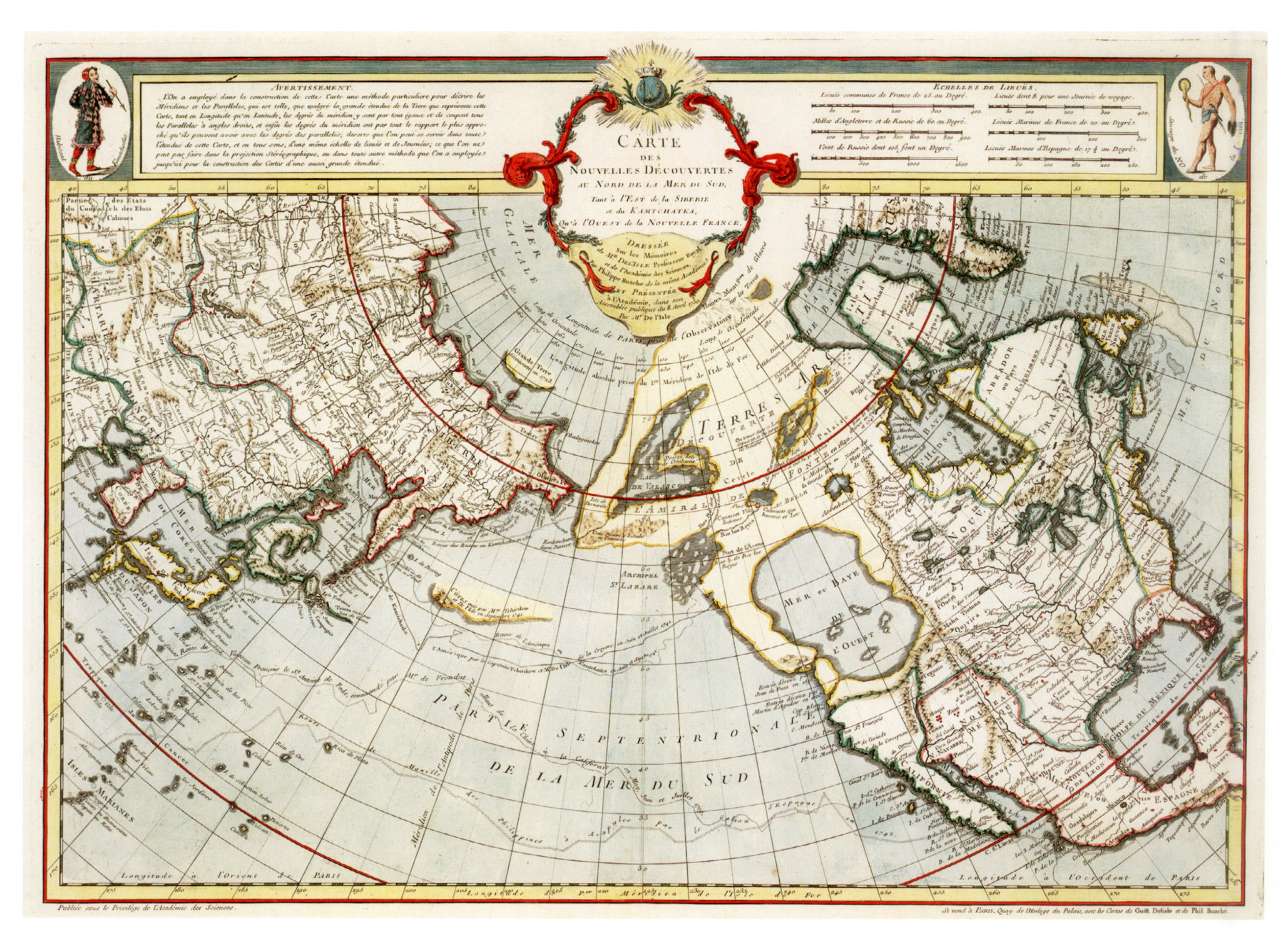

***Carte des Nouvelles Découvertes au Nord de la Mer du Sud* by Joseph-Nicolas Delisle and Philippe Buache, 1752**
In the mid-eighteenth century, the northwestern part of North America was one of the last unknown territories on the planet. Although some mapmakers, such as Jacques-Nicolas Bellin, chose not to speculate about the shape of this coast by leaving the space blank, other geographers saw an opportunity to explore new geographic theories. For instance, the cartographers and Académie des sciences members Joseph-Nicolas Delisle and Philippe Buache published a map that is stunning to twenty-first-century eyes. They presented a gigantic western sea enclaved within North America, the entrance to which had supposedly been discovered by Juan de Fuca in 1592. Farther north, Delisle and Buache drew a passage linking Hudson Bay to the Pacific Ocean, supposedly discovered by Admiral Bartholomew de Fonte in 1640. In reality, de Fonte's report was a fictional text published in London in 1708.

of Lake Superior to the Nipigon River, where they stopped at Kaministiquia, the westernmost post. They then had to cross the "great portage" to reach Lac La Pluie (Rainy Lake) and Lac des Bois (Lake of the Woods). It was not its length that made this "great portage" so formidable, since, as the crow flies, it was only a dozen kilometres in length. Rather, it was the challenge posed by the spectacularly rough terrain, with marshes, rapids, waterfalls, and cliffs, some more than 100 metres high. The convoy had to make the crossing without damaging their canoes and carrying all of their equipment and provisions. This took seven to 10 days of constant, exhausting trudging back and forth with heavy loads. Some of the recruits, terrified by the danger awaiting them, refused to go any farther. La Vérendrye decided to stay with the mutineers but sent 25 men, led by his eldest son, Jean-Baptiste, to erect Fort Saint-Charles at Rainy Lake. The following year, Jean-Baptiste went to Lake of the Woods, where he built Fort Saint-Pierre, intended as a resupply post; in 1733, he built Fort Maurepas on Lake Winnipeg.

Armed with the information brought back by Jean-Baptiste, La Vérendrye set out in search of the "western river," which should flow into the western sea. From Lake Winnipeg, two routes were possible: via the sources of the Missouri River, in today's North Dakota and Montana, and via Rivière Blanche (the Saskatchewan River). From Fort Maurepas, where he found himself in 1738, La Vérendrye chose the first route. His caravan included 25 Frenchmen and 27 Assiniboines, to whom were added, over the 2,000 kilometres travelled, about 600 Assiniboines and, at the end of the voyage, 30 Mandans. It stopped near the mouth of the Little Knife River (in North Dakota), in the middle of Mandan territory, where La Vérendrye and his party were welcomed with open arms. In addition to the relief of finding a nation that was well disposed toward them, the explorers found horses, which the Mandans allowed them to use, and this greatly facilitated the rest of their expedition. La Vérendrye believed that the Missouri was very likely the "western river," but he prudently returned to Fort La Reine and had his son Louis-Joseph explore the environs of Lake Winnipeg. This round trip established the connection among lakes Winnipeg, Manitoba, and Winnepegosis and the nearby rivers.

After 1739, La Vérendrye undertook no new explorations, though he remained commander of the western posts. His sons Louis-Joseph, known as "le Chevalier," and François consolidated the French presence by constructing Fort Bourbon (1739) and Fort Dauphin (1741). The most spectacular campaign, making the La Vérendrye name the stuff of legend, was led by Louis-Joseph and François to the foothills of the Rocky Mountains in 1742–43. Guided by Mandans, whose territory they reached in August 1742, the explorers intended to find the "People of the Bow" (northeast Wyoming), who had objects of Spanish origin and apparently were in contact with white people living in brick structures on the shores of the western sea. They arrived at this destination in November 1742 and, at the request of the Chief of the Bow, agreed to support a military operation against an enemy nation, the "People of the Snake" (Shoshones), whose villages lay at the foot of high mountains on the route to the western sea. But when the explorers went with a war party of more than 2,000 People of the Bow, they found the villages of the People of the Snake deserted. Suspecting that they had gone to attack their own villages, the People of the Bow fled, despite their chief's loud entreaties. "I was greatly mortified not to be able to climb the mountains as I had wished," Louis-Joseph wrote in his journal – and with good reason, as he might have been able to glimpse the Pacific Ocean. During the return journey, on March 19, 1743, the explorers marked with a lead plaque their passage through an Arikara village, today Pierre, the capital of South Dakota.

La Vérendrye's sons had explored for 14 months without finding solid proof that the western sea existed, and the old commander was disappointed when they returned to Fort La Reine in June 1743. The comte de Maurepas, minister of the navy, already poorly disposed toward the La Vérendrye family, accused them of seeking profit by trading with the Indians instead of completing the expedition. Maurepas was conveniently forgetting that the Court of France had left the entire financial burden of the expedition to its members. Governor Beauharnois defended the head of the clan, who then drew up a plan to explore the Saskatchewan River to its mouth. La Vérendrye died in 1749 without realizing his dream, but new trading posts were erected along this watercourse, at Paskoya (Le Pas) in 1748 and Fort La Corne in 1753. In addition, his sons continued to trade in the West and built a number of new trading posts, notably at Grand Portage (Minnesota), Chagouamigon (Ashland, Wisconsin), and Baie des Puants (Green Bay, Wisconsin). The fur trade in the West was secure for the moment, and French domination of all routes in the interior of the continent seemed complete.

Map of a part of Lake Superior and the western territories by Christophe Dufrost de La Jemerais and Gaspard-Joseph Chaussegros de Léry, 1734
In 1731, Louis-Joseph Gaultier de La Vérendrye, commander of Fort Kaministiquia, was asked to find a route to the "western sea." Accompanied by his sons and his trading partners, La Vérendrye ventured west of Lake Superior. Two years later, de La Jemerais presented the governor of New France with a map of the country that had been explored, recopied by Chaussegros de Léry. In "watercolour," the cartographer records the route taken to Lake of the Woods with one of La Vérendrye's sons. Shown are the first two forts built there: Fort Saint-Pierre at Lake TEKAMAMIOUEN (Rainy Lake) and Fort Saint-Charles at Lake of the Woods (in today's Minnesota). The heavy rains at Lake of the Woods led de La Jemerais to believe that he was near the Pacific Ocean (he was in fact 2,000 kilometres away as the crow flies). The crosses along the route represent portages that the travellers had to undertake. Farther west are three watercourses possibly leading to the Pacific: RIVIÈRE ST-PIERRE (Minnesota River), RIVIÈRE ST-CHARLES, and RIVIÈRE DE L'OUEST (Missouri River). These contours were supplied by Assiniboine and Cree informants.

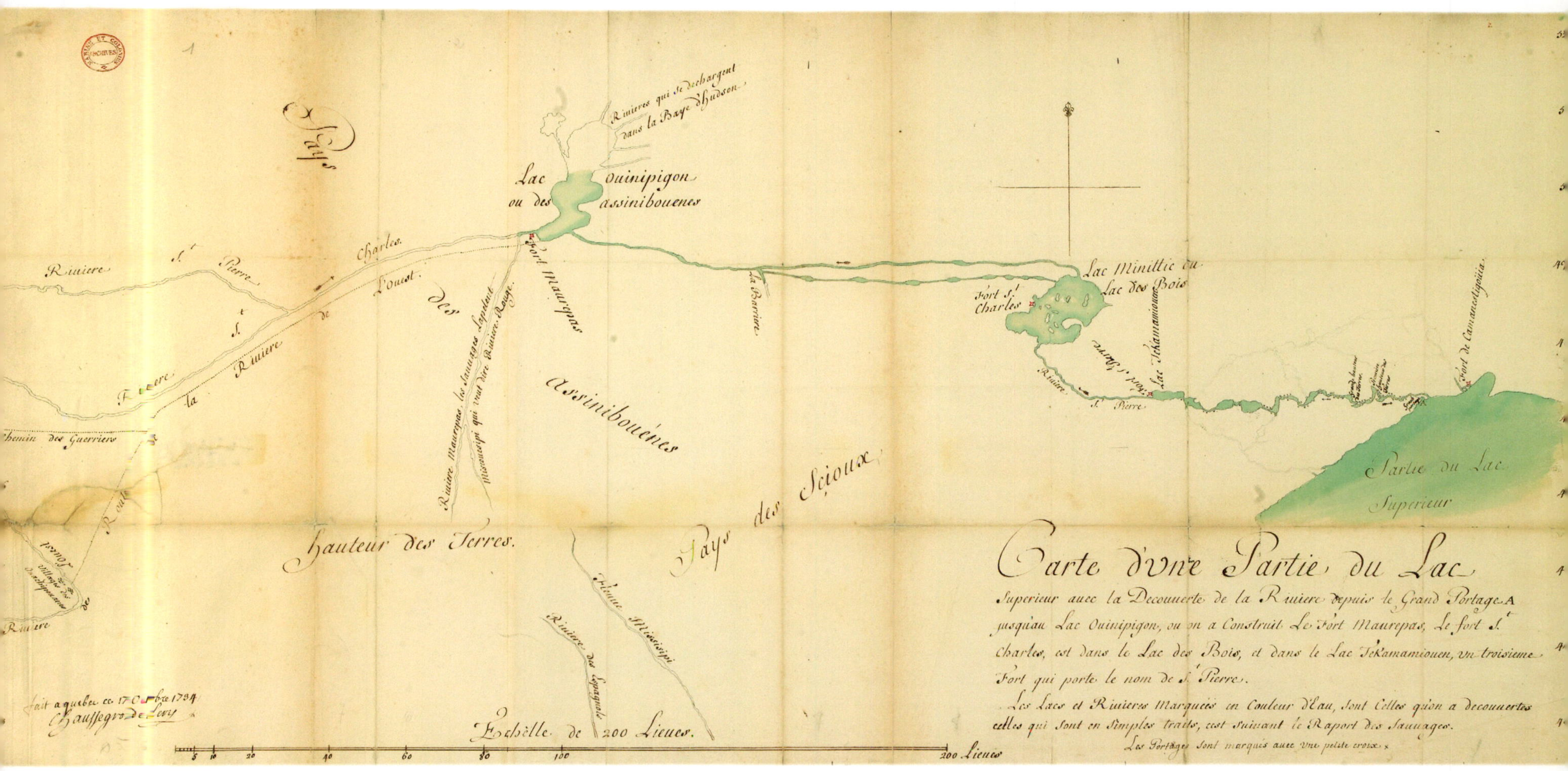

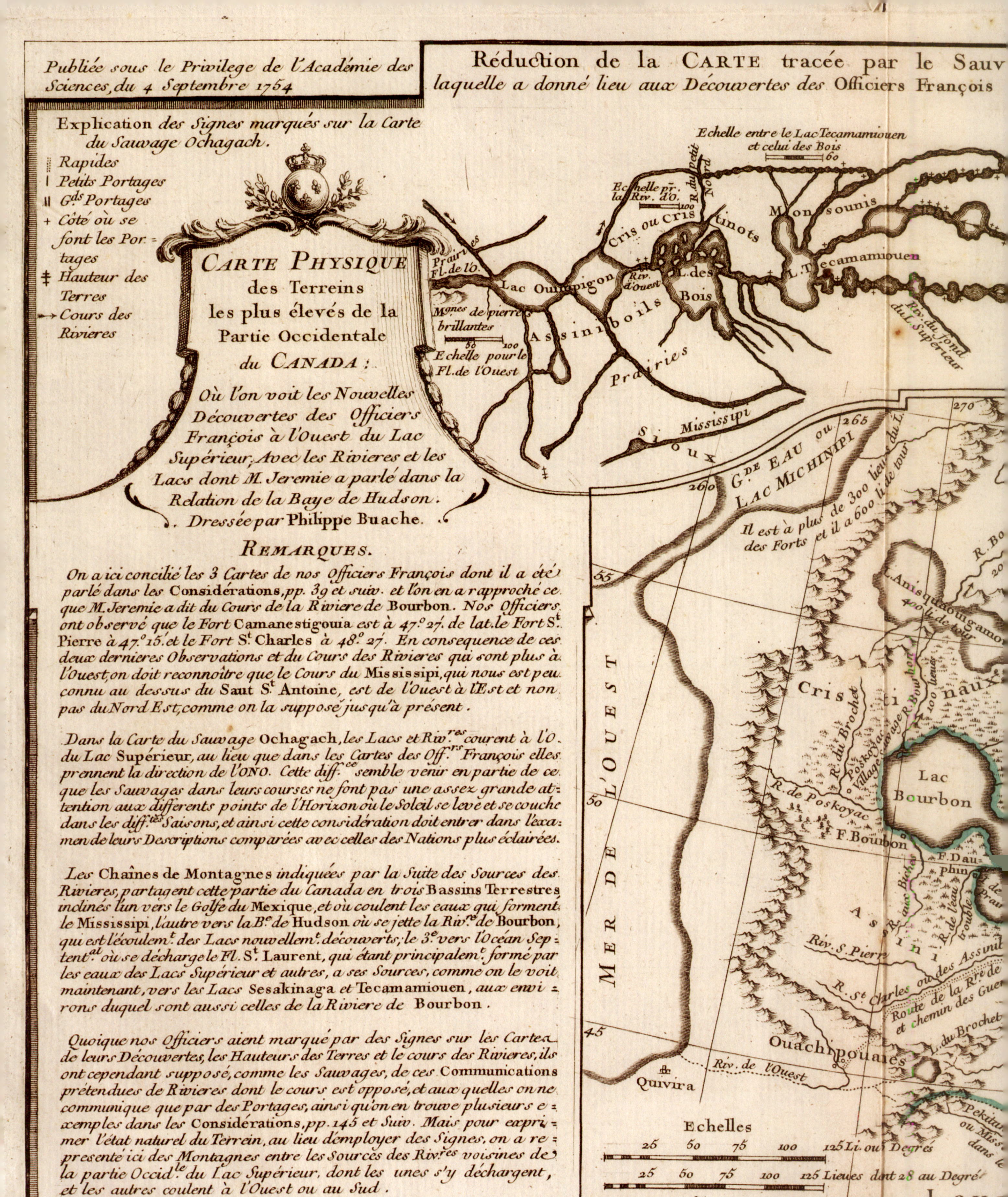

Publiée sous le Privilege de l'Académie des Sciences, du 4 Septembre 1754
Réduction de la CARTE tracée par le Sauv
laquelle a donné lieu aux Découvertes des Officiers François
Explication des Signes marqués sur la Carte du Sauvage Ochagach.
Rapides
Petits Portages
Gds Portages
Côté où se font les Portages
Hauteur des Terres
Cours des Rivieres
CARTE PHYSIQUE des Terreins les plus élevés de la Partie Occidentale du CANADA : Où l'on voit les Nouvelles Découvertes des Officiers François à l'Ouest du Lac Supérieur, Avec les Rivieres et les Lacs dont M. Jeremie a parlé dans la Relation de la Baye de Hudson. Dressée par Philippe Buache.
REMARQUES.
On a ici concilié les 3 Cartes de nos Officiers François dont il a été parlé dans les Considérations, pp. 39 et suiv. et l'on en a rapproché ce que M. Jeremie a dit du Cours de la Riviere de Bourbon. Nos Officiers ont observé que le Fort Camanestigouia est à 47°27'. de lat. le Fort St. Pierre à 47°15'. et le Fort St. Charles à 48°27'. En consequence de ces deux dernieres Observations et du Cours des Rivieres qui sont plus à l'Ouest, on doit reconnoître que le Cours du Missisipi, qui nous est peu connu au dessus du Saut St. Antoine, est de l'Ouest à l'Est et non pas du Nord Est, comme on la supposé jusqu'à présent.
Dans la Carte du Sauvage Ochagach, les Lacs et Rivres courent à l'O. du Lac Supérieur, au lieu que dans les Cartes des Offrs François elles prennent la direction de l'ONO. Cette diffce semble venir en partie de ce que les Sauvages dans leurs courses ne font pas une assez grande attention aux differents points de l'Horizon où le Soleil se leve et se couche dans les difftes Saisons, et ainsi cette considération doit entrer dans l'examen de leurs Descriptions comparées avec celles des Nations plus éclairées.
Les Chaînes de Montagnes indiquées par la Suite des Sources des Rivieres, partagent cette partie du Canada en trois Bassins Terrestres inclinés l'un vers le Golfe du Mexique, et où coulent les eaux qui forment le Missisipi, l'autre vers la Be. de Hudson où se jette la Rivre de Bourbon, qui est l'écoulemt. des Lacs nouvellemt. découverts; le 3e. vers l'Océan Septental. où se décharge le Fl. St. Laurent, qui étant principalemt. formé par les eaux des Lacs Supérieur et autres, a ses Sources, comme on le voit maintenant, vers les Lacs Sesakinaga et Tecamamiouen, aux environs duquel sont aussi celles de la Riviere de Bourbon.
Quoique nos Officiers aient marqué par des Signes sur les Cartes de leurs Découvertes, les Hauteurs des Terres et le cours des Rivieres, ils ont cependant supposé, comme les Sauvages, de ces Communications prétendues de Rivieres dont le cours est opposé, et aux quelles on ne communique que par des Portages, ainsi qu'on en trouve plusieurs exemples dans les Considérations, pp. 145 et Suiv. Mais pour exprimer l'état naturel du Terrein, au lieu d'employer des Signes, on a représenté ici des Montagnes entre les Sources des Rivres voisines de la partie Occidle. du Lac Supérieur, dont les unes s'y déchargent, et les autres coulent à l'Ouest ou au Sud.
Echelle entre le Lac Tecamamiouen et celui des Bois
Echelle pr. la Riv. d'O.
Cris ou Cris
tinots
Monsounis
L. Tecamamiouen
Riv. du fond du L. Supérieur
Prairies
Fl. de l'O.
Lac Ouinipigon
Riv. d'Ouest
L. des Bois
Mgnes de pierres brillantes
Echelle pour le Fl. de l'Ouest
Assiniboils
Prairies
Sioux
Missisipi
Gde EAU ou LAC MICHINIPI
Il est à plus de 300 lieues des Forts et il a 600 li. de tour
L. Anisquaouigamou
400 li. de tour
Cris ti naux
Lac Bourbon
R. de Poskoyac
F. Bourbon
F. Dauphin
R. aux Biches
Riv. S. Pierre
R. St. Charles ou des Assiniboils
Route de la Rre de et chemin des Guer
L. du Brochet
MER DE L'OUEST
Ouachipouanes
Quivira
Riv. de l'Ouest
Echelles
25 50 75 100 125 Li. ou Degrés
25 50 75 100 125 Lieues dont 28 au Degré
55
50
45
260
265
270
LOU
Delahaye l'Aîné Sculpsit

***Carte physique des terreins les plus élevés de la partie occidentale du Canada* by Philippe Buache, 1754**
In 1754, the cartographer Philippe Buache published a memorandum and maps on the subject of cartography in North America. Continuing the work of his father-in-law, Guillaume Delisle, he demonstrated, through analysis of maps and voyage accounts, the existence of a vast inland sea, called the MER DE L'OUEST (western sea). Among other documents, Buache used Russian and Japanese maps, accounts by Henry Ellis, and maps and memoranda issued by "our French officers" – that is, the La Vérendryes. In an inset at the top of the map, Buache placed Ochagach's map showing the succession of lakes and rivers west of Lake Superior and the many rapids and portages to be crossed (see p. 207). To stitch his geographic theories together, he used Nicolas Jérémie's *Relation du détroit et de la baie d'Hudson*, published for the first time in 1720 in Amsterdam. Buache found one passage in this account particularly encouraging: according to Indian sources, there was, a few months by foot to the west, a sea on which could be seen "large canoes or ships with bearded men, gathering gold on the shore of the sea."

Map containing the new discoveries in western Canada, attributed to Louis-Joseph Gaultier de La Vérendrye, 1737 (detail)
In 1737, Pierre La Vérendrye and his men had advanced westward, but not far enough to please the minister of the navy, the comte de Maurepas. If he did not want to be recalled, La Vérendrye had to promise to reach the country of the Mandans. That same year, he distributed a new map of the territories that he had visited; this may have reassured Maurepas. The map shows the Rivière Blanche (Saskatchewan River) leading to a mountain range (the Rockies) and a river flowing into the Pacific Ocean. Along this "sunset river" was a "white man's fort" and at its mouth was a town.

Travelling thousands of kilometres between the Great Lakes, the central plains, and the Rocky Mountains during the search for the western sea between 1730 and 1750, explorers were fully immersed in the natural environment of the Indians. Pierre Gaultier de Varennes et de La Vérendrye, himself the product of a family of explorers, was firmly convinced of the handicaps facing foreigners encountering unknown men and unknown territories. He knew that long-term trade and colonization could succeed only with the continuously reaffirmed agreement of the Indians. Thus, La Vérendrye's undertaking in the West was marked by a clear and meticulously applied diplomatic policy of not only establishing alliances with nations but also maintaining peace among them. Peace was necessary both to the survival of the fur trade and to the collection of information on resources and on the territory to be travelled to reach the western sea.

In addition to providing useful information, the Indians showed the explorers how to grow corn, wild rice, squash, and peas. They supplemented the explorers' provisions with fish and game and provided logistical support in terms of means of transport, workers, dogs and other travel necessities. Women joined the expeditions and took charge, as was their custom, of carrying the heaviest loads, raising tents, preparing meals and mending clothing. Historians date the beginning of French–Indian mixing of blood in Manitoba to the 1720s and 1730s; this would lead to tumultuous times in the nineteenth century.

With more than two centuries of familiarization with North America behind them, French explorers trailblazed through the great plains of the West over some two decades, employing very different means and strategies of travel. Canadian explorers, who had grown up in the eastern part of the country, spontaneously adapted to a new geographic context and gradually adopted land travel, a mode of travel rarely used by their predecessors. Skilful at the tiller of their sailing ships, then good paddlers in canoes, they sometimes had to learn to forget their ships and canoes as they proceeded across the continent in search of the western sea. The frequent need to portage led Europeans voyaging in temperate climates to appreciate winter, when, among other advantages, the spongy, marshy ground froze, making for easier walking. Horses supplied by the Mandans began to be used by explorers of the plains only in 1741. The men who had come from Europe thus profoundly changed their technologies, practices, and strategies in order to thrive in North America and continue their search for the utopia of the western sea.

Main sources

Champagne, Antoine. *Les La Vérendrye et le poste de l'Ouest.* Quebec City: Les Presses de l'Université Laval, "Les cahiers de l'Institut d'histoire" collection, 1968. A still-relevant book on the explorers of the plains. — La Vérendrye, Pierre Gaultier de Varennes de. *À la recherche de la mer de l'Ouest: mémoires choisis de La Vérendrye/In Search of the Western Sea: Selected Journals of La Vérendrye*, edited by Denis Combet, Emmanuel Hérique, and Lise Gaboury-Diallo. Saint-Boniface/Winnipeg: Éditions du Blé/Great Plains Publications, 2001. The reconstruction of voyages on contemporary maps is very enlightening. — La Vérendrye, Pierre Gaultier de Varennes de. *Journals and Letters of Pierre Gaultier de Varennes de La Vérendrye and his Sons: With Correspondence Between the Governors of Canada and the French Court, Touching the Search for the Western Sea.* Edited by Lawrence Johnstone Burpee. Toronto: Champlain Society, 1927. A large number of original documents assembled for this book are now accessible in the database at www.archivescanadafrance.org — Ruggles, Richard Irwin. "The Historical Geography and Cartography of the Canadian West, 1670–1795: The Discovery, Exploration, Geographic Description and Cartographic Delineation of Western Canada to 1795." Ph.D. dissertation, University of London, 1958. The exploration, geographic description, and cartographic delineation of the Canadian West before 1795.

Louisiana

LAND OF PROMISE

French possession of Louisiana was hanging by a thread at the end of the War of Spanish Succession. During the negotiations, which dragged on from 1711 to 1713 as the many agreements that would result in the Treaty of Utrecht were being formulated, the young king of Spain, Philippe V, almost wreaked havoc by claiming the vast territory abutting New Spain in the name of the papal bull of 1493 and the 1494 Treaty of Tordesillas. To the relief of Philippe's grandfather, Louis XIV, not only did Louisiana remain French, but the plenipotentiaries required that the king of Spain renounce his claims to the French throne.

In the early eighteenth century, France had great stakes in Louisiana, as this territory was intended to contain the English colonies east of the Appalachians and serve as a buffer against the Spanish possessions south of the Rio Grande. Louis XIV still hoped to find what Canada had not offered him: the metals constantly prospected for since the earliest explorations. "The great concern is the discovery of mines," wrote the minister of finance, Jérôme de Pontchartrain, to Pierre Le Moyne d'Iberville in 1699, when the latter was leaving for Louisiana. Pontchartrain recommended that d'Iberville attack neither the English nor the Spanish, but "study the resources of the country, and more particularly if it is possible to make wool from the cattle of the country. He is to bring back several skins and even several living beasts if he can. It is said that mulberries grow there; if they do, find out if silkworms could be raised profitably. He is to study the country from the point of view of mines, and so forth."

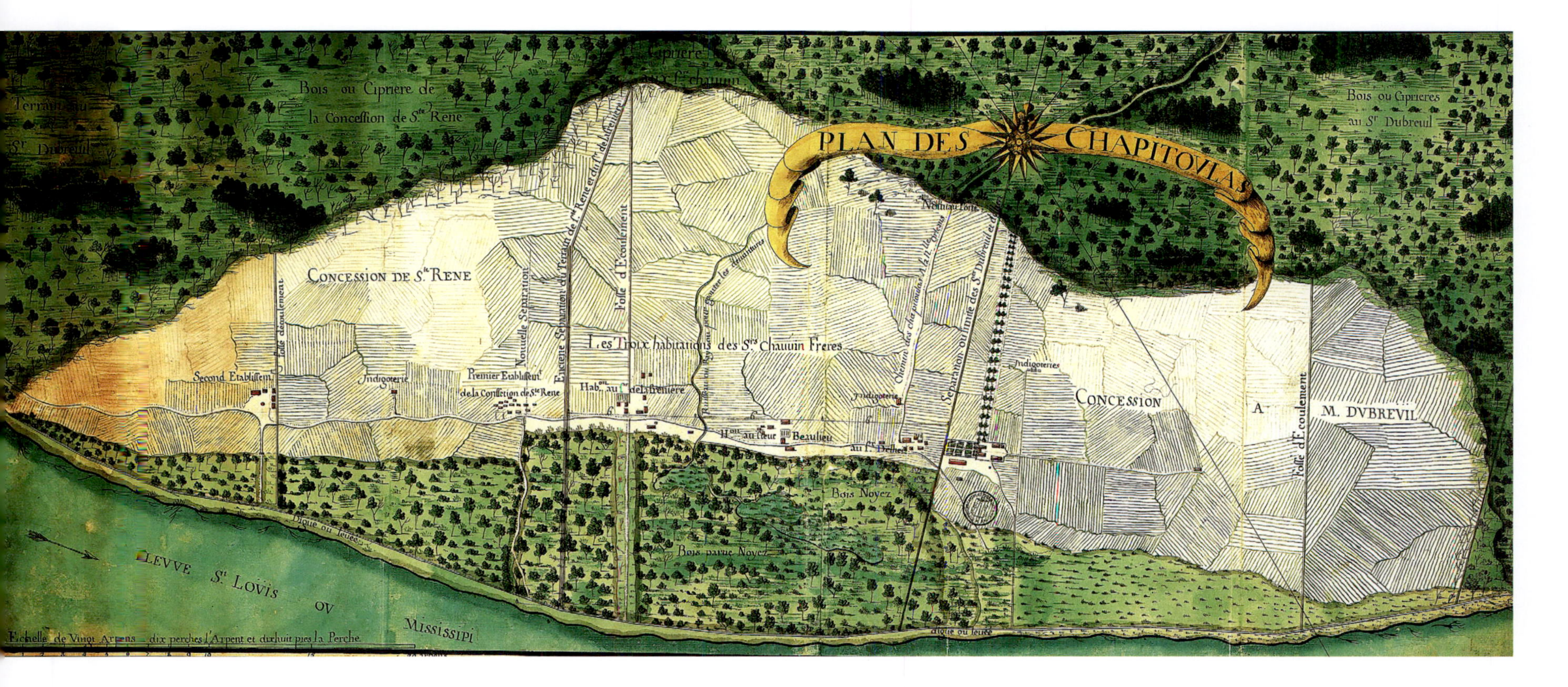

Plan of Chapitoulas, circa 1726
In 1726, Chapitoulas (a town situated some 10 kilometres upstream of New Orleans) had a population of 385 African slaves and 42 Europeans; its main crop was indigo, a plant that produces a natural blue dye. This plan accurately outlines the indigo plantations in Chapitoulas and shows the main buildings, roads, dikes, and drainage canals. Among the main landowners were the Chauvin brothers, born in Montreal.

***Carte du Mexique et de la Floride* by Guillaume Delisle, 1703**
Settled in forts along the shores of the Gulf of Mexico, the French quickly sought to form alliances with the peoples of the interior living along the Mississippi and its tributaries. In Louisiana's early years, Commander Le Moyne de Bienville made overtures to the Natchez and Taensas, as well as to the Natchitoches and Cadodaquious on the Red and Ouachita rivers. The Indians found in the French powerful allies to help them repulse the Chicachas' slave raids. To make this map, the *géographe de cabinet* Guillaume Delisle received information from d'Iberville and Pierre Le Sueur. Delisle placed a border between the colony of Canada and that of Louisiana. The stakes were high: to establish who had authority over the Great Lakes basin, the largest reservoir of fur-bearing animals in North America.

D'Iberville made a list of the resources that might justify colonization. Following the instructions that he had been given, he had Fort Maurepas erected at Biloxi (Ocean Springs, Mississippi) and Fort Saint-Louis at Mobile (Alabama). With his brother, Jean-Baptiste de Bienville, he sought to gain the friendship of all the tribes in the region in order to keep the English of Carolina and Georgia from winning them over to their side. Thus, Bienville began his career in Louisiana in 1701, when he was 21 years old. He was the commander of Biloxi, the first in a long series of appointments that would culminate in his serving as governor from 1732 to 1743. It was to be a difficult assignment: d'Iberville would have to negotiate peace with Indian tribes, regain the trust of the colonists, and defend Louisiana against the English and Spanish. Although he succeeded with the last two tasks, he met total failure with what had been, up to then, his strong point: Indian politics.

Louis XIV had realized that it would be a long and costly process to prospect this immense territory, especially since the founding period coincided with a succession of wars that were exhausting and costly for France, leaving the colony short of resources. The few settlers that there were had to count on the Indians for basic supplies. Thus, there was a return of sorts to the "trading company" formula that had been used to establish Canada. After founding Detroit in 1702, Antoine Laumet, known as Lamothe Cadillac, was appointed governor of Louisiana (1710); in 1712, he and the banker Antoine Crozat started a company that obtained a 15-year trade monopoly in the colony. In a lucky coincidence, peace had been restored, and many French families ruined by the war were encouraged to emigrate to the new colony. But Crozat did not invest enough, and in 1717 his monopoly was transferred, at his request, to the Compagnie d'Occident, owned by the banker John Law. Law changed the company's name to Compagnie des Indes in 1719 and had its monopoly extended to Illinois so that he could exploit the mines there. He conducted a publicity campaign to attract investors. A crowd of shareholders rushed to place their assets with the company, which saw capital, colonists, builders, and craftsmen flow in. Bienville, a member of the company's board of directors, founded New Orleans in 1718. But the good times did not last. In 1720, excess speculation led to the spectacular bankruptcy of Law's bank, although the Compagnie des Indes, reorganized in 1722, enjoyed another 10 years of prosperity.

The 1720s were the years of the Black Code, the objective of which was to protect slaves from mistreatment by their masters and to define the conduct of both parties. The code forbade the practice of any religion but Catholicism and provided explicitly, when applicable, for the expulsion of Jews from Louisiana. Thus the obligations and roles of social groups were very clearly defined. This was to have a long-term effect on the evolution of the region.

In 1726, Louisiana had a population of 3,987, of whom 44 percent were slaves. The population of Canadian or French origin lived mainly in Illinois country, around Mobile and New Orleans, and in Natchez territory. Between these groupings in the north and the south were hundreds of kilometres dotted with small trading posts surrounded by Indian tribes, few of which were happy with the French presence. The Natchez revolt against the French sowed fear in the colony, mistrust of the Compagnie des Indes directors, and disastrous doubt among the company's shareholders. In 1731, therefore, Louis XV revoked the company's privileges, Louisiana

***Carte de la Louisiane et du cours du Mississippi* by G. Delisle, 1718**
Inspired by a map drawn by Father François Le Maire, Guillaume Delisle published this map of Louisiana and the course of the Mississippi in 1718. Delisle's map shows the river's many tributaries, including the Missouri, presented as a possible route to the "western sea." It also indicates the French forts existing within various allied nations. There is a "Los Teijas mission established in 1716" – the first mention of Texas on a printed map. For almost 50 years, this map was the main reference used for representations of the Mississippi Basin.

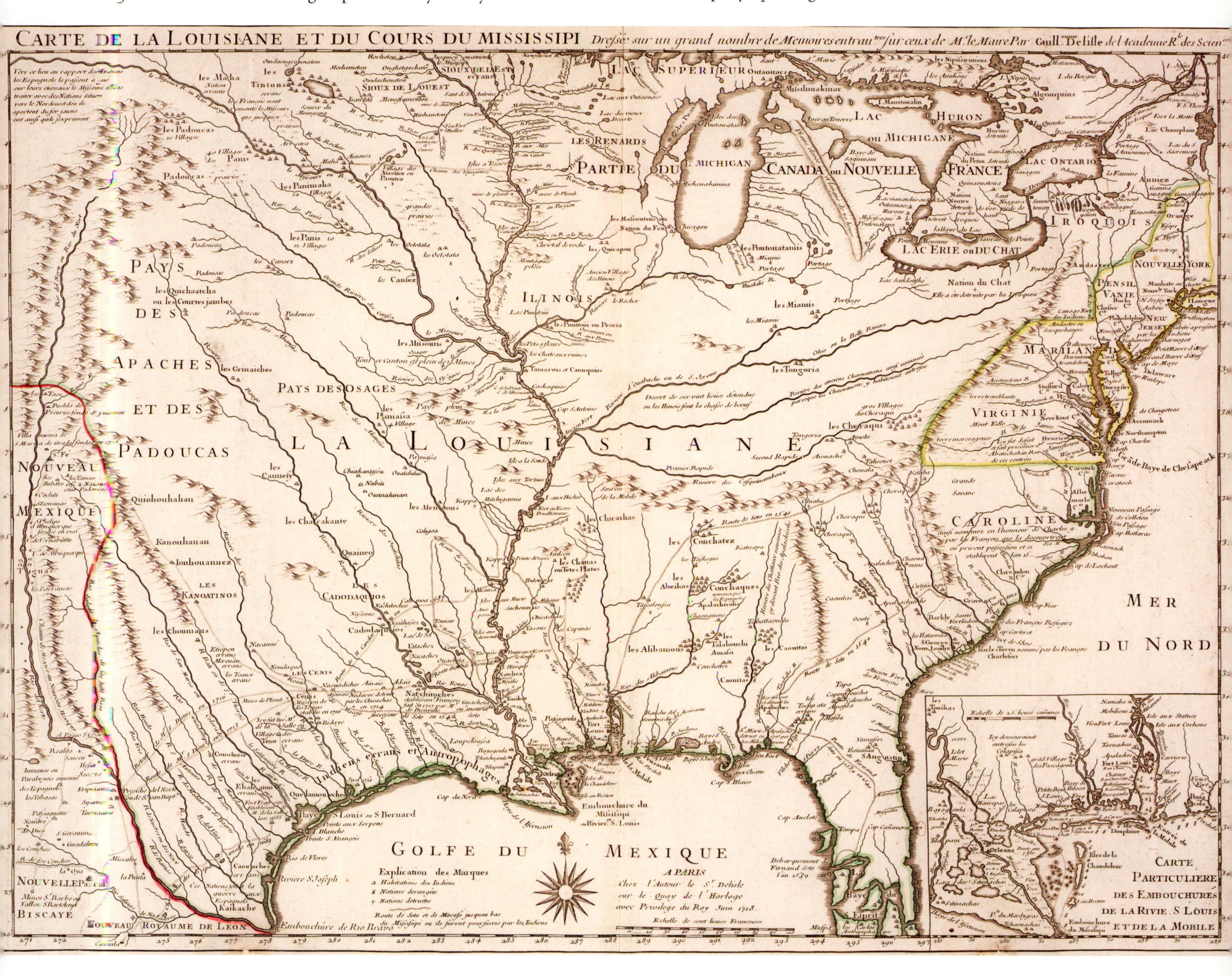

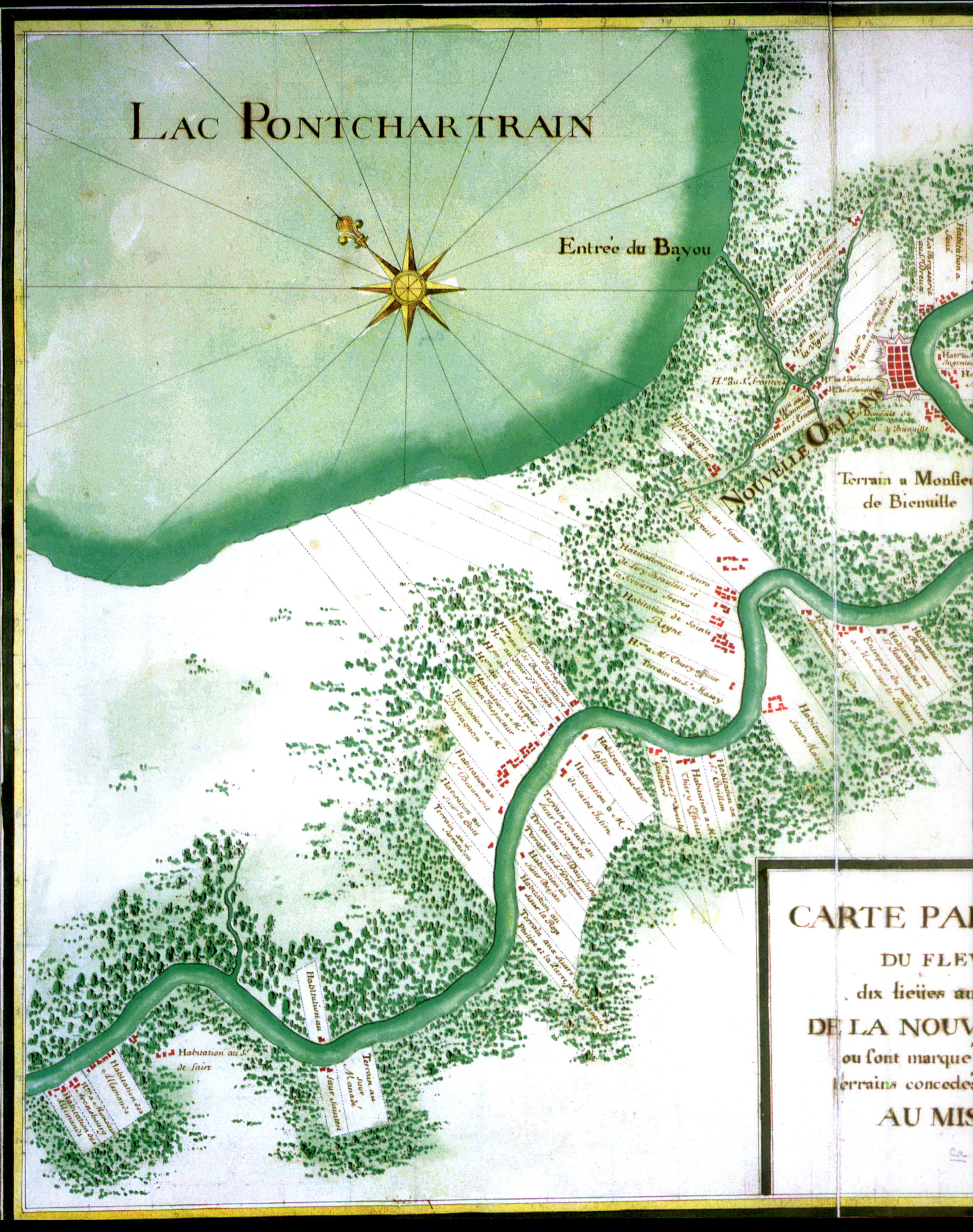

LAC PONTCHARTRAIN
Entrée du Bayou
NOUVELLE ORLEANS
Terrain a Monsieu
de Bienuille
CARTE PA
DU FLEV
dix lieües au
DE LA NOUV
ou sont marque
terrains concedes
AU MIS

Map of the Mississippi and the environs of New Orleans, circa 1723
In 1720, the situation in Louisiana was as precarious as ever. The capital had moved no fewer than four times in Louisiana's 20-year history. Then it was decided to build a city a few kilometres inland, in the fertile Mississippi Valley: New Orleans. This map produced by an anonymous mapmaker shows New Orleans and environs just a few years after its founding. Already fortified, it was nestled into a meander of the Mississippi River. There were only two ways of approaching New Orleans: by taking the Mississippi from its mouth, or by taking Lake Pontchartrain and then the watercourse through the St. John bayou. The site had another advantage: it was elevated and thus relatively invulnerable to flooding. (Hurricane Katrina, in August 2005, was proof of this; it caused the flooding of all of New Orleans except the original French Quarter.) The map also shows that most of the land bordering the Mississippi is already occupied. The names of the owners are written in, including those of Governor Bienville and the engineer Adrien de Pauger. These two men took the best land around the city – and for good reason: one was the governor of Louisiana and the other had drawn up the plans for New Orleans. On the lower left is the "German settlement," referring to 300 colonists recruited by John Law, who settled there in 1722.

Plan of the Tamarois mission in Illinois country by Jean-Paul Mercier, 1735

This sketch of the Tamarois seigneury, located east of the Mississippi a few leagues south of the mouth of the Missouri, was drawn in 1735 by the missionary Jean-Paul Mercier for the superior of the Quebec seminary. It shows that the missionary lived in a house 84 feet long, surrounded by a warehouse, a bakery, a grange, a stable, and a house for black slaves. Mercier also situates the half-dozen French colonists to whom the missionaries conceded land without a contract. These settlers farmed land purchased from the Cahokias, who lived on the site in order to benefit from the armed protection of the French.

was returned to the direct authority of the king, and free trade was instituted. Louisiana was run similarly to the other French colonies, with a governor, a commissioner-director, and a Superior Council. Although it was part of New France, its links of dependency with the general government in Quebec were tenuous. As it was following an economic policy that was vastly different from that of Canada, and as the authorities drew little profit from it, Louisiana was the object of government intervention only when its security was threatened and concerted action was deemed necessary.

Louisiana began to prosper only in the late 1740s: its trade with France, the French Antilles, and the Spanish colonies suddenly grew prodigiously due to its export of wood, indigo, wax, flax, pitch, tar, and furs. Governor Vaudreuil arrived at the best possible time, in 1743, ushering in an era of peace, prosperity, and festivities that left a mark of indulgence and refinement. Under Vaudreuil, the sugar industry was launched and grew in importance. After his departure in 1752, however, rivalries arose between Canadians and French, between Jesuits and Capuchins, and among the Chickasaws, Choctaws, Natchez, and Alibamons. The new governor, Louis Billouard de Kerlérec, had great difficulty managing such unrest in the midst of the Seven Years' War.

By 1746, Louisiana was home to around 9,000 people. More than half of them were slaves, both indigenous and black, though the latter were in the great majority. Many Canadians settled in Upper Mississippi, which had almost 800 colonists in 1752. They supplied flour, dried meats, and animal skins to Lower Louisiana and to Canada. Starting in 1755, another group of French

descendants was added – Acadians expelled from Nova Scotia. These Catholic exiles were given land along the banks of the Mississippi, not far from where the German colonists brought by John Law had settled.

With the fall of Quebec in 1759 and Montreal in 1760, New France was reduced to the isolated territory of Louisiana. Merchandise rotted in the stores for lack of ships; an "elite" accustomed to corruption, trafficking, and gaming was ill prepared for the scrupulous honesty of Governor Kerlérec (1752–62). Only the immensity of its territory and its low trade potential allowed Louisiana to escape the English domination reserved for Canada; the English, having taken Fort Duquesne (Illinois), could have extended their reach southward.

But the court of Versailles was not convinced of the potential value of this remote land. On November 3, 1762, Louis XV decided to make a gift of Louisiana to his cousin in Spain, under a "family pact." The arrival in power of Napoleon Bonaparte was to change everything, for the colony was of strategic interest to him. In 1800, under a secret treaty, Louisiana was returned to the French. But the American president, Thomas Jefferson, feared that France would one day close access to the Mississippi and thus control trade. An American emissary was sent to Paris to propose that the United States purchase the small parcel of land in the Mississippi Delta providing access to the river for $15 million. Surprise! For this very price, Bonaparte, in 1803, ceded the entire territory of 1.6 million square kilometres. The United States doubled its area in one fell swoop. The territory was then divided into 14 states, most of which overflowed the borders of the former Louisiana. France officially disappeared from North America. The path was open for the United States to expand westward.

Main sources

La Louisiane: de la colonie française à l'État américain. Paris: Somogy/Mona Bismarck Foundation, 2003. Exhibition catalogue. — LANGLOIS, Gilles-Antoine. *1682-1803: la Louisiane française.* Paris: Ministère de la Culture et d la Communication, 2003 [www.louisiana.culture.fr]. Virtual exhibition produced for the bicentennial of the cession by Napoleon I of Louisiana to the United States in 1803.

Plan of New Orleans the Capital of Louisiana **by Louis Pierre Le Blond de La Tour, published in London in 1759**
Made from whole cloth on a site chosen by Governor Bienville, New Orleans was named in honour of the regent Philippe d'Orléans. The engineers Leblond de la Tour and Adrien de Pauger used the urban plan in vogue in the eighteenth century. As this map shows, the new capital of Louisiana featured a grid of parallel streets forming perfectly square blocks. In the centre is Place d'Armes (today, Jackson Square), around which were erected the parish church, the Capuchin convent, and the governor's house. Even though the original population was too small to occupy all of the blocks, the city was built over a large area (88 hectares) – in anticipation of population growth.

Fortified Towns and Posts in New France

To survive on North American soil, the French erected a number of defensive structures. In the early years of colonization, New France was at war, and it tried to fend off Indian raids with picket palisades made of trees and intertwined branches. Such structures were not new; they had been used in Europe for centuries. In the late seventeenth century, Quebec, Montreal, and Trois-Rivières had fortified walls to protect against English artillery. A number of king's engineers were sent to the colony to orchestrate their erection or repair. Initiated into the science of fortifications during their service in the army, these men also knew the art of drawing. The first to be sent to Canada, in 1685, was Robert de Villeneuve, an excellent drawer who made a number of beautiful plans of the settlement of Quebec and its environs. He was, however, a mediocre engineer, and the local authorities considered him to be "mad, a libertine, debauched, and a spendthrift." In the end, his fortification plans were never used by the Court. Other, more competent engineers followed, notably Boisberthelot de Beaucours and Levasseur de Néré, but the one who had the greatest influence in New France was Chaussegros de Léry, who was sent to the colony in 1716 to inspect the fortifications at Quebec and Montreal. While a stone wall had been built around Montreal, Quebec was neglected in spite of its status as colonial capital. Only in 1745 were its fortifications finally completed, and this without the king's authorization.

In North America, most of the walls that impeded the expansion of cities were eventually destroyed. Only Quebec City has preserved some of the fortifications dating from New France. They are now a world heritage site protected by UNESCO.

In the second half of the seventeenth century, the French erected other fortifications to protect Canada from English invasion, notably on the Richelieu River (see map p. 186). Following the signature of the Treaty of Utrecht, which ratified, among other things, the loss of Newfoundland and Acadia, they built Louisbourg, an immense seaside fortress, to protect the lucrative fishing industry and ensure better control of the St. Lawrence estuary (see map p. 185). In the interior of the continent, in the Great Lakes and Mississippi regions, the French erected a number of forts that were smaller but strategically placed on main travel routes. The forts were placed as links in a long supply chain linking Montreal and New Orleans to the most inland points farthest west. The French thus ensured that their colonies were well supplied with furs and at the same time blocked English expansion into the interior of the continent. The English responded by building forts that consolidated their position, notably in the Ohio Valley.

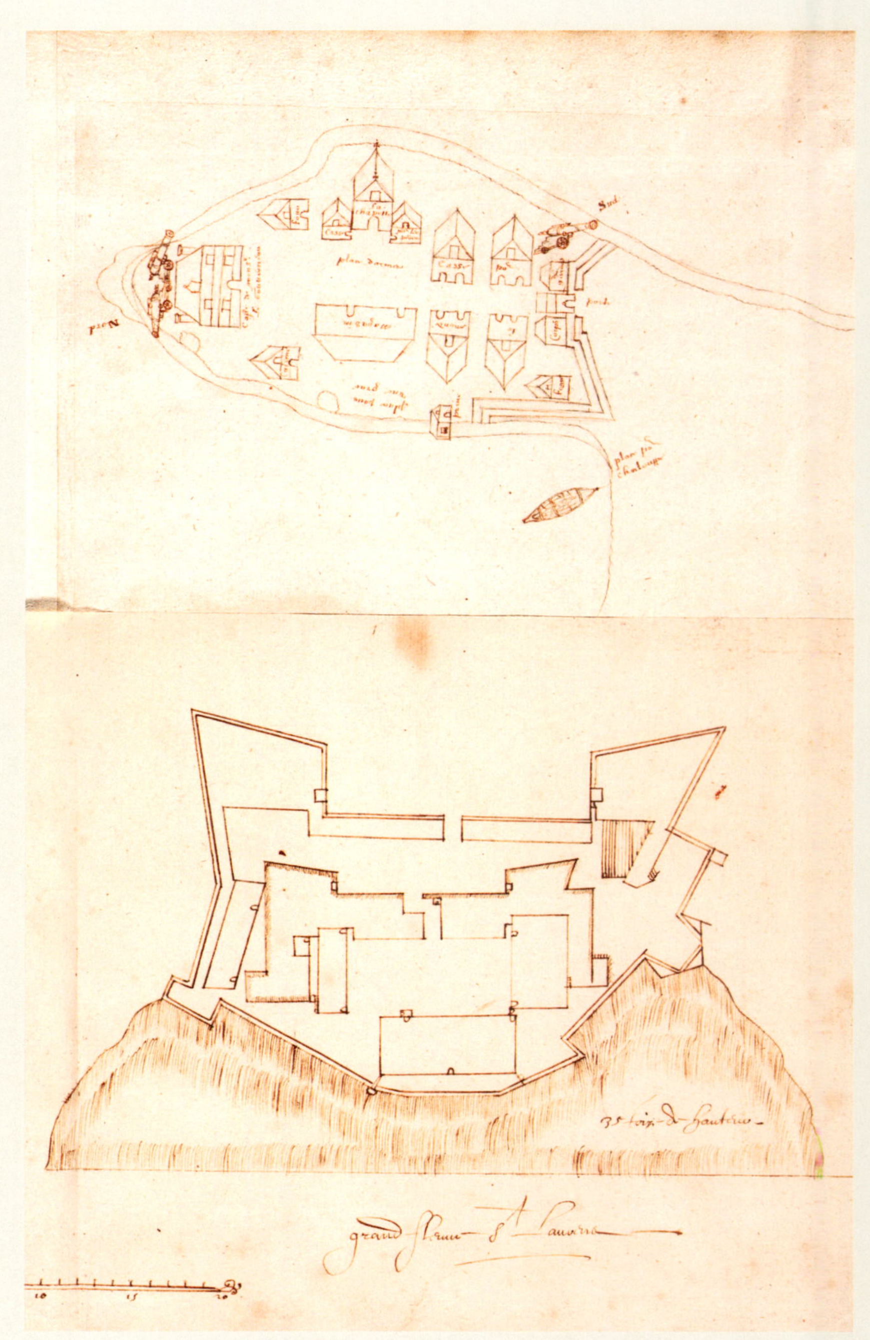

Presumed plan of the Ville-Marie fort, attributed to Jean Bourdon, circa 1647
This is likely the first plan for the Ville-Marie fort on the island of Montreal. When it was founded, five years earlier, the settlement was intended as a centre of spirituality and devotion, a point of departure for missionaries setting out for the western regions. The idea of this mystical utopia was quickly abandoned, and the settlement became instead an indispensable trading centre in New France. The fort shown in the plan was situated on the shores of the St. Lawrence River, in today's Old Port opposite Pointe à Callière, where the eponymous museum now stands. This plate was part of a group of nine plans attributed to the surveyor Jean Bourdon that are conserved in the library of McGill University.

Plan of Boston, 1693
In 1689, war broke out between France and England. After withstanding the aborted attack by William Phips, Governor Frontenac planned to counterattack by invading Boston and Manhattan. He therefore sent Jean Baptiste Louis Franquelin and Lamothe Cadillac on an espionage mission (1692). This map, the result of their reconnoitring, shows the main defence elements of Boston. The routes leading to the city, by land and sea, are highlighted. The city is located in a deep bay, well protected by a barrier of rocks and islands, including Isle du Fort, known as Castle Island. The map also shows that at Nantucket (Noutarquet) there are pilots to bring ships into Boston. Linked to the mainland by a narrow isthmus, Boston was defended by a palisade and well-positioned batteries that kept enemy vessels from anchoring in the vicinity. Fort Andros (Fort Hill), situated on a hill, also served to defend the city. Franquelin said that he worked for five months on this map, until it was approved by those who had been present at the site – French, English, and Indian. Comparing this map with a modern cartographic image, one observes that Boston and Charlestown had more pronounced peninsulas in the seventeenth century.

Plan of Louisbourg by Étienne Verrier, 1735
After ceding Acadia to England, the French built a port fortress at Louisbourg on Île Royale. By defending this site, which was ice-free year round, the French facilitated travel to and from the metropolis, supported the fishery, and secured the entrance to the St. Lawrence River. The engineer Verger de Verville chose the site on the peninsula at the south end of the port, and then set up a checkerboard plan reflecting the classical aesthetic of the time. His successor, Étienne Verrier, drew up a number of town plans, including this one, which shows the three main batteries defending the city. In spite of all these precautions, the French twice had to surrender Louisbourg to English attack.

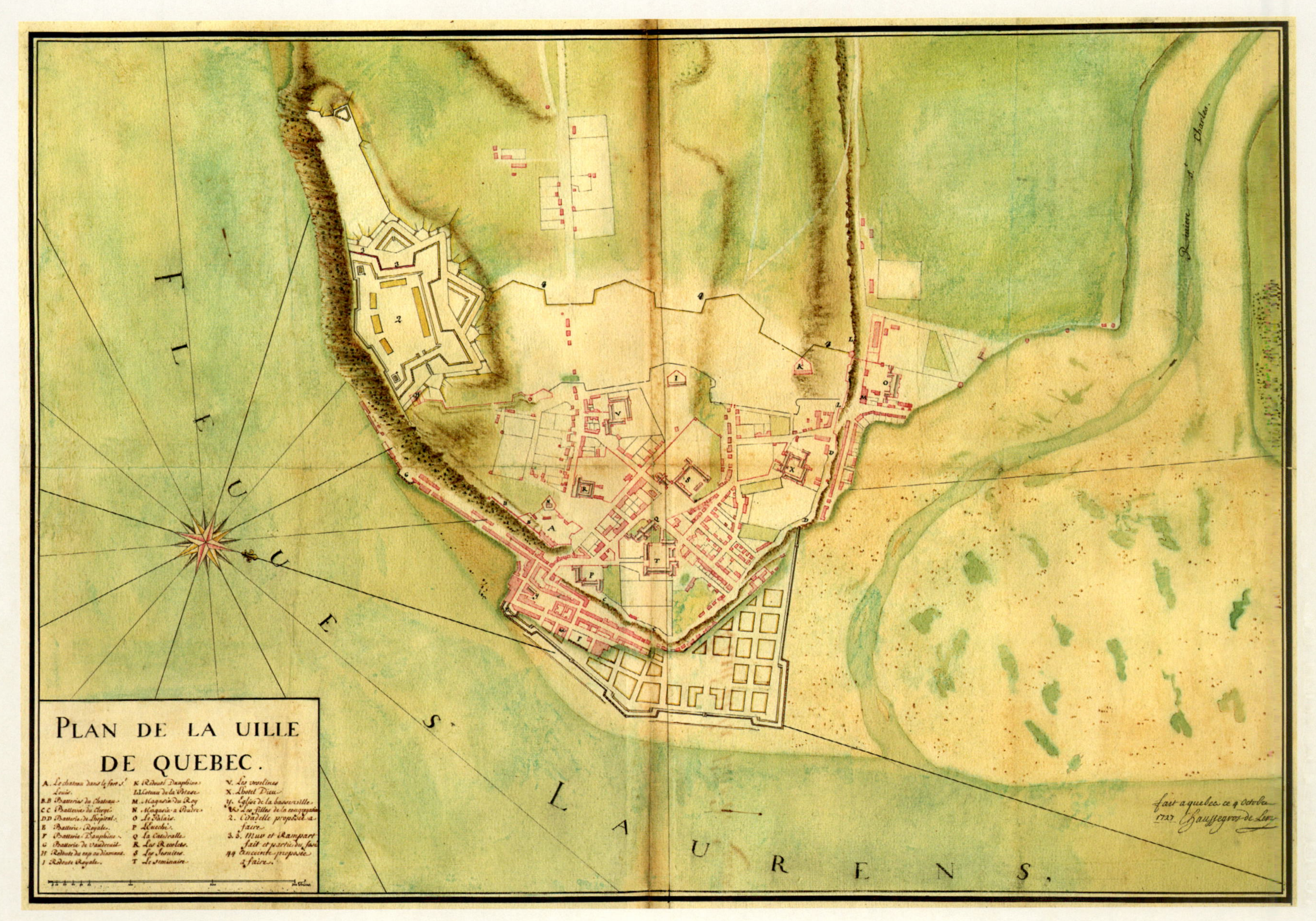

***Plan de la ville de Québec* by Gaspard-Joseph Chaussegros de Léry, 1727**
When he arrived in Canada, in 1716, the king's engineer Chaussegros de Léry was mandated to improve the fortifications of the town of Quebec. He drew up an impressive number of construction and repair plans for presentation to the king. This one, dating from 1727, offers a solution to congestion in the Lower Town. Taking up an idea that had been advanced in the late seventeenth century, he proposed to open the urban space onto the river. Construction of a dike would allow for recovery of a good part of the land that was exposed at low tide. Although this project was approved by the authorities, it did not come to fruition. The plan also proposed construction of a citadel on the heights of Cap Diamant. This large building would overlook the wall that surrounded the city on one side and the passage on the St. Lawrence River on the other. The project was rejected by the king on the grounds that it would be too expensive and would take too long to complete. The idea was taken up again after the Conquest by the English officers, who had a citadel erected to forewarn Quebec against an American attack or a popular uprising.

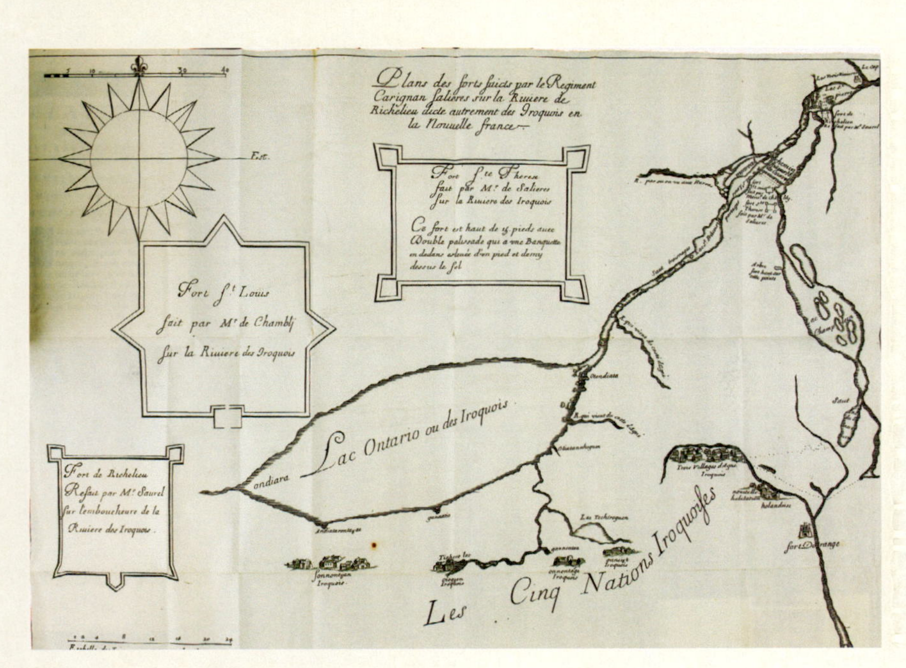

Plan of the forts on the Richelieu River, published in 1666
"Never will New France cease to bless our great Monarch, having undertaken to surrender its life to him and pull it from the fire of the Iroquois." The objective was to crush the Iroquois, as the Romans had crushed the Barbarians, in order to build a "great Christian empire" in North America. The author of the Jesuit relation published in 1666 had no patience for those who impeded the fur trade. Such military concerns were also manifested in the publication of a map of three recently constructed forts – Richelieu, Saint-Louis, and Sainte-Thérèse – on the Richelieu River, also called Rivière des Iroquois, which was the main route between Iroquoisia and the St. Lawrence Valley. According to the author of the relation, a fourth fort was planned deeper into Iroquois country, where the French might "make continual sorties on the enemies, if they do not see reason." This chain of fortifications was the work of the Carignan-Salières regiment, composed of 1,200 soldiers and officers sent to Canada in 1665 to invade Iroquoisia.

View of Quebec taken from the *Atlantic Neptune*
Published for the first time in 1781 in one of the volumes of the *Atlantic Neptune*, this portrayal shows the town of Quebec perched on a high promontory. The height of the cliffs and the main buildings is exaggerated to make it easier for navigators to locate. The artist highlights Fort Saint-Louis, the redoubt on the cape, the batteries in the port, and two warships, giving the impression, in the midst of war between the English and the Americans, that the city was invulnerable.

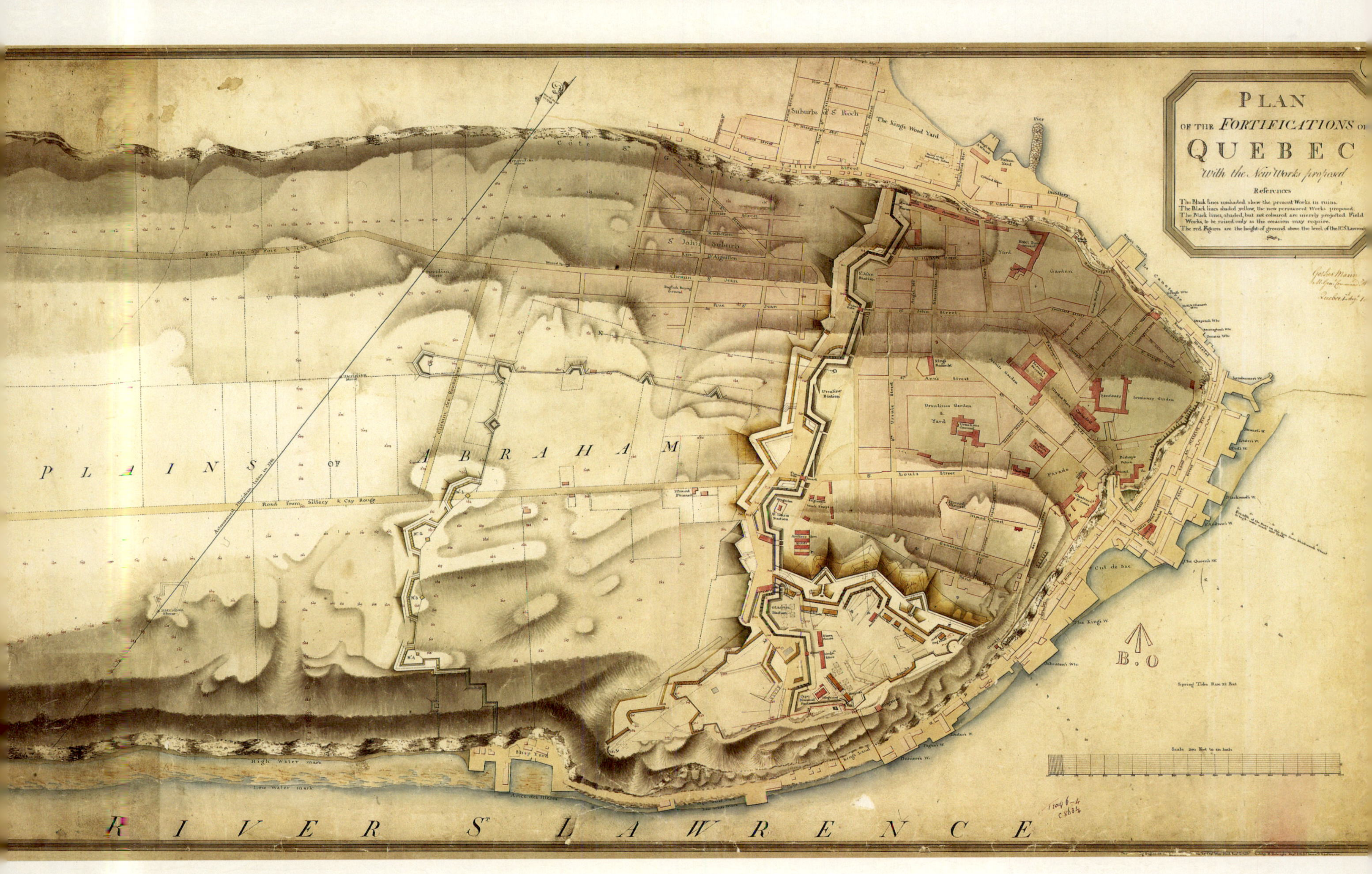

***Plan of the Fortifications of Quebec*, 1804**
Having conquered New France, the English continued the work begun by the French engineers to strengthen the defensive system of the town of Quebec, one of the keys to British possessions in North America. The American War of Independence was to demonstrate the importance of the town and its fortifications to preserving the integrity of Canada. After the blockade of Quebec by Montgomery's American troops, the British authorities decided to repair Chaussegros de Léry's wall, known for its effectiveness, and build a temporary citadel. In 1804, the engineer Gother Mann drew up a new defence plan in which he proposed to complete the wall around the town, occupy the heights of the Plains of Abraham with an entrenched camp, and replace the temporary citadel with a permanent one.

Bird's-eye view of Fort Chambly, 1721
In 1709, during the War of Spanish Succession, Governor Pierre Rigaud de Vaudreuil ordered the construction of a stone fort at Chambly, on the Richelieu River. Inspired by the fortifications of Sébastien Le Prestre de Vauban, the engineer Boisberthelot de Beaucours produced an imposing structure that would become one of the main points in New France's defensive system.

***A Plan of the Town and Fort of Carillon*, 1758**
During the Seven Years' War, the French built Fort Carillon at the southern end of Lake Champlain to defend Canada against a British invasion. This map shows the environs of the fort besieged by General James Abercrombie's enormous 15,000-man army in the summer of 1758. Anticipating the imminent danger and considering the fortifications unlikely to resist an assault, the Marquis de Montcalm had retrenchments hastily built on the heights west of Carillon. Tree trunks were piled up to a height of eight feet, impeding the progress of the British army toward Carillon. The English attack of July 8 turned into a fiasco, with almost 2,000 British soldiers killed or wounded, compared to 377 on the French side.

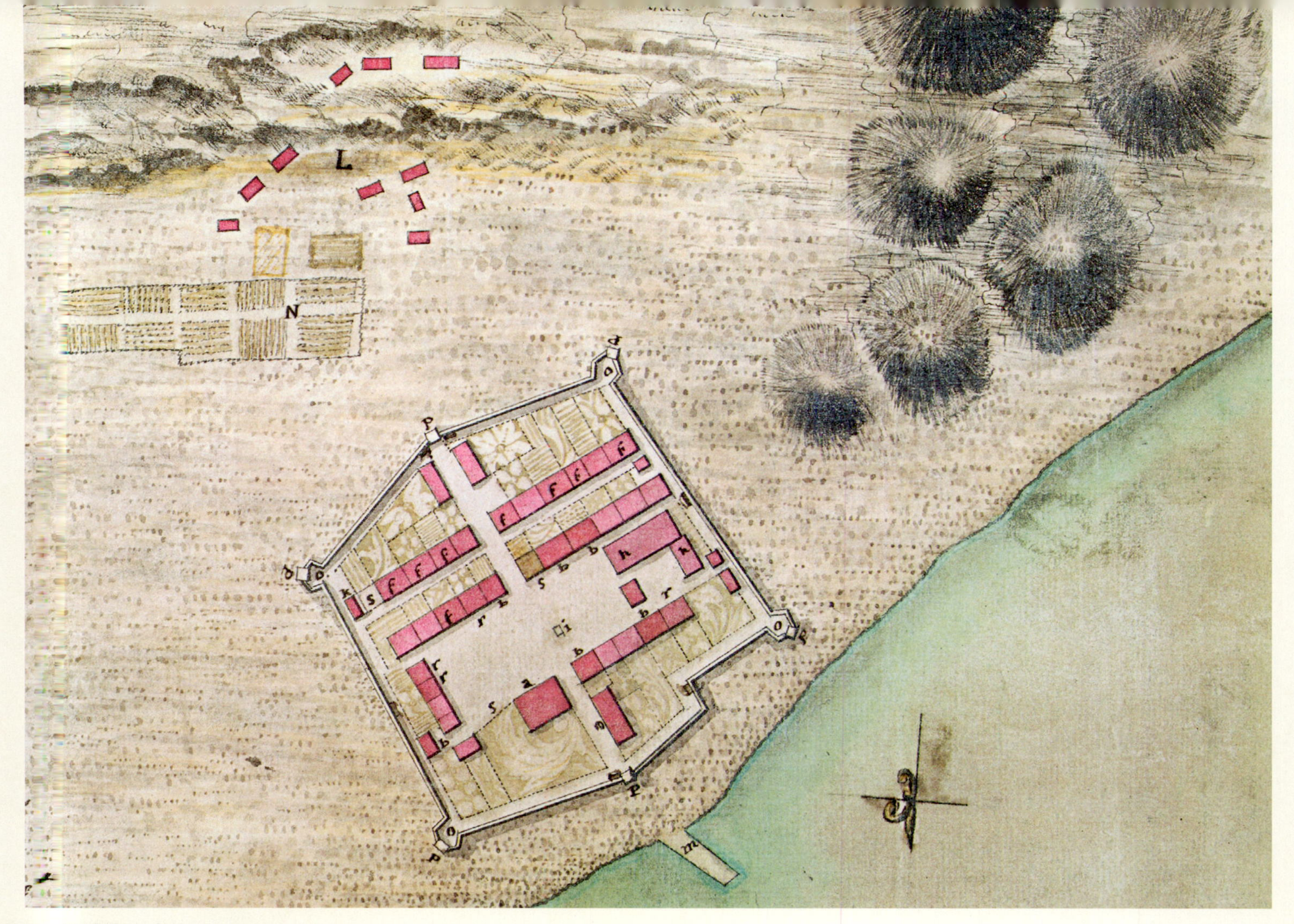

Fort Michillimackinac, 1766
A small military and trading post strategically located at the intersection of lakes Huron, Superior, and Michigan, Fort Michillimackinac was an indispensable stopover for Canadian merchants venturing farther into the interior of the continent. Abandoned in 1696, the fort was rebuilt some 20 years later and became an important meeting place for Indians and French and the centre of the fur trade in the upper country. After the surrender of Montreal (1760), Michillimackinac was turned over to the British. This plan, made by Lieutenant Perkins Magra, shows the state of the fort a few years later, in 1766. Surrounded by a wooden palisade and six bastions (p), the village consisted of several buildings surrounding the well (i): the commanding officer's house (a), officers' barracks (b), soldiers' barracks (f), traders' barracks (r), provisions store (e), and chapel (h). Outside the fort, there were a small wharf (m), stables (L), and a garden (N). In 1781, the site was abandoned in favour of a new fort built on the nearby island of Mackinac.

Fort Prince of Wales
In the seventeenth and eighteenth centuries, Hudson's Bay Company agents built several permanent posts on the shores of the bay. This engraving, taken from Samuel Hearne's travel accounts, shows Fort Prince of Wales, located at the mouth of the Churchill River. Built in stone starting in 1731, this fort, with walls more than 5 metres high and 12 metres thick, was intended to protect the fur trade and prevent attacks by the French. In addition to conveying the monumentality of this structure in the middle of the northern tundra, the engraving depicts the wigwams of the Chippewyans, the main trading partners living around the settlement.

NOUVELLE FRANCE OU CANADA
LAC SUPERIEUR
I. Royale
I. Philippeaux
I. Maurepas
I. Pontchartrain
OUTOUACS
LAC HURON
Isle Manitouahn
LAC MICHIGAN
LAC ERIE ou OSWEGO
Lac S.te Claire
Detroit
SIOUX DE L'EST
ASSINIBOILS
OUTAGAMIS
MASCOUTENS
PAYS DES RENARDS
LES MIAMIS
ILLINOIS
INDIENS DE L'OHIO
Sauteurs ou Messesague
Prairies Immenses
remplies de Buffles
Fort Lullier
Sault S.t Antoine
Quadoghe
Quicapous
Fort des Miamis
Fort Orleans
Missouri Riviere
Osages
Pays rempli de Mines
Kaskakies
VIRGINIE
Twightwees
Delaware
Shawnoes
Grande Plaine tres fertile selon le raport des Indiens et des Anglois
CAROLINE SEPTENTRIONALE
CHERAKIS
AKANSAS
CHICACHAS
Alliés et soumis aux Anglois
CATAWBAS
Anson C.
les Anciens Eriez ont été detruits par les Iroquois il y a 100 ans et c'est depuis ce tems qu'ils possedent le Lac Erie.
Limites Septentrionales de la Nouvelle Angleterre par Lettres Patentes du 3 Novembre 1620, jusques a la Mer du Sud.

Confrontation in Ohio

THE WASHINGTON–JUMONVILLE AFFAIR

Under the influence of Samuel de Champlain, the French had started their march to the interior of the continent in the early seventeenth century. Ceaselessly, they traversed, explored, surveyed, and mapped North America. Unimagined distances did not put them off. At Lachine, they quickly boarded canoes and set out for Michillimakinac, Kaministiquia, Baie des Puants, Kaskaskia, Cahokia, Vincennes, and, one fine day, New Orleans – in this case, a nine-month journey. The country of the Illinois, or Upper Louisiana, was given the role of breadbasket. Its vast natural prairies were ready to receive colonists; some came from Lower Louisiana via the Mississippi River, while the others came from the St. Lawrence Valley via the Ottawa River, then possibly the Great Lakes – that is, through the Ohio region. Pacification of the Iroquois made this route particularly attractive.

Although they did not feel crowded between the Appalachians and the sea, the Americans began to gaze westward, toward the Ohio River, in the early eighteenth century. The Treaty of Utrecht gave them a golden opportunity. Article 15 placed under the protection of the British the "Five Nations," which it qualified as "under the dominion of Great Britain," while, on the other hand, the reverse was planned "for the *Ameriquains* [the Indians, of course!] who are subjects or allies of France." In fact, free circulation was authorized "for the reciprocal advantage of trade, with no molestation or impediment by one party or the other." The text was quite circuitous, allowing for just about any interpretation. The country of the Iroquois, the Five Nations – the Six Nations after the integration of the Tuscaroras in 1722 – was fair game for greedy hands.

The Ohio – the "Beautiful River" – flowed through the backyards of Virginia and Pennsylvania. These two colonies had the advantage of proximity, but they were not in agreement. Their interests diverged. The Iroquois skilfully played them off each other, as was their wont. In 1742, the Treaty of Easton confirmed the Iroquois' cession to the Penn family of 600,000 acres, which belonged in reality to the Delawares. Two years later, at Lancaster, they received gold and silver in exchange for the Ohio region, fulfilling the dreams of European speculators. A solidly fortified trading post was built on the north branch of the Potomac River at Wills Creek; a second fort followed on the Monongahela River, 35 miles up from its confluence with the Allegheny River.

The French kept up the pressure. On the orders of Governor La Galissonière, Pierre-Joseph de Céloron de Blainville, at the head of an expedition of some 250 men, patrolled the region, warmed the hearts of their Indian allies, distributed gifts, and buried lead plaques marking France's taking possession of the Ohio and its tributaries. In November 1749, Blainville returned to report to a new governor, La Jonquière. Pennsylvanian traders were visiting the region in greater numbers, and Blainville recommended that permanent settlements be established. The authorities hesitated. In 1752, Governor Duquesne took command, and he was not afraid to make decisions. The following spring, a line of forts was planned and construction was begun: Fort Presqu'île on the south shore of Lake Erie, Fort le Bœuf on a tributary of the Allegheny River, Fort Machault a bit farther south, near a Delaware village (Venango), and Fort Duquesne at the confluence of the Ohio, Allegheny, and Monongahela rivers. This last was a true fortress flanked by four bastions, as well as a village with a bakery, a forge, and a hospital. The French were definitely there to stay. The stage was set for a serious confrontation. This was when George Washington entered the history books with the assassination of Joseph Coulon de Villiers de Jumonville.

The Washington–Jumonville affair was the type of frontier incident that was bound to set the tinderbox alight. The French officer had been sent on a diplomatic

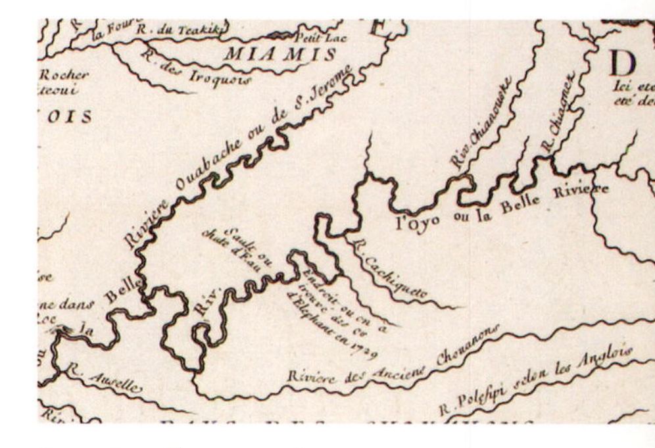

Remains of mammoths on the Ohio River
Along the Ohio River, Jacques-Nicolas Bellin noted the presence of elephant bones. This reference recalls the armed expedition of Baron Charles Le Moyne de Longueuil, who (in 1739) was probably the first European to explore the environs of what is now Big Bone Lick State Park, known for its remains of mammoths.

The Ohio River according to John Mitchell, 1756
Like the French, the English coveted the Ohio Valley for its fauna and its fertile land. Outraged by the French strategy of surrounding British colonies, John Mitchell made a map focusing on the Ohio River. By incorporating this territory into Virginia, Mitchell sought to prove that the English had been in the region longer than the French. The map shows the land routes taken by merchants, as well as the main reserves of salt and beaver. In 1755, when this map was first published, the situation was explosive. The previous year, the future president of the United States, George Washington, had been accused of participating in the assassination of the French officer Joseph Coulon de Villiers de Jumonville (near today's Jumonville, Pennsylvania).

CANADA
LOUISIANE
ET
TERRES ANGLOISES
PAR LE Sr D'ANVILLE
de l'Académie Rle des Inscriptions et Belles-Lettres, et de celle des Sciences de Petersbourg,
Secrétaire de S.A.S. Mgr LE DUC D'ORLEANS
Novembre MDCCLV
Sous le Privilége de l'Académie
Chez l'Auteur, aux Galeries du Louvre.
Longitude Occidentale du Prémier Méridien
ECHELLE
Lieues Françoises, de 3000 Pas Géométriq. ou de 2500 Toises.
Lieues Marines Françoises et Angloises, de 20 au Degré.
Lieues Cannadiennes, évaluées à 2000 Toises.
Milles Anglois, fixés par Henri VII à 1760 Verges ou 5280 Pieds Anglois.
Milles communs et d'usage en Angleterre, de 50 au Degré.
JAMES BAY
Lac Alemipigon
Gens des Terres
CANADA
NOUVELLE
LAC SUPERIEUR
SIOUS
OUTAOUACS
ALGOMQUINS
OUTAGAMIS ou Renards
MASCOUTENS ou Gens du Feu
LAC MICHIGAN
LAC HURON
IROQUOIS DU NORD
LAC ONTARIO
LAC ERIE
IROQUOIS
ILINOIS
MIAMIS
LONG-ISLAND
LOUISIANE
NORT CAROLINA
CAROLINE
TCHACACHAS
Abeyka
HAUTES RIVIERES
TCHATAS
BASSES RIVIERES
FLORIDE
APALACHE
GEORGIE
Catawbas
Congarees
Lac des Abitibis
TIMISKAMIOUETS
Lac Temiskaming

Canada, Louisiane et Terres angloises by d'Anville, 1755

Appointed king's geographer at age 22, Jean-Baptiste Bourguignon d'Anville (1697–1782) was one of the great cartographers of the eighteenth century. Known first for his maps produced for Father Pierre-François-Xavier Charlevoix's *Histoire de Saint-Domingue,* d'Anville made more than 200 maps during his career. This one, *Canada, Louisiane et Terres angloises,* was published in 1755, on the eve of the Seven Years' War, when the French and English were disputing rights to the Ohio Valley. It shows the immensity of the territory, extending from the St. Lawrence River to Louisiana via the Great Lakes, claimed by France before the fall of New France. The map includes an impressive number of Indian and French names, eloquently indicating that a large part of the continent had been occupied by Indians and frequented by the French. This map represented an improvement over previous ones. The contours of the Missouri, for example, are more accurately drawn thanks to the expeditions of Étienne de Véniard de Bourgmond. D'Anville used scientific rigour, describing in a memorandum the method and sources employed to draw this map. He made use of observations from various voyages: those of Marquis Chabert de Cogolin for Île Royale (today, Cape Breton Island); Jean Deshayes, Gabriel Pellegrin, Jean-François de Verville, and Michel Chartier de Lotbinière for Quebec and the St. Lawrence River; Louis Jolliet and Father Pierre-Michel Laure for the region north of the St. Lawrence; the Jesuits, notably Father Francesco Giuseppe Bressani, for the Great Lakes; and Pierre Baron, Adrien de Pauger, and Pierrre Le Sueur for the Mississippi River. D'Anville had a diverse network of correspondents, among them Father Joseph-François Lafitau, Marshal d'Estrées, Intendant Jacques Raudot, and the general controller of finance, Philibert Orry. All of the cartographic sources that he obtained from these men are conserved in the maps and plans department of the Bibliothèque nationale de France. Because of these diverse sources, d'Anville was able to form a relatively true image of North America, which was to have a profound influence on his successors, including the English cartographers of the second half of the eighteenth century.

A Scale of Miles
Part of Lake Erie
The French are now coming from their
Forts on Lake Erie & on the Creek, to Venango
to Erect another Fort – And from thence
they design to the Fork's of Monongehele
and to the Logs Town, and so to continue
down the River building at the most con-
venient places in order to prevent our
Settlements &c.
NB. A Little below Shanapins Town in
the Fork is the place where we are going
to Build a Fort as it commands
the Ohio and Monongehele –
The Logs Town
Ohio
River
Aligany
Mountains

French forts on the Ohio River
This 1754 map is interesting mainly for the role played by its maker, George Washington, in the Jumonville affair. Washington's actions might have been the end of him. Luck was with him on this occasion, however, as it was when he miraculously escaped death during the defeat of General Edward Braddock in 1755. The previous year, Washington had explored the Ohio from Monongahela to the shores of Lake Erie, to anxiously observe a strong French presence. Without naming them, he notes the existence of a number of forts and recommends that this progress be halted immediately.

mission to formally demand that the English leave a territory claimed by France. Caught in an ambush, Jumonville tried to read the order that he had brought. The Americans were not listening; they had already opened fire. Jumonville was killed by a tomahawk strike to his head by the Mingo Indian Tanaghrisson (Tanacharisson, sometimes known as the Half King, born a Catawba Indian and raised by the Senecas), according to some, or by gunfire, according to others. The news was brought to Fort Duquesne by a Canadian named Monceau who had escaped the massacre. An Indian also came to give his version. Louis Coulon de Villiers, Jumonville's younger brother, took command of a large detachment of more than 500 men and went to track Washington down. At the site of the ambush, the corpses of 10 Frenchmen had been abandoned without burial and had been preyed upon by wolves and crows. The Americans had fled. Coulon de Villiers cornered them in Fort Necessity. On July 4, the crestfallen young Washington – he was barely 20 years old – surrendered and signed a statement acknowledging his responsibility in the "assassination" of Jumonville. The text was in French, and Washington claimed later that the interpreter, Van Braam, had translated "assassination" as "death" or "loss." In any case, Washington's morale was at a low ebb. Convinced that the French would avenge their losses and that they were as good as dead, his men proceeded to get drunk. When he realized that the Indians with the French "were all our own Indians, Shawnesses, Delawares and Mingos," Washington literally collapsed. He would have signed anything.

Coulon de Villers became famous for the restraint that he had shown. Washington would be allowed to explain himself; emergencies, however, temporarily rescued him from this embarrassment. Chance smiled on him: he was one of the survivors of the disaster that awaited General Edward Braddock in July 1755, a few kilometres from Fort Duquesne on the Monongahela River.

Despite their superior numbers, the Anglo-Americans were on the defensive. In Europe, hostilities had spread. On May 17, 1756, declaring war on France, George II of England denounced "the encroachments and usurpations of the French." In North America, this declaration was a formality. Until the battle of Carillon (1758), the French had victories, but then everything fell apart: Louisbourg, Fort Frontenac, Fort Duquesne, Fort Niagara, Fort Carillon, and Fort Saint-Frédéric fell. In 1759, the English were on the Plains of Abraham. In 1760, Montreal surrendered.

William Pitt, the great victor of the Seven Years' War, was ready to surrender Canada to France. He felt that the presence of this enemy at the borders of the Thirteen Colonies could have salutary effects, and he would have wanted instead to keep the "sugar islands" away from the grasp of the French. The rich planters, however, did not want commercial competition from Guadeloupe and Martinique. Pitt was denied. Canada was wiped off the map. A new colony was formed, the Province of Quebec, and the British authorities got busy drawing borders. Over 50 years, English cartographers – Herman Moll, John Senex, Henry Popple – extended the borders of the English colonies to the south shore of the St. Lawrence River. Pennsylvania and Virginia claimed that their territories extended to the shores of lakes Ontario and Erie.

London decided otherwise. West of the Alleghenies became Indian territory; the border of the Province of Quebec was not the south shore of the St. Lawrence but the watershed between the river and the ocean.

Main sources

Anderson, Fred. *Crucible of War: The Seven Years' War and the Fate of Empire in British North America, 1754–1766.* New York: Alfred A. Knopf, 2000. A monumental work. — Frégault, Guy. *Canada: The War of the Conquest.* Translated by Margaret M. Cameron. Toronto: Oxford University Press, 1969. — Vaugeois, Denis. *The Last French and Indian War: An Inquiry into a Safe-Conduct Issued in 1760 that Acquired the Value of a Treaty in 1990.* Translated by Käthe Roth. Montreal: McGill-Queen's University Press, 2002.

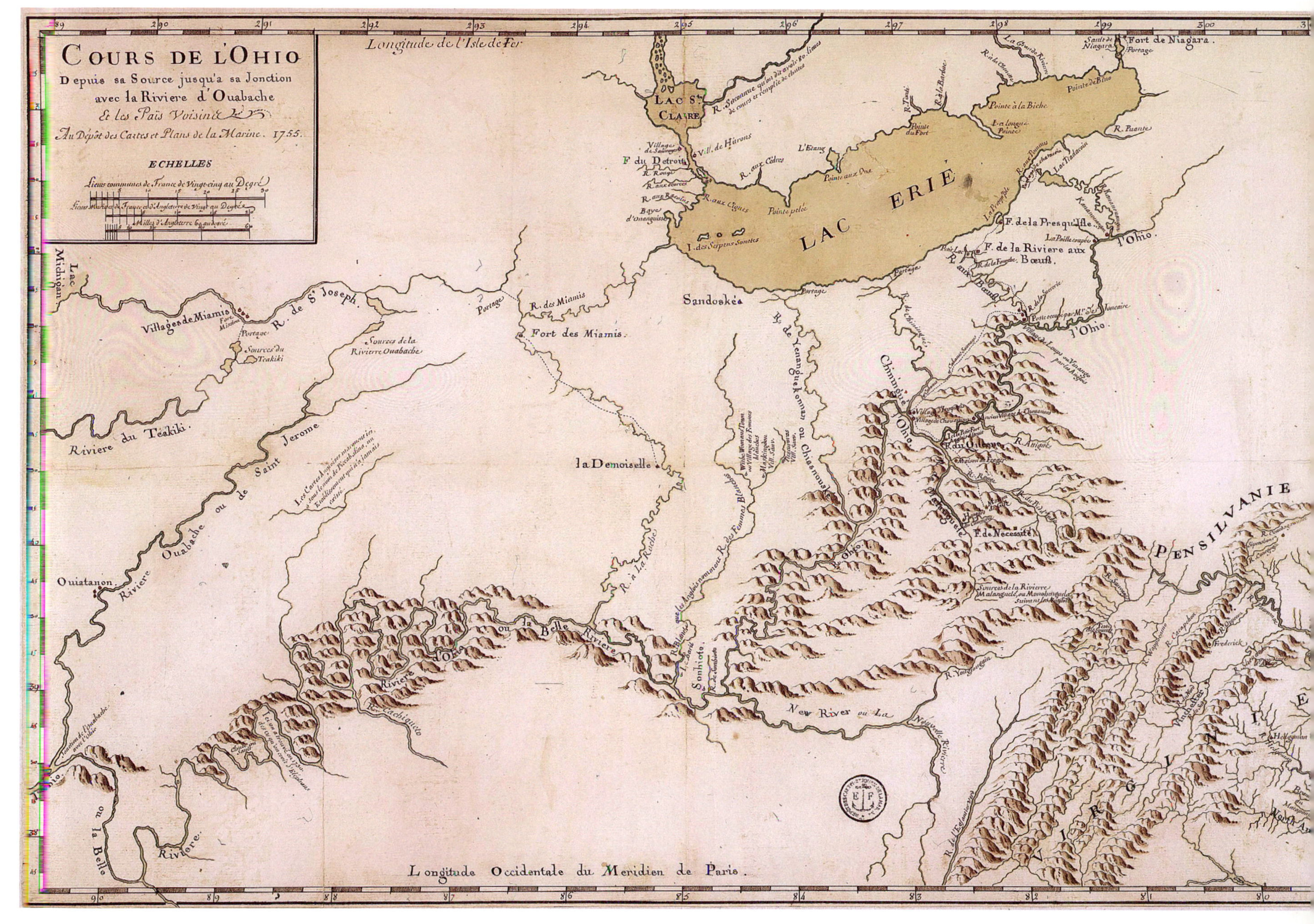

Cours de l'Ohio, 1755
Up to the 1740s, maps of the Belle Rivière, as the Ohio River was known, were relatively approximate. With information garnered from various military expeditions, however, cartographers were able to improve their representations of the river's contours. At the time, French-Indian alliances, so crucial to containing the British, were being put to the test. A few Indian chiefs moved to the Ohio Valley to escape French control and trade more with the British. Among these was the Miami chief Memeskia (also known as La Demoiselle), who established a village called Pickawillany on the Rivière à la Roche (Great Miami River), a tributary of the Ohio. In 1749, the officer Céloron de Blainville was sent to meet with the chief and convince him to return to the French fold, but this attempt was unsuccessful. In 1752, the French adopted a more forceful approach. The Métis Charles-Michel Mouet de Langlade, leading 210 Outaouais warriors and 30 French soldiers, took a number of prisoners, including La Demoiselle, who was boiled and eaten. The French then retook control of the Ohio from the British. This map shows the village of Chief La Demoiselle, which was attacked by the French. It also depicts the routes between Lake Erie and the Ohio, which the French were trying to control by building forts: Fort de la Presqu'île, Fort de la Rivière au Bœuf, Fort Machault ("post occupied by Mr. de Joncaire"), and Fort Du Quesne. This region, key to the continent for both French and English, was openly disputed even before the official start of hostilities in Europe in 1756.

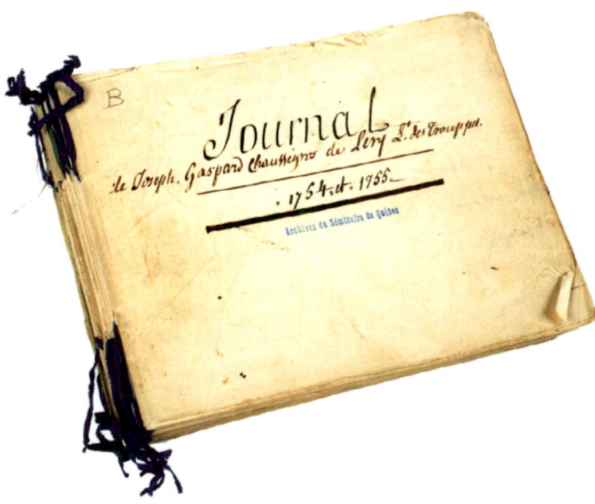

Journal of Chaussegros de Léry, 1755
Following in his father's footsteps, Gaspard-Joseph Chaussegros de Léry (1721–97) was a military engineer in Canada from an early age. Like many officers, he was in the habit of writing about the events he experienced day by day. This journal, covering the period March 7, 1754, to August 5, 1755, describes French military activities on the eve of the Seven Years' War. Sent to Detroit by General Duquesne, Chaussegros de Léry explored the border zone disputed by France and England and was occupied with construction of fortifications and transport of supplies and trade merchandise. His journal not only recalls the difficulties of transportation in the colony but includes beautiful cartographic sketches, including a map drawn from information supplied by an Oneida informant (see p. 207).

Hydrography of the St. Lawrence

Along with the Hudson River, Hudson Strait, and Mississippi River, the St. Lawrence River was one of the main routes to the interior of North America. By founding a permanent settlement at Quebec in 1608, the French assumed tight control of access to this perilous river, whose shoals, rocks, and sandbars struck fear into the hearts of the most experienced sailors. In fact, the St. Lawrence swallowed up more than one ship and took many human lives; its currents, fogs, and storms turned it into a cemetery full of watery graves.

Faulty navigation and poor climatic conditions on the St. Lawrence were the cause of one of the worst catastrophes in maritime history: en route to besiege Quebec in the summer of 1711, the English admiral Hovenden Walker lost eight ships opposite Île aux Œufs, and 900 men and women drowned or froze to death on the shore. But the difficulties of navigating the river also presented a major advantage: it was a formidable rampart for the colony.

The French themselves had much trouble negotiating the St. Lawrence and making it more accessible to the ships that linked Canada, the West Indies, and France. In 1665, Intendant Jean Talon sent a memorandum to the Court underlining the hazardous nature of the river and expressing his wish that this be communicated to the pilots from La Rochelle and Normandy. To contain the losses of sailing ships and human lives, the colonial authorities decided to provide better training to pilots. The main ports in France already had their own hydrography schools, and so it was decided to set up a similar school in Quebec. The first courses in navigation were given by Martin Boutet de Saint-Martin, a lay teacher at the Jesuit college. At Talon's urging, Boutet opened the doors of his classroom to aspiring navigators. Pleased with this initiative, Talon wrote, a few years later, "Young men in Canada are dedicating themselves and rushing to the schools for sciences, arts, trades, and especially sailing, such that if this inclination is further fed, there is reason to hope that it will become a breeding grounds for navigators, fishermen, sailors, and workers, all having a natural disposition for such jobs."

In spite of these efforts, the river seemed as perilous as ever. In a letter sent to the minister of the navy in 1685, Louis Jolliet, seigneur of Mingan and Anticosti, wrote, "It is not for no reason that at all times, those who have come to this country of New France have been apprehensive about the entrance to the Gulf of St. Lawrence and all the passages from Anticosti to Quebec, about 30 leagues apart. We know, Sir, that a number of ships sent by His Majesty and by merchants have perished in said river, for lack of maps on which they may navigate." These few lines introduced an immense map of the estuary of the St. Lawrence from Quebec to Newfoundland, accomplished after some 50 trips by bark and canoe (see pp. 196–97). This map does not show

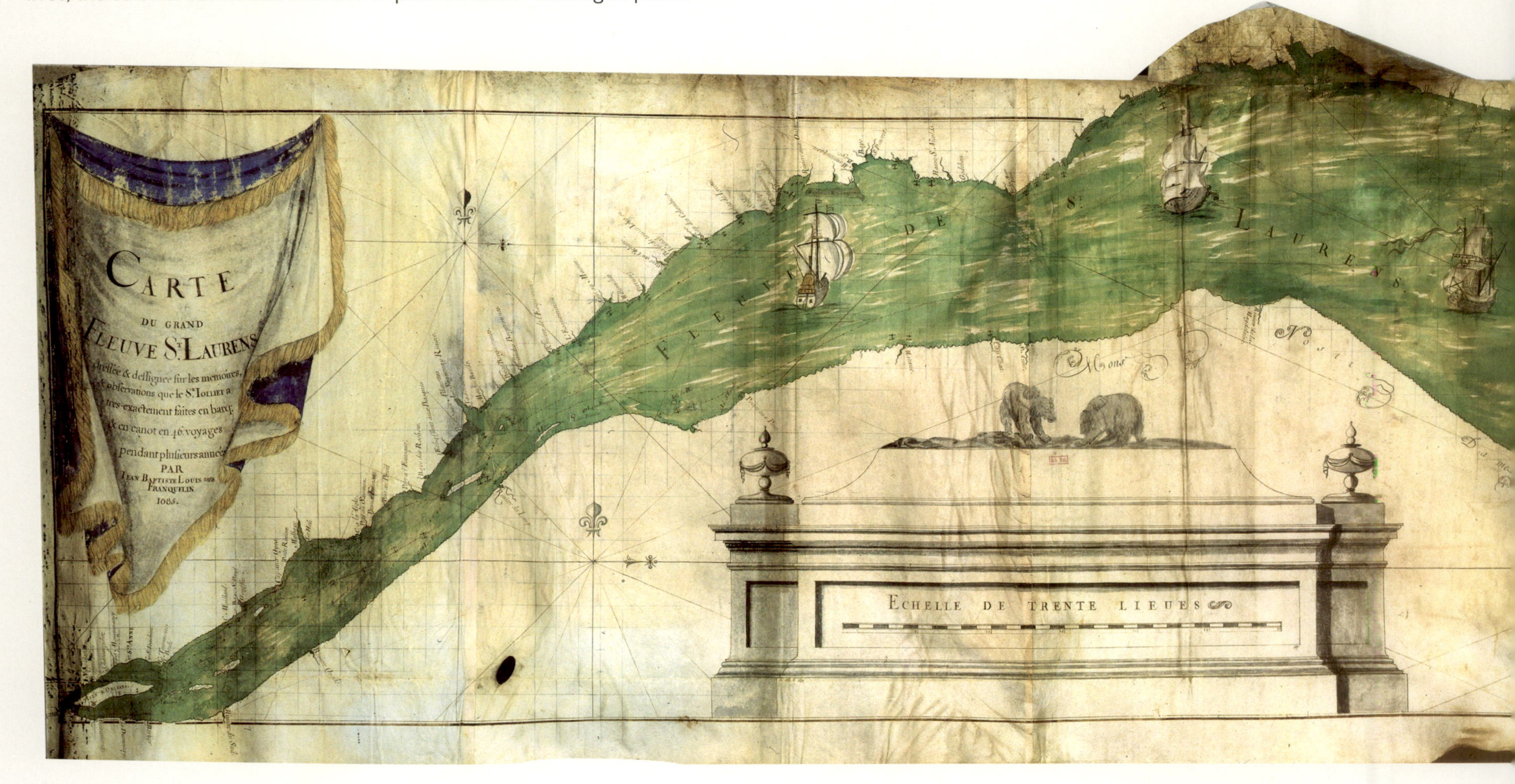

The St. Lawrence River according to Jean Deshayes, 1686

Map of the St. Lawrence River by Louis Jolliet and Jean Baptiste Louis Franquelin, 1685

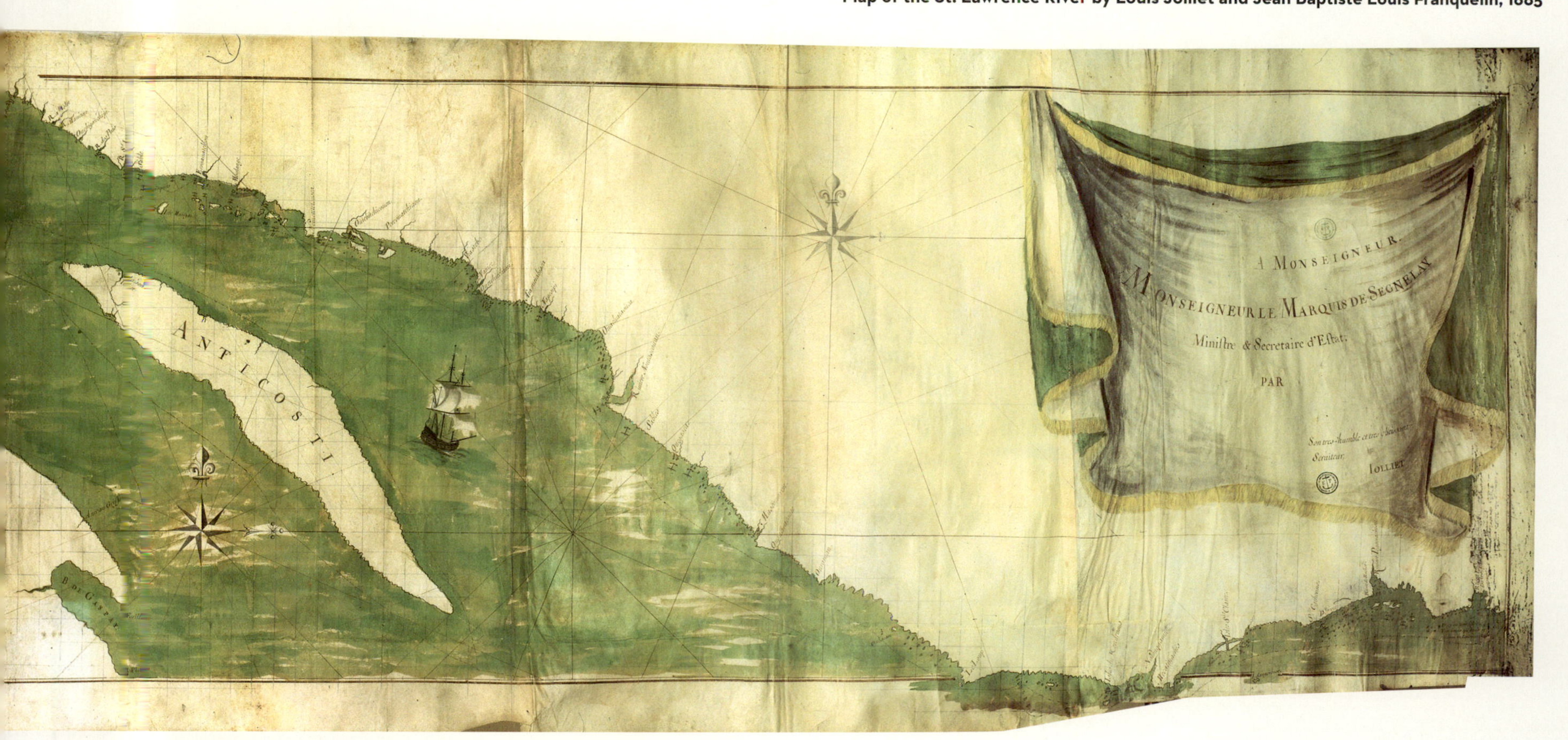

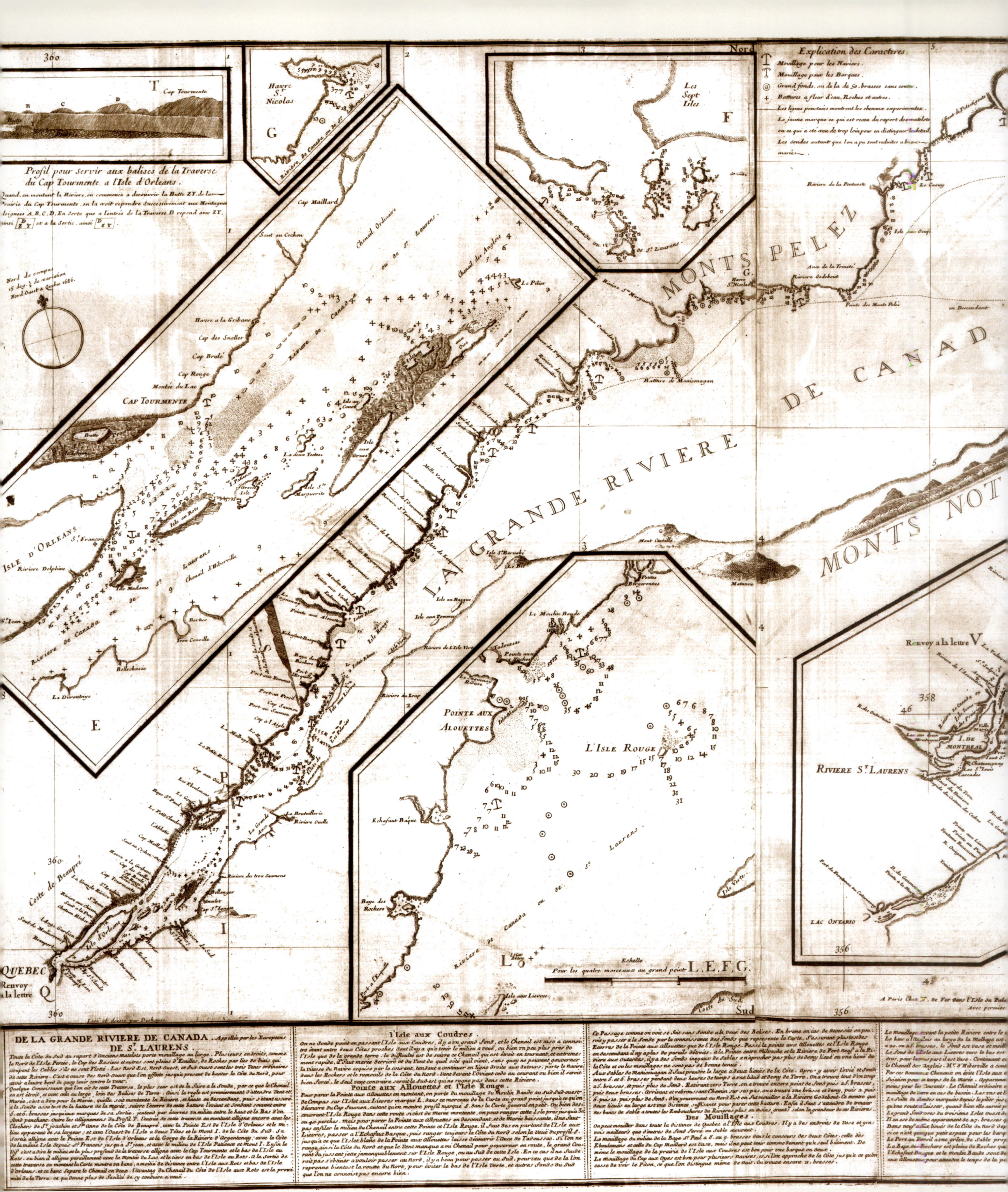

Explication des Caracteres
Mouillage pour les Naviers.
Mouillage pour les Barques.
Profil pour servir aux balises de la Traverse du Cap Tourmente a l'Isle d'Orleans.
Cap Tourmente
Havre St. Nicolas
Les Sept Isles
MONTS PELEZ
LA GRANDE RIVIERE DE CANAD
MONTS NOT
Cap Maillard
Cap Tourmente
Isle d'Orleans
Isle aux Coudres
Isle aux Grues
Chenail Ordinaire
Chenail des Anglois
Pointe aux Alouettes
L'Isle Rouge
Echafaut Basque
Baye des Rochers
Isle aux Lievres
Riviere St. Laurens
I. de Montreal
Lac Ontario
Renvoy a la lettre V
Quebec
Renvoy a la lettre Q
Coste de Beaupré
Isle Verte
Mont Camille
Matanne
Le Moulin Baude
Riviere du Loup
La Bouteillerie
Kamouraska
Batture de Manicouagan
Echelle
Pour les quatre morceaux au grand point L. E. F. G.
Nord
Sud
DE LA GRANDE RIVIERE DE CANADA. Appellée par les Europeens DE St. LAURENS.
L'Isle aux Coudres.
Pointe aux Allouettes et l'Isle Rouge.
Des Mouillages.

water depths, but it does show capes, islands, sandbars, anchorages, and courses to follow to avoid reefs and shoals.

In the same year, 1685, New France received a visit from an esteemed guest, sent by the Académie royale des sciences: Jean Deshayes. The Académie, founded in 1666, had the mission of redrawing the map of the world according to the most accurate measurements taken from astronomical observations. When he saw the new, corrected map of France, Louis XIV remarked, "These gentlemen of the Académie have taken part of my kingdom away from me." Deshayes was among the French scientists who travelled overseas to places seen as laboratories. He went to Gorée, Senegal, then to the Antilles (Guadeloupe and Martinique), before arriving in Quebec in 1685 to take astronomical observations and a survey of the St. Lawrence River. Although his health was delicate, Deshayes sailed with Governor Denonville to Fort Cataracoui, at the mouth of Lake Ontario. On December 10, 1685, he took advantage of a lunar eclipse to calculate the longitude of Quebec. The following year, he mapped and took soundings of the estuary of the St. Lawrence downstream from Quebec. He used the most technologically advanced instruments, including a combination cross-staff, compass, and telescope. He travelled using typically Canadian conveyances, including canoes and snowshoes.

Deshayes stayed in Quebec for only one year on this trip, but he returned in 1702 as king's hydrographer to teach navigation and piloting, which he did until his death in 1706. His most valuable legacy was, without doubt, his maps of the river. One of his manuscript maps, conserved at the Bibliothèque du Service historique de la Marine in Vincennes, shows the section of the river between Quebec and Lake Ontario (see pp. 196–7). It is obvious that Deshayes did a masterful job, going so far as to show houses and mills along the river. After the Académie royale des sciences deemed it of great usefulness for navigation, the map was published on a reduced scale by Nicolas de Fer in 1702, then republished in 1715 (see opposite). This map is accompanied by a detailed text and teems with instructions for navigators: bearings, directions, nature and depth of the bottom, anchorages, recommended distance from the shore, sandbars, shoals. According to some, this map was the authoritative reference for pilots and ship's captains until the end of the French regime. Others, such as the cartographer Jacques-Nicolas Bellin, stated that it was drawn at too small a scale to be useful to navigation.

In fact, navigators quickly discovered the map's limitations and pointed out the need for new surveys. By this time, cartography projects were no longer being initiated by the hydrography school in Quebec, run by the Jesuits, but by the engineer-hydrographer of the Dépôt des cartes et plans de la Marine, Nicolas Bellin, based at a central location in Paris where all of the ministry's maps were kept. Although he accumulated new information there, he admitted that he delayed making it public. The French thus fell behind the English, who, after conquering Canada, quickly published new and accurate maps of the river. Talented officers such as Samuel Holland, James Cook, and Frederik Wallet DesBarres continued the work begun by their rivals. With publication of the *Atlantic Neptune,* England set out to position itself as sovereign of the Atlantic and master of a rebel river.

Opposite
La Grande Rivière de Canada **by Jean Deshayes, 1715**

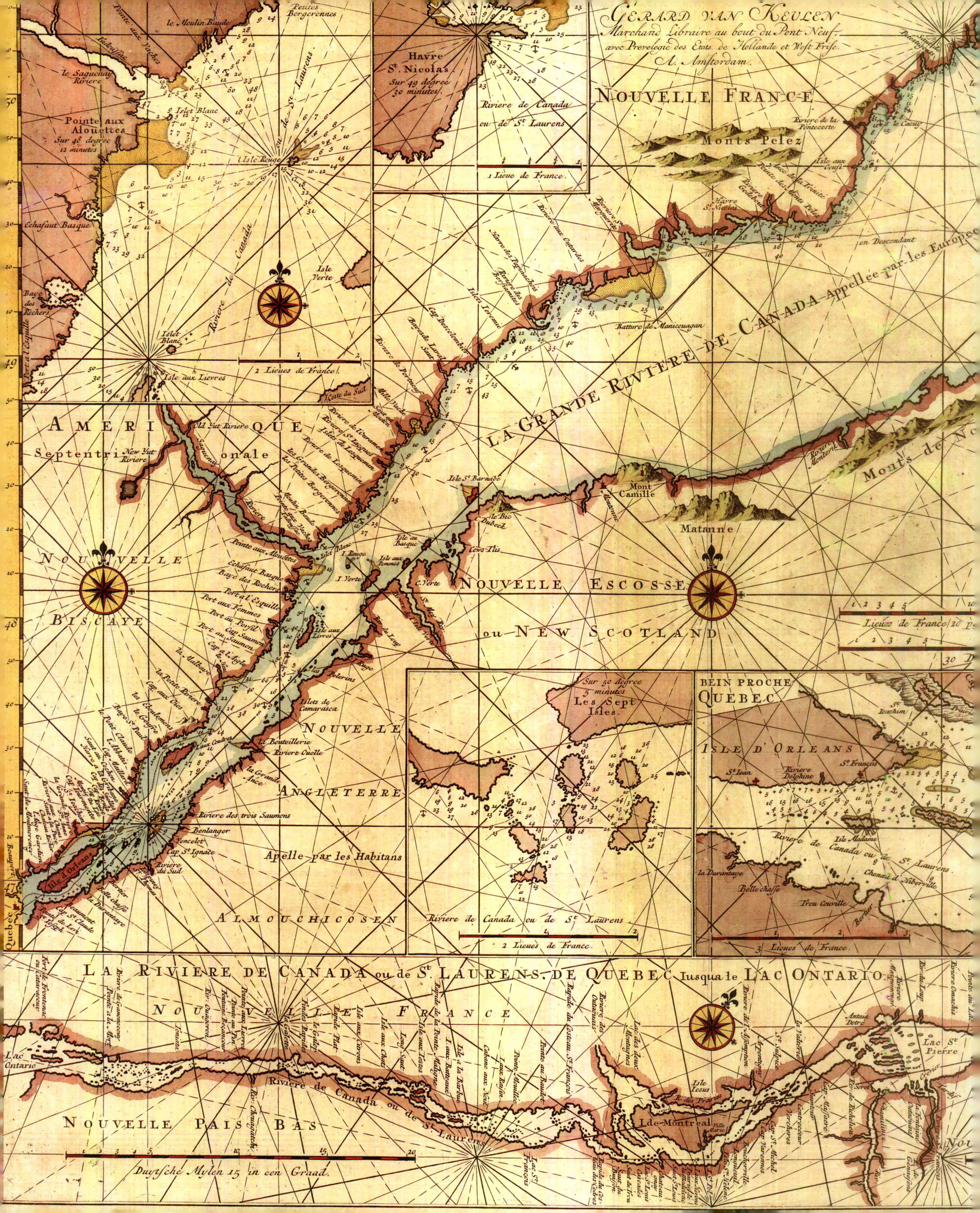

Gerard van Keulen
Marchand Libraire au bout du Pont Neuf.
avec Previlegie des Etats de Hollande et West Frise.
A. Amsterdam.
Nouvelle France
Monts Pelez
La Grande Riviere de Canada Appellée par les Europe
Havre St. Nicolas
Sur 49 degree 30 minutes.
Riviere de Canada ou de St. Laurens
1 Lieue de France.
Pointe aux Alouettes
Sur 48 degree 12 minutes
le Saguenay Riviere
Isle Rouge
Isle Verte
Echafaut Basque
Baye des Rochers
Isle aux Lievres
2 Lieues de France.
Ameri que
Septentri onale
Nouvelle Biscaye
Nouvelle Escosse
ou New Scotland
Mont Camille
Matanne
Lieuës de France
Nouvelle Angleterre
Apelle par les Habitans
Almouchicosen
Isle d'Orleans
Sur 50 degree 5 minutes
Les Sept Isles.
Riviere de Canada ou de St. Laurens
2 Lieues de France.
Bein Proche Quebec
Isle d'Orleans
Lieues de France.
La Riviere de Canada ou de St. Laurens, de Quebec Iusqua le Lac Ontario.
Nouvelle France
Lac Ontario
Riviere de Canada ou de St. Laurens
Nouvelle Pais Bas
Ile de Montreal
Lac St. Pierre
Duytsche Mylen 15 in een Graad.

Nouvelle carte de la Rivière de Canada by Gerard Van Keulen, Amsterdam, circa 1717
Before the founding of the Dépôt des cartes et plans de la marine, French officers and pilots consulted Dutch nautical charts. However, this map by Van Keulen is a copy of a map by Jean Deshayes. It was a blow to the pride of the French, who did not lack talented cartographers. Not until the Dépôt was founded were maps by French pilots finally used, and seen for their true worth.

Pages 202–03
Joseph Frederick Wallet DesBarres's _Atlantic Neptune_
During the American War of Independence, the British army had an urgent need for accurate nautical charts to help them combat the American colonial troops. The *Atlantic Neptune* responded to the demand. This maritime atlas was produced by Joseph Frederick Wallet DesBarres, a Swiss native who had enlisted in the British army and was sent to Canada during the Seven Years' War. After the conflict, the British Admiralty asked DesBarres to map the coasts of the Gulf of St. Lawrence, Nova Scotia, and New England to facilitate colonization and safe navigation. Supported by a team of assistants and labourers, DesBarres conducted the required surveys over a 10-year period. In 1774, he returned to London to correct and compile his own maps and those made by other officers, such as James Cook and Samuel Holland. Individual plates were published that year and the four-volume collection was published between 1777 and 1784. In all, the *Atlantic Neptune* contained some 250 nautical charts covering the entire North American coast from the St. Lawrence River to the Gulf of Mexico. The atlas also contained beautiful views of ports, harbours, and bays (see p. 187), some of which were useful to navigators reconnoitring these areas. This map of the St. Lawrence is an example of the level of detail found in the *Atlantic Neptune* hydrographic surveys. The excerpt (the environs of Quebec and Île d'Orléans) shows the depth, the channels, the tidal flats, and the reefs in the river, as well as the roads, hills, villages, buildings, and cleared land that navigators could use as reference points.

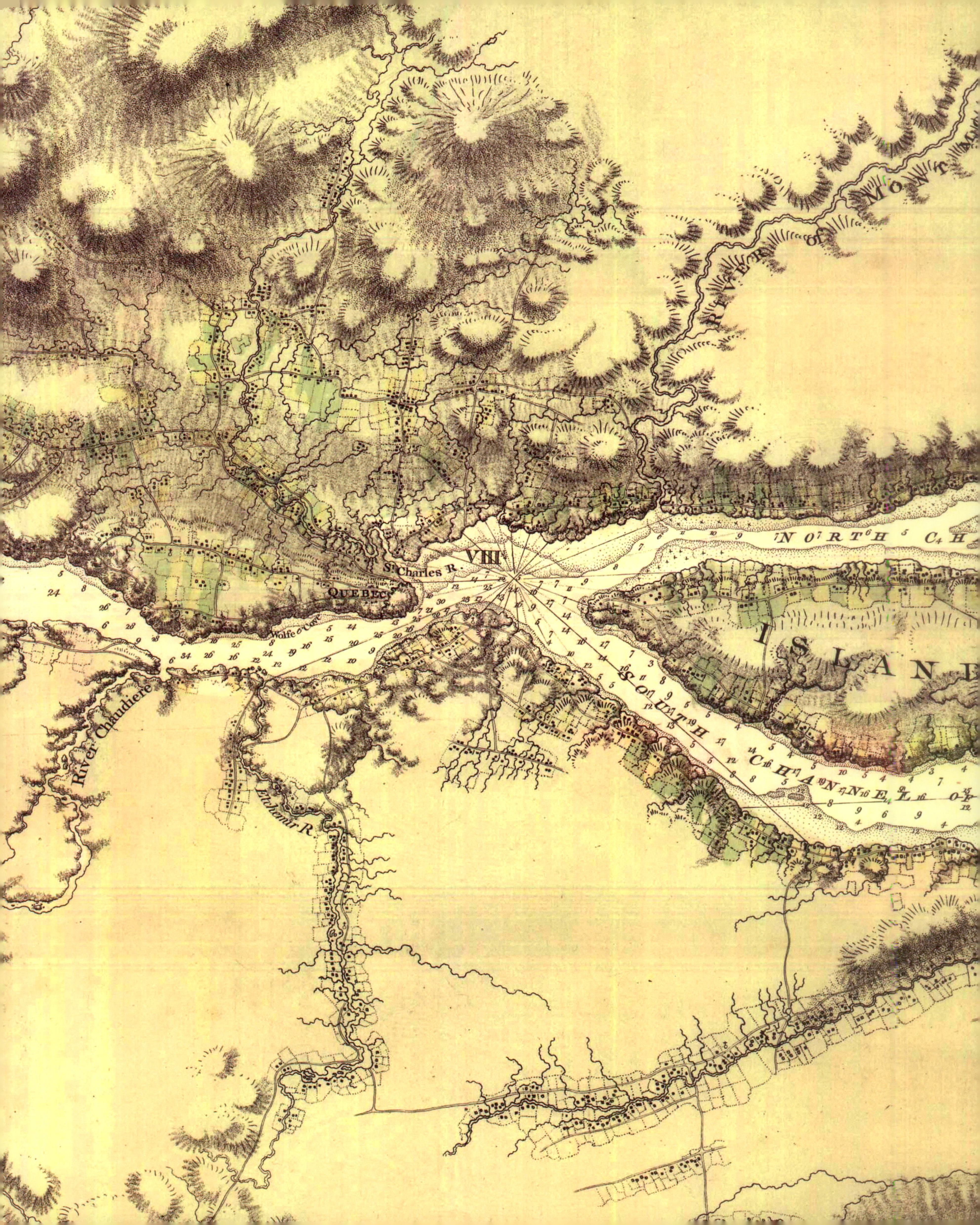
RIVER OF MONT M
St Charles R.
VIII
QUEBEC
Wolfe Cove
NORTH CH
ISLAND
SOUTH CHANNEL OF
River Chaudiere
Etchemin R.

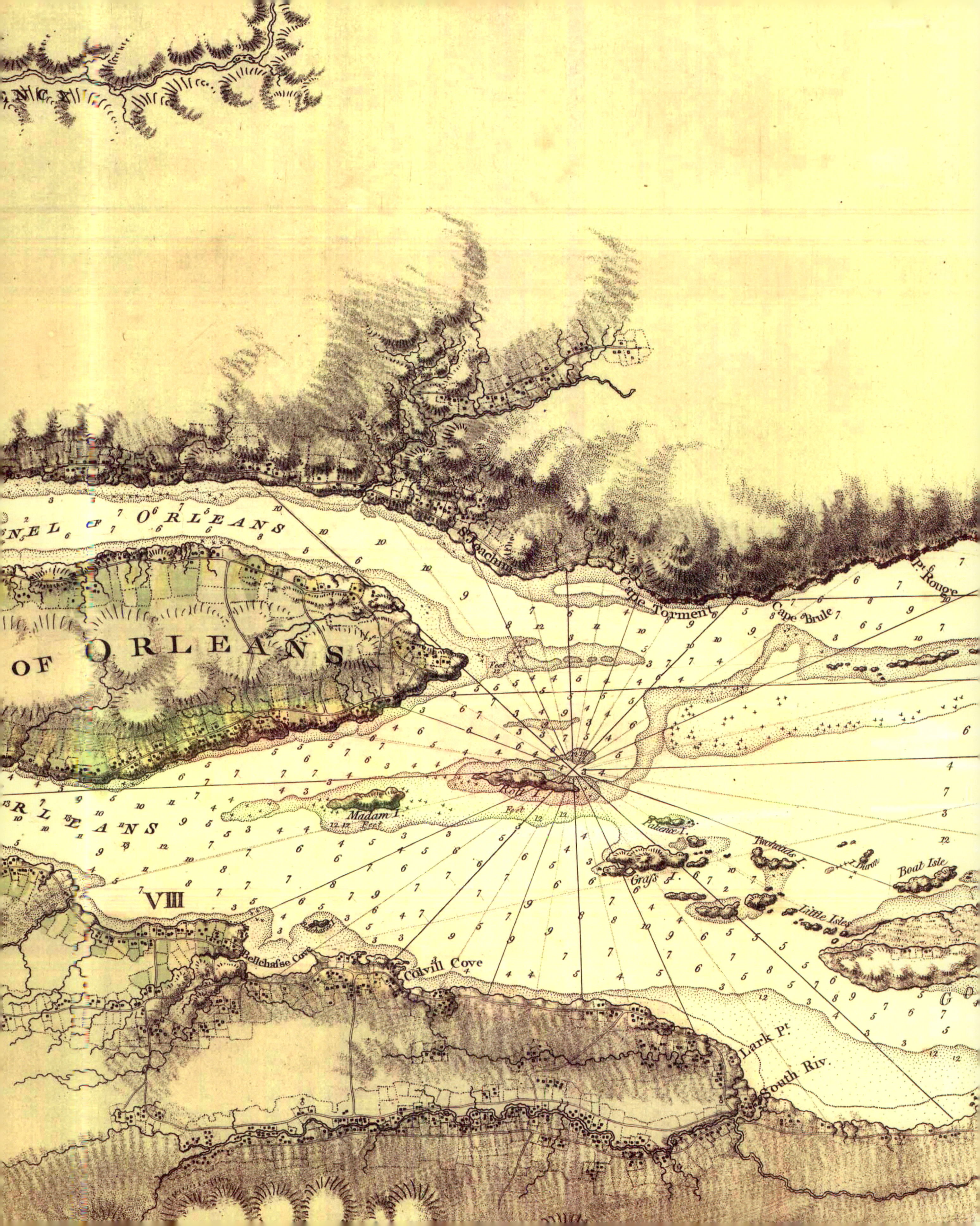

NEL OF ORLEANS
OF ORLEANS
RLEANS
VIII
Cape Torment
Cape Brule
Pt. Rouge
Madam I.
Bellchasse Cove
Colvill Cove
Lark Pt.
South Riv.
Boat Isle
Little Isles

JAMES BAY
KRIS INDIANS
ASSENIPOELS
CANADA
LAKE SUPERIOR
LAKE HURON
LAKE MICHIGAN
LAKE ONTARIO
LAKE ERIE
Algomkins
Outagamis
Outowaks
Uttawas
The IROQUOIS or Six Nations
Illinois
VIRGINIA
Albemarl County
NORTH CAROLINA
SOUTH CAROLINA
GEORGIA
Upper Creeks
Middle Creeks
Lower Creeks
LONG ISLAND
MASSACUSET BAY
BUNDY BAY
Cape Cod
PHILADELPHIA
Charlestown
St Augustin
ATLANTIC O
GULF OF MEXICO
British Statute Miles 69 to a Degree

Indians and Cartography

"THEY MARK THE TRUE NORTH"

At the time of the arrival of the Europeans in North America, the Indians had, over the ages, woven a vast trade network of both water routes and well-worn footpaths that stretched throughout the continent.

The best makers of paths were large animals, which preferred to walk on gentle slopes and knew how to get around obstacles. They reliably took the same trails over and over, and humans naturally imitated them. The law of least effort applied universally (as it does today) – in fact, it led to material and technical progress. When the whites came, they used the Indian paths and natural corridors. Even today, a number of highways and superhighways follow those same paths.

Indian farmers exchanged the fruits of their harvests and their labour, such as hemp products (string and rope), for tobacco and skins, some of which they resold. Hunters and fishermen offered furs, mats, dried berries, and moose antlers. They gathered shells on the Atlantic coast and on the shores of the Gulf of Mexico, and they obtained flint from Ungava Bay, copper from Lake Superior, and pipestone (now also known as catlinite in honour of the painter George Catlin) from Dakota.

One day, the Hurons saw knives, hatchets, and cauldrons arriving with traders from the east. Positioned at the centre of major networks, their territory formed a major crossroads. From Huronia, many roads led in all directions.

To these land routes were added water routes, which, for some nations, were even more important. From Tadoussac, for example, one could go either to James Bay via the Saguenay River, Lake Saint-Jean, the Ashuapmushuan River, Lake Mistassini, and the Rupert River, or to the Great Lakes via the St. Lawrence River, the Ottawa River, the French River, and Lake Nipissing, or via the "copper route," which went farther north.

The Europeans were interested in trading and did not, at first, consider the geographic knowledge of the indigenous peoples. It was obvious that the Indians found their way easily and could travel long distances. "These Savages," wrote Intendant Jacques Raudot in the early seventeenth century, "understand the ways through the forest and know them like we know the streets of a city." The French readily recruited them as guides and adopted their means of transport. "In the thickest forest and in the darkest weather," wrote Father Joseph-François Lafitau, "they never lose, as they say, their Star. They go straight where they want to go, in untravelled country, where there is no marked path."

The Europeans, however, needed maps. Some of the Frenchmen who travelled the continent knew how to render astonishingly accurate depictions of the places they visited. Sometimes they questioned the Indians and asked them to make maps, or at least drawings. At the request of Jacques Cartier, who wanted to push westward from Hochelaga, his young Iroquoians guides put sticks on the ground to represent the river and placed twigs over them to indicate rapids and waterfalls, explains Richard Hakluyt in an annotation to Cartier's account of his third voyage.

The practice of "making drawings" for the Europeans became customary. The Indians got into the habit. "They drew roughly, on bark or on the sand, accurate maps on which only the distinction of the degrees was missing," wrote an enthusiastic Lafitau. The early explorers had no scruples about asking questions, as John Smith did for his map of Virginia published in 1612. Usually quick to boast, Smith confessed that he had not explored all of the territory depicted. He even indicated, by placing

Shawnee Savage
Drawn from life in Illinois country by Joseph Warin.

***A New and Accurate Map of North America* drawn according to d'Anville by Peter Bell, 1771**
This map by the English geographer Peter Bell shows an American territory, rich with Indian names, in which the English colonies stretch far to the west. Three years later, these same colonies would be frustrated in their ambitions with the passage of the Quebec Act of 1774, which reserved for Indians the territory between the Appalachians and the Mississippi. Up to that time, few English colonists had ventured into this region. The contours and toponymy of the map are inspired by the work of the famous cartographer Jean-Baptiste Bourguignon d'Anville (see pp. 192–93). Most cartographic information had followed a long trajectory from Indian and coureur des bois to the drawing tables of cartographers in Quebec, then Paris, and finally London.

Carte de la Rivière Longue
by Baron de Lahontan
In September 1688, Baron Louis-Armand de Lahontan left for Michillimakinac heading not for the town of Quebec, as the governor had requested, but for the Mississippi. Guided by a band of Ottawas and Outagamis (Foxes), Lahontan and 10 soldiers took an unknown watercourse, which they called Rivière Longue or rivière Morte, and passed a number of densely populated Indian villages: Eokoros, Essanapés, and Gnacsitares. Historians long considered this expedition a fabrication. Guillaume Delisle believed that the Rivière Longue was an extension of the rivière Moingona (Des Moines River). Others believed that it was the Missouri River. Today, some experts are of the opinion that it corresponds to the Minnesota River and that the Indians encountered were Dakotas. Lahontan was very well received; the chief of the Gnacsitares offered him not only female companionship but also a map drawn on deerskin. Tired and hungry, Lahontan refused the women. He accepted the map, however, and published it in his accounts. The engraving is a diptych, with the two panels separated by a central line and a lily. On the east side, Lahontan portrayed the known territories. On the west side, he transcribed the Indian map. Included are, among other things, a mountain range corresponding to today's prairie hills, separating the country of the Gnacsitares from that of their enemies, the Mozeemleks.

small Maltese crosses here and there, where he had worked "by information of the Savages."

Samuel de Champlain, for his part, constantly questioned his Indian guides. He admired both the Indians and their interesting boats: "But with the canoes of the savages one may travel freely and quickly throughout the country, as well as up the little rivers as up the large ones. So that by directing one's course with the help of the savages and their canoes, a man may see all that is to be seen, good and bad, within the space of a year or two."

Other explorers sent the data that they collected to *géographes de cabinet,* who carefully studied everything that came from North America. It was the era of the mapmakers Nicolas Sanson and Guillaume Delisle. Delisle, in particular, produced, among other things, superb maps of the Mississippi Basin (1703 and 1718) based on work done by Jean Baptiste Louis Franquelin and information gathered by many voyagers, including Louis Jolliet, Father Jacques Marquette, Cavelier de La Salle, Henri de Tonty, Father Louis Hennepin, and Pierre Le Moyne d'Iberville. D'Iberville and his brother, de Bienville, questioned the Indians frequently. During his 1699 voyage, d'Iberville asked his guide to make him a drawing on at least two occasions. One of these was on March 22, when he questioned a Taensa, and to be certain that he had understood – considering the language barrier – he asked the Taensa to make him a map. D'Iberville does not say what the map was drawn on. If it was a tanned hide, he might have carefully preserved it; otherwise, he certainly would have made a copy. In any case, his information was a delight to mapmakers such as Delisle.

During the same period, the Recollet Chrestien Le Clercq noted, in his *Nouvelle relation de la Gaspésie,* "They have much industry to make on bark a sort of map that marks precisely all the rivers and streams of a country of which they want to give a description: they mark all the places exactly." The Indians of the Mississippi and Missouri rivers did not have the same materials, but they did have similar skills. Jean-Baptiste Trudeau gave a glowing report in his account of his voyage on the Upper Missouri in 1794–96: "They make the most correct maps of the countries that they know, which lack only the latitudes and longitudes of locations. They mark True North according to the North Star . . . and count distances by warriors' days and half-days; each day is worth five leagues. They make these particular chorographic maps on birchbark."

In spite of their contributions, the Indian informants were forgotten or ignored, while their copyists were celebrated. There are two exceptions: Ochagach and Ackomokki, also called Acaoomahcaye or Old Swan. The former was a Cree whose map, copied by La Vérendrye, was to guide Jacques-Nicolas Bellin in producing a map of northern North America published in 1755; as well, Philippe Buache wrote at the top of his *Carte physique des terreins les plus élevés de la partie occidentale du Canada* (Physical map of the highest lands of the western part of Canada, 1753), "Une Réduction de la Carte tracée par le Sauvage Ochagach et autres" (A reduction of the map drawn by the savage Ochagach and others).

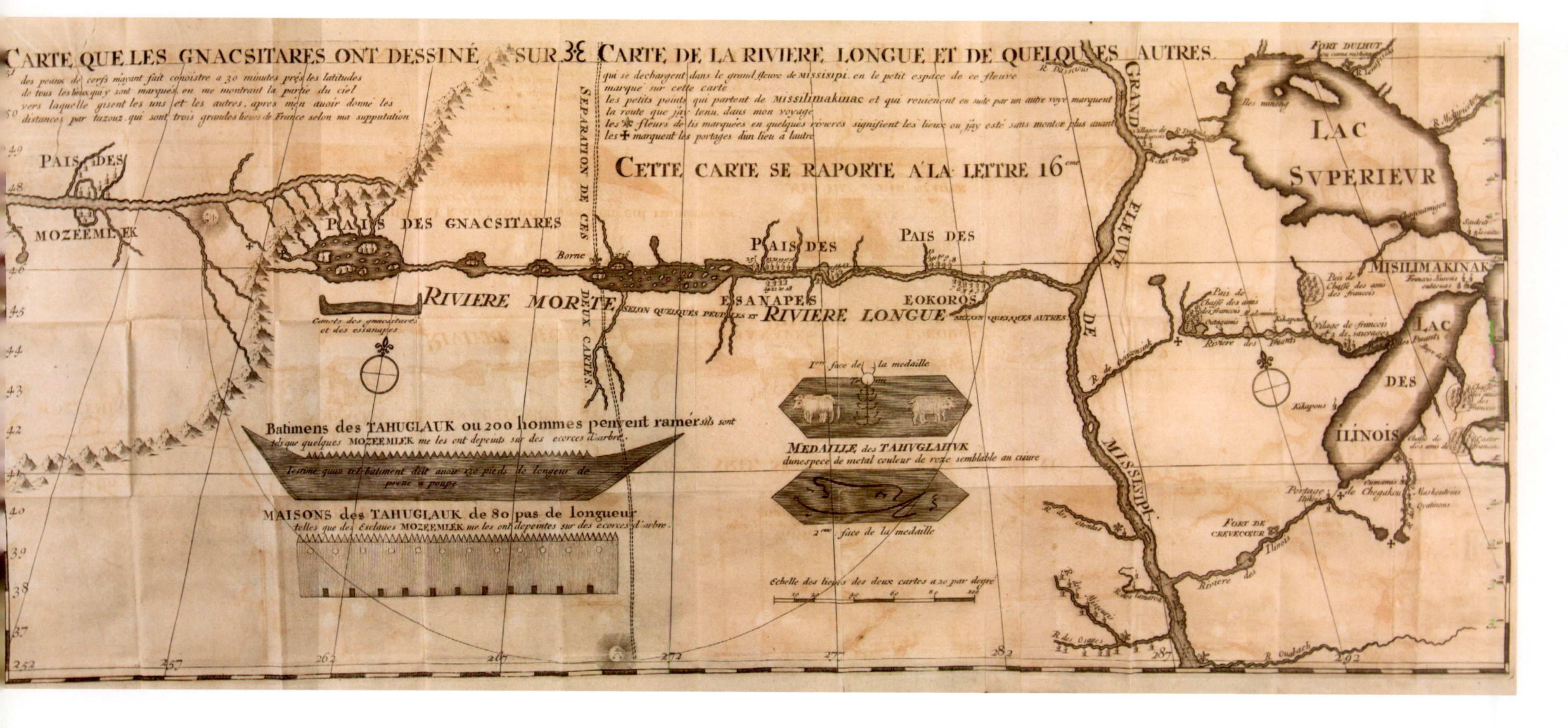

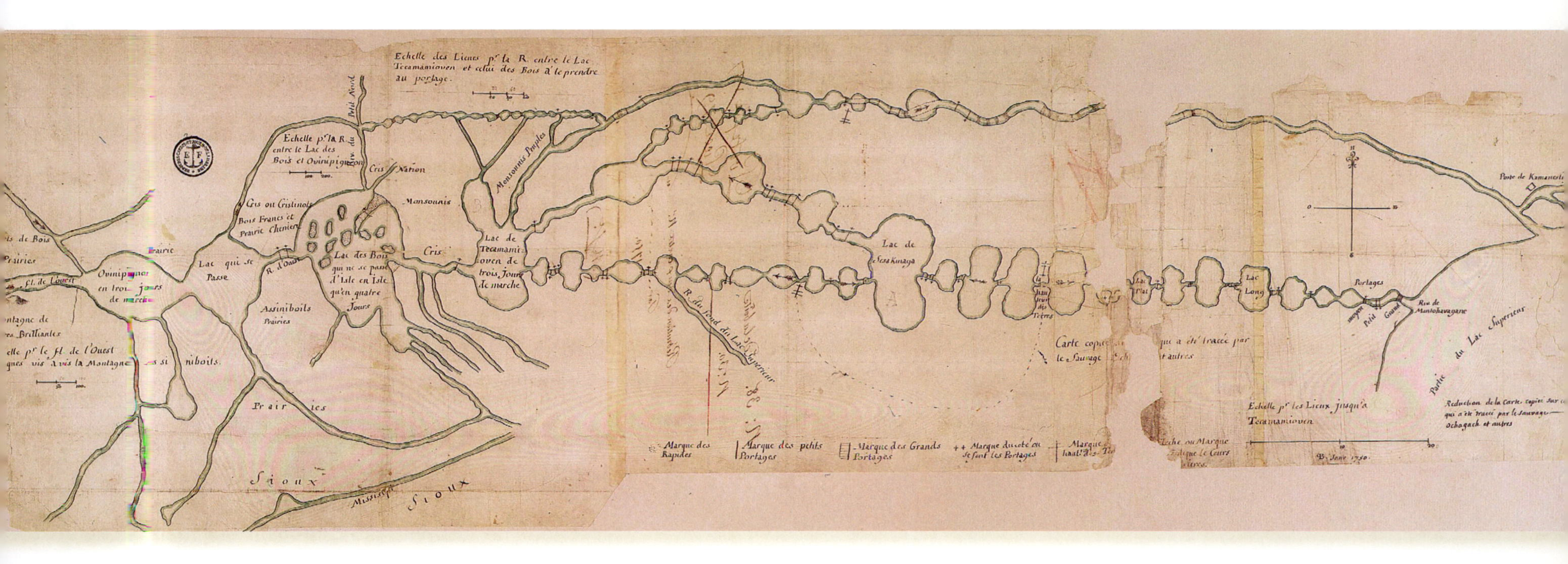

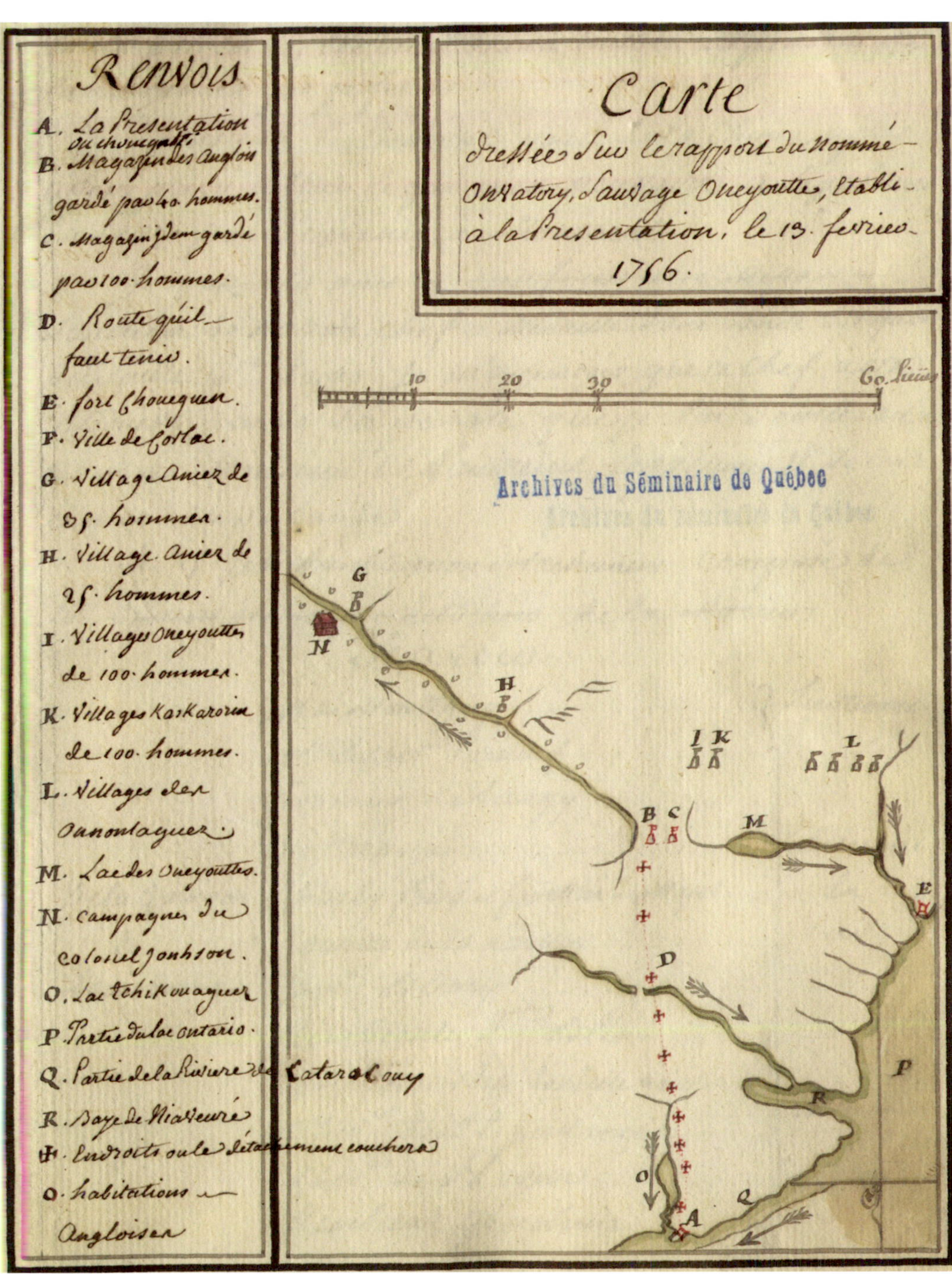

Above
Map by the Cree Ochagach, circa 1729
European travellers wishing to explore unknown territories could not do so without the assistance of Indian guides and translators. The greatest explorers were certainly aware of this, but they did not always consider it a good idea to introduce these guides and translators into their accounts. The Cree Ochagach (or Auchagah) was an exception. Around 1729, this fur trader crossed the path of Pierre Gaultier de La Vérendrye at Kaministiquia and became his guide for an expedition, via terra incognita, to the "great western river." Ochagach became known by drawing the drainage basin west of the Great Lakes in charcoal on birchbark. At least two other Indian chiefs, Tacchigis and La Marteblanche, also supplied maps to La Vérendrye. All of these original maps are likely lost forever, but La Vérendrye took pains to compile them and send them to the authorities as documentary proof. This map shows two ways to reach Rainy Lake from Lake Superior: one from Kaministiquia and the other from Grand Portage. Farther west, a river flows west from LAC OUINIPIGON to a mountain of glittering stone. The intention was to show that it was very close to the southern sea so that the authorities would not be discouraged. In reality it is likely that the western river was the Nelson River, flowing into Hudson Bay. Although this map does not meet the standards of European cartography, it did serve the needs of voyagers by highlighting and amplifying the key elements: lakes, watercourses, and portages presented in succession. Also, the scale is not constant, as it represents more often the travel time than the real distance. A number of European cartographers used Ochagach's map, among them Jacques-Nicolas Bellin (see the map on p. 233), Jean-Baptiste Bourguignon d'Anville, and, late in the century, Jonathan Carver and Aaron Arrowsmith. Philippe Buache published it unaltered in 1754, under the auspices of the Académie des sciences (see pp. 174–75).

Left
Carte dressée sur le rapport d'Onouatoury
by Chaussegros de Léry, 1756
At the beginning of the Seven Years' War, most of the Indians were allied with the French. This support, which compensated for France's numerical inferiority, was particularly useful for attacking enemy positions. The map shows how Commander Chaussegros de Léry received information from an Oneida Iroquois named Onouatoury in preparing an attack on Fort Bull from Fort La Présentation.

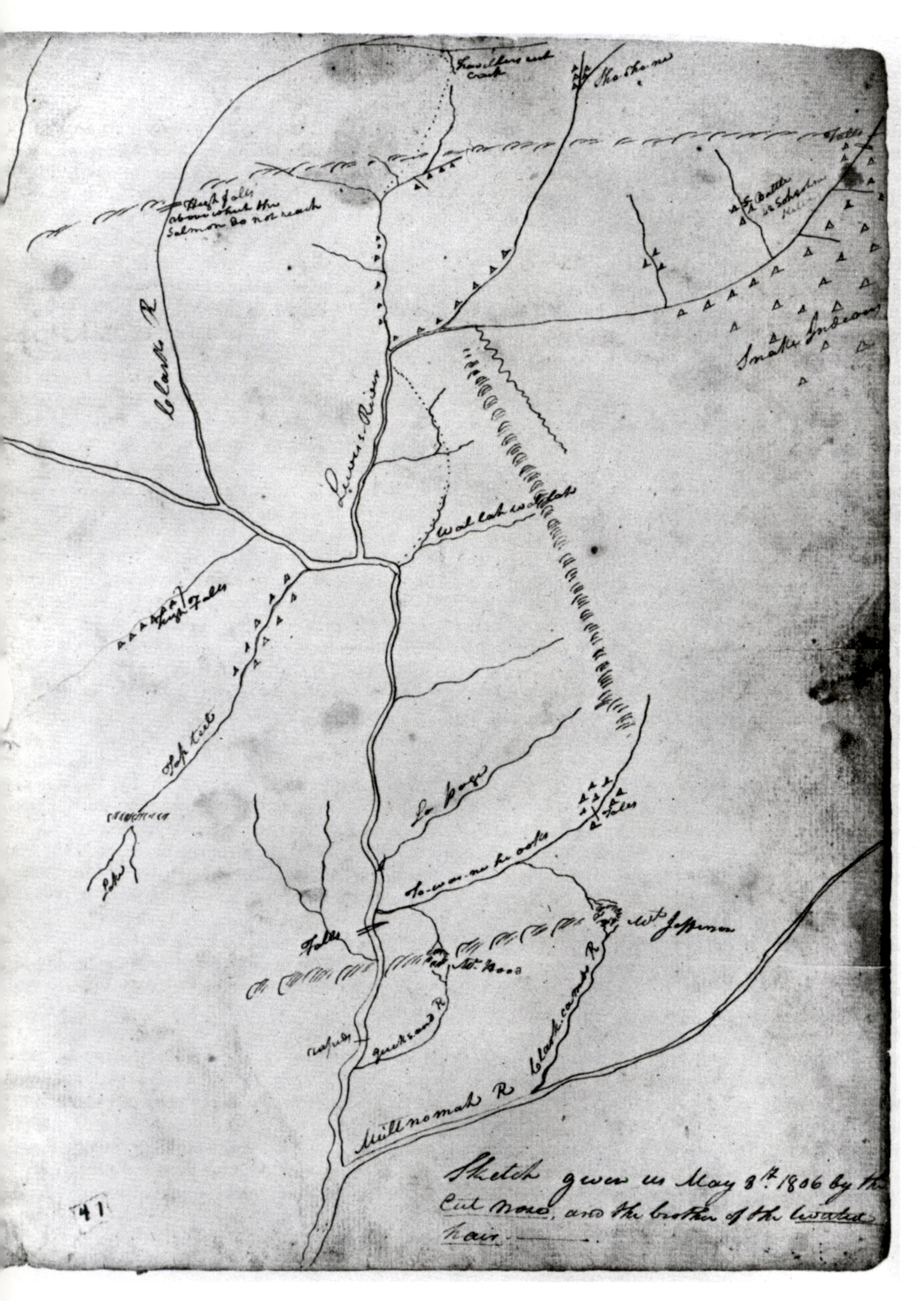

Map by the Indian named Cut Nose
In May 1806, Lewis and Clark were on their way home. A Nez Percé Indian drew a map to show them the best way to reach the Rockies. The distances were in time and not space. For instance, the trip on the Columbia River between the Snake River (Lewis River) and the Spokane River (Clark River), presenting few obstacles, appears to be much shorter than it really is, while the preceding stretch, involving several waterfalls, is portrayed as longer.

It appears that maps were provided by two other Indians: Tacchigis and La Marteblanche. A map of the Upper Missouri by Ackomokki, a Blackfoot chief, was copied by Peter Fidler and incorporated into Aaron Arrowsmith's version of his map of North America published in 1802, just in time to be given by President Thomas Jefferson to Meriwether Lewis on the eve of Lewis's famous expedition with William Clark.

Drawn in snow or sand, sometimes in the ashes of a fire, the maps made by the Indians were quickly erased. Others, drawn on birchbark, tanned hide, or a rock, might last longer, but not long enough to be with us today. The copies that have survived testify to the contribution of indigenous people to the exploration of their continent. The guides – whether Indian, Métis, or French Canadian – played an essential role. Even more important were the Indian and Métis women who became the companions of great explorer-cartographers such as Peter Fidler and David Thompson. Taking up the ways of the new land, Fidler married a Cree woman named Mary in 1794, while Thompson married a mixed-blood woman of Cree origin, Charlotte Small, in 1799. The Fidlers had 14 children; the Thompsons, at least five. When they returned to "civilization" after many years of surveying and mapping the continent, each had his marriage sanctified by the church. The mixing of blood, once a solely French–Indian phenomenon, was now taken up by the English. Many Métis people were able to make the best of both worlds by preserving their Indian roots, something that has been often observed but rarely studied.

Main sources in order of consultation

Malcolm, Lewis G. *Cartographic Encounters: Perspective on Native American Mapmaking and Map Use.* Chicago: University of Chicago Press, 1998. — Malcolm, Lewis G. "Communiquer l'espace: malentendus dans la transmission d'information cartographique en Amérique du Nord." In Laurier Turgeon, Denys Delâge, and Réal Ouellet (eds.), *Transferts culturels et métissages Amérique/Europe xvie-xxe siècle, pp. 357–76.* Sainte-Foy: Presses de l'Université Laval, 1996, — Malcolm, Lewis G. "Misinterpretation of Amerindian Information as a Source of Error on Euro-American Maps." *Annals of the Association of American Geographers,* vol. 77, no. 4 (1987), 542–63. — Warhus, Mark. *Another America: Native American Maps and the History of our Land.* New York: St. Martin's Press, 1997. — Le Clercq, Chrestien. *New Relation of Gaspesia.* Translated and edited by William F. Ganong. New York: Greenwood Press, 1968. — Hayes, Derek. *Historical Atlas of Canada: Canada's History Illustrated with Original Maps.* Vancouver: Douglas & McIntyre, 2002. — Lahontan, Louis Armand de Lom d'Arce, baron de. *Œuvres complètes,* vol. 1, edited by Réal Ouellet and Alain Beaulieu. Montreal: Presses de l'Université de Montréal, "Bibliothèque du nouveau monde" collection, 1990, — Trudeau, Jean-Baptiste. *Voyage sur le Haut-Missouri, 1794-1796,* edited by Fernand Grenier and Nilma Saint-Gelais. Sillery: Septentrion, "V" collection, 2006. — Bressani, Francesco Giuseppe. *Relation abrégée de quelques missions des pères de la Compagnie de Jésus dans la Nouvelle-France,* translated from the Italian and supplemented with a foreword, biography of the author, and many notes and engravings by R.P.F. Martin. Montreal: John Lovell, 1852. The missionary insists on the Indians' sense of orientation. — Champlain, Samuel de. *The Works of Samuel de Champlain.* Vol. 1, translated and edited by H. H. Langton and W. F. Ganong. Toronto: Champlain Society, 1922.

Bison skin
This beautifully painted bison skin was part of a valuable collection at the Musée de l'Homme in Paris; today, it is housed at the Musée du quai Branly. Collected for use in educating princes of the Maison de France, the skins were usually made in memory of a chief or an important event. In spite of a rather hermetic iconography, each detail of shape and colour had meaning. This skin, one of the most beautiful in the collection, has symbolic significance that is difficult to describe. To read it, one should look at it horizontally. To the right, two sticks with feathers (perhaps peace pipes) open toward the moon, in the centre of which is a man in motion. A magnificent sun shines to its left side. At the bottom, the inhabitants of three villages have gathered for a dance. The story begins in the right-hand village and leads us toward European-style houses. In the corner, an Indian armed with a bow is crouching, while in the middle of the houses two other Indians smoke the pipe. A road leads to another village, in front of which a battle is taking place. The combatants on the village side are naked and apparently very excited by the confrontation. All have guns, except for one drawing a bow. Outside, a duck hunt is taking place. According to experts at the museum, the action takes place on the lower course of the Arkansas River.

Maps Full of Place Names

Like the act of mapping, the act of naming is a form, however symbolic, of appropriation. To name a territory, a place, a *topos*, is, in a sense, to baptize it, to remove it from a barbarian limbo and civilize it. For more than three centuries, the French covered with place names the colonial space that they were carving out in North America, and this heritage is still visible on today's maps of Canada and the United States. In New France, creative minds had an opportunity to give free rein to the imagination as they stamped the territory with meaningful names. Each of the thousands of toponyms, invented or borrowed from the Indians, had a specific significance, a history of its own.

It was natural for European explorers and geographers to take inspiration from their kingdoms of origin. New France, New England, New Castile, Nova Scotia, New Belgium, not to mention the short-lived New Sweden and New Denmark, all appeared on maps. By forging these territories from whole cloth, European explorers paid tribute to their sovereigns, who saw their kingdoms expanded with very little effort. America was perceived as a vast virgin territory onto which could be transposed a name, and then the concept of civilization.

The origin of the name New France illustrates the idea of utopia in North America. The name appeared for the first time – in its Latin form, *Nova Gallia* – on a map made in 1529 by the Italian explorer and cartographer Gerolamo Verrazzano. This name, coined by his brother, Giovanni, during his expedition to the North American coast, became solidly anchored in the cartographic landscape of the time. And yet, New France did not really exist. The French had attempted to occupy the territory of course, but unsuccessfully. It was not until the seventeenth century and the arrival of Champlain in Canada that reality finally caught up with the name.

Before the French settled on the continent, the North American coast was nevertheless familiar to the fishermen and traders who visited the Grand Banks off Newfoundland. The earliest names to figure on maps were Spanish, Portuguese, Basque, Breton, and French. A good number of them have withstood the test of time. In Newfoundland, certain names of French origin that appeared in the sixteenth century survive to the present day, including Cape Degrat, Cap Blanc, Fichot Island, Saint Julien's Island, Groais Island, Cap Rouge, Belle Isle, Baie Blanche (White Bay), Cap Saint-Jean (Cap St. John), and Plaisance. This is also true for a number of islands south of Newfoundland: Saint-Pierre, Miquelon, Saint-Paul, Brion. Even before Cartier arrived in Canada, Europeans were landing at Blanc-Sablon and Havre de la Baleine (Baie Rouge or Baie Forteau). Some place names recall more specifically the presence of Bretons – notably Cape Breton and Cape Breton Island. A few maps also emphasize the seniority of Breton discoveries in 1504 (*Britonibius primum detecta*). When Jacques Cartier sailed into the Strait of Belle Isle in 1534, he noted the existence of a harbour named *Karpont* (Quirpon). Near the entrance to the Gulf of St. Lawrence, he reported another harbour, Brest. Both were named after the homeports of Breton fisherman who visited these sites in summertime.

Other European fishermen and traders also navigated on the St. Lawrence very early. A 1529 map by the Spaniard Diego Ribeiro showed a certain Ochelaga River on the North American coast. This sixteenth-century name for the St. Lawrence, referring to the village of Hochelaga, indicates that Europeans knew about the river before Cartier arrived. In the 1540s and 1550s, very accurate Norman maps showed the Gulf of St. Lawrence and the St. Lawrence River, territories explored by Cartier, Roberval, and Alfonce between 1534 and 1543. Some place names seem to have been inspired by these explorers (Baie des Chaleurs, Blanc-Sablon, Canada, Gaspé, Honguedo, Île d'Orléans, Île aux Coudres, Saguenay, Stadaconé, Sept-Îles, and so on), while others are of unknown origin: Aquachemida, Baie Sainte-Marie, Mecheomay (Mirimachi), Saint-Malo, Lac d'Angoulême (Lac Saint-Pierre), Chateaubriant, Grosse femelle, la Bastille, and others. A later Norman map (Guillaume Levasseur, 1601) is also an excellent testimony to the French presence on the St. Lawrence. This very valuable document precedes Champlain's arrival in Canada and shows the names Quebec and Trois-Rivières even before these became sites for European settlements.

In proportion to the importance of their voyages, Cartier and Champlain did not leave that many toponymic traces. Most of the names mentioned in their voyage accounts have not survived. One factor that increases the chances that a place name will survive is whether people use it. In the St. Lawrence Valley, geographic names stabilized only around the mid-seventeenth century, following the first waves of French

***Il disegno del discoperto della Nova Franza* by Paolo Forlani, 1566**
This Italian map shows North America as it was imagined by Europeans in the second half of the sixteenth century. Evidently, the continent is now separated from Asia by the Strait of Anian (Bering Strait). New France, although it does not really exist yet, occupies an enviable place – on paper, at least. It also figures in the title, on the upper left-hand side, which is translated as "The drawing of discoveries in New France." Although the continent's contours are more or less recognizable, the interior is quite confusing. The mapmaker was mistaken about the placement of the St. Lawrence River, which is drawn too far south and should actually irrigate the villages of Stadacone, Ochelai, and Ochelaga.

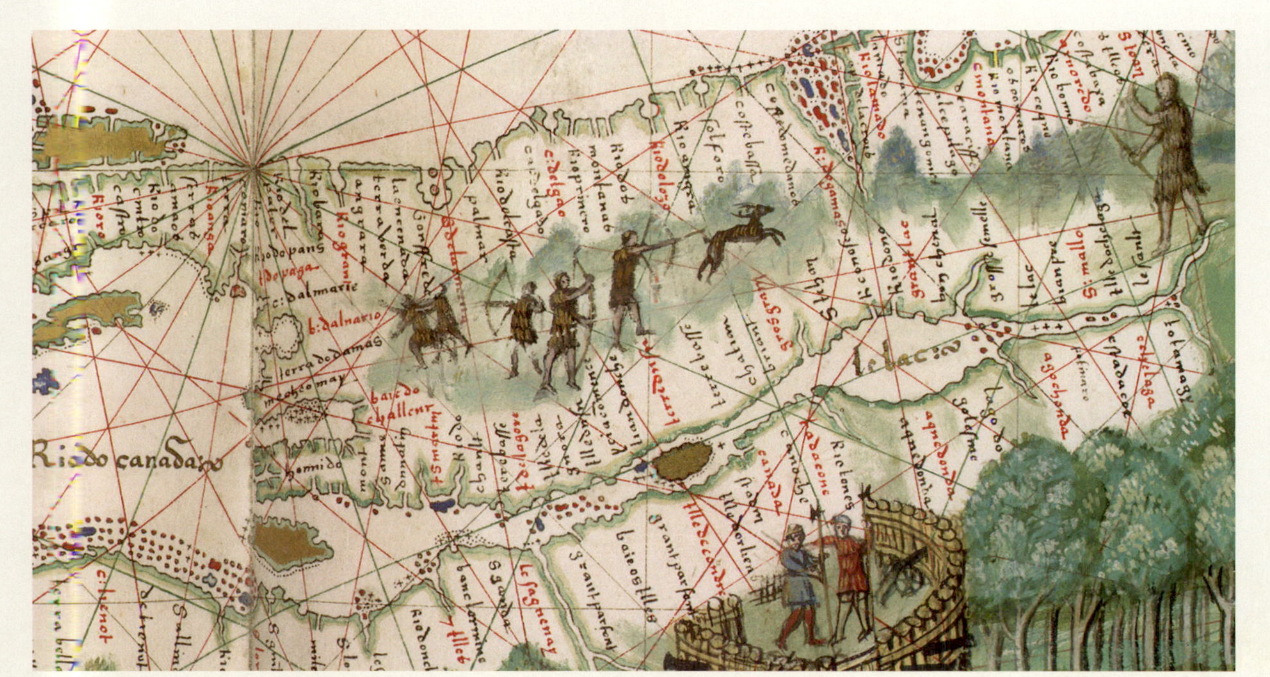

Map by Nicolas Vallard, 1547 (detail)
This map, attributed to Nicolas Vallard, is remarkably rich in place names. Dated 1547, it shows, among others, Newfoundland, the Gulf of St. Lawrence, and the St. Lawrence River as far as Saut Saint-Louis (LE SAULT), just after the voyage of Cartier and Roberval. Most of the geographic names are Portuguese (RIO DO CANADA, R. CONOSCO, TERRA DE DAMAS, RIO DE PARIS, ARCABLANC, etc.), giving the impression that Vallard was himself Portuguese or had received information from Portuguese navigators. The map also includes a number of French names (BREST, ST-LORENT, 7 ISLES, ISLE D'ORLEANS, ISLE DE COUDRE, ST-MALLO, CHATEAUBRIANT, GROSSE FEMELLE, etc.); the origin of these names is uncertain, since many of them do not appear in the accounts of Cartier or Jean Alfonce. A number of names also testify to the presence of Indians, with whom Europeans traded not only materials but also geographic information (MECHEOMAY, AGNEDONDA, AGOCHONDA, OCHELAGA, TOTAMAGNY, CANADA, STADACONE, etc.).

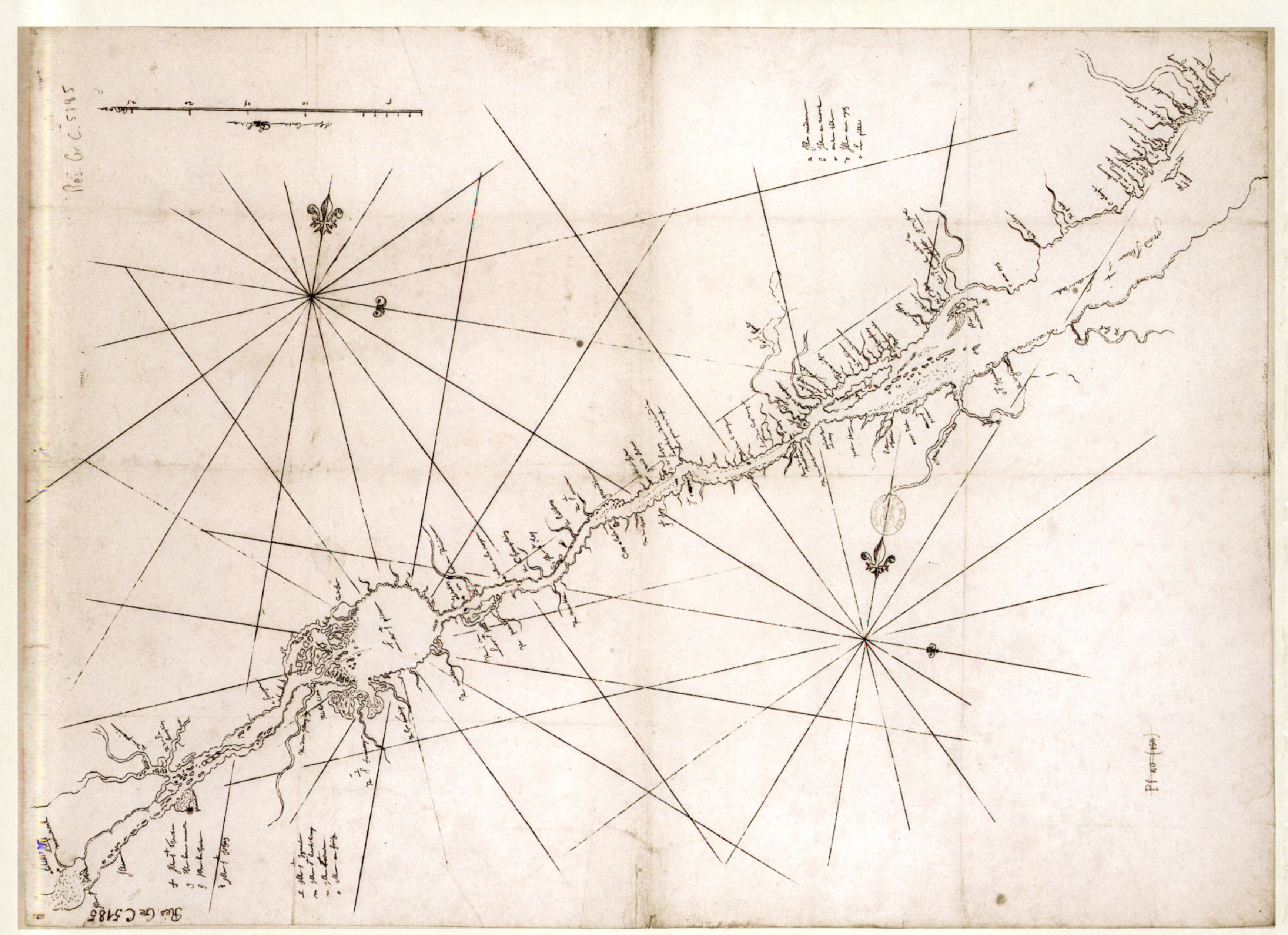

***Rivière de St Laurens depuys Montréal jusqu'à Tadoussac*, attributed to Jean Bourdon, circa 1641**
This little-known map shows the St. Lawrence River from Montreal to Tadoussac. It was probably drawn in the 1640s by the surveyor and trader Jean Bourdon, thus offering a link between the cartography of Samuel de Champlain and that of the mapmakers of the late seventeenth century. On this map appear, for the first time, names that are familiar today: Cap-Rouge, Rivière Portneuf, Rivière Sainte-Anne, Rivière de l'Arbre à la Croix, Rivière de l'Assomption, Rivière Nicolet, Rivière Saint-François, Chutes de la Chaudière, Rivière du Sud, Île aux Oies, Cap-Saint-Ignace, Château-Richer, Beaupré, and Baie-Saint-Paul.

CARTE
DU DOMAINE DU ROY
en Canada
dressée par le P Laure miss. J.
et dediée en 1731
a
MONSEIGNEUR LE DAUPHIN
augmentée de nouveau revüe et corrigée
avec grand soin par le meme en
attendant un exemplaire
Complet l'automne
1732
Guyot f.
1733
Lieües Communes
10 20 30 40
BAYE D'HUDSON
L. OURINAGAU ou les Anglois vont en traite
les Mistassins assurent quil est presque aussi
grand que leur lac.
Fort et R. Rupert
NEMISKAU
La grande decharge
Terre des Gessé
Montagnes du Nord
TIMISKAMING
Miskouaskne
Les MAMIOUETZ
TIMISKAMING
Domaine du chef des Mistassins
L. ALBANEL
L. DAUPHIN
Tous les lacs et RR au dessus de
cette ligne se degorgent dans
le grand lac des Mistassins
L. CHOMONCHOUANE
L S JEAN
ou
Pekouagami
L. St. PIERRE
FLEUVE

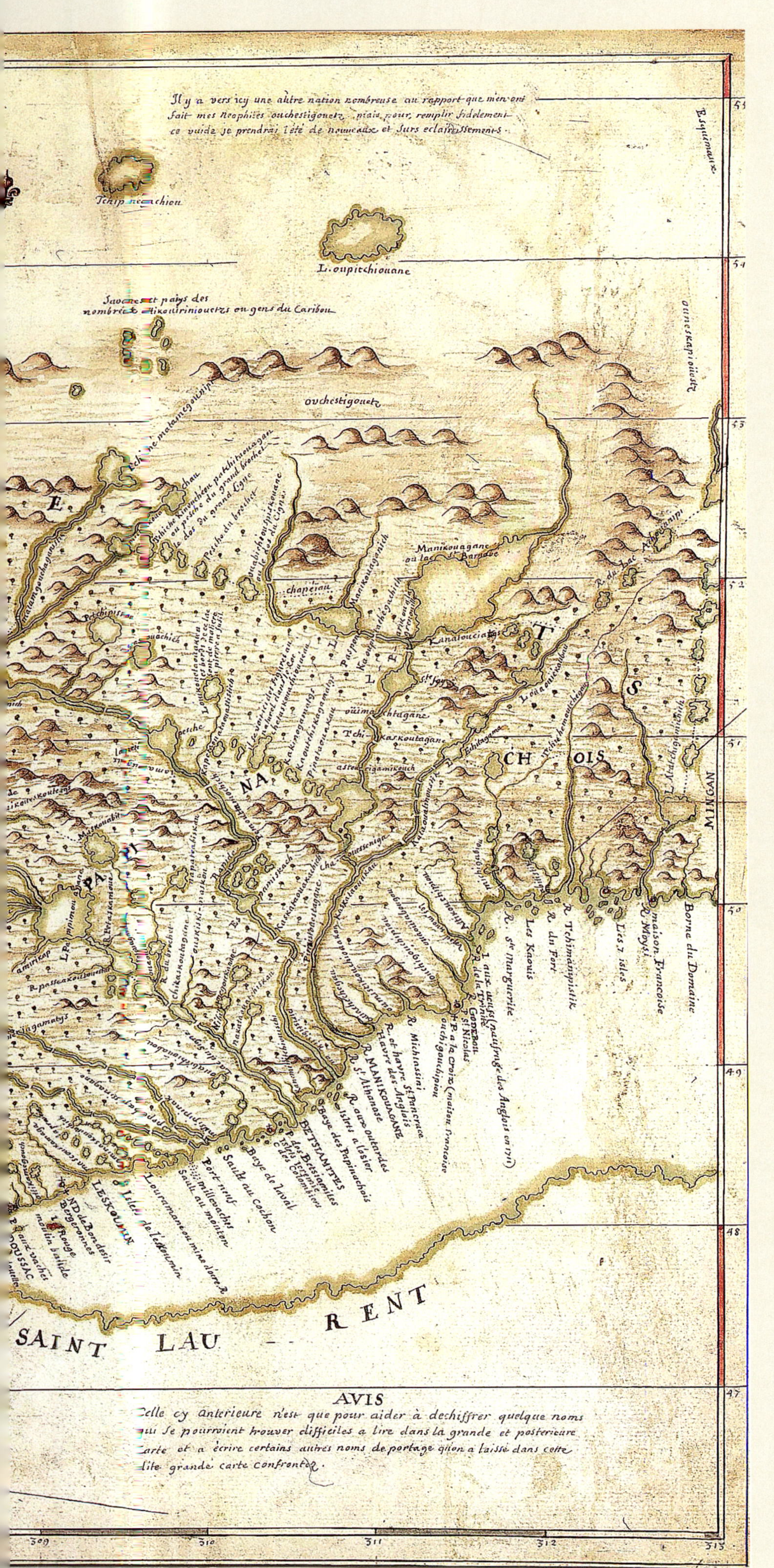

Carte du Domaine du Roi en Canada by Father Laure, 1733
Made in August 1731 in Chicoutimi, this map by the Jesuit missionary Pierre-Michel Laure describes the king's domain north of the St. Lawrence River. Laure reveals a large territory in which revenue from trade was reserved for the holder of a monopoly, who, for his part, had to pay an agreed-upon rent to the king. Given the important source of revenue that this territory represented, it is not surprising that it was desirable to have it mapped and its borders defined. The domain stretched along the north shore of the St. Lawrence from Île aux Coudres and La Malbaie to the Mingan seigneurie. It included the trading posts at Sept-Îles, Rivière Moisie, Chicoutimi, and the Lac Saint-Jean region. This richly detailed map includes a multitude of indigenous names. The mapmaker had travelled around the region as dictated by apostolic needs. He had learned Algonquin and met a number of people who introduced him to their culture, including Montagnais, Papinachois, and Mistassins. At Lac Albanel, he discovered a "marble lair in the form of a chapel." In a relation written one year earlier, Laure explained that only the jongleurs (the shamans) could enter this "house of great spirit" to communicate with the gods. He was the first to report the existence of rock paintings in what is today the province of Quebec: upstream from Tadoussac, opposite the Baie de Mille-Vaches, he noted the Indian name PEPÉCHAPISSINAGANE, below which he wrote, "one can see on the rock there figures naturally painted."

In 1730, Laure wrote a substantial memoir on the state of the King's Domain. The following year, he prepared the map shown here, while intendant Hocquart delegated Louis Aubert de La Chesnaye (1731) and Joseph-Laurent Normandin (1732) to explore the King's Domain. The information gathered was likely used by Hocquart to prepare his 1733 ordinance. See Russel Bouchard, *L'exploration du Saguenay par J.-L. Normandin en 1732: Au cœur du Domaine du Roi*. (Septentrion, 2002).

Following pages (214–15)
A Map of the British Empire in America by Henry Popple, London, 1733
After the signing of the Treaty of Utrecht in 1713, both French and English colonies entered an era of unprecedented prosperity. Their goals were to increase the area of their respective territories while restricting, though peacefully, the development of rival colonies. In order to get a better idea of the borders between the colonies, and also to better plan their expansion on the continent, the English needed more accurate geographic descriptions. *A Map of the British Empire in America* by Henry Popple, published in 1733, met these needs. Once the 20 sheets were assembled, the document, measuring about 2.5 x 2.5 metres, covered a territory bordered by Hudson Bay to the north, Panama to the south, Newfoundland to the east, and the Mississippi to the west. Most of the major American administrators had a copy, including George Washington and Thomas Jefferson. As his brother was a member of the Board of Trade and Plantations in London, the office that determined the fate of the colonies, Popple had little difficulty gaining access to the best English cartographic sources. For the territory west of the Appalachians, he used French sources, notably Guillaume Delisle's *Carte de la Louisiane* (see the map on p. 179). Popple thus used a number of Indian place names. The French names were translated into English or simply written in French.

MEXICO
Part of Hudson's Bay
Christianaux Lake
L. Alepimigon
L. Nemico
KILISTINONS
Heads of the Misisipi
LAKE SUPERIOUR
LAKE HURONS
LAKE ILLINOIS
LAKE ERIE
Bay de Toronto
Algonquins
Nation des Renards
Lead mine
Paris
Padouca
N. des Illinois
F. Crevecoeur
Copper mine
Tamarois
Mechigamias
Misouris R.
Hohio R.
Fall of Niagara
Hurons
N. du Chat
PENNSYL
APALACHEAN MOUNTAINS
NUEVO MEXICO
Nabiri
Kappa
Chaovanons
Charokees N.
A fit place for an English Factory
LOUISIANA
Mobile
CAROLINA
GEORGIA
SOUTH
Cape Carteret
Charles Town
St. Helena Sound
FLORIDA
Ayavalla
Palachucola F.
Echete N.
Apalache B.
I. au Vaisseau
Mississipi R.
Vera Crux to the Havana to avoid the Tra
GULF
NUEVO
Guardaloupa
Presidio del Norte
The Southward Bo

THE
WESTERN
OR
ATLANTICK
TERRA de LABRADOR or NEW BRITAIN
NEW FOUND LAND
The GRAND FISHING BANK of NEWFOUND LAND
Gulf of St Lawrence
Embouchure of the R. St Lawrence
NOVA SCOTIA
ACADIE
BAY of FUNDY
GASPESIE
NEW ENGLAND
NEW YORK
Cape Cod
Boston
Long I.
Cape Charles
Cape Henry
Chesapeak
Cape Hatteras
Bermuda
C. Charles
C. des Chateaux
Belle I.
C. Hambleton
C. Race
C. Canso
C. Sables
NEW YORK
Placentia
Annapolis Royal
Boston
New York
Charles Town
Bermuda
Providence
CAROLINA by the last Charter.

Map of Manitounie
The Illinois country, newly discovered by the Frenchmen Louis Jolliet and Jacques Marquette, afforded an opportunity to invent various names that vied for the prize for originality: Colbertie, Frontenacie, and Manitounie. This last place name was unusual, to say the least, because it referred to an Indian cultural artifact: a statue titled Manitou, adored as a divinity by the inhabitants of the region. This anonymous map seems to be the inspiration for the map of the Mississippi published by Thévenot a few years later (see p. 101).

immigration. The first truly permanent place names appeared on maps made by Jean Bourdon in the 1640s and by Franquelin in the 1680s and 1690s. A number of Quebec locales bear the names of the early seigneurs of New France, whether or not they participated in the settlement of Canada. For instance, maps of the time depict the seigneuries of Berthier, Bécancour, Deschambault, Île-Bizard, l'Île-Perrot, Lauzon, Lavaltrie, Longueuil, Lotbinière, Neuville, Tilly, and so on. A number of these seigneurs were engaged in the fur trade; for instance, Charles Aubert de la Chesnaye, the wealthiest man in the colony, was seigneur, among others, of La Chesnaye (which became Lachenaie), strategically placed at the confluence of the Mille-Îles and Des Prairies rivers.

To seek or solidify royal patronage, explorers immortalized their sovereigns or their senior advisors by naming new discoveries after them. There are many examples. Christopher Columbus, sponsored by Isabella of Castile, named what is today Cuba *Isabella*. The English who settled in Virginia named their territory in honour of Queen Elizabeth I, nicknamed the "Virgin Queen" because she did not have children. In Brazil, the French built Fort Coligny (1555) to acknowledge their patron, Gaspard de Coligny. In Canada, Île d'Orléans was named after the duke of the same name, the future Henri II. Today's Lac Saint-Pierre was called Lac d'Angoulême, after the dynasty of King François I, a Valois-Angoulême. Samuel de Champlain, who needed the support of the Court of France to maintain a trade monopoly, also played this game. On his maps there is a Rivière du Pont (Rivière Nicolet), a Cap de Chaste, the Chutes de Montmorency, a Rivière du Gast (also Rivière Nicolet), a Cap de Condé (on Île d'Orléans), and a Lac de Soissons (Lac des Deux-Montagnes). All of these names pay tribute to men who, through their administrative functions or commercial activities, influenced the destiny of New France: the merchant François Gravé Du Pont, Lieutenant Aymar de Chaste, Admiral Charles de Montmorency, the merchant Pierre Dugua de Monts, Prince de Condé, and Comte de Soissons. Champlain also lent his own name to three places on the continent: Lake Champlain and two rivers (one in the Mauricie region, which still bears his name; the other in Massachusetts, now called Mashpee River). Later, the names Richelieu, Colbert, Maurepas, and Pontchartrain also appeared again, bearing witness to the importance of ministers of the navy in decisions influencing the fate of New France. When he returned to Quebec, the explorer and trader Louis Jolliet gave Governor Frontenac a map illustrating a highly anticipated geographic discovery, the Mississippi River. With an eye on a concession in Illinois country, Jolliet baptized the Mississippi Fleuve Colbert and the surrounding region *Colbertie*. In spite of these original place names, the powerful minister of the navy was not swayed and refused Jolliet the coveted concession. "We must increase the number of inhabitants of Canada," he responded, in essence, "before we consider other lands." The toponymic creativity of Cavelier de La Salle met with better luck: in 1682, he named the territory that he took possession of Louisiana, in tribute to King Louis XIV, who had supported his expansionist intentions.

Other names recall more modest characters who nevertheless marked the territory. The Pere Marquette River, for example, flows into

Lake Michigan at the spot where the Jesuit priest died. The Perray River, which flows into Hudson Bay, evokes the passage of the explorer Jean Péré. Port-Rossignol in Acadia was named for ship's captain Jean Rossignol of Rouen, taken prisoner by Dugua de Monts for having contravened the trade monopoly granted to the king. Saut au Récollet recalls the Recollet priest Nicolas Viel, who was murdered and thrown into the rapids of Rivière des Prairies.

In general, names that were European in origin had more success on the Atlantic coast, where the French could live in a state of almost total autarky, without the help of the Indians. The name St. Lawrence is a good example of this; it first designated a bay on the north coast of the river explored by Cartier on the day of the festival of St. Lawrence. Perhaps by mistake, Mercator and other cartographers applied this name to the gulf. Long called Rivière d'Hochelaga or the River of Canada, the great watercourse became the St. Lawrence, supplanting the other names in the seventeenth century.

But the trend reversed when the French left the shores of the Atlantic and the St. Lawrence to venture to the interior of the continent, where names of Indian origin became more prominent. The farther the French advanced, the more indispensable the Indians were to them. As a minority in North America, they could not allow themselves the luxury of renaming square kilometre after square kilometre of territory – especially because they were entirely beholden to the Indians for transportation and for supplies of furs, the economic cornerstone of New France. To make themselves understood, the explorers had no choice but to use Indian languages and names.

There were a few vague attempts at "toponymic acculturation." In 1688, Vincenzo Coronelli published a map of the western part of Canada on which he described the most important geographic entities by their Indian and French names, among them Lac des Hurons/Lac d'Orléans, Lac Érié/Lac de Conty, Lac Ontario/Lac Frontenac, Lac des Illinois or Lac Michigami/Lac Dauphin, Fleuve Missisipi/Rivière Colbert, Rivière de Illinois/Rivière Seignelay, Rivière Nantounagan/Rivière Talon, and Lac des Issati/Lac Buade.

The following year, perhaps inspired by Coronelli, the cartographer Franquelin proposed to divide New France "into provinces that can be given set borders and stable, permanent French names, along with rivers and particular places, abolishing all the savages' names, for these only cause confusion because they change very often and each nation names places and rivers in its own language." We can sympathize with the cartographer, who must have been confused by the profusion of Indian names, each one more difficult to transcribe than the last. But Coronelli's attempt and Franquelin's proposal bore no fruit; the French continued to adopt Indian names when they made their maps.

The map by Father Laure (1731), portraying the king's domain north of the St. Lawrence, is a good example. In this domain, which was royal in name only, there were practically no French people because trade there was banned. The Indians were masters of the territory, as evidenced by the impressive number of Indian place names reported by the missionary cartographer. Farther west, an anonymous cartographer, immersed in Indian culture, dubbed the area around the Mississippi *Manitounie*, referring to a statue called Manitou, venerated as a divinity by the Indians.

The Indian toponymic heritage is much more pervasive than we realize, as such names sound perfectly normal in a contemporary context. In the United States, for example, a number of large cities, states, and rivers have names of Indian origin, used by French cartographers in various forms – among them Alabama, Arkansas, Chicago, Illinois, Kansas, Miami, Michigan, Minnesota, Mississippi, Missouri, Ohio, and Wisconsin.

Canada is not left out – its name means "village" in Algonquin. A number of province and city names are also of Indian origin, including Saskatchewan, Winnipeg ("dirty water" in Winnipi), Manitoba, Toronto ("trees standing in water" in Mohawk), and Nunavut ("our country" in Inuktitut). In Quebec, whose name means "receding waters" in Algonquin, Abitibi, Anticosti, Batiscan, Chicoutimi ("where the deep water ends" in Montagnais), Hochelaga, Kamouraska, Manicouagan, Natashquan, Rimouski, Saguenay, Témiscaming, Témiscouata, Mascouche, Maskinongé, Mistassini, Nemisco, Tadoussac, and Yamachiche are just some of the Indian names that were assimilated by the French. Appearing for the first time on maps made by French explorers and geographers, they are now an integral part of the North American landscape, evidence of a successful crossbreeding of place names.

This detail of a map by Franquelin (also see detail on p. 105) shows the area inhabited by the French around Montreal. Still with a relatively small population, the colony extended barely beyond the shores of the St. Lawrence. Upstream of Montreal were seigneuries entrusted to officers of the Carignan-Salières regiment to defend the Quebec–Montreal corridor (Lavaltrie, Contrecœur, Saint-Ours, Varennes, Verchères). This detail also shows the La Montagne mission (on the southern flank of Mount Royal) and the mission at Sault Saint-Louis (Caughnawaga, now Kahnawake), which housed Iroquois, Hurons, and Algonquins who had converted to Christianity.

Following pages (218–19)
Map of North America showing the borders resulting from the 1783 Treaty of Paris by Carington Bowles
This map by the London publisher Carington Bowles shows the borders that resulted from the Treaty of Paris of 1783 and testifies to the birth of the United States, a country that was built on a territory occupied by Indians and frequented by French voyagers. The Great Lakes and Mississippi regions are crammed with Indian names, many of which were used again to name American states, cities, and counties (Michigan, Akansas, Ilinois, Huron, Myamis, Alibamous, Ft Alabama, Chicago R., Ouisconsin, etc.). For a number of years after the conquest of New France, the English had to rely on French cartography to describe this immense territory that now belonged to them. Maps by Bellin and d'Anville were partially copied, with many of the geographic names translated from French to English. A number of names thus refer to the French presence in the American West. These include Cristal de Roche, les Deux Mamelles, les Pots à Fleur, St. Jerome R., St. Joseph Fort, R. du Rocher, Puants Bay, St-Peter R., Cap St-Anthony, Marquette R., R. du Raisin, Lacs de Sel, Thuillier, I. Maurepas, Pontchartrain I, Lac des Vieux deserts, and Vermillon R.

The Southern Boundary of Hudsons
NORTH SEA
HUDSONS BAY
NEW NORTH WALES
NEW SOUTH WALES
EAST MAIN
LABRADOR, OR
NEW BRITAIN
HUDSONS STRAIT
Resolution I.
Davis Inlet 1586
C. Farewel
British Miles 69 to a Degree
French Leagues 25 to a Degree
TEGUAYO GRANDE
or TEGUAS NATION
Apaches de Navayo
Apaches Vaqueros
Bounds of Virginia
New York
VAST TRACT OF LAND UNKNOWN
NEW MEXICO
Zuni or Cinola
Apaches de Xila
APACHES NATION
Apaches of Seven Rivers
Hojomes and Janos
Sumas
PIMAS
Tinamares
MAHA and TIN
WHITE PADUCAS NATION
BLACK PADUCAS NATION
PANIS NATION
Canecys
Canoatinos
TICAS
Wandering Indians
Parts very

Companys Territories Settled by Commissaries
ALGONQUINS
ADIRONDAKS
of ADIRONDAKS
LAKE HURON
OUTAOWAS
CATARAQUI
LAKE ERIE or OSWEGO
PENSILVANIA
CHIKTAGHEEKS
Maspoutens
NORTH
CAROLINA
CHICASAWS
AKANSAS
GEORGIA
C. Hatteras
C. Lookout
Old Barnegat Beach
Charles Town
Extirpated by the Carolinians

Volcano
C. Douglas
Mt. St. Augustine
C. Elizabeth
COOKS INLET
Russian Factory
Anchor Pt.
Port Chatham
C. St. Hermogene
Whitsuntide Bay
Pt. Greville
Rose I.
PR. WILLIAMS SOUND
C. Hinchinbrook
Comptrollers B.
Cape Suckling
Mt. St. Elias
Admiralty Bay
C. Phipps
Port Mulgrave
Saw Mt. St. Elias 40 Leagues distance
Port Francois
C. Fairweather
Mt. Fairweather
Lynn Canal
CROSS SOUND
Cross Cape
Portlocks Har.
Bay of Islands
C. Edgecumbe
Norfolk Sound
C. Banks
Port Snettisham
Holkham Bay
Port Houghton
C. Ommaney
Saw the Land for the first time since leaving Nootka
C. Addington
Pto. Bucareli
Foresters I.
Dixons Entrance
North I.
Dundas I.
Observatory Inlet
Stephens I.
Hippa I.
Rennells Sound
QUEEN CHARLOTTES Is.
Skidegats
Princess Royal Islands
Cumshewaes Sound
Ibbertsons Sound
Kingsett
C. St. James
Lucerna Bay
Salmon R.
Village of Rascals
SLOUACUS IND.
NAGIL IND.
ATNA IND.
Queen Charlottes Sound
Calverts I.
Fitzhugh Sound
Scots Is.
Scots Bay
C. Split
Port Brooks
Woody Pt.
Quadra and Vancouvers Island
WAKASH NATION
Nootka
Breakers Pt.
supposed Str. of Juan de Fuca
Nitinat
Green I.
Pinnacle
Claaset
Queen Hythe
Whitbeys Har.
Saddle Hill
Birch Bay
MOUNTAIN
ROCKY MOUNTAIN INDIANS
Mackenzies River
The Indians say they sleep 5 Nights in descending this River to the Sea
N O R T H P

North America, West Coast

THE PACIFIC LITTORAL

The west coast of North America was an enigma for European navigators. The expanse of water separating Alaska from Asia, called the Strait of Anian in the sixteenth and seventeenth centuries, was presumed to lead to the Northwest Passage between the Pacific and Atlantic and thus offer Europe a shorter route to Asia. The navigators' intuition came closer to reality with the expeditions of Bering, who gave his name to the strait between Asia and North America. Farther south on the west coast, certain basic geographic "facts" were overturned. For instance, the peninsula of California had been defined accurately during the voyage by Francisco de Ulloa in 1539. But de Ulloa's account was barely glanced at by other voyagers – so that for almost two centuries, California appeared on maps as an island.

In reality, a vast portion of the western littoral of North America was considered by Europeans to be the exclusive preserve of Spain, which had taken possession of it as part of the division instituted in the Treaty of Tordesillas (1494). After reaching the "Spanish Sea" (the Pacific Ocean), the Spanish went in a straight line to the Philippines, the Moluccas, China, and India – in fact, to all of Asia, just as Christopher Columbus had dreamed. They were not interested in the west coast of North America itself, and they turned their eyes to Mexico and the other territories with mineral resources. The Strait of Anian and the Northwest Passage also lost their appeal, until competitors appeared.

In 1577, Queen Elizabeth I of England commissioned Francis Drake to undertake a major expedition, which became a circumnavigation of the globe. Drake sailed to the Strait of Magellan, against the current up the Pacific coast to 48°N, then around Asia and Africa, finally returning to England after more than three years. Spain now realized that it needed to strengthen its presence on the California coast. In 1602, Sebastián Vizcaino, a Basque seaman living in Mexico, scoured the inlets of the coast in a vain search for gold and pearls. He went as far as 43°N (Cape Mendocino), and then returned empty-handed to the west coast of Mexico. Meanwhile, villages and religious missions sprang up in the interior of New Spain, north of the Rio Grande. The explorations of the Jesuit Eusebio Francisco Kino, from 1685 to 1702, covered a vast territory from the Gulf of Mexico to the Gulf of California. It was not until 1747, after the voyages of Fernando Consag, that the Spanish authorities once again recognized Lower California as a peninsula.

The real threat of conquest of the Pacific coast arose where it was least expected – not from the sea, but from the land facing the North Pacific. During a stay in Western Europe in 1716–17, the czar of Russia, Peter the Great, was impressed by the accounts of explorations bringing new jewels to the crowns of the colonial powers. At the eastern edge of his kingdom, beyond Kamchatka, was a mysterious region, the Strait of Anian, where he believed that exploration should take place. In 1725, he called upon the Dane Vitus Bering to undertake a mission: Bering was to cross all of European and Asian Russia, and then sail as far north as possible. Bering went to 67°18'N in the strait that now bears his name and returned to St. Petersburg by the same land route in 1730, after five full years of voyaging. The results of the expedition were received triumphantly: Russia had, in its turn, discovered America. The czar asked Bering to undertake another mission in 1733. Much of the geographic information that Bering brought back from that voyage challenged the assumptions of geographers such as the German Gerhard Muller and the Frenchman Joseph-Nicolas Delisle, who was living in the czar's court with his brother, Louis Delisle de La Croyère. Now, an

Nootka Indian

The northwest coast of North America by Aaron Arrowsmith, London, 1802
The London cartographer and publisher Aaron Arrowsmith gathered information from explorers and voyagers to make one of the most complete cartographic syntheses of the North American west coast to appear in the early nineteenth century. His sources were explorers from various European countries and included surveys by the sea-otter trader James Colnett, George Vancouver's careful contours, and Lapérouse's "Port aux Français," "Pamplona Rocks of the Spaniards," and "Russian factory." The line meandering through the ocean near the coast depicts the route of Vancouver, thanks to whom the cartography of the coast was so accurate. This map was the first to successfully delineate the west coast of North America and to account for the main geographic realities and human habitation of the region around the Rocky Mountains.

CH
N.W. COAST of AMERIC
EXPLORE
1778
Prepared by Lieut. Henr. Roberts
THE FROZEN OCEAN
PART OF ASIA
THE ARCTIC CIRCLE
SEA OF OKOTSK
KORIAKS
TSHUKUTSKOI
GULF OF ANADIR
SEA OF KAMCHATSKA
THE SOUTH
Omoloweja Ghooba
Cape North
Mouth of the Anadir
C. Serdze Kamen
East Cape
Cape Lisburn
Cape Mulgrave
NORTON
CLERKES ISLAND
Anderson's Isle
SHOAL WATER
BRISTOL BA
Cape Upright
PENCINSK GULF
INGEKINSK GULF
SEA of OLJUTORSKOI
St. Thadeus Noss
BEHRING'S ISLANDS
BEAVER
SEA
Track in the
SEA
Return
Atako
Atghka
Amluk
OONALASHKA
Longitude East from the Meridian of Greenwich.
170
180
190
LONDON: Published by Wm. FADEN, Geogr

R T
nd the N.E. COAST of ASIA,
the YEARS
1779.
mediate Inspection of CAPT. COOK.
250
75
70
240
230
220
THE SEA
Coppermine River
BUFFALO LAKE
NORTHERN
INDIANS
COPPER INDIANS
DOBAUNT LAKE
PART OF NORTH AMERICA
DOG RIBB'D INDIANS
THE ARCTIC CIRCLE
ARATHAPESCOW LAKE
ARATHAPESCOW INDIANS
CHESTERFIELD INLET
EXPLANATION.
The Figures annexed to the Ships Track denote the Depth of Water in Fathoms.
Var. Variation of the Compass Easterly.
The Ships Track in 1778.
The Ships Track in 1779.
Supposed direction of such parts of the Coast as were not seen.
PRINCE WILLIAM'S SOUND
Mount St. Elias
Mt. Fairweather
CROSS SOUND
Bay of Islands
Mt. Edgecumbe
C. Edgecumbe
Bristol River
Foggy I.
NOOTKA or KING GEORGES SOUND
GREAT SEA
210
220
230
the KING, Charing Cross, July 24, 1784.
Engraved by W. Palmer

Preceding pages
The northwest American coast and the northeast Asian coast, in 1778 and 1779, by James Cook
Cook was chosen by the British Admiralty to lead a scientific expedition to the Pacific. He was sent to Tahiti to observe the transit of Venus, a rare astronomical event during which Venus passes between Earth and the sun. On a third voyage, his mission was to explore the northwest coast of North America and find a passage to the Atlantic. In Britain, there was great interest in this passage: an act of Parliament promised a reward of 20,000 pounds to the discoverer. This map, the product of Cook's single voyage to the North Pacific, contributes completely new information on the region explored. His investigations, concentrated on Nootka Bay and the northern coast, resulted in exact longitudinal measurements to north of Bering Strait. Cook's surveys of Bering Strait up to 65°N, the Alaska Peninsula, and the Aleutian Islands were relatively accurate, but from there to Nootka he simply made a linear contour revealing a continuous shoreline, without an opening toward the Atlantic. As a direct outcome of his longitudinal calculations, the distance between North America and Asia became clear, and for the first time the entire width of North America could be assessed.

extremely profitable trade began, involving the hunting of sea otters, whose fur was highly prized in China. Over the protests of Aleuts, who were frequently recruited to take part in the slaughter, the species was almost completely killed off in less than a century.

Meanwhile, England was extending its empire to all parts of the globe. Admiral George Anson's circumnavigation, from 1740 to 1744, alerted other European nations to the risks of English settlement in the Falkland Islands, a provisioning site in the South Atlantic and key to reaching the Pacific Ocean. In 1764, after losing Canada, France took its revenge by sending Louis-Antoine de Bougainville with more than 150 people, including several dozen Acadians, to found a colony in the Falklands. At almost the same time, the Englishman John Byron explored the islands and established a base at Port Egmont. Spain objected strenuously and claimed the territory in the name of the Treaty of Tordesillas; France withdrew from the Falklands and repatriated most of the colonists, but England remained and strengthened its hold on the islands. This first Falkland Islands crisis concluded with a British victory.

Spain was now fully aware of the threat. In 1769, it began to send naval expeditions to the northern part of the California peninsula. In 1774, Juan Pérez Hernandez sailed up to 55°30'N, within sight of what is today called the Alexander Archipelago, and came into contact with the Haida Indians when he anchored at Nootka. The following year, Bodega y Quadra found the mouth of the Columbia River after reaching 58°30'N. In 1779, he went even farther, up to 60°N, within view of the St. Elias Mountains, without seeing any Europeans except Russians hunting sea otters.

Then, James Cook inaugurated a new era in the history of navigation. The scientific progress made during the Enlightenment made for better sailing conditions, more exact bearings at sea with the use of Harrison's chronometer, and vastly improved sanitation on board ships. European exploration ships, led by those of the British Royal Navy, drew 400 to 600 tons instead of the 100 tons drawn by the older ships. They became considerably stronger and more comfortable with the advent of copper-lined hulls, which were particularly useful for travelling in the northern seas, where ice posed a constant danger for wooden hulls.

Leaving England with two ships, Cook crossed the Atlantic from north to south. In July 1776, he reached the Cape of Good Hope; he crossed the Pacific toward New Zealand and the Sandwich Islands, and then returned to the American coast, first sighting land at 44°33'N, the latitude of Oregon. It was March 7, 1778, and Cook had been at sea for two full years. Nootka Bay offered an ideal harbour for repairing the ships and taking on supplies. He and his crew were stunned by the design of the Nootka Indians' houses and boats, their skill at trading, and their social organization, reflected in fabulous totems whose meaning the Europeans could not decipher. The drawings of John Webber, a member of the expedition, accurately portrayed the English visitors' perspective on this nation of Siberian origin.

The ships continued northeast, sailing along the coast to conduct hydrographic surveys. Having sailed almost

all the way around the immense Alaska Peninsula, up to 70°N, on August 16 the explorers found themselves surrounded by ice. Cook was forced to turn back: "The season was so far advanced, and the time when the frost is expected to set in so near at hand, that I did not think it consistent with prudence, to make any farther attempts to find a passage into the Atlantic this year, in any direction; so little was the prospect of succeeding." In a single season of navigation, Cook had confirmed the existence of the Bering Strait, defined the Alaska Peninsula, and sketched out hundreds of kilometres of extremely jagged coastline.

Jean-François de Galaup, comte de Lapérouse, followed up on Cook's explorations. France wanted to regain its international stature after losing almost all of its possessions in North America. The frigates the *Boussole* and the *Astrolabe* were fitted out with the greatest of care, and the operation was followed closely by Louis XVI himself. The king intervened directly, as no other sovereign of France before him had done, on many aspects of organizing the voyage, particularly the choice of the 10 scholars who were to be part of the expedition and the technical equipment needed for their research. On August 1, 1785, the two frigates set sail from Brest for Cape Horn and then Easter Island, which they reached on April 9, 1786; they then headed for the Sandwich Islands. On June 23, less than a year after the expedition had left, Lapérouse was within sight of Mount St. Elias. He surveyed the shoreline from north to south as far as Monterey, with an eye to the possibility of founding settlements, similar to the English "factories" at Hudson Bay but under much more favourable climatic conditions even though the two sites were situated at the same latitude.

Lapérouse's report of his voyage is that of both a sailor careful to elucidate all technical aspects and a humanist faithfully conveying the mentality of his times. His hydrographic surveys were presented with the modesty of an experienced navigator who is aware that something inevitably slips past even the most meticulous observation. Although he was troubled by differences in the level of progress, Lapérouse's vision of the territories and the men he encountered reflected the rationalist optimism of his century, as he had great faith in human intelligence and the progress that it was capable of making. Thus, he believed that people who had not yet been touched by the Enlightenment had everything to gain by opening their minds to it. He had developed a scientific curiosity as free of prejudice as possible and with a spontaneity of thought that led to several new ideas on the anthropological front and on the presence of foreigners in the lands to be colonized. Lapérouse's poetic, elegant, and enthusiastic account, the work of a skilled author, was the ultimate reference in eighteenth-century exploration literature.

Spain had to face the facts: several other Western powers were present on the west coast of North America. England, France, and the United States had found a

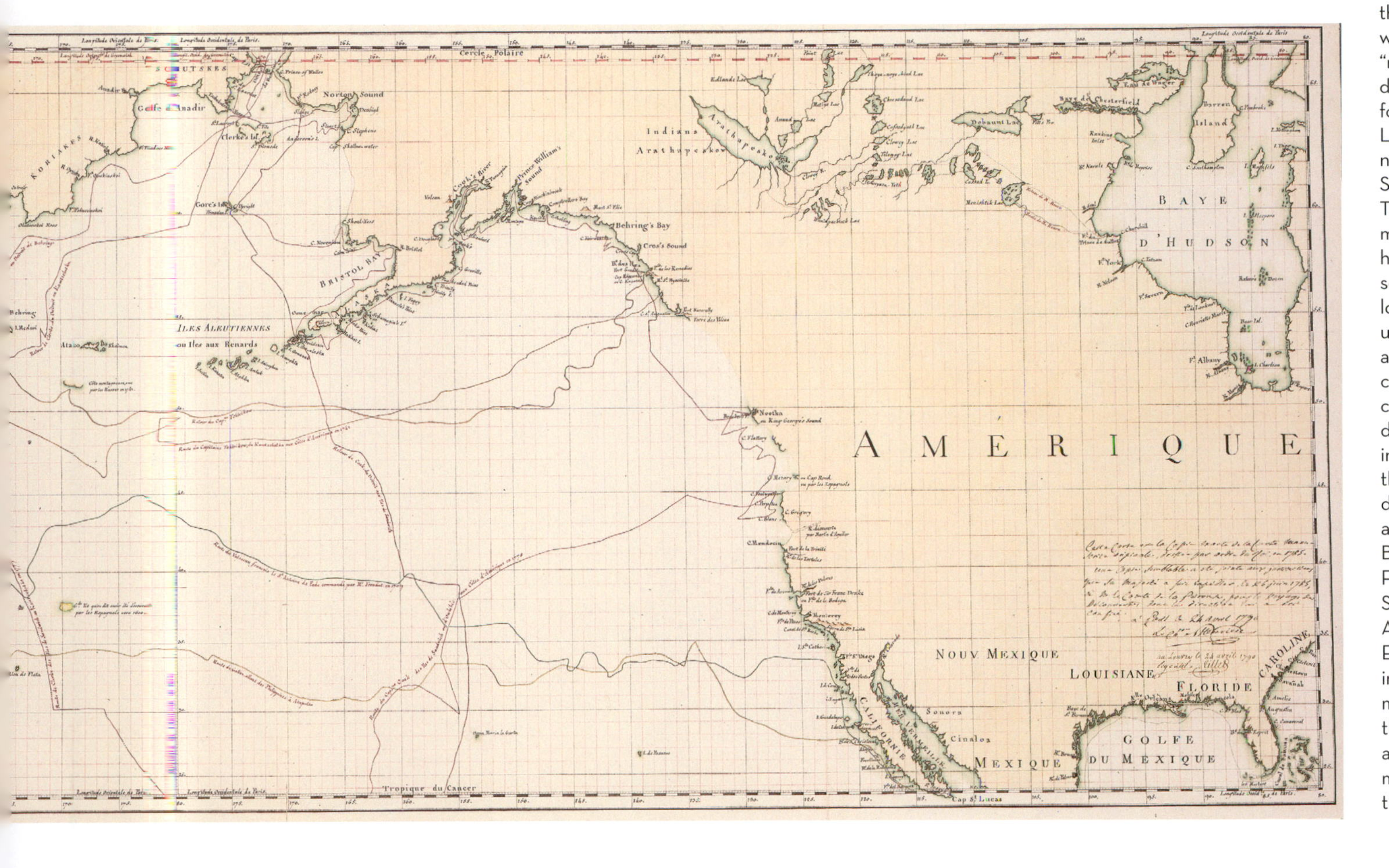

The voyages of Lapérouse (opposite and below)
During the American War of Independence, the naval officer Jean-François de Galaup, comte de Lapérouse, distinguished himself at Hudson Bay, to the detriment of the English ships and settlements. Having entered the good graces of the king, he was entrusted with a scientific mission equal to his talents: to discover a northwest passage by following up on the searches begun by others. The map below, produced by order of the king in 1785, shows the routes taken by navigators up to that time: the Frenchman Nicolas de Frondat in 1709, the Russian Alexei Chirikov in 1741, and the Englishmen James Cook and Charles Clarke in 1778–79. In the summer of 1786, one year after his departure, Lapérouse arrived at a beautiful, peaceful port on the coast of Alaska, more than 100 kilometres north of Los Remedios, the final stop for Spanish voyages of exploration. It was the perfect site, he believed, for a "factory" (trading post). To assert French rights on the coast, he named it Port aux Français. The remainder of the expedition showed dramatically that the professions of cartographer and hydrographer were not devoid of risk. In July, two small boats that had gone to sound the entrance to the port were carried away by the tide (see opposite). The bodies of the 21 drowned sailors were never recovered. To commemorate this tragic event, Lapérouse and his men erected a monument on the main island of Port aux Français, which they named Île Cénotaphe. Two weeks later, the French departed, "moved by the bad fortune, but not discouraged." This disaster was followed by another. In March 1788, Lapérouse was shipwrecked and died near the island of Vanikoro, in the Salomon Islands archipelago. Throughout the expedition, he had made nautical charts of the coasts he had visited. An astronomer on board set the coordinates in latitude and longitude of all the places visited, using the best instruments available at the time (sextant, quadrant, marine chronometer). The problem of calculating longitude had been definitively solved, thus permitting improvements in the cartography of the Pacific. Fortunately, before his death Lapérouse had given reports and maps to the interpreter Barthélémy de Lesseps (uncle of Ferdinand de Lesseps, builder of the Suez Canal), who crossed North America by land and returned to Europe bearing one of the most important documents in French maritime history. In 1797, the journal of the voyage was finally published, accompanied by an atlas containing maps and illustrations resulting from the expedition.

The northwest coast of North America reconnoitred by Captain Vancouver
Between 1772 and 1774, while accompanying the famous explorer James Cook, George Vancouver began conducting hydrographic surveys in the North Pacific. Sent again to the North American coast in 1792, Vancouver demanded that the Spanish apply the terms of the Nootka Convention, which recognized the territorial rights of the English. At the same time, he explored and mapped the entire coast between the 30th north parallel (today's Mexico) and the Bering Strait in Alaska. This long, drawn-out task lasted three years and the results were published in 1798. While many found the account tedious, the hydrography and cartography work was acclaimed. On this point, Vancouver may be considered the worthy successor to his master, Cook. This map shows the route taken by Vancouver's crew facing what is today the border between Canada and the United States. After entering Juan de Fuca Strait (long believed to be the entrance to a vast inland sea), Vancouver explored Puget Sound, around the location of today's Seattle. He then discovered an immense island, which he named Island of Quadra and Vancouver, in honour of the two negotiators of the Nootka Convention. This map and the others published in the same book finally laid to rest all allusion to a northwest passage at this latitude. It was the end of a myth that had motivated numerous explorers.

formidable market for furs there. Remembering that the Nootka site had been explored 15 years earlier by Spaniards, Madrid had no intention of letting an area that it considered Spanish territory slip through its fingers. Thus, between 1788 and 1793, Spain launched no fewer than eight expeditions with the goal of occupying the territory, imposing its authority, continuing explorations, and evangelizing the Indians. Although it had a modern fleet and remarkably well-trained officers and crews, Spain was no match for England, which had become the greatest European maritime force, and it had no choice but to sign the Nootka Convention on October 28, 1790, agreeing to return the British goods that it had seized, and to abandon all claims to property or to exclusive occupation of the North Pacific coast.

But the first country to colonize North America did not abandon its plan to maintain a presence on the west coast. Spain took a deliberately scientific approach. The expedition of Alessandro Malaspina, with ships equipped according to the most up-to-date methods and with artists and scientists on board, was equal in quality to the expeditions of Cook and Lapérouse. In 1791, Malaspina navigated to 60°N and explored the coast and islands of Alaska. He stopped at Nootka to visit with Muquinna, the main chief of the Nootkas, while his collaborators busied themselves making detailed descriptions of the Indians and their environment. Malaspina returned to Cadiz in 1794. His exploration was followed by at least four more missions to the North Pacific. All of these undertakings, which aroused neither the admiration of Cook's voyage nor the anxious expectations of Lapérouse's, nevertheless raised Spain to the ranks of the great exploring nations and demonstrated the abilities of its sailors and scholars, whose original and conscientious research reports made them, too, worthy representatives of the Enlightenment.

England rose to the challenge and commissioned a new expedition to the southern seas and the North Pacific. George Vancouver, an excellent sailor who had travelled with Cook in the Pacific, was the leader, though he was only 33 years old. Vancouver was also charged with negotiating the application of the Nootka Convention with the Spanish post commander. The main mission, however, was to conduct a detailed hydrographic study of the Pacific coast from 30°N to 60°N. When he reached 39°27', Vancouver began to examine, in a sloop, every inlet and island of the coast, foot by foot, in order to make a map. On August 19, 1794, the boats returned from their last anchorage, southeast of the island of Baranof, in a bay that Vancouver appropriately named Port Conclusion. He had travelled some 65,000 miles, and his boats had logged another 10,000 miles, making this one of the longest expeditions of discovery in history. Vancouver fulfilled his mission, although he was not able to arrange for restoration of the British goods seized by the Spanish at Nootka. Indeed, in spite of the excellent relations between Vancouver and Bodega y Quadra, the two men could not agree on the details; they informed their respective governments of this and waited for instructions. More seamen than diplomats, they did not wish to come into conflict, preferring to maintain the scientific tone of their relations during the three summers that they spent together on the coast. Fortunately for the two men, no new instructions arrived during this period from either Spain or England.

In making detailed drawings of the coast, Vancouver vanquished the tenacious myth of a passage, via the Columbia River or a "western sea," to the Pacific Ocean. His painstaking hydrographic surveys were lasting reference works. In 1795, barely a century after it had been virtually unknown territory, the west coast of North America was fully mapped. In this century of scientific conquest, the vanquishing of scurvy was not an insignificant feat; now, long expeditions in cold seas could be undertaken. Led by European countries with conflicting interests, the exploratory missions came very close to ending in armed conflict, but the negotiators wisely avoided both confrontations and attempts at colonization. What did remain, however, were the fur-trading posts, which led to the gradual European invasion of Indian lands and the decimation of the sea-otter population. This trade resulting from exploration was accompanied by serious damage that the expedition leaders – even those inspired by the principles of the Enlightenment and with a humanistic view of the lands they found and the people who lived there – could neither measure nor control.

Main sources in order of importance

Gough, Barry M. *The Northwest Coast: British Navigation, Trade and Discoveries to 1812.* Vancouver: University of British Columbia Press, "Pacific Maritime Studies" collection, vol. 9, 1992. — Lapérouse, *Jean-François Galaup. Voyage autour du monde sur l'Astrolabe et la Boussole (1785-1788), edited by* Hélène Patris. Paris: La Découverte, "Littérature et voyages" collection, 1997. — Poniatowski, Michel. *Histoire de la Russie d'Amérique et de l'Alaska.* Paris: Horizons de France, 1958. — Bernar, Gabriela, and Santiago Saavedra. *To the Totem Shore: The Spanish Presence on the Northwest Coast.* Madrid: Ediciones El Viso, 1986.

Source for translation: Cook, James. *A Voyage to the Pacific Ocean.* London, 1784 [www.americanjourneys.org].

The Imminent Conquest

ENGLISH DOMINATION

THE SEVEN YEARS' WAR (1756–63) was, above all, a struggle between European states for control of territories located both in Europe and on other continents. On the one hand, Austria wanted to retake Silesia from Frederick II of Prussia; on the other hand, England, allied with Prussia, wanted to appropriate the French colonial empire in India and North America. In 1755, the English admiral Edward Boscawen attacked the coasts of Saint-Malo, Rochefort, and Cherbourg and, without a formal declaration of war, captured more than 300 merchant ships and 6,000 sailors. At the same time, in North America, a number of major friction points, independent of what was occurring in Europe, were ignited at the borders between the English colonies and New France, notably along the Ohio River and at Louisbourg. This was all it took to officially launch a conflict involving several continents in the spring of 1756.

Weakened by the long Austrian War of Succession (1740–48), France sought reinforcement by confirming or developing alliances in treaties with Austria, Russia, Sweden, Saxony, and Spain. The inevitable battlefield was Europe, where the French army was deployed on a number of fronts and had some successes in 1756 and 1757. In North America, France sent new battalions to support its naval troops, militias, and Indian allies. French officers, including François Charles de Bourlamaque, Louis-Antoine de Bougainville, and Chevalier de Lévis, had excellent knowledge of the terrain on which the battles were fought, and they devised effective strategies. The governor-general of New France, Pierre Rigaud de Vaudreuil, born in Canada and previously governor of Louisiana, was particularly skilled with regard to the subjects of French–English borders in North America and the political position of the Indian nations. With the French troops led by Louis-Joseph de Montcalm, the Canadians went squarely on the attack and seized Fort Oswego on Lake Ontario (1756), launched "Indian-style" raids against English settlements, took Fort William Henry on Lake Saint-Sacrement (1757), and was spectacularly triumphant in the battle of Carillon on Lake Champlain (1758), with 3,500 men against the 15,000-man English army.

For France, North America was only part of the front that it wanted and needed to defend. But, beyond Europe, it was the most coveted, especially once William Pitt the Elder took command in England, in 1757, with the main goal of appropriating the French colonial empire. Pitt devoted huge resources to this end, committing troops, military equipment, and provisions of all sorts. England's main asset was its fleet of 107 large warships with which it undertook a pitiless privateering campaign on the Atlantic. With some 50 ships, France was a minor maritime force. The minister of the navy commenced naval construction again in 1748, but due to a lumber shortage only 15 vessels were launched. In addition, aging assets, a paralyzed administration, and omnipresent nepotism considerably impeded the fitting out of the warships.

In spite of the sad state of its navy and the harassment by English corsairs on the Atlantic, France sent a number of ships loaded with troops and provisions to Canada each year. It was a considerable effort, given the country's obligations in Europe. These expeditions, however, had little effect; the English fleet waiting in the Gulf of St. Lawrence and the St. Lawrence River often intercepted convoys before they arrived at Quebec. A tragic event unrelated to the war further weakened the French navy: in 1757, the squadron of Dubois de La Motte, who had repelled the English attack on Louisbourg, was decimated, upon its return to Brest, by a typhus epidemic.

Following pages
Plan of the environs of Quebec during the 1759 siege by Samuel Holland and Joseph Frederick DesBarres (above: le Foulon. Following pages: detail showing the battleground)
This map in its entirety accurately depicts the locations of the French and English troops during the siege of Quebec in 1759. The capital of New France is under fire from enemy batteries installed on the other side of the river, at Pointe Lévis. To the east is the landing place of the army of General James Wolfe (Foulon Cove). The surveying and drawing were done by prolific cartographers: the Dutchman Samuel Holland, who became the surveyor general of Canada, and Joseph Frederick DesBarres, a Huguenot from France who had published a remarkable atlas titled *Atlantic Neptune* (see p. 201).

HAUTEURS D'ABR
FOULOU COVE
ANCE DES MERES
Road from St Fois
Quebec
Languedoc
Bearn
La Guienne
Montreal
Trois Riv

RIVER S^T. CHARLES
QUEBEC
HAM
CAPE DIAMAND
Dock Yard
Dock
POINT DES P
S
T
F
General Hospital

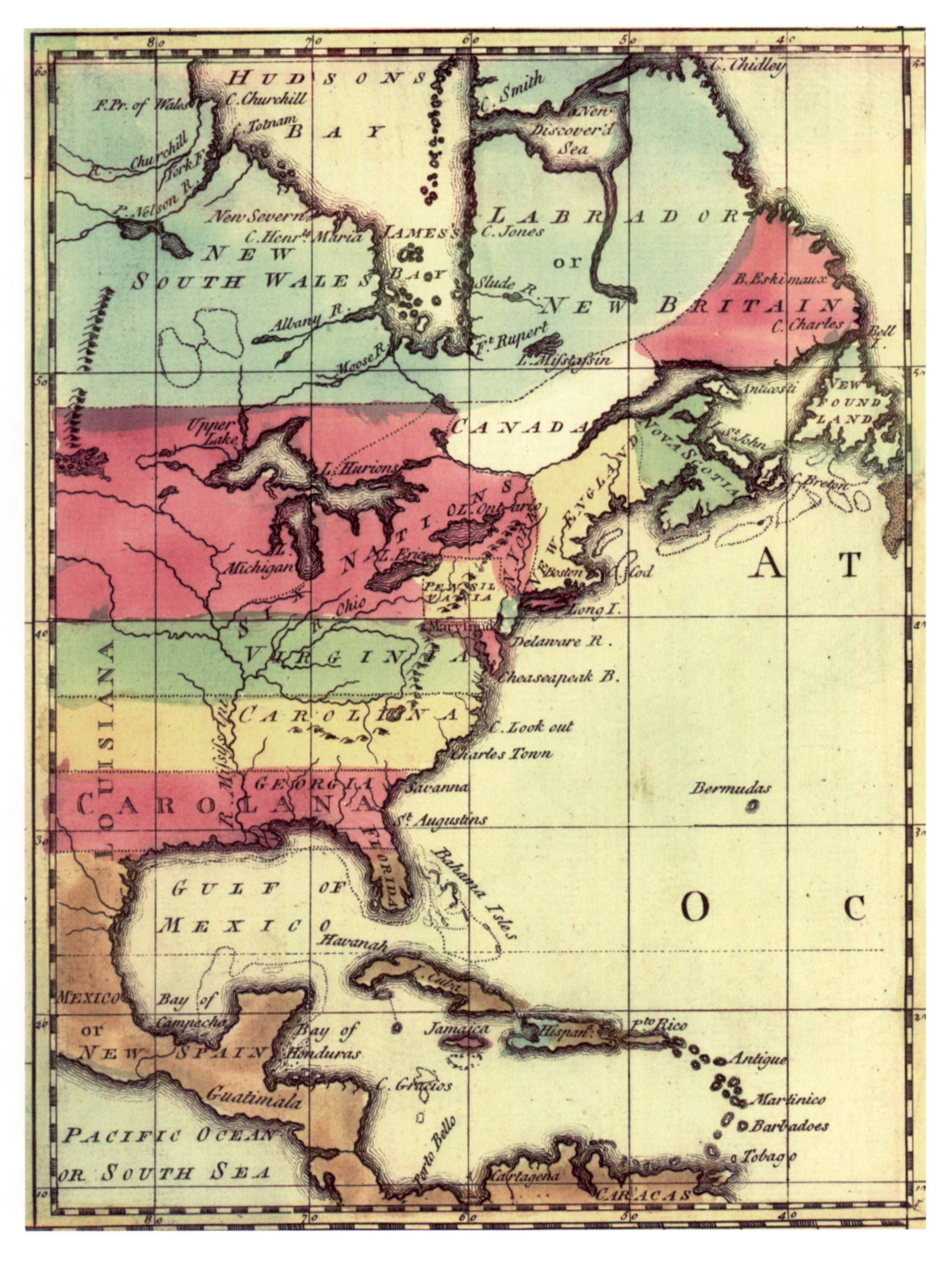

An anti-French map published in London in 1755 (detail)
A number of maps were produced by the Society of Anti-Gallicans on the eve of the Seven Years' War, aiming to show that French claims to North America infringed on territory already occupied by the British or by their Indian friends and allies. Most zones of friction were located in the Ohio Valley, where France had built a chain of forts. On this map, colours are used to illustrate the English colonies as they were described in their founding charters. Canada, for the British, was limited to the north shore of the St. Lawrence River, Cape Breton Island, and other islands in the gulf, as well as the coast of Newfoundland later known as the French Shore.

Six thousand seamen died. Supplies of provisions became ever more unreliable.

It is remarkable that the 85,000 inhabitants of New France were able to resist a million and a half English colonists as long as they did. In the British Parliament, the leader of the opposition, George Grenville, was of the opinion that "it was not the war in Germany but the lack of sailors that kept the French from continuing operations in America and landing in Great Britain." Given that the French secretary of state for the navy had a budget of 30 million pounds and his British counterpart 150 million pounds, there were no illusions about France's capacity to win. After the famous English victory of September 13, 1759, on the Plains of Abraham at Quebec, the capital of New France, and the French victory by Chevalier de Lévis in the spring of 1760, the arrival of English ships in front of Quebec led to the retreat of French troops to Montreal. The numbers spoke for themselves: all the French maritime forces together could not have countered the Royal Navy and defended France's possessions dispersed throughout North America, the West Indies, Africa, and the Indian Ocean.

The war in North America ended with the general surrender of Montreal on September 8, 1760, under the terms of which Canadians were granted protection of their goods and the free exercise of their religion but were required to give up their arms. As they waited for hostilities to end in other parts of the world, the English instituted a temporary military regime to run the country under the leadership of General Jeffery Amherst, from 1760 to 1763. During this three year occupation, the custom of Paris remained in force, the seigneurial system was maintained, the militia captains continued to serve as liaison between the authorities and the population, the Catholic Church continued in its role, internal administration of the colony was conducted in French, and the economy, which had been severely stifled by years of war, gradually recovered. The three administrative regions of Canada were retained but were made more autonomous: Ralph Burton was appointed gover-

nor of Trois-Rivières; Thomas Gage, governor of Montreal; and James Murray, governor of Quebec. The new English leaders evaluated their assets. Thomas Jefferys, Richard Short, James Cook, and Samuel Holland described, drew, surveyed, and mapped the conquered territory.

The Indian nations were also affected by the war. A number of tribes living close to the St. Lawrence Valley negotiated neutrality with the English. The Hurons of Lorette, in a certificate signed in September 1760 by James Murray, were "allowed the free Exercise of their Religion, their Customs and liberty of trading with the English Garrisons." A few days later, others met with Sir William Johnson, superintendent of Indian affairs, at Caughnawaga to discuss peace. At Detroit and the other posts in the Great Lakes region, English settlers arrived and changed the rules of the game for the tribes, most of which had been allied with the French. They ended the custom of exchanging "gifts," refused all credit, required that trade take place only at the posts, and eliminated the traffic in rum. An alliance then formed under the Ottawa war chief Pontiac, the goal of which was to "destroy all the English that they find on the land that they had permitted their brothers and good friends the French to occupy." After a number of Indian victories in 1763, in the end the English took over.

Meanwhile, diplomats were engaged in lengthy peace negotiations. France, the loser, was hardly in a position to make demands and abandoned all thoughts of returning to war. "I am like the public," wrote Voltaire. "I like peace more than Canada and I believe that France can be happy without Quebec." The British negotiators were concerned primarily with defending England's existing colonies against future attack, and for this they needed Canada. They also wanted to preserve territories that offered consumer markets for products made in England. The French wanted to preserve the European market, which was fond of tropical products – especially West Indian sugar, production of which increased tenfold during the eighteenth century, generating large revenues for merchants and for France.

Finally, through the Treaty of Paris of February 10, 1763, England, Spain, France, and Portugal ended the Seven Years' War. France lost Hindustan, except for five trading posts, and Senegal, except for Goree Island, which it retained for the slave trade. It recovered Martinique, Guadeloupe, St. Lucia, and the western part of Santo Domingo (Haiti). France had already given Louisiana to Spain as compensation for Florida, which Spain had ceded to England. On February 15, 1763, Maria Theresa of Austria, Frederick of Prussia, and their allies signed a peace treaty at Hubertusburg, taking as its basis the status quo ante bellum. They had thus fought for seven murderous years in Germany to find themselves, at the end of the war, exactly where they had started. France was bled dry; it had lost its navy, its best post on the west coast of Africa, many of the Lesser Antilles, and all of its possessions in North America except for Saint-Pierre-et-Miquelon, "to serve as a shelter for French fishermen."

England came out of the struggle impoverished but victorious, with enormous gains in territory and prestige that enabled it, until the U.S. War of Independence, to act as the arbiter in Europe. In North America, it exerted total control over the territory, including the former possessions of France and Spain. The southern and western parts of the continent were now open to British settlement. North America's fate was sealed: it would be English.

Main source

Taillemite, Étienne. *L'histoire ignorée de la marine française.* Paris: Librairie académique Perrin, "Passé simple" collection, 1987.

Murray's map showing the Mille-Îles River between Lachenaie and Sainte-Rose, 1761
After the British troops conquered Quebec and Montreal, General James Murray had a map made of the newly conquered country. Between February and November 1761, Murray assigned this task to the most talented engineers in his army, including Samuel Holland, John Montresor, and William Spry, who produced more than 40 large-scale maps covering most of the seigneurial area from Les Cèdres, upstream of Montreal, to Île aux Coudres. Thanks to this meticulous and monumental work, British military officers had intimate knowledge of the topography of the conquered territories, which would be useful if Canada were ever to be retroceded to France. In addition to being formidable management and planning tools, these maps ensured tight military control of the colony. This excerpt shows the Mille-Îles River between the villages of Lachenaie and Sainte-Rose via Terrebonne. It demonstrates a concern with cartographic detail: the mapmakers have portrayed not only the main topographic elements – watercourses, roads, churches, mills, relief features – but also each inhabitant's house and cultivated land.

Nicolas Bellin and the Dépôt des cartes et plans de la Marine

French cartography was long deprived of a central archive for travel accounts and maps, such as existed in Spain and Portugal. Fortunately, this gap was filled in the eighteenth century with the institution of the Dépôt des cartes et plans de la marine, which contributed greatly to knowledge about French North America. The origins of the archive went back to Jean-Baptiste Colbert, who in 1669 established a department of the navy responsible for overseas colonies. In 1682, an engineer was put in charge of collecting and preserving the maps sent to the minister. Various other types of documents were collected over the years, testifying to the French colonial adventure under the Ancien Régime. These included general maps, marine maps, cadastral plans, travel accounts, reports, and other memoranda. To facilitate the management and preservation of the many portfolios and boxes that accumulated, and to encourage the production of new maps, this corpus was separated from the administrative archives. In 1720, the Dépôt des cartes et plans de la marine was created.

The following year, Jacques-Nicolas Bellin started working at the archive. As a hydrographer and engineer, he began to gather, organize, and use the maps that he was receiving. He held the position of hydrographer, which today would be qualified as curator and cartographer, for more than 50 years. Because of the variety of documents that came to him, Bellin was able to compare different sources, both texts and maps. He profited from his contacts with pilots and captains to countercheck information and update imprecise or inaccurate maps, and he used data from several scientific sea voyages to draw correct maps of the St. Lawrence River and the Gulf of St. Lawrence. Useful observations were contributed by a number of navigators, including the Marquis de l'Étenduère and the Marquis de Chabert de Cogolin in Cape Breton, Testu de La Richardière in the St. Lawrence region Gabriel Pellegrin in Newfoundland, and Jacques L'Hermitte at the Baie des Chaleurs. One of Bellin's missions was to address the dearth of nautical charts and to replace the Dutch maps that the French pilots were still using. Although he had the title of hydrographer, Bellin drew not only nautical charts but, by virtue of the department's cartographic sources, maps of the interior of the continent. He was inspired, among other things, by La Vérendrye's exploration voyages west of the Great Lakes. Everything was grist for his mill, and this was reflected in his often innovative maps.

Bellin became even more interested in North America when *Histoire et description générale de Nouvelle-France*, by the Jesuit Pierre-François-Xavier de Charlevoix, was published. In fact, the two worked together to enhance this historical synthesis, the most complete and popular of its time, with 28 maps. But the work that brought Bellin international renown was *Histoire générale des voyages* by Father Antoine François Prévost. This book, which included a number of Bellin's maps, was translated into several languages and distributed beyond France. Later, two other large-scale publishing projects kept Bellin occupied for a number of years. These were *Hydrographie française* and *Petit Atlas maritime*, an impressive collection of almost 600 maps and plans.

A scientist specializing in marine and cartography issues, Bellin was a proud representative of the Enlightenment. He also wrote an impressive number of articles (1,400) in the *Encyclopédie ou dictionnaire raisonné des sciences, des arts et des métiers* by Diderot d'Alembert. Bellin never tried to capitalize unduly on the information that he collected. An honest cartographer, he always took care to cite his sources in memoranda accompanying his main maps. His reputation was flawless, and great explorers such as Louis-Antoine de Bougainville consulted him before undertaking long voyages. For all of these reasons, Bellin was one of the most prominent cartographers of his time.

The seal of the Dépôt de la marine
This seal identified maps that were in the collection of the Dépôt des cartes et plans de la marine, founded in 1720. In the nineteenth century, the Dépôt became the Service hydrographique de la marine. The valuable collection is now divided among three conservation centres: the Département des cartes et plans at the Bibliothèque nationale de France, la Bibliothèque centrale of the Service historique de la marine (Château de Vincennes), and the Centre historique des Archives nationales. It is the richest collection of maps linked to New France.

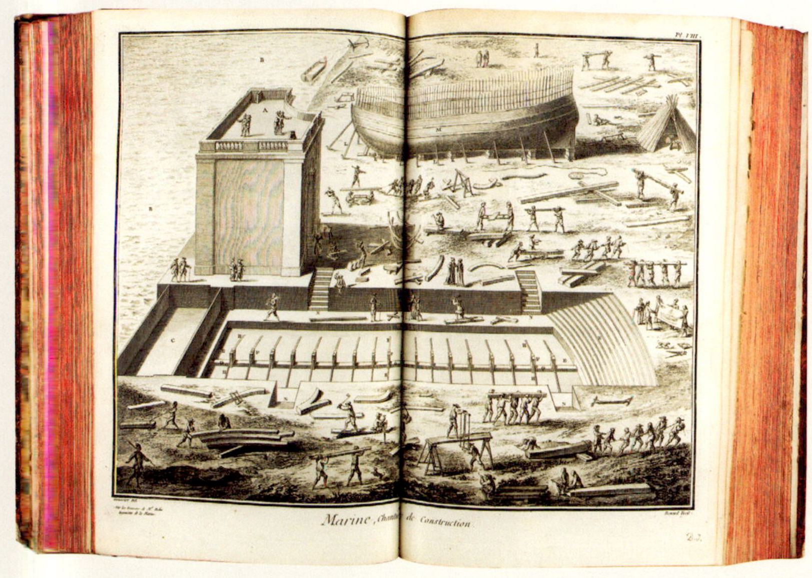

Illustrations accompanying an article by Jacques-Nicolas Bellin, *Encyclopédie des sciences*
The cartographer Jacques-Nicolas Bellin was closely tied to the scientific community of his time. He was responsible for more than 1,400 definitions that appeared in the *Encyclopédie ou dictionnaire des arts et des sciences*, which he signed with the letter Z. This plate, taken from the *Encyclopédie*, was engraved from a drawing by Bellin.

***Carte de l'Amérique septentrionale pour servir à l'Histoire de la Nouvelle France* by Jacques-Nicolas Bellin, 1743**
For more than two centuries, the French dreamed of a route across North America to China. This map by Jacques-Nicolas Bellin, published in 1743, conveys these hopes. It shows, facing the 50th parallel, a river flowing westward from Lac TECAMANIOUEN (Rainy Lake) to the Pacific, not far from a "mountain of glittering stones," where "according to the savages' report begins the flow and ebb." Based on cartographic information obtained by Pierre Gaultier de La Vérendrye from guides and Indian chiefs (see p. 207), this map indicated that there was an easy route from one ocean to the other. There was reason to be optimistic: the La Vérendryes thought that they were not far from their goal. In fact, at Fort la Reine they were about 2,000 kilometres (as the crow flies) from the Pacific coast. How could a cartographer as meticulous as Bellin make such an error? The main problem was the absence of longitude measurements. Bellin had to rely on estimates of distance, which were altered by the large number of detours and portages that the voyagers had to take. In addition, the information supplied by the Indians, orally or in written form, was often poorly interpreted. It is likely that the westward-flowing river on Bellin's map is the Nelson River, which, in reality, flows into Hudson Bay!

PARTIE OCCIDEN
de la
NOUVELLE FRA
ou du
CANADA
Par Mr. Bellin Ingenieur du Roy
Pour servir à l'Intelligence des Affai
present en Amerique, communiquée
les Heritiers de Homan, en
Echelles
Lieues Communes de France de 2
Lieues Marines de France et d'Anglete
LAC ALIMIPEGON
On ne connoit point le Cours de toutes ces Rivieres.
LAC SUPERIEUR
Isle Royale
Isle Maurepas
I. Pontchartrain
I. Philippeaux
I. Ste. Anne
Iles du Tonere
I. S. Ignace
I. S. Roch
Iles de la Carpe
Lac Long
Baye de Michipicoton
C. Chaillon
Ance aux Sables
Minabaujou
Cap Hoquart
I. Hocquart
I. Beauharnois
Pointe de Kioueounan
La Grande Isle
Isles aux Princesses
Portage
Pointe aux Thessalons
LES SAUTEURS
Isle Manitoualin
Mission S. Marie
Fort et Mission abandonnés
Michillimakinac
I. aux Bois Blanc
Ance au Tonerre
LAC HURON
PAYS DES RENARDS
Baye des Noquets
I. du Castor
Lacs aux Outaouacs
Lac des vieux deserts
Mines de Cuivre
Lac Pesin
Fort abandonné
Baye des Puants
Village des Malouines
LES SAKIS
OTCHAGRAS
Mission de S. François Xavier
LAC MICHIGAN
Baye de Saguinam
La Grande Riviere
R. du Detroit
Lac Ste. Claire
Le Detroit
I. des Serpens Sonnettes
Mississagués
PAYS DES MASCOUTENS
Mine de Plomb
R. et Port de Checagou
Portage des Chenes
Village des Miamis
Village de Pouteouatamis
Le Fort
R. S. Joseph
Sources de la Riviere Ouabache
Sources du Theakiki
PAYS DES MIAMIS
Lac
PAYS DES ILINOIS
La Charbonière Fort des Miamis
Le Rocher
Village d'Ilinois
Lac Pimiteouy
Fleuve Missisipi
Le Missouri R.
Les Pots à fleurs
Chateaux ruinés
Kaskasquias et Tamarouas
Montagnes Pelées
Mine de Plomb
Riviere Ouabache ou de S. Jerosme
l'Oyo ou la Belle Riviere
LAC
PAYS
Qui ont été detr
Toute cette coste n'est pas

The Great Lakes according to Jacques-Nicolas Bellin, 1755
In the seventeenth century, the Jesuits had rendered a very good cartographic description of the Great Lakes (see the map by Bressani on p. 94). In the next century, exploration of the environs continued, spurred on by the coureurs des bois. In Paris, Jacques-Nicolas Bellin gathered voyage accounts and other geographic observations that were sent to the Ministère de la Marine and used new information to enhance the accuracy of his maps. The French in fact reconnoitred an impressive number of lakes and rivers. This map shows the few forts that enabled France to keep control of the region on the periphery of its empire. In spite of its high quality, the map is not without errors. With information from a coureur des bois named Louis Denys de La Ronde, Bellin added gigantic islands in the middle of Lake Superior, named in honour of ministers of the navy (Pontchartrain, Phélypeaux, and Maurepas). These islands, said to contain copper ore, proved to be fictional.

Following pages
Map of North America by Jacques-Nicolas Bellin, Paris, 1755
With the Treaty of Utrecht (1713), Hudson Bay officially became English. Using information provided by the English pilot Christopher Middleton, the French cartographer Jacques-Nicolas Bellin made a relatively accurate map of the region. The mapmaker admitted that he lacked information on the east coast of the bay (today in the Nord-du-Québec region) but he refuted the hypothesis of another passage south of Hudson Bay. Even though the English governed the region, Bellin wrote the names of forts in French. On the west coast of Hudson Bay, he summarized more than a century of searching in just one short sentence: "The English sought a passage in this part, but it does not exist." In the interior, Bellin integrated the information regarding La Vérendrye's expeditions. Lac Ouinipigon (Lake Winnipeg) is at the centre of a rather disorganized basin, mainly because a watershed line is poorly placed. At the edge of the continent, Bellin sketched in the western sea that explorers had dreamed about, but it does not have solid contours, and he admitted his ignorance about most of the landmasses around it.

REMARQUES
Les Possessions des Francois sont colorées en Bleu, Celles des Anglois en Jaune, Et celles des Espagnols en Rouge
Fort Francois
Ancien Fort Francois abbandoné
Fort Anglois
Fort Espagnol
ECHELLES
Lieues communes de France de Vingt cinq au Degré
Lieues Marines de France et d'Angleterre de Vingt au Degré
Milles d'Angleterre de Soixante au Degré
Ces Parties sont entierement inconnues
On ignore si dans cette Partie ce sont des Terres ou la Mer
On ne scait si ce sont des Isles ou le Continent
Terres Découvertes par les Russes en 1741. Sans y avoir abordé
LA MER DE L'OUEST
On n'a aucune Connoissance de ces Parties
On peut placer ici les Provinces de QUIVIRA et TEGOUAIO dont on n'a aucunes connoissances certaines
Cap Blanc ou de S. Sebastien
Cap Mendocin
Port du nouvel An
Port de Pinos
PAIS DES ASSINIBOILES
NATION DU SERPENT
CRISTINAUX DES LACS
LAC BOURBON
F. Bourbon
Lac Ouinipique ou Ouinipigon
LES CRIS OU CRISTINAUX
Fort la Reine
F. Maurepas
F. S. Charles
F. S. Pierre
Lac des Bois
Lac Rouge
Grande Etendue de Pays entierement Inconnue
LAC SUPERIEUR
LAC MICHIGAN
LAC ALIMIPIGON
BAYE HUDSO
Baye Chesterfield
Port Nelson
R. Bourbon
R. Churchill
Padoucas
Panis
les Panimalia
Ces Nations Sauvages et les Villages des Panis sont sujets a changer
Octatas
les Kansez
Osages
Missouri R.
Ohio
NOUVEAU MEXIQUE
APACHERIA
Apaches
Moqui
Janos
Sobaipuris
Cocomaricopas
Apaches des Sept Rivieres
Ces Contrées et les Nations Sauvages sont peu connues
LOUISIANE
les Cadodaquios
les Choumans
les Natchitoches
PAIS DES CENIS
Nlle Orleans
les Chicachas
les Tchatas
Alibamons
FLORI
NOUV. NAVARRE
SONORA
CALIFORNIE
MER DU SUD
Cap S. Lucas

CARTE DE
L'AMERIQUE SEPTENTRIONALE
Depuis le 28. Degré de Latitude jusqu'au 72.
Par M. Bellin Ingenieur de la Marine et du Dépost des Plans, Censeur Royal, de l'Académie de Marine, et de la Societé Royale de Londres.
M.DCC.LV.
Avec une Description Géographique de cette Partie de l'Amerique
Nota qu'on na point marqué de Limites
GROENLAND
TERRE VERTE
ISLANDE
DETROIT DE DAVIS
ISLE DE BONNE FORTUNE
Baye de Cumberland
DETROIT DE HUDSON
LABRADOR
nomée anciennement par les François
NOUVELLE BRETAGNE
BAYE JAMES
PAYS DES ESQUIMAUX
GOLPHE S.T LAURENT
ISLE DE TERRE NEUVE
Grand Banc de Terre-neuve Sur le quel on fait la Peche de la Morue
le Bonnet Flamand
Banc Jaquet
Cap Farewell
Cap Cod
Baye de Chesapeak
Charles Town
Toute cette Partie est entierement Inconnue
DEPOT DE LA MARINE

IV

CROSSING NORTH AMERICA

NINETEENTH CENTURY

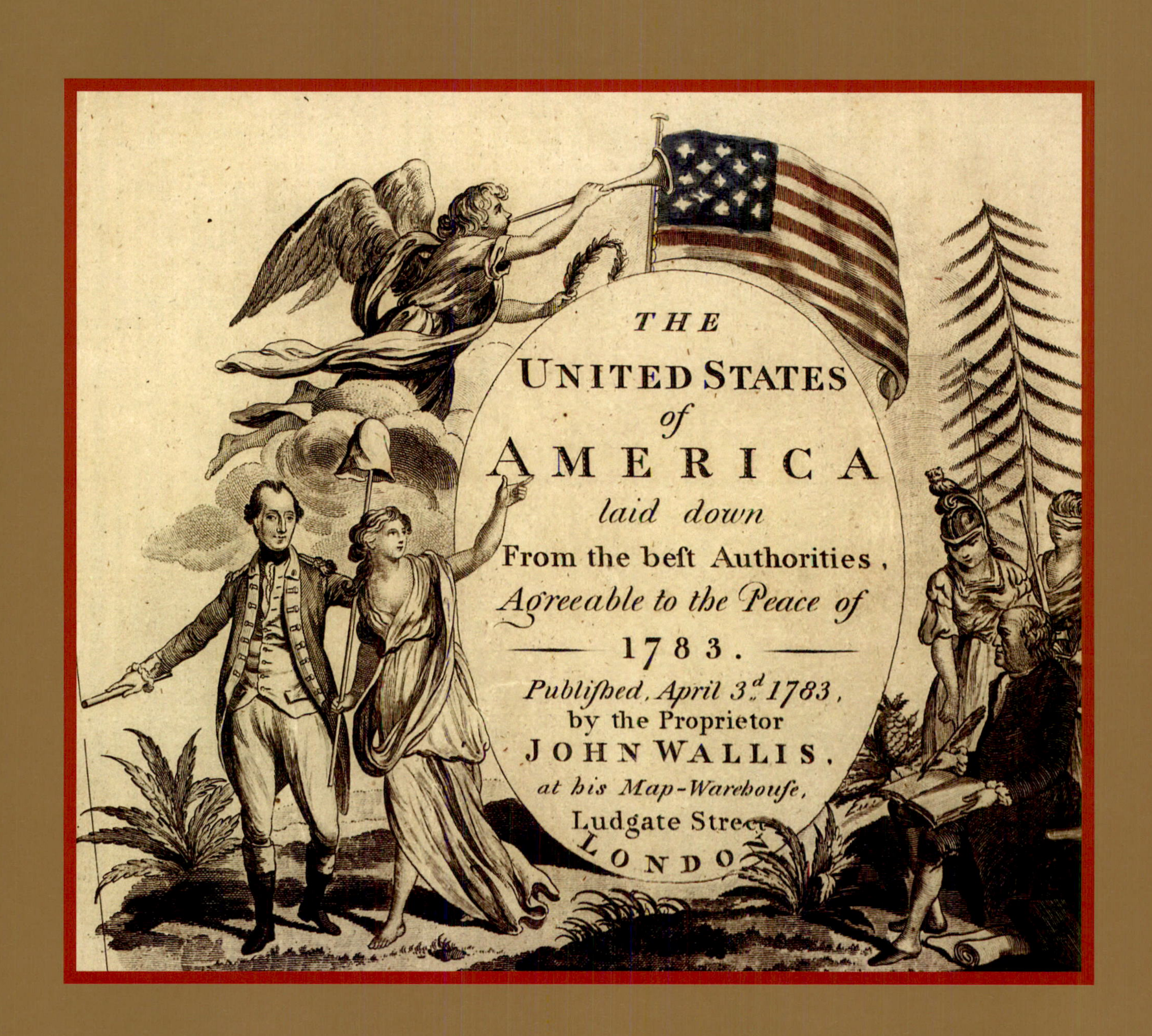
THE
UNITED STATES
of
AMERICA
laid down
From the best Authorities,
Agreeable to the Peace of
1783.
Published, April 3d 1783,
by the Proprietor
JOHN WALLIS,
at his Map-Warehouse,
Ludgate Stre
LONDO

N O R T H
P A C I F I C
O C E A
WILLIAMS SOUND
Admiralty Bay
CROSS SOUND
QUEEN CHARLOTTES Is
Quadra and Vancouvers Island
Pt Vancouver
Mt Hood
COAST OF
NEW ALBION
INDIANS
DOG RIB'D
INLAND IND.
BEAVER
SLAVE LAKE
CHEPAWYANS
BLOOD IND
BLACKFOOT IND.
FALL IND
CREE INDIANS
Long Hair Nation
Blue Mud Nation
SNAKE NATION
TEGUAYO
CUTAGANAS
YUMAS
TAOS
APACHES
NEW MEXICO
SONORA

HUDSON'S STRAIT or FROBISHER'S MISTAKEN STRAIT
SOUTH BAY
HUDSONS BAY
LABRADOR
EAST MAIN
JAMES'S BAY
NEW SOUTH WALES
NEWFOUNDLAND
GULF OF ST LAWRENCE
ISLE OF ANTICOSTI
NEW BRUNSWICK
NOVA SCOTIA
UPPER CANADA
LOWER CANADA
LAKE SUPERIOR
LAKE HURON
LAKE MICHIGAN
LAKE ERIE
LAKE ONTARIO
UNITED STATES
MISSOURIS
LOUISIANA
ATLANTIC OCEAN

A New Map
OF THE
PROVINCE OF LOWER CANADA,
Describing all the
Seigneuries, Townships, Grants of Land, &c.
Compiled from Plans deposited in the
PATENT OFFICE QUEBEC;
By Samuel Holland, Esqr. Surveyor General
To which is Added
A PLAN of the RIVERS, SCOUDIAC and MAGAGUADAVIC,
Surveyed in 1796, 97, and 98,
by Order of the Commissioners, appointed to ascertain the true
RIVER ST. CROIX intended by the TREATY of PEACE,
BETWEEN HIS BRITANNIC MAJESTY,
and the
UNITED STATES of AMERICA.
LONDON.
Published by WILLm. FADEN, Geographer to His Majesty,
and to His Royal Highness the Prince Regent.
No. 5, Charing Cross, April 12th. 1813.
PROVINCE OF LOWER CANADA
DISTRICT OF QUEBEC
DISTRICT OF THREE RIVERS
DISTRICT OF MONTREAL
QUEBEC COUNTY
HAMPSHIRE
ST. MAURICE
BUCKINGHAMSHIRE
WARWICK
LEINSTER
SURREY
KENT
RICHELIEU
BEDFORD
HUNTINGDON
NORTHUMBERLAND
HERTFORD
DORCHESTER
ST. JOHN'S LAKE
LAKE ST. PETER
LAKE CHAMPLAIN
MONTREAL
QUEBEC
BOUNDARY LINE of the UNITED STATES
NEW YORK
VERMONT
NEW HAMPSHIRE
THE UNITED

New Frontiers

HISTORICAL FOUNDATIONS

"THANK GOD, MY SON," said Sully Prudhomme, "for having placed rivers near cities." This ironic comment points out that cities did not arise by chance. They were formed when a stop was made along a route, often related to an obstacle: a mountain range, desert, river, or feature of a river such as a waterfall, large meander, or tributary. When voyagers arrived at these places, they stopped to change their means of transport, recruit a guide, rest and relax, pick up equipment, or gather provisions. And thus, a place for services developed.

A city might also come into existence near a mine, a mineral deposit, or, best of all, a combination of factors. Some cities arose as a result of human beings challenging nature, as was the case with Mexico City, which was built on a lake; others, such as Washington, Ottawa, and Brasilia, were a manifestation of political will.

In North America, the locations of Quebec City, Montreal, Detroit, New York, Chicago, and New Orleans can be explained by geography. The same is generally true for borders that define nations. Sometimes, the geographic aspect is not so obvious. For example, how can one explain the choice of the 49th parallel to establish the longest part of the border separating the United States and Canada? What are the geographic specificities? They are certainly not obvious. What caused this apparent compromise?

In October 1763, following the signing of the treaty that crossed New France off the map, London upset the expectations of the Thirteen Colonies regarding – not to mention their claims to – a region that stretched from Baie-des Chaleurs to Lake Champlain. Instead of setting the border at the south shore of the St. Lawrence, the British chose a line running along "the highlands that separate the rivers that flow into said St. Lawrence River from those flowing into the sea." Once it reached 45°N, near the source of the Connecticut River, this line left the watershed and went toward the St. Lawrence River, and from there to the southwest end of Lake Nipissing. (The 45th Parallel marked the northern edge of the territory ceded in 1606 by James I of England to the London Company and the Plymouth Company.)

Ten years later, at the time of the passage of the Quebec Act (1774), the British kept the line following the watershed up to 45°N, but, once it reached the St. Lawrence, they drew it along the south shores of lakes Ontario and Erie, then, in the middle of Lake Erie, turned it to meet the Ohio River, which it followed to the Mississippi River. The border of this new "Province of Quebec" then roughly followed the Mississippi to the territory already conceded to the "merchant-adventurers of England who trade at Hudson Bay."

This border did not survive the second Treaty of Paris, which recognized the independence of the United States of America (1783). Cut off from the territory east of the Sainte-Croix River, which flowed into the Bay of Fundy, the Province of Quebec kept its border south of the St. Lawrence, which returned to that river along 45°N as far as Lake Ontario, but this time the line crossed the lake and bisected lakes Erie, Huron, and Superior. From there, a choice had to be made between two points: Grand Portage and, farther north, Kaministiquia. Based, for better or worse, on the famous map made by John Mitchell (1755), west of Lake Superior the border followed Pigeon River and a series of lakes, made another zigzag to Rainy Lake, and reached its northwest culmination at Lake of the Woods, then turned to find the Mississippi. Beyond that point, border disputes were frequent and would no doubt have continued but for the role played by the distant city of New Orleans.

***A New Map of the Province of Lower Canada* attributed to Samuel Holland, 1813 (detail)**
First published by William Faden in London in 1802, this map by Samuel Holland was republished at least seven times, and each edition was enhanced with data supplied by the Office of the Surveyor General of Lower Canada under the direction of Joseph Bouchette. With the Quebec Act of 1774, the British conquerors finally decided to retain the seigneurial regime while introducing a new system that would lead to the creation of townships. These appeared west of Montreal and were scattered at the edges of seigneuries on the north shore of the St. Lawrence – for instance, Rawdon and Brandon and, farther east, Tukesbury and Jersey. Below the seigneuries on the south shore the Eastern Townships were born, extending to the border of Lower Canada as defined by the 45th parallel. The 1802 map, however, gives no indication of the headlands planned to mark the eastern border. Both Holland and Bouchette would always be concerned with the contours of this border. In this version, dated 1813, the 45th parallel stops at the Connecticut River, which forks to the east and is joined by two tributaries, Halls Stream and Indian Stream. To the west, the 45th parallel meets the St. Lawrence at the mouth of the Saint-Régis River. On the Canadian side is noted, "Lands possessed by the St. Regis Indians." Holland's map, constantly updated by Bouchette and his son, became one of the most valuable sources of toponymic information on Lower Canada.

Preceding pages: Arrowsmith, 1802; see p. 249

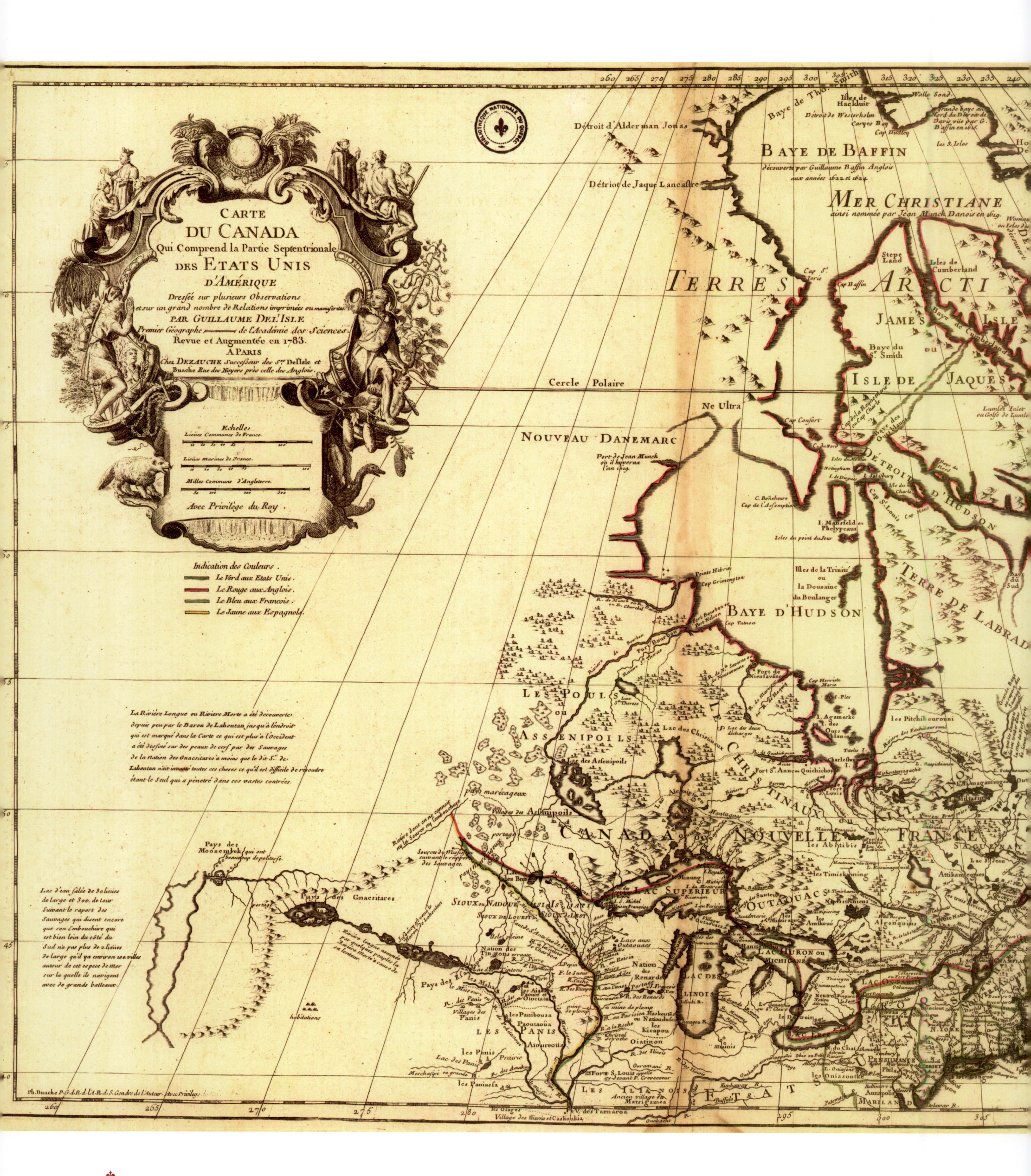
CARTE
DU CANADA
Qui Comprend la Partie Septentrionale
DES ETATS UNIS
D'AMÉRIQUE
Dressée sur plusieurs Observations
et sur un grand nombre de Relations imprimées ou manuscrites.
PAR GUILLAUME DEL'ISLE
Premier Géographe de l'Académie des Sciences.
Revue et Augmentée en 1783.
A PARIS
Chez DEZAUCHE Successeur des Srs Del'Isle et
Buache Rue des Noyers près celle des Anglois.
Echelles
Avec Privilège du Roy.
Indication des Couleurs.
Le Verd aux Etats Unis.
Le Rouge aux Anglois.
Le Bleu aux François.
Le Jaune aux Espagnols.
La Riviere Longue ou Riviere Morte a été découverte depuis peu par le Baron de Lahontan jusqu'à l'endroit qui est marqué dans la Carte ce qui est plus à l'Occident a été dessiné sur des peaux de cerf par des Sauvages de la Nation des Gnacsitares à moins que le Sr. de Lahontan n'ait inventé toutes ces choses ce qu'il est difficile de résoudre étant le Seul qui a pénétré dans ces vastes contrées.
Lac d'eau salée de 30 lieues de large et 300 de tour Suivant le raport des Sauvages qui disent encore que son Embouchure qui est bien loin du côté du Sud n'a pas plus de 2 lieues de large qu'il ya environ 100 villes autour de cet espece de mer sur la quelle ils navigent avec de grands batteaux.
Détroit d'Alderman Jonas
Détroit de Jaque Lancastre
BAYE DE BAFFIN
découverte par Guillaume Baffin Anglois aux années 1622 et 1624.
MER CHRISTIANE
TERRES ARCTI
Cercle Polaire
Ne Ultra
NOUVEAU DANEMARC
ISLE DE JAQUES
DÉTROIT D'HUDSON
BAYE D'HUDSON
TERRE DE LABRAD
LES POULS ou ASSENIPOILS
Lac des Assenipoils
Pays des Gnacsitares
CANADA ou NOUVELLE FRANCE
LAC SUPERIEUR
LAC HURON ou MICHIGANE
LAC DES ILINOIS
LES PANIS
LES ILINOIS
Ph. Buache P. G. d. R. d. l'A. R. d. S. Gendre de l'Auteur. Avec Privilege.

Well before the start of the Seven Years' War, the Thirteen Colonies had begun to eye the territory south of the Great Lakes, which stood between them and the Mississippi. Through the Royal Proclamation of October 7, 1763, Great Britain had created an Indian reservation, but this decision did not stand long given the appetites of the Anglo-Americans. They quickly settled in the region, sketching out the states of Kentucky (1792), Tennessee (1796), and the future Indiana, which were, however, cut off from the Atlantic coast by the Appalachian Mountains. The other route to the sea was via the Mississippi. Indeed, at the time of its independence, the United States had England cede to it the territory located between the Appalachians and the Mississippi, thus acquiring access to this gigantic river.

Under the Treaty of Paris of 1763, it was agreed, in theory, that navigation on the Mississippi River would be "equally free, as well to the subjects of Great Britain as to those of France, in its whole breadth and length, from its source to the sea . . . as well as the passage both in and out of its mouth." Why "as well as"? Simply because the border did not follow the Mississippi to its mouth; the line of demarcation between the two powers followed "the middle of this river, from its source to the river Iberville, and from thence, by a line drawn along the middle of this river, and the lakes Maurepas and Pontchartrain to the sea." This specification of the "river Iberville" had great consequences. New Orleans and "the island in which it is situated" remained France's (and eventually Spain's). The text of article 7 is stunningly clear: from the mouth of the small "river of Iberville," the Mississippi flowed through non-American territory. The English were trying to reduce the dimensions of what was considered to be an island, but they finally agreed to renounce possession of the city itself. Nevertheless, they wanted to introduce a clause forbidding the erection of any fortification on either shore of the river where it ran by New Orleans. The French protested that such a requirement could be misinterpreted. Finally, the parties returned to the agreement between the two for free navigation on the full stretch of the Mississippi, "from its source to the sea." The English took the ultimate precaution of having it "farther *[sic]* stipulated, that the vessels belonging to the subjects of either nation shall not be stopped, visited, or subjected to the payment of any duty whatsoever." It seems that everyone understood very well the strategic importance of New Orleans.

In 1800, Napoleon demanded that Spain retrocede Louisiana to France. The future emperor had plans for the West Indies, particularly Santo Domingo, and he needed a base of operations on the mainland. Did he also realize that possession of New Orleans offered him

Map of Canada and the United States, published in Paris in 1783
Guillaume Delisle was among the most respected cartographers of his time. Many see him as the torchbearer of modern cartography. Trained by his father, Claude Delisle, a great professor of history and geography, then by the astronomer Giovanni Domenico Cassini, he was the first geographer to be admitted to the *Académie des sciences*. His renown earned him the title of King's First Geographer, created especially for him. Delisle's sources were many and varied. He attached great importance to scholars' observations, many of which enabled him to improve the contours in his maps. He had in his possession a large number of published voyage accounts, notably the Jesuit *Relations*, and met a number of explorers in Paris, including Pierre Le Moyne d'Iberville and Pierre Le Sueur. Examination of the *Carte du Canada ou de la Nouvelle France*, originally published in 1703, reveals many influences, notably that of Baron de Lahontan, who published his *Nouveaux Voyages* that year. Delisle was therefore able to incorporate, at the last minute, the RIVIÈRE LONGUE (see p. 206) in the extension of Rivière MOINGONA (Des Moines River), although he noted that Lahontan had perhaps "invented all of these things, which is difficult to resolve as he was the only one to visit these vast lands." This map was republished a number of times before 1783, when it was used to show the new borders between English, American, and Spanish territories determined by the Treaty of Paris. To the northeast, the American territory is bordered by Rivière Sainte-Croix, then by the headland as far as the St. Lawrence, thus cutting Lake Champlain in half. Farther west, the border runs through the middle of lakes Ontario, Erie, Huron, and Superior, then follows the course of the Mississippi.

an extraordinary means of blackmail? In any case, events occurred more quickly than planned. The Haitians put up resistance to the French, and the Americans, irritated by the controls that had been placed on them around New Orleans, were determined to acquire the city – whether at an agreed-upon price or by force. For Thomas Jefferson, this could even have become a reason to wage war against his beloved France. But Napoleon offered Jefferson the entire basin west of the Mississippi for a price similar to what he was prepared to pay for the city alone, which provided control of the immense river, and what came to be called the Louisiana Purchase was completed in 1803.

Suddenly, the border of the United States extended as far as the northern edge of the enormous Gulf of Mexico basin. The source of the Mississippi – yet to be found for certain – was surely a bit farther north than that of the Red River, which flowed northward. This zone was no doubt at the edge of the northern slope of North America laid bare by the retreat of the ice sheet – that is, the Hudson Bay basin. It was a thorny problem for makers of borders! Would the line be extended in interminable zigzags, as it was west of Lake Superior? In the meantime, the Pacific coast was beginning to reveal its secrets: news of Lewis and Clark's expedition, and the ensuing founding of Astoria at the mouth of the Columbia River, caused great excitement.

The Napoleonic wars had repercussions in North America. Britons and Americans confronted each other in 1812. The result was a draw. Reason finally prevailed. Contrary to what some have stated, the Treaty of Ghent (1814) left open the question of the border west of the Great Lakes. In 1816, Great Britain agreed to cede the part of the Red River basin south of the probable source of the Mississippi; two years later, it recognized, in an honourable compromise, 49°N as the border from Lake of the Woods to the Rocky Mountains. In short, Great Britain accepted the northern limits of the Louisiana Purchase. Later, in 1846, the United States and Great Britain extended this border (49°N) to the Pacific Ocean.

Today, Sully Prudhomme might say to the Americans, "Thank God for having placed New Orleans on the eastern shore of the Mississippi and the French for having refused to let the English control it at that moment in time." It was the location of the city that incited the

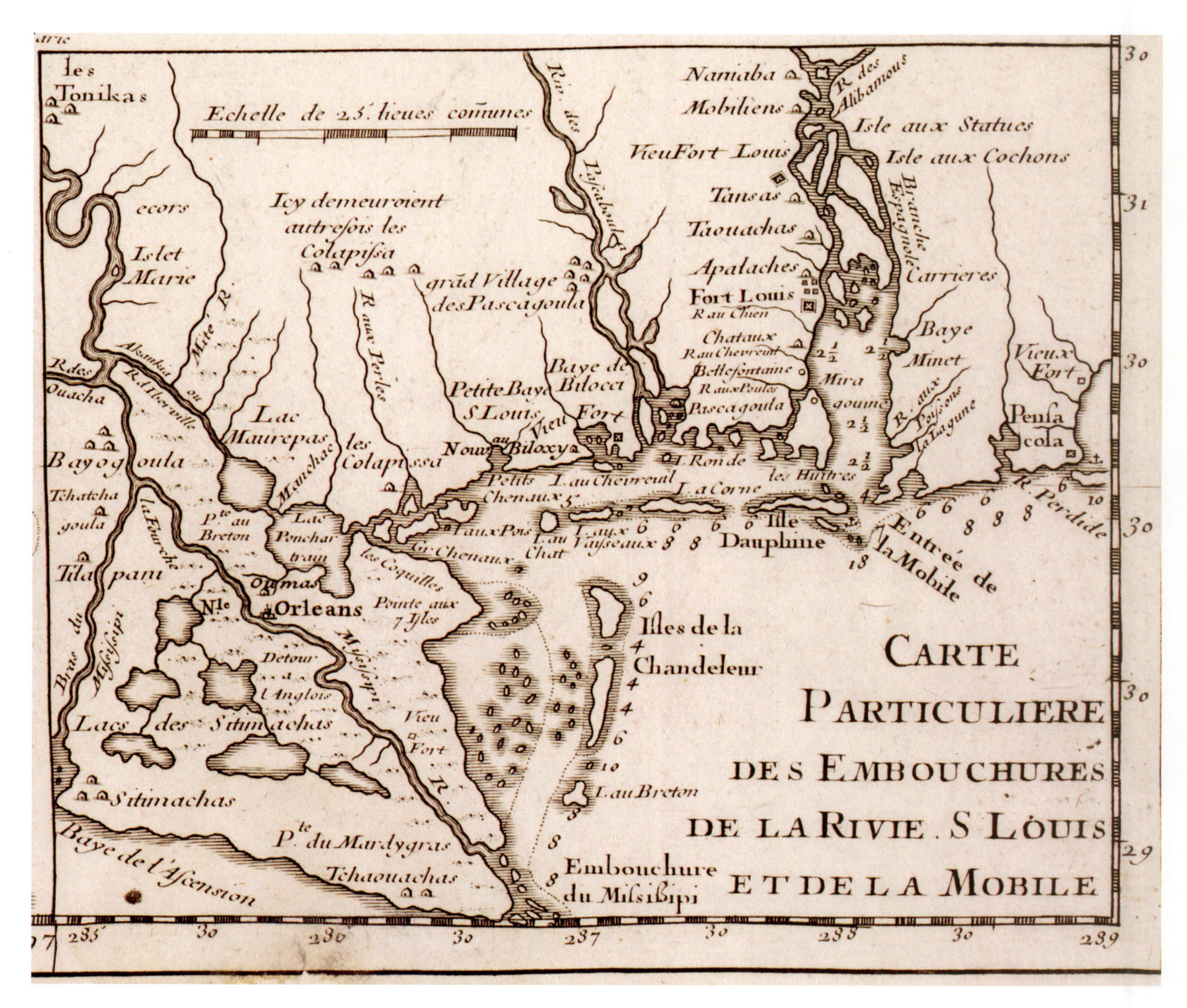

Map of the mouth of the Mississippi by Guillaume Delisle, 1718
In the lower right-hand corner of his 1718 *Carte de la Louisiane et du Cours du Mississippi* (see p. 179), Guillaume Delisle indicates that he considered it important to show a *Carte particulière des embouchures de la rivière S. Louis et de la Mobile.* For a *géographe de cabinet* such as Delisle, a map this significant, the fruit of years of work, was created in steps, and from various sources. The details on the mouth of the Mississippi, added in 1718, were noted by the authorities and evidently were to guide the French negotiators during the writing of the Treaty of Paris in 1763. It is no exaggeration to say that the terms used in article 7 of the treaty changed the course of history. Under the conditions agreed to in the treaties of both 1763 and 1783, the Americans were at the mercy of a foreign power for using the Mississippi, at least as far as the Gulf of Mexico. Thomas Jefferson made this a *casus belli*; he pressed for the French to sell the city of New Orleans. Napoleon took everyone by surprise by offering to cede the immense territory that he had just recovered from the Spanish. In 1803, in what came to be known as the Louisiana Purchase, the United States doubled in area. This gigantic leap had been triggered by the fact that the Iberville River was not really navigable. Who would have suspected that the founder of the French colony of Louisiana would give his name to so modest a watercourse? In 1763, the French negotiators mocked their counterparts, but they were pushing their luck, as history would show.

The mouth of the Mississippi according to Collot and Warin, 1796

Was General Victor Collot on a secret mission, in 1796, along the Ohio and Mississippi rivers? When informed of Collot's presence in these regions, some Americans considered eliminating him, but it was the Spanish who finally had him arrested in New Orleans. Collot's activities were indeed disquieting. The very nature of his topographic surveys made him suspect, as they had a clear military connotation.

A former governor of Guadeloupe, "with no money or navy or army," Collot was forced to surrender his island to the English, who deported him to the United States. There, he was threatened with lawsuits by a merchant who held him responsible for personal losses. His service record under the command of the comte de Rochambeau, who had come to the rescue of the Americans at the height of the War of Independence, had been forgotten. Forced to stay within reach of American justice, Collot agreed, at the request of the new French ambassador to Washington, Pierre-Auguste Adet, to undertake a reconnaissance mission concerning the western borders of the United States. In March 1796, he and a valued colleague, Joseph Warin, began to examine closely the courses of the Ohio and Mississippi rivers. Aware of the persistent border problems in both the north and the south, the Americans and Spanish could not help but be annoyed. Both suspected France of wanting to retake the western part of Louisiana and perhaps support a secessionist movement west of the Appalachians. The supreme stake, however, was free navigation on the Mississippi as far as the Gulf of Mexico. In fact, under article 7 of the 1763 Treaty of Paris, the western border of the English possessions followed the middle of the Mississippi River not to its mouth on the Gulf of Mexico, as the English had hoped during the negotiations, but to the Iberville River (roughly the 30th parallel), which led to Lake Maurepas. Because this article was not modified in the Treaty of Versailles of 1783, Collot drew the eastern shore of the river in such a way as to show the Iberville River leading to Lake Maurepas and Lake Pontchartrain (only the western shore of which is shown), as well as the link with New Orleans. To make his point even clearer, at the end of the meander located at the mouth of the Iberville River, on the western shore he indicated a marshy area flowing into the Plaquemine River. This terrain dried out at certain times of the year, his report mentioned, as did the Iberville River. In other words, navigation was possible only on the Mississippi, and was thus controlled by the New Orleans authorities. Collot portrayed the centre of the city, indicated the surrounding marshes, and specified the depths of the river's three main outlets into the Gulf of Mexico. As usual, the map showed only the shores and the principal tributaries; in fact, an observer unfamiliar with this region might wonder where exactly the gulf was. Somewhat negligently, the cartographer simply noted that the shore was a "marsh covered with reeds" beyond which was the "high-water line." A question persisted: This map is known as "Collot's map," but did he draw his maps himself? Were the observations and surveys the result of teamwork, and the maps drawn by the engineer Joseph Warin? At the time that the *Carte générale du cours de la rivière Ohio* was printed, Collot asked the designers to prepare a cartouche with the following dedication: "To the spirit of Joseph Warin." Seriously injured by two Chicacha Indians and arrested with Collot in late October 1796, Warin died, according to found documents, on October 27 at the latest. All indications are that the originals were drawn by Warin; although the map of the Mississippi is not signed, the map of the Ohio clearly bears his signature.

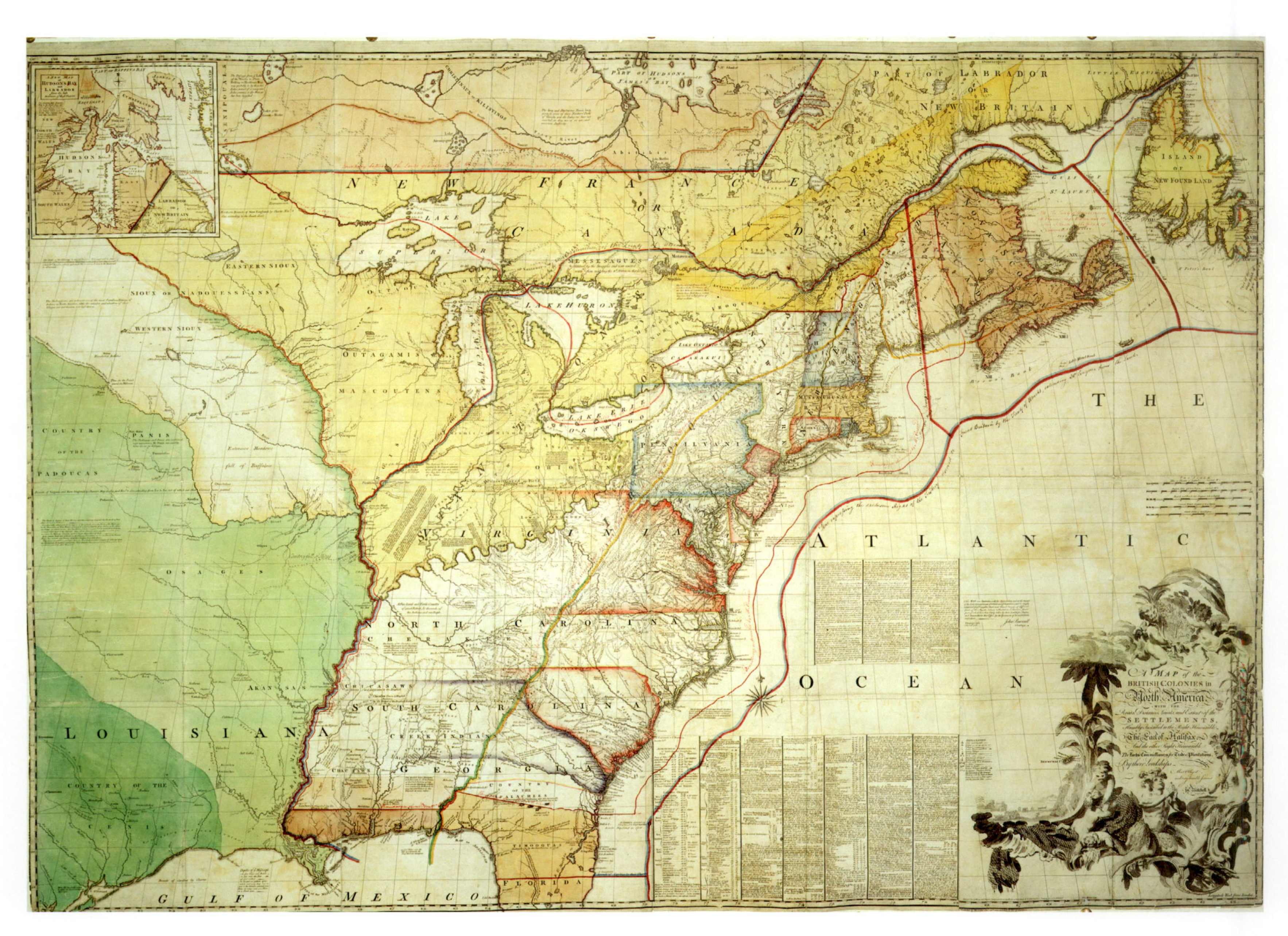

***A Map of the British Colonies in North America* by John Mitchell, London, circa 1783**
This invaluable map by the American John Mitchell has been dubbed the "red-lined map" because of the red lines drawn during negotiations leading to the 1783 Treaty of Paris. These lines indicate the borders of the new country created by the treaty: the United States of America. The map, now conserved at the British Library, is also known as King George's Map, as it once belonged to George III. The colours used on the map are related to the new borders established after the conquest of New France. Louisiana, ceded by France to Spain in 1762, is coloured green. The borders of the "Province of Québec (1763–1774)" are indicated in dark yellow and its expansion according to the Quebec Act of 1774 in pale yellow – a real provocation for the English colonies in America.

Americans to demand ownership of it, and ownership was offered to them on a silver platter, with the western basin of the Mississippi as a bonus. Their appetite whetted, the Americans demanded more, threatened the British to the north, repulsed the Mexicans to the south, and massacred the Indians for good measure. In 1853, having assumed control of a large part of the continent, they appropriated – with "manifest destiny" on their side – the entire continent. The word "America" became synonymous with "United States of America." In Europe, "America" came to mean "United States"; after all, the inhabitants of the United States were already called Americans.

Main sources

For the texts of the Treaty of Paris (1763), the Royal Proclamation, the Quebec Act, and the second Treaty of Paris (1783), see Adam Shortt and Arthur G. Doughty (eds.), *Documents Relating to the Constitutional History of Canada, 1759–1791* (Ottawa, 1911). I am grateful to Éric Bouchard of the Bibliothèque et Archives nationales du Québec for bringing to my attention the work of Theodore Calvin Pease on the negotiations for the 1763 Treaty of Paris, published by the Trustees of the Illinois State Historical Library, "Illinois State Historical Library" collection, 1936, vol. 27, *Anglo-French Boundary Disputes in the West, 1749–1763*. [D. V.]

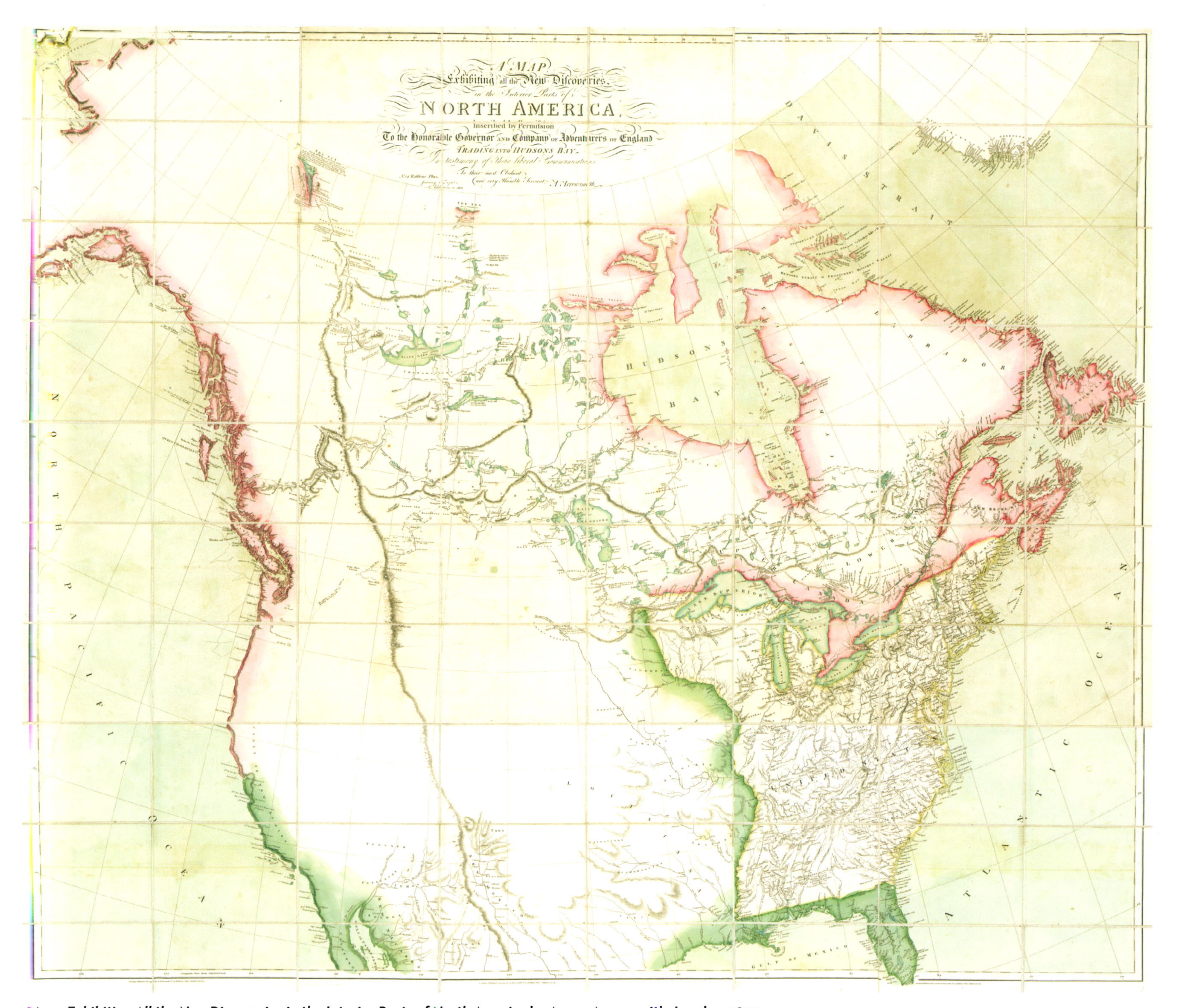

A Map *Exhibiting All the New Discoveries in the Interior Parts of North America* by Aaron Arrowsmith, London, 1802
In the early nineteenth century, President Thomas Jefferson of the United States was fascinated by the western territories. He sent two of his military officers, Meriwether Lewis and William Clark, to find a practicable route to the Pacific. One map in particular was of great use to them: *A Map Exhibiting all the New Discoveries in the Interior Parts of North America*, by the London cartographer and publisher Aaron Arrowsmith (1750–1823). Originally published in 1795, this large wall map was copied, improved, and republished numerous times up to 1850, and it was seen as authoritative by both explorers and the British government. Jefferson had in his possession the second edition, dated 1802. He also commissioned a copy for Lewis and Clark, who noted on it their own route across the continent. The map shows the rivers used by traders from Montreal and Hudson Bay and highlights the route taken by Alexander Mackenzie to reach the Pacific in 1793. It also uses a number of other sources from the Hudson's Bay Company and the North West Company, notably the work of Samuel Hearne, Philip Turnor, and the trader Peter Fidler, which included a map by a Blackfoot chief named Ackomokki. The map approximately situates the territories of the various Indian nations and shows the rivers from their sources in the Rockies. Farther south, between the Mississippi, known since the seventeenth century, and the Pacific, explored by George Vancouver shortly before, was still virgin territory. The mapmaker nevertheless draws in a dotted line extending from the Missouri, crossing the Rockies, and meeting the Columbia River. This probably meant that the route was still hypothetical. It is possible, however, that by making the dotted line and including a note indicating that the ocean could be reached in eight days ("The Indians say they sleep 8 nights in descending this river to the Sea"), the mapmaker was portraying an easy route from the mountains to the Pacific Ocean. This led Lewis and Clark to believe, mistakenly, that a relatively uncomplicated passage to the Pacific existed.

DISTRICT OF MONTREAL
DISTRICT OF THREE RIVERS
DISTRICT
COUNTY of HAMPSHIRE
COUNTY of St MAURICE
COUNTY of WARWICK
COUNTY of LEINSTER
COUNTY of QUEBEC
BUCKINGHAM
COUNTY of RICHELIEU
COUNTY of BEDFORD
HUNTINGDON
SURREY
KENT
LAKE St PETER
St LAWRENCE
RIVER
ISLAND of MONTREAL
ISLAND of JESUS
LAKE St LOUIS
Isle Perrault
Peterborough
Caxton
Hunterstown
Brandon
Kildare
Rawdon
Wexford
Kilkenny
Chertsey
Radnor
Acton
Maddington
Blandford
Somerset
Stanfold
Bulstrode
Aston
Horton
Wendover
Simpson
Upton
Grantham
Wickham
Kingsey
Tingwick
Ham
Weedon
Wotton
Shipton
Durham
Windsor
Dudswell
Lingwick
Melbourne
Brompton
Stoke
Westbury
Bury
Milton
Roxton
Ely
Granby
Shefford
Stukeley
Orford
Ascot
Eaton
Newport
Farnham
Brome
Bolton
Hatley
Compton
Clifton
Auckland
Standbridge
Dunham
Sutton
Potton
Stanstead
Barnston
Barford
Hereford
Hemingford
St Amand
Sabrevois
Noyan
Foucault
Bleury
Delery
Seigneurie of Lotbiniere
Gentilly
Seigneurie of Courval
Seigneurie of Bourg Louis
Fief of St Gabriel
Seigneurie of Batiscan
Seigneurie of Cap la Magdelaine
Seigneurie of Lachenay & St Ours
Seigneurie of Terrebonne
Seigneurie of Sauls St Louis
Seigneurie of La Salle
Seigneurie of Lacolle
Seigneurie of Villechauve or Beauharnois
St Giles
St Croix
Lake St Francis
Maskinongé
Sorel
Yamaska River
St Francois River
Lake Memphremagog
River Sorel or Richelieu
East Bay
CHECK LINE RUN BY
JEREMIAH McCARTHY
CHECK LINE RUN BY Wm VONDENVELDEN in 1795
Line between the late Provinces of Quebec & New York run by John Collins Esquire in the Years 1771, 1772 & 1773.

Aftermath of the Conquest

TWO CANADAS

THE TOWN OF QUEBEC was far from the ocean – almost 1,500 kilometres away! The St. Lawrence River was not only very long but also extremely wide, and therefore believed to be easily navigable. In truth, however, the gigantic river had a succession of hazards that offered natural protection against invasions. The French were well aware of this and tightly controlled the production of nautical charts.

Experienced Pilots Wanted!

Since they did not have maps, the Americans and English captured as many French and Canadian pilots as they could at the beginning of the war that broke out in 1756. When the time came, these pilots would have to guide ships, and they would be allowed to make no errors.

The best known of these pilots was Théodose-Matthieu Denys de Vitré, who was in the employ of Abraham Gradis, the wealthy partner of Intendant François Bigot. Vitré was taken prisoner in 1757. He explained later that Vice-Admiral Charles Saunders had demanded that he choose between death by hanging and a generous pension. Of course, he chose to serve the English – more precisely, Rear Admiral Philip Durell, who recruited by force, at the entrance to the estuary, almost 20 French and Canadian pilots, to whom he entrusted a strong squadron of 14 warships charged with cutting off resupply to Quebec City. The French blithely outmanoeuvred them. On board the *Chézine,* Louis-Artoine Bougainville passed under their noses in the spring of 1759, as did the small fleet of Kanon, the famous frigate lieutenant. Durell followed them closely and dropped anchor opposite Bic, just upstream of Rimouski and about 260 kilometres from Quebec City.

Saunders entrusted his own 90-cannon ship, the *Neptune,* to Augustin Raby, a pilot whose skill was widely recognized, and who had been captured at Bic. The English had their spies. They aimed for the best. The role played by Vitré and Raby was known to their compatriots. Vitré decided to go and live in England, while Raby, known as "the Ugly," returned to work in Quebec. There he rubbed shoulders with, among others, Martin de Chennequy, who had also been taken prisoner and put into the service of the English fleet. Chennequy was the grandfather of the renowned Charles Chiniquy (1809–99), apostle, then apostate, of temperance, who claimed that his forebear was of Spanish origin, that his real name was Etchiniquia or Etcheneque, and that he had been born in a parish near Bayonne, France.

In the summer of 1759, the ships of a formidable armada glided, one after another, up to Quebec. As they went, Saunders ordered that the crews make as many observations as possible. The British had very talented topographers, including the Huguenot Joseph DesBarres and the Englishman James Cook, whom Samuel Holland had taken under his wing. They tirelessly and ceaselessly surveyed the shallows and channels of the St. Lawrence. The dangerous crossing from Île d'Orléans was made quietly. "Sixty of the enemy's warships passed," noted a French observer, "where none of our 100-ton vessels ventured, neither day nor night."

The Surrender of Quebec

Soon, General Montcalm and Governor Vaudreuil had before them some 2,000 cannons on 40 warships escorted by 80 troop ships and 50 to 60 boats and schooners. On June 23, 1759, the English fleet, with a total of 30,000 sailors and 9,000 soldiers, took up a position off Île d'Orléans. Their general, the 32-year-old James Wolfe, had no doubt that he would be victorious. And yet Quebec resisted. The summer went by. Wolfe was on edge. He

Lower Canada according to Samuel Gale and Jean-Baptiste Duberger, 1795 (detail)
In 1795, Samuel Gale and Jean-Baptiste Duberger made an immense map showing the borders of districts, seigneuries, and townships, as well as the border with the United States via the headland as set under the Treaty of Paris in 1783. Since 1790, many townships had appeared between the river and the border, where it was hoped that a population could be settled to impede American ambitions. This map shows how the division into (square-shaped) townships was juxtaposed against the division into (rhomboid-shaped) seigneuries, radically transforming the landscape of the province of Quebec.

Following pages
Murray's map, town of Quebec sector (detail)
Conserved at the Centre d'archives de Québec of the Bibliothèque et Archives nationales du Québec, this document is the sixth manuscript version, on a smaller scale, of Murray's map (see also p. 231). The drawing by the English engineers John Collins and Samuel Holland shows how the clearing of land and forest was proceeding along the river at the time of the conquest of New France. Visible are not only watercourses, relief features, and the edge of the forest, but also the sites of the villages and houses. In the seventeenth century, the French had begun to colonize inland, following the St. Charles River to the north, but also along other, smaller watercourses, to form a number of villages.

Property
Cap Rouge
St. Augustin
St. Nicolas
The Seigniory of Pe
Ley the Propert

ISLAND OF ORLEANS
Mr. Murray's Seignory
North
Channel
South Channel
Falls of Mountmorancy
Fief De Vencent
Belonging
to Mrs. Pechon

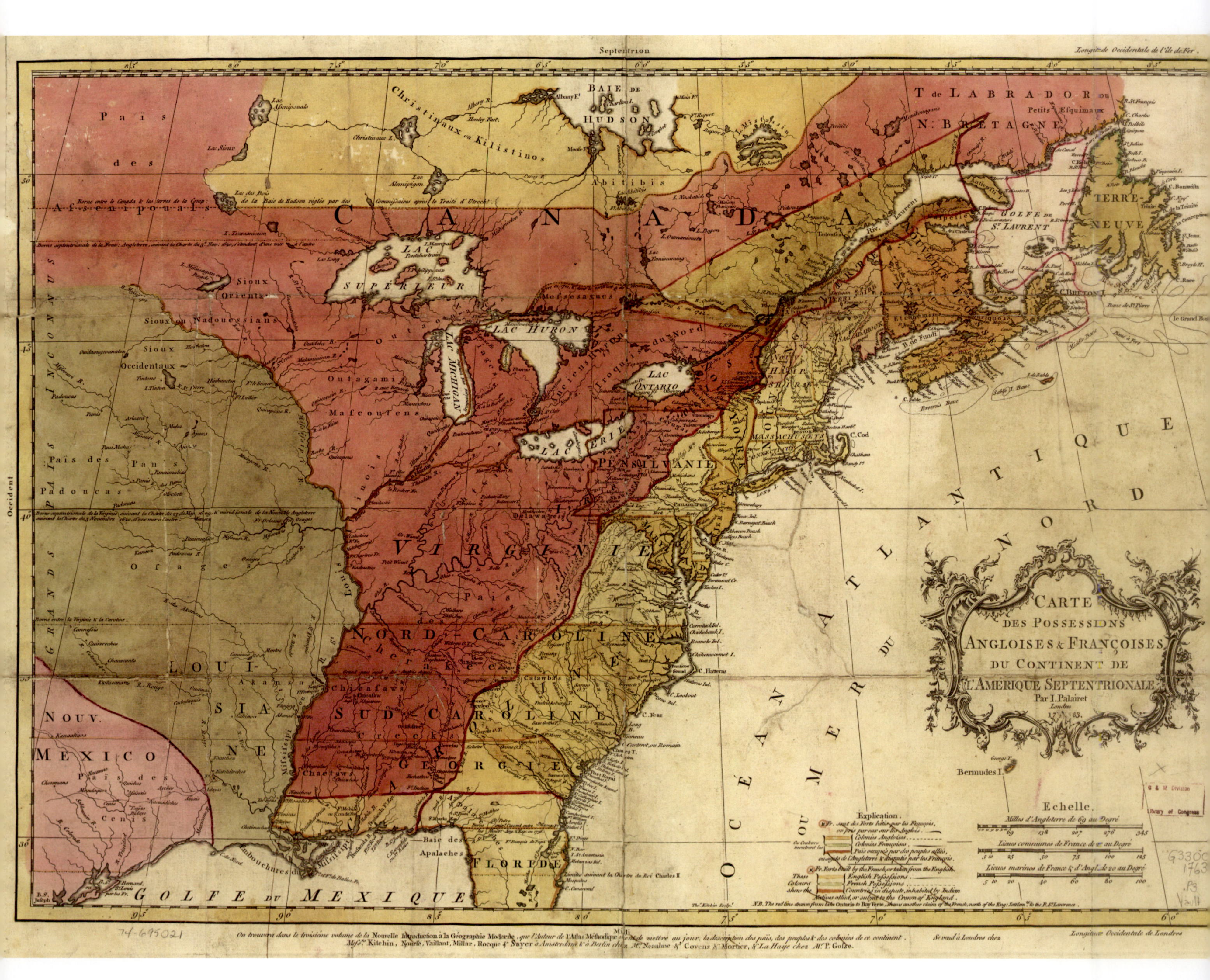

***Carte des possessions angloises & françoises* by Jean Palairet, 1763**
In February 1763, the fate of New France was sealed. Under pressure from American merchants, new borders were drawn cutting the St. Lawrence Valley off from its hinterland and the fur trade that took place there. A new territory was created, the Province of Quebec, enclosed in a square restricted to the middle St. Lawrence, as illustrated in this map by Jean Palairet.

looked for a breach through which to attack. On September 13, he decided to roll the dice. He managed to land part of his army on the heights near Quebec. Montcalm fell into the trap. Without waiting for possible reinforcements, and in spite of the fact that the English were cornered, he rushed forward. Five days later, the town surrendered. Saunders's fleet, with relief, sailed home.

James Murray inherited the command. His superior was General Jeffery Amherst, who was held up somewhere in New England. It was important to follow up on operations. What would the next step be? Guided by Athanase La Plague, a Huron from Jeune Lorette, Lieutenant John Montresor set off through the forests of Maine. On the way back, he took advantage of the advance of English troops toward Montreal to make notes. A lover of cartography, he also made a sketch of the river from the Baie des Mille Vaches to the vicinity of Montreal, continuing the work begun by Saunders's men. Impressed by the result, Murray decided to put his engineers to work. For now, the war was over; Montreal had surrendered in September 1760. The British therefore had to learn more about the territory. "Never again," Murray wrote to William Pitt, "will we be incapable of attacking and conquering this country in a single campaign." Pitt, though, was thinking of returning Canada to France. Would it not be better, he wondered, to leave a foreign presence north of the Thirteen Colonies, and in this way encourage and maintain a strong Anglo-American relationship?

The engineers took their job seriously. They described routes, streams, copses, marshes, hills, and farmland. They located villages, counted the inhabitants, and made lists of men who were likely to bear arms. The discipline of the battlefield was not found, however, in the drawing studios. The officers argued easily. Montresor and Holland were jealous of one another; Montresor and Murray hurled abuse at each other. Nevertheless, data collection remained prolific. Under the expert hands of the drawers, a huge map took shape. Known as "Murray's map," it measured 13.7 x 11 metres. Some copies were taken to London; others were kept on site, though they were not used for a new military campaign. In effect, Pitt, who was seriously thinking of returning Canada to France in order to keep the French sugar-producing islands in the West Indies, was denied the role of negotiator. The sugar lobby, made up of wealthy English planters worried about the prospect of competition, had successfully pressed for the recalling of Pitt and for restitution of the sugar islands to France, mainly Guadeloupe, Martinique, and St. Lucia. The French planters demanded much lower prices than did the English planters. Lovers of sweet tea, the English were ready to pay dearly for their sugar. Canada became British not because of furs but because of sugar.

Gone Was Canada, but Not the Canadians

On February 10, 1763, the Treaty of Paris formalized the British conquest. Étienne-François Choiseul, the French

***The United States of America* by John Wallis, 1783**
In 1783, a republic was born in North America: the United States, whose borders are traced in this map by John Wallis a full five months before the Treaty of Paris was signed. The title cartouche celebrates the two main figures of the American Revolution, George Washington and Benjamin Franklin, along with personifications of liberty, justice, and wisdom. Floating above is the American flag with 13 stars and 13 stripes, portrayed for the first time on a British map.

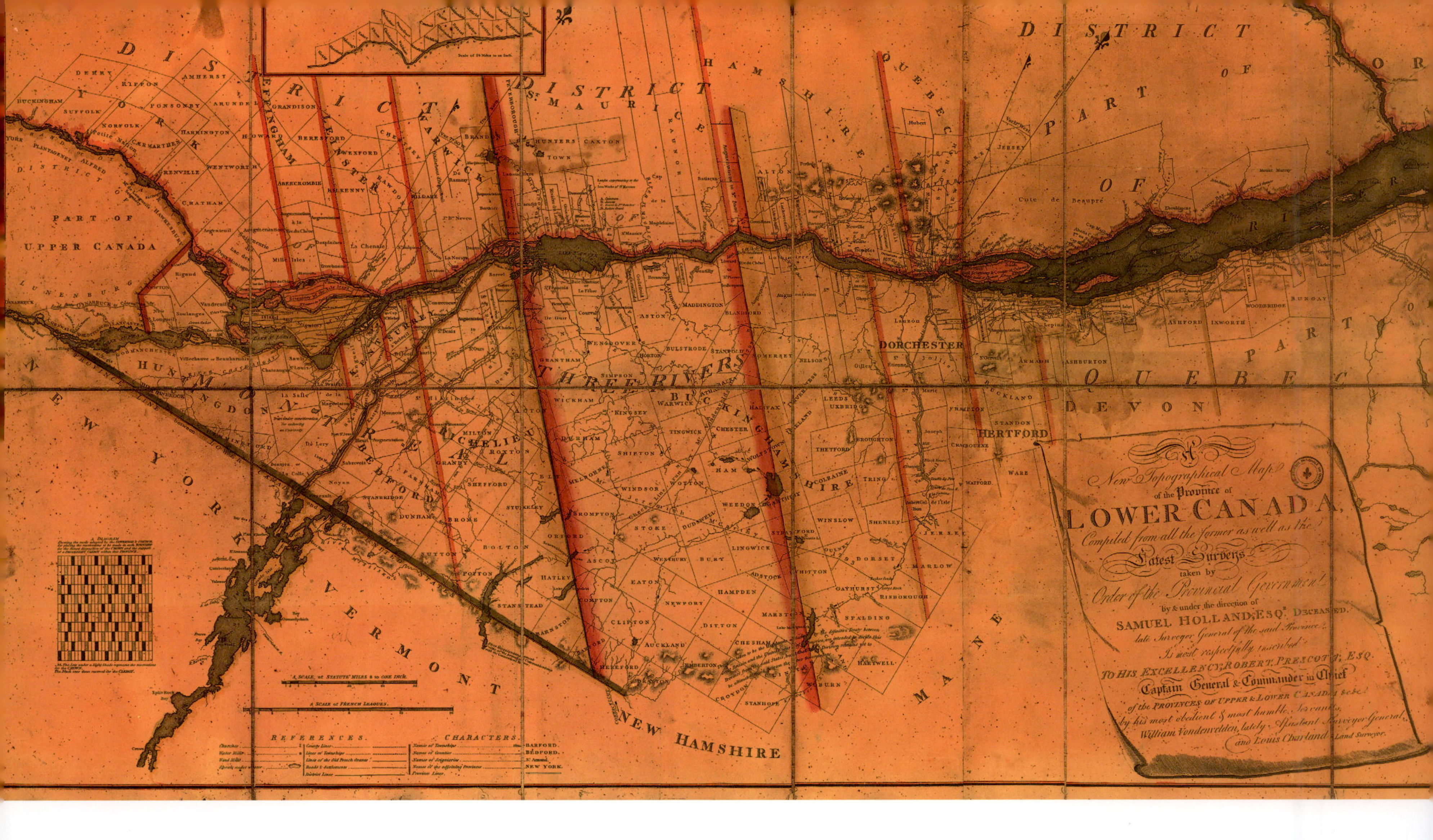

A New Topographical Map of the Province of Lower Canada **by William Vondenvelden and Louis Charland, London, 1803**
William Vondenvelden, who immigrated to Quebec in 1776, had many skills. He worked as a translator, then as a surveyor and a printer. Vondenvelden formed a partnership with a Canadian born in the town of Quebec, Louis Charland, to publish a detailed map showing the seigneuries and townships of Lower Canada, largely inspired by Samuel Gale and Jean-Baptiste Duberger's map (see p. 250), in 1803. Vondenvelden's map was accompanied by a book containing excerpts of titles of former concessions made before and after the conquest of New France. These two publications gave readers of the time a clear idea of the state of landownership in Lower Canada.

minister of the navy, was rubbing his hands with glee. His thinking was in line with that of Pitt: without a French threat at their door, the Thirteen Colonies would become more independent of London; without an enemy at the border, the Americans would no longer need England's protection. History would bear Choiseul out. Twenty years later, a new Treaty of Paris was signed, this one recognizing the independence of the United States. It was sweet revenge for the French.

Immediately after the signing of the Treaty of Paris (1763), the Canadians, descendants of French people "accustomed to the country," were abandoned. France could do nothing more for them. They had 18 months to leave a colony that was now British. If they chose to leave, they could sell their assets, but only to British subjects. The leaders had no alternative; they left, as did some of the elite. Families wept over their dead and rebuilt their houses. Peasants reconstituted their herds. French money was worthless; people had to start with nothing. Businesses changed hands and public office was reserved for former subjects – that is, the British. "In 1763, there were still Canadians," wrote the historian Guy Frégault in the conclusion of his short essay *Canadian Society in the French Régime*, "but there was no longer a Canada."

Their Indian allies of the French felt the same way. The news of an English victory was devastating. They lost the balance of power that they had held between the two colonial forces. Their complicity with the French was accompanied by the possibility of negotiations; they could go to the English in the case of disputes. Already, war had slowed trade. The blockade applied by Pitt had deprived the French of their trade merchandise – the gifts that their Indian partners so appreciated. The Indians also feared for their land. The French, few in numbers, posed no threat. But the English were a different story. An Indian prophet got mixed up in things. It took no more for revolt to break out, skilfully led by Pontiac, the Ottawas' war chief. Amherst did not know what to do. Desperate, mistrustful, he was not far from proposing germ warfare. If necessary, he ordered his colonel, Henry Bouquet, pass out contaminated blankets and infected handkerchiefs. At Fort Pitt, the trader William Trent had already taken the initiative. Two Delaware chiefs had warned the English of an imminent attack, pleading with them to leave the area. The English responded that they would defend themselves against "all the Indians in the Woods," if necessary, that reinforcements were on the way and "you should take care of your women and children." But the Indians' victories were disturbing. "We gave them two Blankets and a Handkerchief out of the Small Pox Hospital. I hope it will have the desired effect," Trent wrote in his journal on May 24, 1763, several weeks before Amherst and Bouquet came up with the idea.

In London, no one understood how it was that a few thousand Indians were standing up to the conqueror of New France. Amherst was recalled. The "vermin," as he called the Indians, had gotten the better of him. Pitt was well aware that it was William Johnson, superintendent of Indian Affairs, who had made the difference in the conquest of Canada, but there was no time to consult Johnson. They had to move quickly. George III could not decide who, among Lords Grenville, Halifax, and Egremont, should prepare a draft constitution. From July to late August 1763, the news from North America was increasingly alarming. The king was still having trouble making up his mind when fate intervened: Lord Egremont, the man responsible for colonial policy, died suddenly. This led to more hesitation, which ended with the confirmation of the roles of Lords Halifax and Grenville. They chose to concentrate on the short term. Starting from a plan that had been prepared by Lord Egremont, Lord Halifax hurriedly sketched out what was to become the Royal Proclamation of October 7, 1763.

A New Colony: The Province of Quebec

South and west of Georgia, two new provinces were laid out: the two Floridas. To the north, the heart of the old French colony in the St. Lawrence Valley became the Province of Quebec. The remainder, from the Great Lakes to the Florida basin, from the east bank of the Mississippi River to the Appalachian Mountains, was reserved – "for the present," according to the text of the proclamation – for the use of the Indians.

The Americans were deeply disappointed by this territorial division. George Washington, a major landowner, was one of those who coveted the land to the west. He complained to one of his partners that he considered the proclamation "a temporary expedient to quiet the Minds of the Indians." For Washington, it would be just a few years before the situation changed, especially because the Indians themselves were willing to let the Americans acquire this land. London had moved quickly, but without taking even the slightest risk with regard to the Thirteen Colonies. They were blocked to both the west and the north. They wanted access to the St. Lawrence River and hoped to annex its south shore. They would have to contain their ambitions to the watershed between the river and the Atlantic Ocean.

As if these measures were not provocative enough, Grenville decided to levy taxes in the colonies to settle the war debt and pay for the troops that had to be maintained to prevent a new uprising. London understood that Canadian agitators had supported Pontiac's revolt and that it was not out of the question that the Canadians themselves would start an uprising.

The Americans barely had time to digest these affronts when London radically altered the borders of the

Treaty of Paris of February 10, 1763
Curiously, the treaty that opened the St. Lawrence Valley to English colonization was written in French only. Twenty years later, in 1783, a second treaty was signed in Paris, and it was through this that the United States gained its independence. On the same day, September 3, the Treaty of Versailles, which ended the war among France, Spain, and the United Kingdom, was also signed. The Americans quickly came to see themselves as the masters of North America: The United States became "America" – the word used by the cartographer Martin Waldseemüller from Saint-Dié, France.

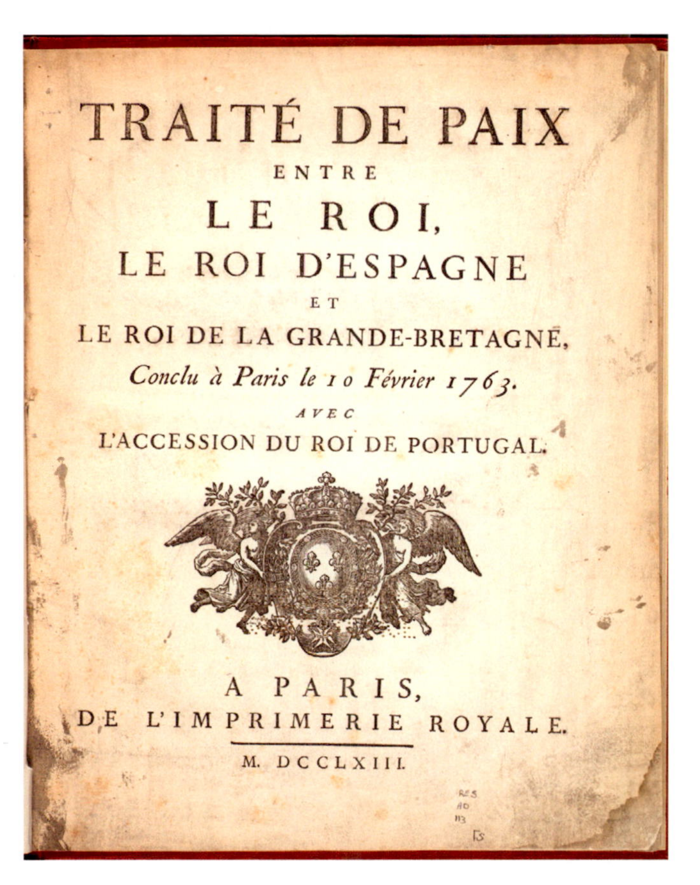
TRAITÉ DE PAIX
ENTRE
LE ROI,
LE ROI D'ESPAGNE
ET
LE ROI DE LA GRANDE-BRETAGNE,
Conclu à Paris le 10 Février 1763.
AVEC
L'ACCESSION DU ROI DE PORTUGAL.
A PARIS,
DE L'IMPRIMERIE ROYALE.
M. DCCLXIII.

Province of Quebec. The huge Indian reservation created in October 1763 was uncontrollable. The Canadians, far from letting themselves be shut within the restrictive borders of 1763, returned to their old habits and guided English merchants ever farther into the interior of the continent. In 1774, with the Quebec Act, the Province of Quebec returned almost to the borders of the New France that had existed following the Treaty of Utrecht; it included the Great Lakes up to the confluence of the Ohio and Mississippi rivers. At this point, the Thirteen Colonies revolted. In 1774, it was estimated that some one million Americans – about one third of the population of the Colonies – were ready to consider independence. Guided by powerful leaders, supported by France, they triumphed. North America, which had become British in 1763, was split into two in 1783: to the south, a republican nation was born; to the north, British North America subsisted.

While the United States regrouped, British North America was split up. The arrival of thousands of Loyalists led in 1784 to the creation of New Brunswick, which was carved out of the large province of Nova Scotia that had been formed in 1713. Soon, the Province of Quebec would also be split in two.

The Province of Quebec Gives Birth to Two Canadas

In the St. Lawrence Valley, the British conquest was not followed by a strong wave of immigration. For the same reasons that had impeded French immigration, English settlers were slow in coming. "The industrious portion of the English people are not going to the Colonies," observed the novelist Frances Brooke, who fervently wished that the new power would do all it could to keep the Canadians. Otherwise, what would this new colony be worth? Nevertheless, it would not be a bad idea to encourage them to speak English!

For the moment, the Canadians had the numerical advantage. In 1763, some 50,000 of them were making a living from farming. They occupied about a million arpents of land divided among 200 seigneuries or fiefs. They had a high birth rate, and they easily integrated more than 1,000 German mercenaries who had come to support the British army and chose to stay after the American Revolution. The Loyalists who poured in were determined to continue living in English under English law. They succeeded in having the Province of Quebec split into two Canadas. The Constitution Act of 1791 provided for the creation of "Upper Canada for the English or American colonists and Lower Canada for the Canadians." The British prime minister, William Pitt the Younger, admitted that he would have preferred "a merger into a single body such that national distinctions might disappear forever," but he did not want to rock the boat. The American Revolution had led to new wisdom: "It will have to be experience that teaches the Canadians that the English laws are the better ones." London also took this opportunity to give in to both British and Canadian pressure and create parliamentary institutions. The Canadians had secured the illusion of power. How long would it be before they noticed?

To sum up: in one generation, from 1760 to 1792, Canadians had three constitutions (1763, 1774, and 1791) and lived through the repercussions of three revolutions (1763, 1775, and 1789). Following a major Anglo-Saxon schism, they survived in the weaker entity that resulted, British North America.

Main sources

ANDERSON, Fred. *Crucible of War: The Seven Years' War and the Fate of Empire in British North America, 1754–1766.* New York: Alfred A. Knopf, 2000. — FILTEAU, Gérard. *Par la bouche de mes canons: la ville de Québec face à l'ennemi.* Sillery: Septentrion, 1990. — MURRAY, Jeffrey S. *Terra Nostra: The Stories behind Canada's Maps, 1550–1950.* Georgetown, ON: McGill-Queen's University Press, 2006. — VAUGEOIS, Denis. *The Last French and Indian War: An Inquiry into a Safe-Conduct Issued in 1760 that Acquired the Value of a Treaty in 1990.* Translated by Käthe Roth. Montreal: McGill-Queen's University Press. Cross-reference research in the *Dictionary of Canadian Biography* complemented the information.

The Northwest

RIVAL COMPANIES

For many years, the Europeans sailed up and down the North American coast hoping to find a passage to China. Over the years, they reconnoitred the Mississippi Delta, the mouth of the Hudson River, the estuary of the St. Lawrence River, and Hudson Bay. Alas, none of these routes into the continent offered the passage that they were looking for. They did lead, however, to the heart of a continent full of surprises of all sorts. Essentially, furs offered the prospect of great profits. Otter, fox, ermine, and mink fulfilled the dreams of the most elegant dressers. A rather strange fashion, however, made the beaver the most sought-after animal, not for its coat but for its undercoat, which milliners used to make hats of all sorts, admired for their sheen and sturdiness. These hats were so popular that the Baltic beaver had become endangered, and the American beaver was discovered in the nick of time.

The Montreal fur traders first had the Albany traders as rivals, and then went into full competition with the traders of the Hudson's Bay Company, which was created in 1670 on the initiative of Pierre Radisson and Médard Chouart Des Groseilliers. Over the years, even the merchants of New Orleans made their presence felt as far north as Illinois country.

Considering the distances involved, these trade rivalries were remarkable. Each of the many fur routes ran thousands of obstacle-strewn kilometres. At first glance, the St. Lawrence axis offered the most advantages, and it led in various ways to west of Lake Superior, where Grand Portage and Kaministiquia served as new points of departure. Curiously, the Hudson Bay basin presented almost as many advantages as did the St. Lawrence. When the north slope of the continent emerged with the melting of the Wisconsin ice sheet, which had covered the northern part of North America, a number of rivers with sources far to the south began to flow into James Bay or Hudson Bay. The fate of this territory, first disputed by the English and French, was sealed by the Treaty of Utrecht (1713). But the merchants of the Hudson's Bay Company remained hesitant for many years. They were content to wait for the Indians to come to their posts – Moose, Albany, Severn, York, Churchill (Prince of Wales), and others – dotting the shores of the two bays. The men of the company were ordered to stay away from the Indian women; at first, mixed marriages were more or less banned.

In spite of the successful trade, the search for a passage westward was not forgotten. In 1742, Christopher Middleton explored west of Hudson Bay, but the real impetus was provided by Moses Norton, the commander of Fort Prince of Wales at the mouth of Churchill River, a strategic post that, although farther from England than Montreal, was nonetheless accessible for about 100 days a year.

In 1767, the Chipewyans told Norton about copper mines somewhere in the northwest. Matonabbee and Idotliaze even made a map of the coast from the mouth of the Churchill River to where the mines in question could be found. This provided an opportunity both to continue the search for a northwest passage and to verify the information provided by the Indians. Samuel Hearne, a young trader noted for his strength and his ability to adapt to the harshness of life in the North, was placed in charge of this mission. Hearne's first two attempts were failures. The third time, he struck up an alliance with Matonabbee, who led a small group composed mainly of his seven wives. The women were indispensable, the Indian explained: they protected the men from the cold, took care of the food and equipment, prepared the food – and ate little themselves – and

Lake Athapuscow (detail)
An engraving taken from the journal by Samuel Hearne in which he describes, in impressive detail, Lake "Athapuscow," sprinkled with small islands on which grew tall poplars, birch trees, and pine trees. Deer abounded, and on the larger islands there were many beaver.

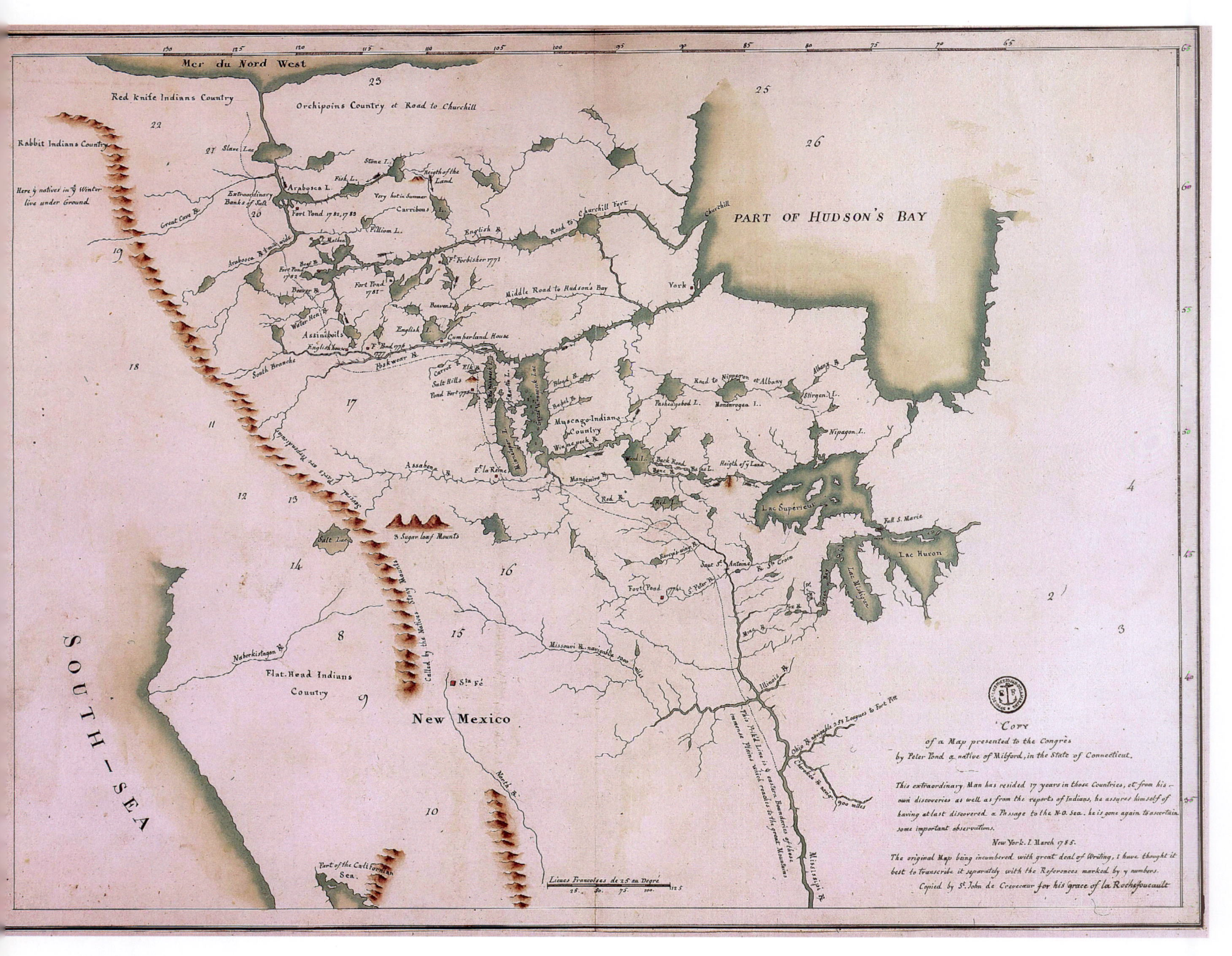

Map of North America by the explorer Peter Pond, "copied by St. John de Crevecoeur for his grace of La Rochefoucault," 1785

Peter Pond's map depicts the territory between the STONY MOUNTS and the Great Lakes (1785). This copy was prepared by St. John de Crèvecœur for the duc de La Rochefoucauld-Liancourt; these two Frenchmen were closely associated, as authors, with the history of the United States. Pond depicted the country west of Lake Superior as full of rivers and lakes, and here and there on his map are entities called Fort Pond, each one dated. There is no doubt that he travelled extensively in the region. Some reference points are useful in reading the map, including WOOD LAKE, CUMBERLAND HOUSE, YORK, AND CHURCHILL. Another landmark is SLAVE LAC, located south of the MER DU NORD WEST, which leads toward Lake ARABOSCA (Athabasca) and the river of the same name ("1/2 mile wide"), located just north of BEAVER River and west of Lake METHEA (Methye or La Loche). In 1778, on the advice of his guides, Pond set out on a gruelling 12-mile portage in order to reach the Athabasca drainage basin more directly. It was in this region that he met Indians loaded down with furs headed for Fort Churchill, a route clearly indicated on his map as "Road for Churchill Fort," via the English River. An arrow indicates the Hayes River route, which goes from the Lake Winnipeg region to Fort York; yet another route follows the WINNEPEEK (Winnipeg) River from Lake of the Woods to south of Lake Winnipeg. In short, finding one's way on this map takes attention, but it can be done. Shrewdly, Pond said nothing about the territory west of the Rockies, leaving hopes alive. Of course, he, too, was hoping to find access to the Pacific, for the route back to Montreal was very long! He ended up confusing his desires with reality, leading his second-in-command, Alexander Mackenzie, into error by presenting a large river flowing from Great Slave Lake to the Pacific; the river, in fact, led Mackenzie to the Arctic Ocean. The intrepid and bellicose Pond had prepared this 1785 map to influence and convince. It was intended for the United States Congress, while a copy was sent to the authorities at Quebec. Pond's strategy was obvious. Although he was American, he was prepared to work for the British if they were willing to provide financial support for his explorations. In fact, his expeditions had been extremely lucrative to date, but he was ambitious; he was so opposed to competition that he had been involved in two murders. Now, Pond had pushed things too far. He became *persona non grata* among fur traders. He drew other maps, but none more successful. Nevertheless, he contributed greatly to establishing the North West Company, the dangerous rival of the Hudson's Bay Company.

transported the supplies; best of all, they could carry twice as much as a man. The expedition that took Hearne and his party to the mouth of the river that would be called Coppermine lasted 19 months (1771–72). In the end, there was little copper, no passage, and the realization that the continent was much larger than had been thought. Hearne wrote an account of his expedition; his observations are now pored over by historians, anthropologists, and naturalists. According to Hearne, Matonabbee was skilful and reliable; Norton was debauched but had a wonderful daughter, Mary, who became Hearne's great love.

After these three expeditions, lasting 32 months in total, and in spite of the hardships endured, Hearne continued to serve in the North. In 1773, he was charged with founding a first post in the interior to contain the traders from Montreal. On the advice of Indian chiefs in the region, he chose a site on Lake Pine Island (Cumberland) linked to the Saskatchewan River and near the Churchill River system. Fort Cumberland was not equal to the task of blocking the westward movement of the Montrealers, but it did send a signal, however weak, from the directors of the Hudson's Bay Company.

Hearne served for some time as a senior agent at Fort Prince of Wales. In 1782, the fort was captured by none other than the comte de Lapérouse, who had been swept northward by the war of independence in the United States. Aware of the superiority of the enemy's forces, Hearne surrendered without a fight. He recounted that Lapérouse, very impressed with his journal and notes, returned them to him intact, on condition that Hearne publish them as soon as possible. Hearne worked on these for years, adding information mainly about the Chipewyans and their environment. He had the passion of a naturalist. The book was published in 1795, three years after his death. It was republished in English a number of times and was translated, over the years, into German, Dutch, Swedish, French, and Danish.

A Rival Company: The North West Company

After the conquest of 1760, the fur trade increased in intensity. For the coureurs des bois of the St. Lawrence

Samuel Hearne at the Coppermine River

On December 7, 1770, Samuel Hearne left Fort Prince of Wales for a third attempt (in red on the map) to reach the copper mines whose existence had been mentioned by the Chipewyans and Crees. This time, he succeeded, with Matonabbee as his main guide. On this map, a green line indicates the second attempt. On July 14, 1771, he reached a river (Coppermine). Three days later, he arrived at its mouth after having witnessed, powerless, a massacre of a band of Inuit. On his return (yellow line), he crossed Great Slave Lake, which he named Athapuscow, returning to his point of departure on June 30, 1772. Although he found little copper, Hearne had achieved a true exploit. In his journal, published in 1795, he noted, "The continent of America is much wider than many people imagine . . ." Elsewhere, he made it clear that he never found a river of any size flowing westward and concluded that there was no northwest passage.

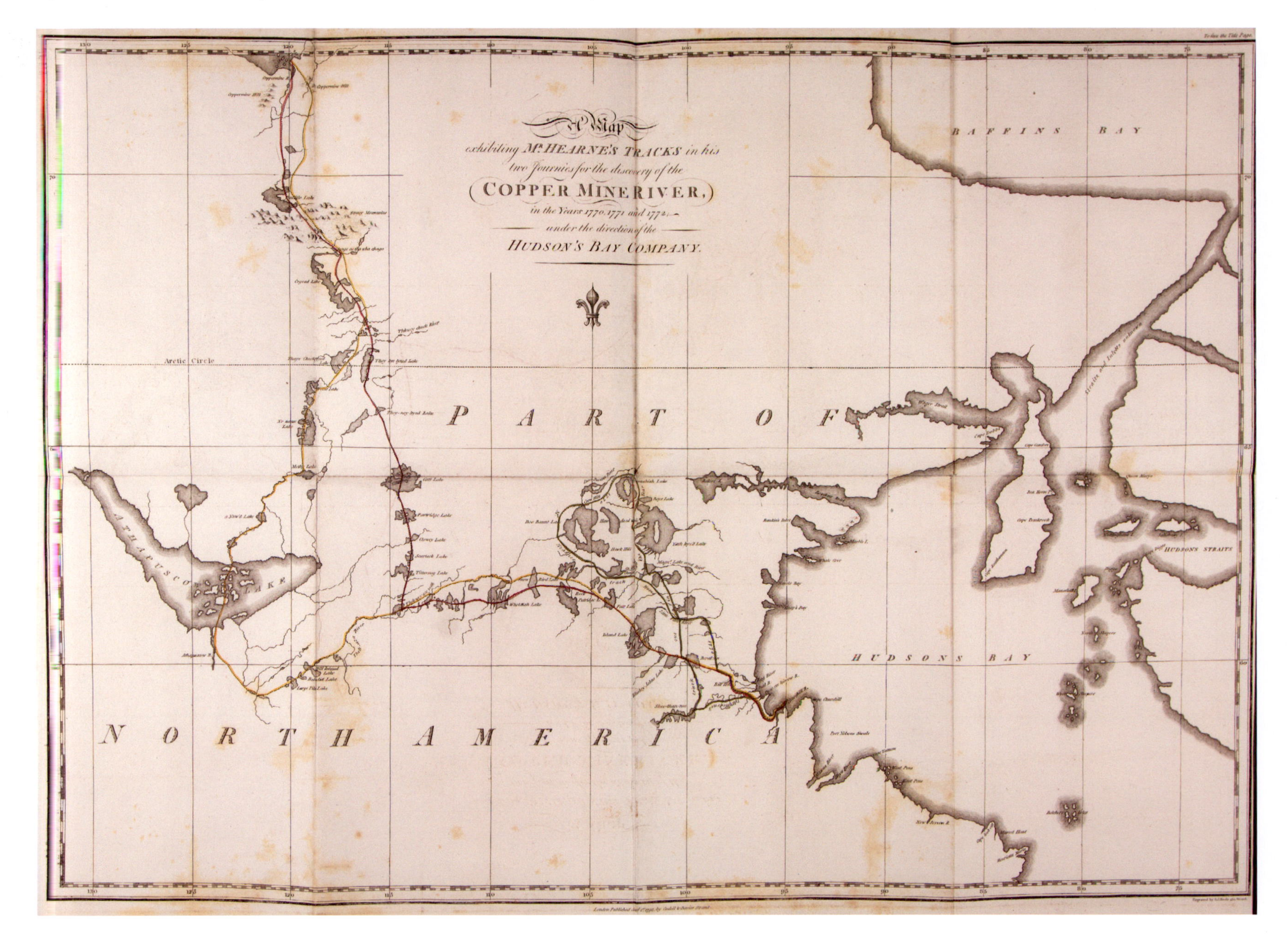

A MAP OF AMERICA,
between Latitudes 40 and 70 NORTH, and Longitudes 45
EXHIBITING MACKENZIE's
From Montreal to Fort Chipewyan & from thence to the
In 1789, & to the West Pacific Ocean in 1793.
NORTH PACIFIC OCEAN
BHERINGS STRAIT
Bristol Bay
STRAIT OF CHELEKHOF
COOKS INLET
Admiralty Bay
QUEEN CHARLOTTES I.
Queen Charlottes Sound
Cape Blanco
Trinidad
THE SEA
Quarrellers
HARE IND.
MOUNTAIN IND
NATHANA IND.
STRONGHOW INDIANS
INLAND IND.
BEAVER IND.
SLAVE LAKE
Fort Chipewyan
COPPER INDIANS
Buffaloe Lake
Chesterfield Inlet
BLOOD IND.
BLACKFOOT IND.
FALL IND.
CATTANAHOWES
LAKE WINNIPIC

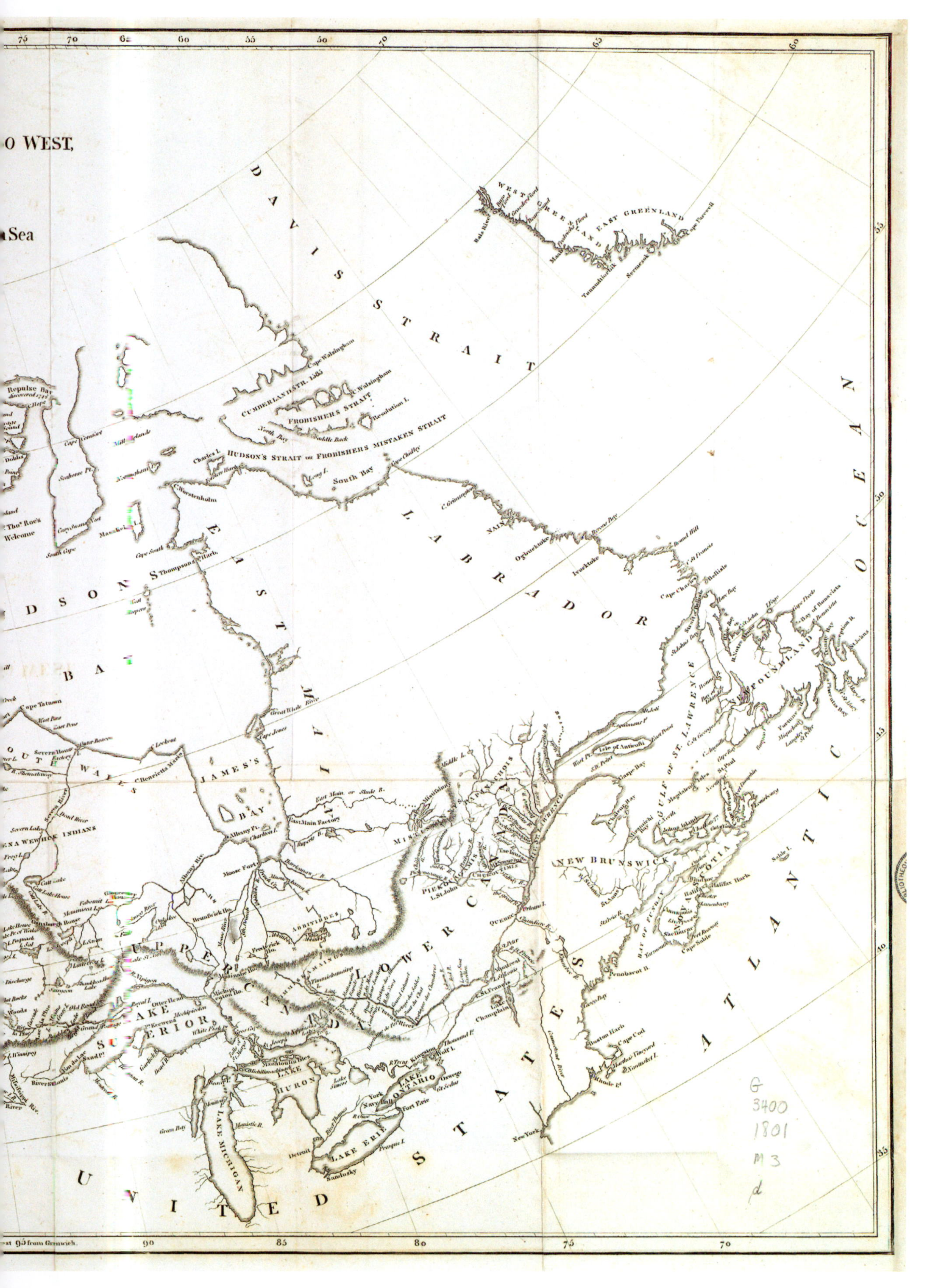

Map of Alexander Mackenzie's voyages, 1801
In June 1789, Alexander Mackenzie began a voyage of exploration on a "Great River," today called the Mackenzie River. Four French Canadians and three Indians accompanied him on the expedition. For a number of days, the explorers made a rapid advance westward. On the 15th day, however, the river suddenly turned northward. The voyagers reached not the Pacific, as they had hoped, but the shores of the Arctic Ocean. Mackenzie was not a man to admit defeat. No doubt planning a new voyage in the west, he went to London to further his knowledge and acquire various instruments for locating his position: compass, sextant, chronometer, ephemeris, and telescope. He returned to Canada with the firm intention of finding a way to the Pacific. This time, he wanted to climb the Rocky Mountains farther south, beyond the Peace River. He left Fort de la Fourche (Peace River Landing) on May 9, 1793, a chronometer in his hand to give the time of his departure. This instrument enabled him to make longitudinal readings so that he would know how much farther he had to go to reach the Pacific, whose longitude had just been calculated. With Mackenzie were two Indian hunters and interpreters who were so useful that he constantly worried that they would abandon the expedition. There were also a number of French Canadians: Joseph Landry, Charles Doucette, François Beaulieu, Baptiste Bisson, François Courtois, and Jacques Beauchamp. These "pork-eaters," as Mackenzie dubbed them, were so skilful at paddling their canoes that they almost never had accidents. But the route that they took, through inhospitable terrain, put their talents to the test. The rivers were tumultuous, full of reefs and waterfalls, while their banks were often steep and rocky. When they were at a very high altitude, near the snow line, the explorers had to leave the river and climb even higher through the forest, carrying canoes and supplies, to find another river flowing west toward the Pacific. They met a myriad of obstacles: heat, cold, mosquitoes, drought, fog, rain, sleet. The secret of their courage was revealed by Mackenzie: "A kettle of wild rice, sweetened with sugar ... with their usual regale of rum, soon renewed that courage which disdained all obstacles that threatened our progress." Fortunately, along the way they met Indians who told them how to proceed; one old man drew, on a large piece of bark, the river running southeast, its tributaries, waterfalls, reefs, and portages through the mountains. Mackenzie described a number of the peoples who he encountered: Cree, Tsattine, Sekani, Kluskus, Bella Coola, and Bella Bella. After days of travel, the exhausted men finally reached the mouth of the Fraser River (which Mackenzie thought was the Columbia). Mackenzie left a mark of their passage on the coast of the Pacific, a sort of graffito in large letters on the rock where he and his men slept: "Alexander Mackenzie, from Canada, by land, 22 July 1793." After he retired, in 1801, Mackenzie published *Voyages from Montreal*, which became a bestseller. Accompanying the book was a map titled *A Map of America* showing his two main voyages in great detail so that readers could better follow his account.

The Northwest Territories by David Thompson, 1814
In 1814, two immense maps were published simultaneously that up the major explorations leading to the Pacific, the ultimate goal for more than three centuries. One, made by William Clark, presented a synthesis of the expedition that he and Meriwether Lewis had led from 1804 to 1806, enhanced with data gathered from later voyagers, including the indomitable Georges Drouillard. Clark's map was engraved and widely distributed. The other, made by David Thompson, was given to the North West Company; it was hung in full view in the Kaministiquia post, where it was used for more than a century by voyagers setting out for the West. The preparation of this gigantic map capped 25 years of exploration west of Lake Superior. Thompson, initiated by Samuel Hearne, trained by Philip Turnor, seconded by the Métis woman Charlotte Small, and accompanied by Indian and Métis guides and French-Canadian voyagers including Augustin Boisverd and Michel Boulard, walked, surveyed, and mapped the northwestern part of North America. On July 15, 1811, he came full circle, reaching the Pacific. When he returned to Montreal in September 1812, "weary of these constant and exhausting voyages," he decided to settle down and devote himself to less tiring activities. Thompson had accumulated a great deal of knowledge, which he distilled in a single map covering the country from Lake Superior to the Pacific Ocean. The reference points were SLAVE LAKE, the Rockies, and the great waterways, including the Athabasca, Fraser, and Yellowstone rivers. Concerned, with good reason, about the problem of the border with the United States, he clearly designated the position of the Red River and situated the region where the source of the Mississippi could be found. Long ignored, today acclaimed, Thompson was the greatest of explorers. His 1814 map is justifiably the pride of the Archives of Ontario. One can only admire everything that it represents: encounters with First Nations, French-Canadian voyagers, and British explorers. It was they who made North America a welcoming and promising land.

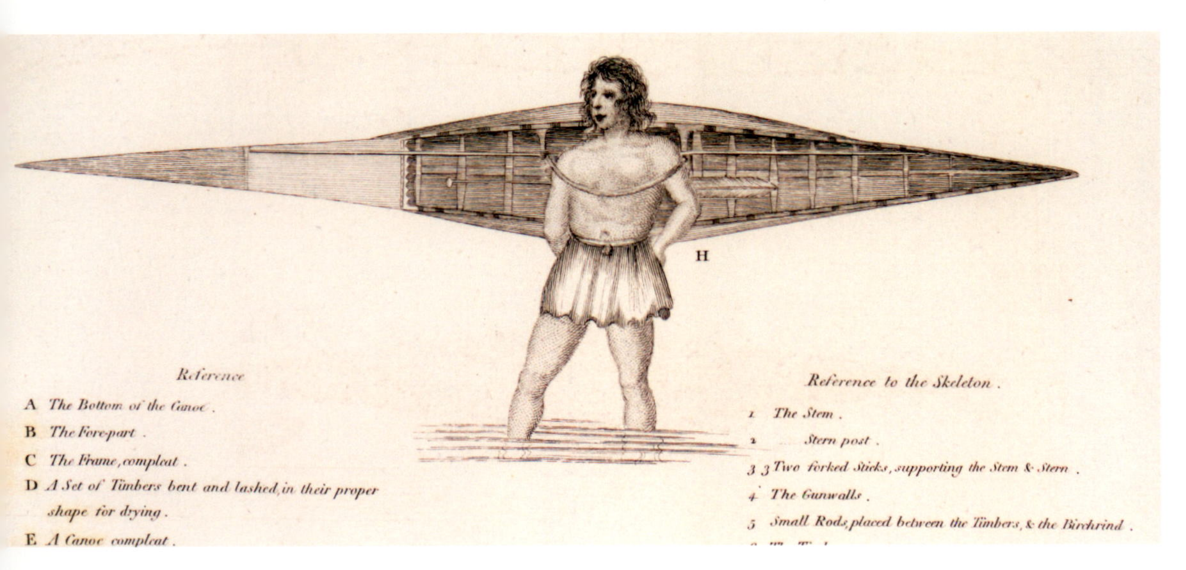

Chipewyan canoe
Engraving taken from Samuel Hearne's journal. The author, who had spent much time observing the Indians, emphasizes here the great flexibility offered by the type of canoe that the Chipewyans used. This boat, light enough to be transported by a single man, could cross rapids and travel long distances, both summer and winter.

Valley, there was no question of being confined within the borders of the small "Province of Quebec" created in 1763. Shaken up by Pontiac's revolt and the jolts of the American Revolution, traders formed a multitude of small companies and then joined forces in the winter of 1783–84 to form the North West Company.

Unaware of the monopoly and rights of the Hudson's Bay Company, a number of Montreal merchants had reached the Northwest region. The impetuous Peter Pond, originally from New England, worked first in the Upper Mississippi, where the merchants of New Orleans were prowling. There, Pond learned about trading, getting along with the Indians, and dealing with the competition. Step by step, he travelled as far as Lake Athabasca, where he crossed paths with Indians heading for Fort Prince of Wales. They still had a long and exhausting way to go. Pond offered to trade for their enormous cargoes of furs. And this was how, in 1779, Pond established a settlement, Fort Chipewyan, west of Lake Athabasca, which would soon serve as general headquarters for operations of the agents of the North West Company.

Once they were there, the agents took the opportunity to enhance their geographic knowledge of the West. One of these men, Alexander Mackenzie, who had been Peter Pond's second in command, launched the first attempt to reach the Pacific Ocean. He reached the Beaufort Sea in 1789, then, leaving from Fort Chipewyan on October 10, 1792, arrived at his intended destination. He returned to the fort on August 24, 1793, bringing all his men back safe and sound. These included Joseph Landry and Charles Ducette, who had been part of the previous voyage.

Another champion of the Northwest reached the Pacific a few years later. David Thompson had received an education through apprenticeship while employed by the Hudson's Bay Company. During the winter of 1789–90, he and Peter Fidler had learned about mathematics, surveying, and astronomy from Philip Turnor, the first surveyor hired by the company to work in the Northwest. Over several years, this trio accumulated a considerable mass of information that was compiled in maps and sent to London. In 1795, the London cartographer Aaron Arrowsmith made a spectacular synthesis of these data in the form of one of the first maps to represent all of North America.

Turnor and Fidler had first been hired by the rival North West Company to find locations for posts and to look for a rapid, navigable route from Hudson Bay to Lake Athabasca and Great Slave Lake. Fidler remained in the region, where he married a Cree woman in 1794. The Hudson's Bay Company had finally relaxed its rules concerning mixed marriages, realizing that such marriages were in fact a major asset for the Canadians, who unhesitatingly moved into Indian communities and found wives. This was a way of building trust, facilitating trade, and making life easier and more pleasant.

In 1799, Thompson followed suit, marrying a young Métis woman whose father had been a partner in the North West Company. Two years before, he had left the Hudson's Bay Company to work for its rival. The North West Company put him to work surveying west of Lake of the Woods. More interested in this work than in trading, Thompson proved to be both skilled and zealous. He came to believe that the source of the Mississippi was Turtle Lake. This information was to become vitally important in 1803 with the acquisition by the United States of the Louisiana territory – that is, from the west bank of the Mississippi River and its western tributaries to their sources, including the Mississippi itself. This is how the border between Canada and the United States came to be established. The 49th parallel had not been drawn by chance; the line roughly followed a watershed in the centre of the continent (between the Gulf of Mexico and Hudson Bay basins) from the Great Lakes to the Rocky Mountains. Farther west, to forestall disputes, it was extended arbitrarily to the Pacific Ocean.

Main sources

HEARNE, Samuel. *Le piéton du Grand Nord: première traversée de la toundra canadienne (1769-1772).* Edited by Marie-Hélène FRAÏSSÉ. Paris: Éditions Payot & Rivages, 2002. — MACKENZIE, Alexander. *Voyages from Montreal on the River St. Laurence through the Continent of North America to the Frozen and Pacific Oceans in the Years 1789 and 1793: With a Preliminary Account of the Rise, Progress, and Present State of the Fur Trade of that Country.* London: R. Noble, 1801. — MASSON, L .R. *Les bourgeois de la Compagnie du Nord-Ouest: récits de voyages, lettres et rapports inédits relatifs au Nord-Ouest canadien, publiés avec une esquisse historique et des annotations*, 2 vols. Quebec City: Imprimerie générale A. Côté, 1889. — NISBET, Jack. *Sources of the River: Tracking David Thompson across Western North America.* Seattle: Sasquatch Books, 1994. — THOMPSON, David, *Columbia Journals.* Edited by Barbara Belyea. Montreal: McGill-Queen's University Press, 1994. Clearly the best edition of Thompson's journal.

The Louisiana Purchase

LEWIS AND CLARK

THE WATERS OF THE MISSISSIPPI RIVER were "beautiful and clear" until it merged with the Missouri, observed Chevalier de Rémonville in the early eighteenth century. Then they became "cloudy and silted."

In 1719–20, the Jesuit Pierre-François-Xavier Charlevoix was sent by the duc d'Orléans to "examine the many rumours regarding the existence and geographic situation of a Western Sea" separating North America from Asia. When he returned to France in 1723, Charlevoix concluded that the best way to find this sea was "to go up the Missouri, whose source cannot be far from the sea. All the Savages ... with one voice have assured me of this." The explorer Véniard de Bourgmond might have been able to correct some of this information; ten years earlier, he had gone up the Missouri about 1,000 kilometres and knew that he had stopped far from its source.

None of the voyagers who descended the Mississippi failed to mention the "considerable river" that flowed from the west. It was all the more fascinating because it led into the unknown, perhaps to the "western sea." On his remarkable map of Louisiana, dated 1718, Guillaume Delisle took obvious pleasure in showing the immense Mississippi river system, a vast region claimed by the French, who had been exploring it in every direction for half a century. After the Treaty of Utrecht (1713) had more or less amputated New France, the cartographer found some comfort in representing the English colonies, crammed between the Appalachians and the sea, east of an immense Louisiana.

Louisiana Divided

In 1762–63, a new conflict was forcing France to retrench for the last time. Étienne François Choiseul, the French minister of the navy, was ready to cede New France to England. He hoped, however, to protect his Spanish ally occupying the southern part of the continent. For his part, Louis XV wanted to please his cousin, Carlos III. At Fontainebleau in November 1762, France secretly ceded to Spain the basin west of the Mississippi, creating a vast buffer zone between its own colonies and those of England.

Peace was officially agreed to at Paris on February 10, 1763, by the four great powers of the time. Article 7 specified that the new border between the French and English possessions would be "set by a Line drawn in the middle of the Mississippi River from its source to the Iberville River, and from there to a line drawn in the middle of this river and lakes Maurepas and Pontchartrain to the sea." The French, who certainly had Delisle's map

Sioux family
Jonathan Carver's voyage account, published in 1781, was an immediate success. His portrait of regions west of Lake Michigan offered exultant prospects: "A region capable of supporting the needs of a multitude of inhabitants," he wrote. As well, he noted, a single mountain range separated this territory from the sources of the four great rivers of North America: the St. Lawrence, the Mississippi, the Bourbon (Nelson), and the Western (Oregon). He also painted a reassuring portrait of the Indians of the region. This illustration is of a man, woman, and child of the Naudowessie nation – his name for the Sioux.

The Missouri by David Thompson, 1798
Thompson was more interested in exploration and surveying than in the fur trade. In 1797, he left the Hudson's Bay Company, where he had been trained by Philip Turnor, to enter the employ of the North West Company, whose directors were more concerned with the border between British North America (the future Canada) and the United States. In the summer of 1797, Thompson was surveying west of Lake of the Woods and trying to situate on maps the various posts of the North West Company when he decided to join a group of French Canadians who were headed for Mandan country. He appreciated their joie de vivre, learned French, and discovered – and was somewhat shocked by – one of the particular attractions of their destination. Following in the footsteps of the La Vérendryes, he learned to reconnoitre and map the great detour that characterized the Missouri, six years before the Americans Lewis and Clark. On the way out, he followed the Assiniboine and Souris rivers; he returned along the Assiniboine, crossed the Red River, and thought that he had found the source of the Mississippi at Turtle Lake. In fact, he was very close to the place generally accepted as its main source, Lake Itasca, named by the explorer and geologist Henry Schoolcraft from the Latin words *veritas caput* – "true head." The map, titled *Bend of the Missouri*, was given to the cartographer Nicholas King, who had been asked by President Jefferson to make a synthesis map to be used in the preparation of Lewis and Clark's expedition. It was the president himself who wrote, "Bend of the Missouri, long 101 25'3, lat 47 32', by Mr Thomson astronomer to the NW Company 1798." In 10 months, Thompson had travelled almost 7,000 kilometres and confirmed that article 2 of the 1783 treaty could not be applied since the source of the Mississippi was quite a bit farther south than thought and the imaginary line traced, the "due west course," from Lake of the Woods, as mentioned in the treaty, could therefore not meet with the river. Finally, the 1803 Louisiana Purchase settled the question by granting the United States the basin west of the Mississippi.

in mind, if not at hand, knew exactly what they were doing. They did not want to cede New Orleans or give the British control over navigation on the Mississippi, even though they said the opposite in the treaty. The English negotiators apparently had no idea that the Treaty of Fontainebleau existed.

With the Treaty of Paris, therefore, the east bank of the Mississippi passed into British hands, at least as far as the mouth of the small Iberville River. Most of the Canadians who lived on the left bank of the great river decided to leave Fort de Chartres, Prairie du Rocher, Cahokia, and Kaskaskia and to settle in Sainte-Geneviève or in Saint-Louis, founded in 1763 at the mouth of the Missouri by two New Orleans merchants, Pierre Laclède and his adopted son, Auguste Chouteau. Other villages sprang up all around, while expeditions to the interior of the continent proliferated. Trade was intense but limited at first to the Lower Missouri region. In 1790, Jacques d'Église reached Mandan country, where he encountered traders from the north who were taking roughly the route followed by the La Vérendryes in 1738.

The Spanish, to whom the Louisiana territory had been ceded in 1762, grew concerned. The time had finally come to get to know this territory better. In 1793, they founded the Company of Discoverers and Explorers of the Missouri. A reward was offered to the first Spaniard to reach the Pacific via the Missouri. The merchants of St. Louis got organized and sent three successive expeditions to the Upper Missouri. The first, in 1794, was led by Jean-Baptiste Trudeau, a Montrealer by birth. The following year, it was the turn of Antoine Simon Lecuyer de la Jonchère, about whom little is known, and then of James Mackay, a former North West Company employee who formed a partnership with the Welshman John Thomas Evans. Trudeau and Mackay left journals, written in French, of their expeditions.

A Source of Major Watercourses in the Mountains

In spite of these laudable efforts, the western sea remained an inaccessible dream. The general surveyor of Louisiana, Antoine Soulard, used the abundant information garnered from the various attempts to extend his maps as far as a mountain range somewhere to the west. It was increasingly recognized that the sources of major rivers were to be found at higher elevations.

Robert Rogers, the indefatigable Ranger whom the Abenakis would never forget for the massacre that he perpetrated at Odanak after the battle of the Plains of Abraham, was more than a simple soldier. He was a curious man, and he dreamed of venturing toward the Pacific. In 1765, during a visit to London, he published *A Concise Account of North America,* in which he tried to convince the government to finance an expedition to the Pacific. He planned to follow a river called the Ouragan (river of the west), which probably had its source in a mountainous region. He even sent the explorer-cartographer Jonathan Carver to make surveys in the Upper Mississippi basin in advance of his mission. Carver was one of the first Americans to explore this region. In 1781 in London, he published *Travels through the Interior Parts of North America in the Years 1766, 1767, and 1768.* He supported Rogers's theory and developed the idea of a watershed. Did Carver make all of the voyages that he mentions in his book? Some doubt it, pointing out the extent to which he draws on the accounts of Hennepin, Lahontan, and Charlevoix. In any case, Carver strengthened the image of the "good Savage" that spread through Europe and popularized the word Oregon.

Jefferson's Great Project

All of these sources and many others, such as those compiled by LePage du Pratz in his *Histoire de la Louisiane,* were carefully studied by Thomas Jefferson. As a Virginian plantation owner, Jefferson naturally looked to the west for land. When it gained independence in 1783, the United States pushed beyond the limits of the Appalachians, wiped out the immense Indian reservation created by the British in 1763, and expanded as far as the Mississippi River.

Jefferson, a key actor at this juncture in history, loved geography and wanted to know more about his new neighbours. He therefore encouraged exploration west of the Mississippi. He personally approached George Rogers Clark, a frontiersman and hero of the war of independence, to lead an expedition, but Clark declined and Jefferson left for Paris without being able to stir up interest in westward exploration. At the time of Jefferson's

Map made in 1803 by Nicholas King, cartographer for the American government
At the request of President Jefferson and under the supervision of the Secretary of the Treasury, Albert Gallatin, Nicholas King prepared for Meriwether Lewis a synthesis of the knowledge acquired not on the projected itinerary but on the territory north of the 45th parallel. More precisely, King had been instructed to portray the territory from 88° W to 126° W (according to the data gathered by Robert Gray, James Cook, and George Vancouver), and from 30° N to 55° N. As references, he relied not only on a synthesis map by Aaron Arrowsmith, but also on the maps by Alexander Mackenzie and David Thompson.

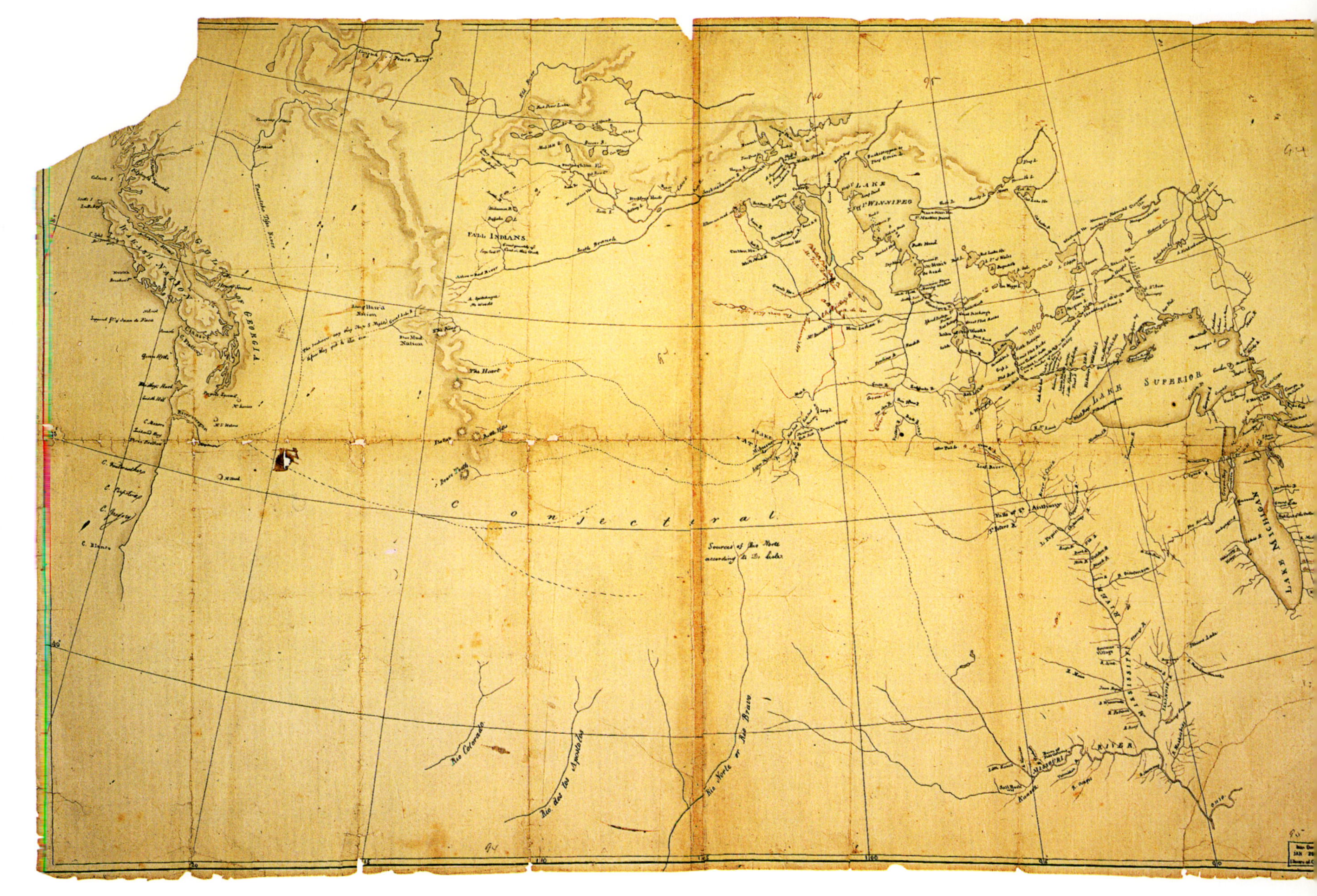

Lewis and Clark's expedition portrayed by Robert Frazer
Although it has major distortions, this map, attributed to Robert Frazer, a member of Lewis and Clark's expedition, gives an idea of the route they took. With his thin mountain range, however, Frazer seems to have forgotten the horrors of the crossing of the Rockies, both going and returning. Clark, for his part, had a tendency to exaggerate somewhat. He placed the mountains well in view (see pp. 272–73). It is worth remembering that Lewis had warned the public against Frazer's publication plans, as he was "only a private . . . entirely unacquainted with celestial observations . . . and therefore cannot possibly give any accurate information on those subjects, nor on that of geography" (Gilman, 2003: 163).

arrival in France as a minister plenipotentiary, the exploration accounts of James Cook on the west coast of North America had just been published (1784). The following year, Louis XVI sent Jean-François Galaup de Lapérouse to follow in Cook's footsteps "to see if there was not some river or some narrow gulf forming a communication, via inland lakes, with some part of Hudson Bay." The search for a northwest passage had not been forgotten.

In 1792, Cook's observations were confirmed by the American captain Robert Gray and then by George Vancouver. The mouth of the Columbia River was situated at 46°N, 124°W. Washington was at 39°N, 77°W. The distance between the two points was some 5,000 kilometres. Jefferson believed that the continent was formed symmetrically. The mountain range in the east, the Appalachians, should, he reasoned, have a counterpart in the west. Between the two, a great river received the water of rivers with sources in one or the other mountain range, while the outer slope of each range fed watercourses that flowed into the Atlantic or the Pacific. It was thus logical to think that the Missouri and Columbia rivers had their sources on one side or the other of the western mountain range.

This was the accepted theory when the exploits of a young Scottish trader ratcheted up the pressure. Alexander Mackenzie journeyed from Montreal to Lake Athabasca, then managed to make a round trip between Fort Chipewyan, on Lake Athabasca, and the Pacific in 1792–93.

Less than a week after assuming the presidency of the United States, on February 17, 1801, Jefferson proposed that a young military officer, Meriwether Lewis, whose family he knew well, join him as a private secretary. Lewis was to help Jefferson fulfil one of his electoral promises: to reduce the army's expenditures without upsetting his followers. During their private meetings, exploration of the West was no doubt a favourite subject. Lewis, after all, had already expressed his willingness to lead an expedition to the West.

In the fall of 1802, Jefferson was ready. In November, he met with the Spanish ambassador to inform him of his intention to send a scientific expedition to find a navigable route to the Pacific via the Missouri River. A few days later, he asked Congress to approve an expenditure of $2,500 "in order to extend trade outside of the United States" – understood to mean exploring west of the Mississippi. At the same time, Jefferson led negotiations to purchase "the island of New Orleans." As the French had foreseen and favoured, this city controlled passage on the Mississippi. As long as New Orleans came under Spanish authority, the Americans could work

around it, but all indications were that the French were in the process of having Louisiana – that is, the basin west of the Mississippi and the city of New Orleans – retroceded to them.

While James Monroe left for France to give a hand to Robert R. Livingston, whose various mandates included negotiating the purchase of New Orleans, Meriwether Lewis, on orders from the president, began to prepare for a major expedition. He bought a new model of gun, the Harper Ferry Model 1803, had a special ship built in Pittsburgh, recruited soldiers, and learned from eminent experts in Philadelphia how to calculate latitude and longitude, surveying methods, observation of the sky, and the rudiments of natural history, medicine, and ethnology. Lewis also gathered gifts and official documents to give to the Indians and provisions for about two years.

As the preparations continued, Lewis assessed the scope of the task ahead and suggested to the president that he take a co-captain. They agreed on William Clark, one of the brothers of George Rogers Clark, to whom Jefferson had previously proposed a similar venture. The partnership of Lewis and Clark functioned wonderfully. The success of their mission depended largely on how they complemented each other, their sense of discipline, and their intelligence and endurance.

From the beginning, luck smiled on them. What had at first seemed to Lewis to be a series of setbacks turned out to be advantages. His keelboat had been promised to him for July 20 but was not ready until August 31. At the same time, things were changing quickly in Paris. Napoleon, humiliated at Santo Domingo, decided to sell not only New Orleans but the entire Louisiana territory. Jefferson received official confirmation of this on July 14, 1803. Under the circumstances, the St. Louis authorities, not knowing which way to turn, refused to let anyone go by the city. Meanwhile, Lewis did not reach St. Louis until December – too late to set sail on the Missouri before winter – and so he found it prudent to obey the authorities and wait until spring.

This delay enabled Lewis and Clark to obtain documents from the St. Louis merchants, to read the voyage accounts of Trudeau and Mackay – the former having been translated into English on Jefferson's orders, the latter by an agent on site – to examine the maps of Aaron Arrowsmith and Antoine Soulard, and to recruit pilot-interpreters such as François Labiche and Pierre Cruzatte. Labiche and Cruzatte were Métis, as was the hunter-interpreter Georges Drouillard, who had already joined the expedition. Added were about 10 employees under the direction of Jean-Baptiste Deschamps who taught the soldiers how to navigate by poling and lining.

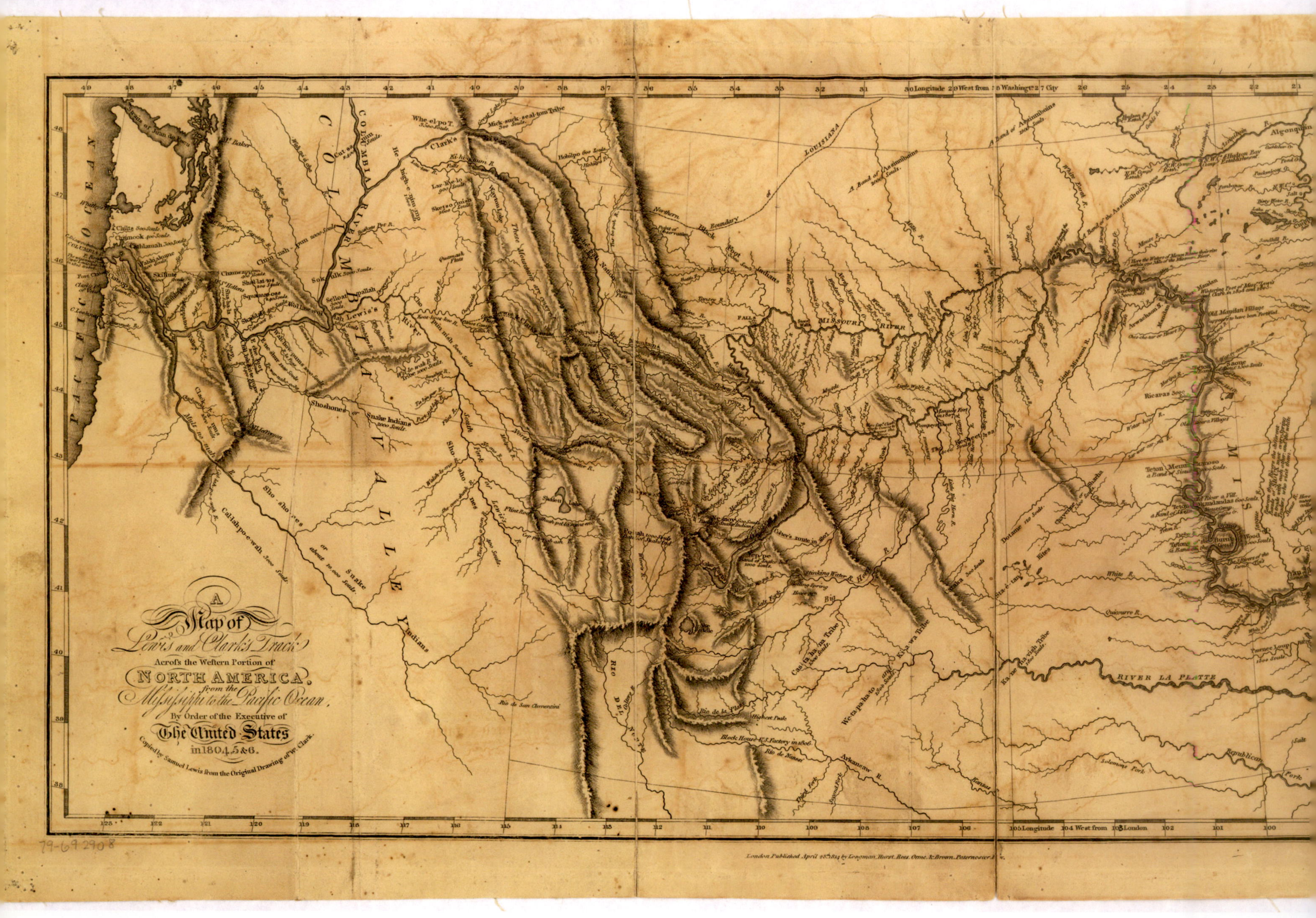

A Map of Lewis and Clark's Track across the Western Portion of North America, 1814

This map, no doubt the best-known one by William Clark, was drawn in the peace and quiet of St. Louis, three years after the return of the famous expedition that he had led with Meriwether Lewis. Intended to accompany the publication of the journals kept by the explorers, it was a synthesis of some 150 maps that Clark had drawn in the field. He now had all the leisure he needed to complement them with information gathered from other sources, including the traders and merchants who frequented the Missouri. "Copied by Samuel Lewis from the original drawing of WM Clark," the map was probably made in 1809–10. It is dated 1814, however – the date of the edition made by Nicholas Biddle and Paul Allen with the collaboration of Clark, since Meriwether Lewis, who was supposed to fill this role, had died in 1809 without completing the work. Just as Lewis had been a talented writer, Clark turned out to be a remarkable cartographer. He had taken extreme care in making his surveys, inscribing his position readings, noting the distances travelled, and recording the altitudes of the mountains and waterfalls encountered. He had only basic military training, but he was inventive. For instance, he developed a method that enabled him to establish with astonishing accuracy the height of the five successive waterfalls on the Missouri. Clark's many maps were filled with information of all sorts. On his 1814 map, he named the main rivers up to the Rockies, blithely baptizing them with American names. In short, he took possession of this part of the country. The Indian tribes were indicated with an assessment of their populations, still extremely small. Abandoned Indian villages were marked. Evidently, epidemics had made their indescribable ravages, as confirmed in the journals kept by the explorers. Clark had not forgotten the punishing crossing of the Rocky Mountains, at the foot of which was the source of the Missouri, fed by three smaller rivers – a location called the Three Forks – which he named Gallatin (the farthest east), Madison, and Jefferson. Naturally, he named the river that looked the largest after the president. Contrary to what Jefferson, a self-taught geographer, was hoping, the source of the Missouri was quite distant from that of the Columbia River, which in fact flowed from the north. Once the many mountain ranges were crossed, the Columbia could, however, be reached via a succession of two large rivers: the Kooskooke (Clearwater) and the Lewis (Snake). His map also suggested that the Rockies, so threatening as a whole, were only one range opposite the source of the Lewis River, while the route taken by the expeditionary party, indicated by a dotted line, had crossed five or six ranges, at least one of which was snow-covered. Clark thus seems to have had the intuition that there was a pass somewhere farther south, and in fact he pointed out the Platte River, which was not navigable but whose shores were firm enough to accommodate wagons in the summertime, making it the point of departure for the famous Oregon Trail. Aware of the importance of the northern border of the Louisiana Purchase, Clark drew a "Northern Boundary of Louisiana" that ran off the edge of the map near the supposed source of the wide Milk River, a tributary of the Missouri. He was not far from the truth: the Canada–United States border would be drawn nearby.

Lewis and Clark's Expedition

Lewis and Clark left St. Charles, a small village slightly northwest of St. Louis, on May 21, 1804, and returned to their starting point on September 2, 1806. Their expedition is now among the founding myths in United States history. It was organized just after the United States doubled in area, from 540 million acres in 1783 to over a billion acres in 1803, due to the Louisiana Purchase. The western route that it opened up, though not passable at first, would soon become synonymous with implacable western expansion, thus feeding the American dream and paving the way for "manifest destiny."

The first part of their voyage took Lewis and Clark to the Upper Missouri, where the Mandans, Hidatsas, and a handful of Amahamis cohabited. The men were to spend their first winter there. After two months of difficult navigation, at a pace of 20 to 35 kilometres per day – sometimes as many as 40 kilometres, sometimes as few as 10 – the expeditionary force of 40 men still had not met any Indians except for the few who were accompanying traders descending the Missouri with cargos of pelts. Exactly two months after leaving St. Charles, having travelled 1,000 kilometres, the party reached the mouth of the Platte River. A few days later, Drouillard, a man trusted by both leaders, returned from the hunt with a Missouri Indian. His people were nearby, but they had been decimated by recent epidemics and the survivors, too few to form a village, had taken refuge with the Otoes. A little farther north, the Omahas had suffered the same fate as the Arikaras and the Mandans, who, it was learned, now inhabited only three villages out of 18 and two out of five, respectively. Lewis and Clark reached these survivors in late October – they had travelled some 2,000 kilometres in five months. They observed that the Indian nations wherever they travelled had been swept away by smallpox epidemics.

During the winter, Lewis and Clark met traders from the North West Company, including young François-Antoine Larocque, who had written an interesting travel journal. After learning, mainly through less-than-friendly contact with the Teton Sioux, of the importance of having interpreters, Lewis and Clark recruited Toussaint Charbonneau and his Shoshone wife, Sacagawea, as well as Baptiste Lepage, who had already travelled slightly farther up the Missouri.

In the spring, a small detachment commanded by Corporal Warfington, carrying specimens of all sorts, was sent back to St. Louis in a keelboat captained by the pilot and interpreter Joseph Gravelines (recruited the previous fall from the Arikaras), assisted by several employees. Meanwhile, Lewis and Clark continued their journey with "seven pirogues," as Larocque put it. They hoped to have an easy time reaching the Great Falls about which the Indians had told them. It took them two months to cross the 1,600 kilometres in question – two months without encountering any Indians. They might well have missed, with relief, meeting up with Crows and Flatheads, even though these tribes should have been in the vicinity, but not the Shoshones. This nation of horse merchants was key to getting through the mountains. In fact, Lewis and Clark had Sacagawea with them precisely to facilitate negotiations. Sacagawea finally did play the role expected of her, though the chain of translation was not particularly simple: from English to French to Hidatsa to Shoshone – that is, from Lewis to Labiche or Drouillard to Charbonneau to Sacagawea and Chief Cameahwait (who turned out to be Sacagawea's brother, kidnapped from his people during a Hidatsa raid long before).

In 1998, the Dollar Coin Design Advisory Committee recommended that the new American dollar bear an image of Sacagawea, "the Native American Woman who accompanied Lewis and Clark on their exploration of the American West." One of the indisputable merits of this Indian woman was that she carried her newborn baby during the long and difficult journey. It was decided to portray her with her child, Jean-Baptiste, the son of the interpreter Toussaint Charbonneau, who had been recruited in the Mandan village during the winter of 1804–05. Father and son were to live astonishing lives.

Missouri Territory 1814

In 1803, Napoleon ceded to the Americans, for a pittance, the basin west of the Mississippi – a vast territory extending to the source of the Mississippi and its western tributaries – in a deal that became known as the Louisiana Purchase. It took years to establish the territory's borders in a climate of dispute and controversy. Neither the source of the Mississippi nor those of its northern tributaries, such as the Milk and Maria rivers, were known for certain. Lewis and Clark had received instructions to find and establish the source, but they did not succeed with this mission. On this 1814 map, Mathew Carey indicated to the north a curved line that started at Lake of the Woods, passed south of the Winnipeg River, skirted several geographic features, and stopped north of Mount Reiner (Rainier), a little north of 47° N, the "probable North Boundary of the Missouri Territory," he noted. The Louisiana Territory had been renamed the Missouri Territory once the state of Louisiana, which can be seen at the mouth of the Mississippi, was created in 1812. Another element in dispute was the western border of the Louisiana Purchase between 30° and 35°. President Jefferson dared to claim as far as the Rio Grande, but the American negotiators quickly retreated to the Colorado River. In reality, the Louisiana Purchase seemed, rather, to resemble the rough sketch on the opposite page and conceded to the Americans a strip of land between the Sabine River and the Mississippi.

In pain and misery, the party, with the help of the Shoshone guides, climbed interminable mountain ranges. The Salish, Flatheads, Nez-Percés, and Wallas Wallas took turns helping the explorers. Once they arrived at the Clearwater River, the Americans learned from the Indians how to make canoes by hollowing out large tree trunks not with an axe but with fire.

Starting in mid-October 1805, Lewis and Clark noted in their journals that the Indians they met had in their possession objects of European origin, obtained no doubt from their counterparts on the coast. The Pacific Ocean was not far off. They reached it in mid-November.

From the Great Falls of the Missouri, they had travelled 1,500 kilometres, including 230 on terrifying mountain trails and another 600 assisted by Wassapans, Yakimas, and Umatillas. They had reached the Pacific Ocean after a gruelling journey that took 18 months and covered some 6,000 kilometres. The winter spent near the Clatsops and Chinooks was long and dismal. Drouillard, a good hunter, provided food, Lewis treated his men's venereal disease, and Clark worked on his maps.

The return journey was much faster, though punctuated by a murderous confrontation between Lewis and a band of Blackfoot Indians. The Missouri River represented about 40 percent of the route. Paddling downstream, the party sometimes travelled 120 to 130 kilometres per day. Lewis and Clark brought back all members of the expedition safe and sound except for Sergeant Charles Floyd, who had died at the beginning of the journey from acute appendicitis.

The route taken by Lewis and Clark, like that taken by Mackenzie, was not really feasible. People would have to come to terms with the idea that a navigable route simply did not exist. Paradoxically, it was a watercourse that indicated what means could be used – one of the tributaries of the Missouri, the Platte River, which flowed from the west. The Platte was shallow and very wide in spots, and toward the end of summer its banks became suitable for wheeled vehicles. This became the Oregon Trail, the route taken by westward-bound pioneers.

Lewis and Clark did not find the passage they sought, but they broke an imaginary barrier and returned with much information. They revealed to enthusiastic naturalists the existence of 178 plants and 122 animals unknown until then. Clark's numerous maps were used as references for many years. The journals of both captains and those of their officers offer a wealth of material that to this day has not been fully exploited. It took a long time for anthropologists and historians to plumb their depths, even though accounts of early contacts between Indians and whites are quite rare. In fact, the journals were not published until much later. Some parts went astray for a time; all the notes concerning the Indians still have not been found.

Nevertheless, an enormous amount of material is available. Reuben Gold Thwaites edited an eight-volume scholarly edition published to celebrate the centenary of the expedition. Gary Moulton recently published twice as much material. These are monumental works, models of the genre and worthy tributes to a remarkable expedition that concluded three centuries of adventure, exploration, determination, and dreams.

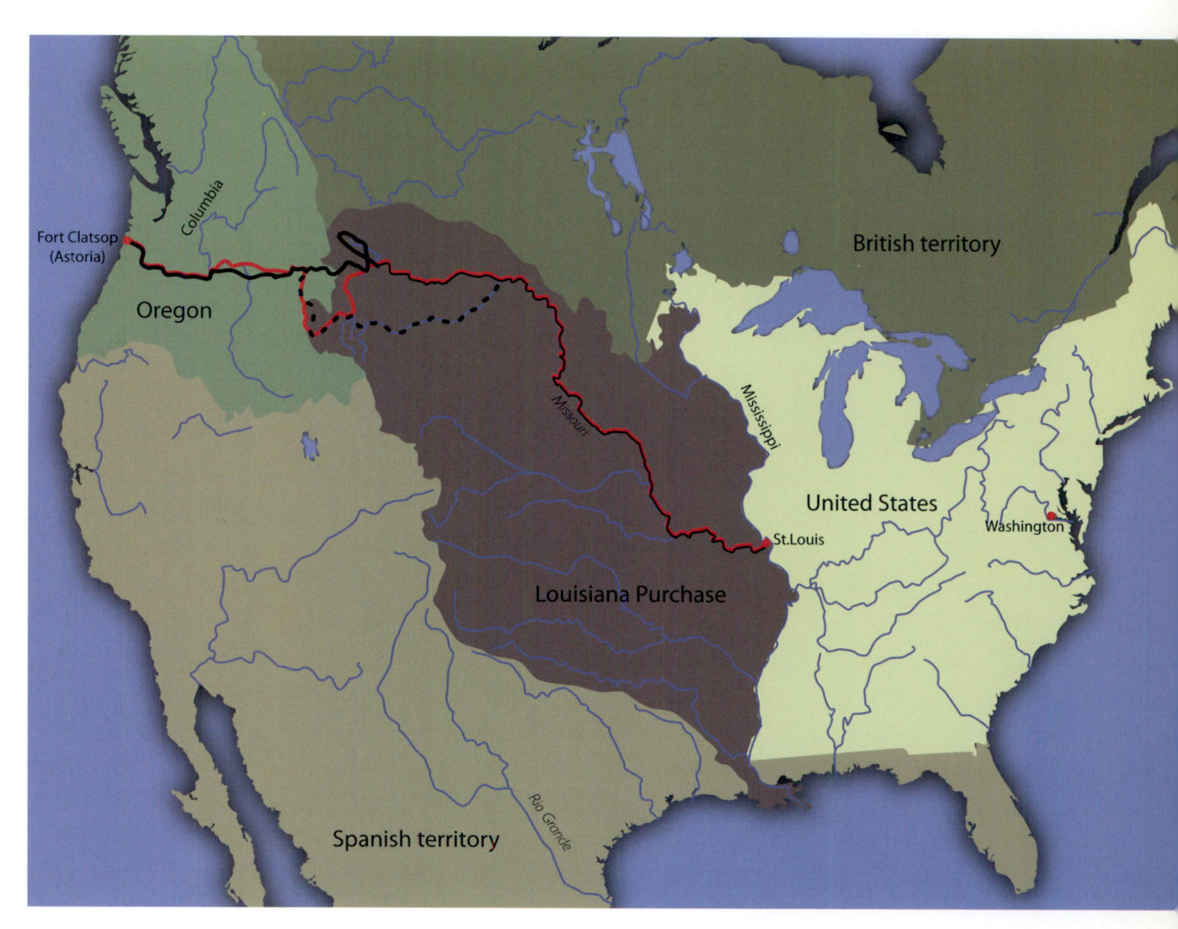

The Louisiana Purchase and the expedition of Lewis and Clark
The United States was expanding with the acquisition of the western basin of the Mississippi. Shown here is the route taken by Lewis and Clark from 1804 to 1806: from St. Louis to Fort Clatsop, then to the mouth of the Columbia River.

Main sources

Appleman, Roy E. *Lewis and Clark: Historic Places Associated with Their Transcontinental Exploration (1804–06).* Washington, D.C.: U.S. National Park Service, 1975. — Chaloult, Michel. *Les Canadiens de l'expédition Lewis et Clark, 1804-1806: la traversée du continent.* Sillery: Septentrion, 2003. — Foucrier-Binda, Annick. *Meriwether Lewis & William Clark: la traversée d'un continent, 1803-1806.* Paris: M. Houdiard, 2000. — Moulton, Gary E. (ed.), *The Journals of the Lewis and Clark Expedition, 13 vols.* Lincoln: University of Nebraska Press, 1983–2001. — Ronda, James P. *Lewis and Clark among the Indians.* Lincoln: University of Nebraska Press, 1984. — Thwaites, Reuben Gold (ed.). *Original Journals of the Lewis and Clark Expedition, 1804–1806, 8 vols.* New York: Antiquarian Press, 1959. Although Moulton's edition, makes obvious progress, Thwaites's edition is still interesting. — Vaugeois, Denis. *America, 1803–1853: The Lewis and Clark Expedition and the Dawn of a New Power.* Translated by Jane Brierley. Montreal: Véhicule Press, 2005. This book contains a list of top-notch essays, including those by James Ronda and John Logan Allen, and collections of documents assembled by Donald Jackson and A. P. Nasatir. As an introduction to Lewis and Clark's epic journey, the stimulating essay by Stephen Ambrose, *Undaunted Courage: Meriwether Lewis, Thomas Jefferson, and the Opening of the American West* (New York: Simon & Schuster, 1996) is highly recommended.

Following pages
Map of the United States with the Contiguous British & Spanish Possessions by John Melish, Philadelphia, 1816
In the nineteenth century, some felt that the American nation had an obvious fate: to expand and conquer the entire continent of North America. This map by the publisher John Melish can be seen as a precursor of the ideology that has been called Manifest Destiny. The United States is portrayed as stretching from ocean to ocean (due in part to the voyages of Lewis and Clark), which Melish considered a notable innovation for a large-format map. Commenting on his work, he noted, "The map so constructed, shows at a glance the whole extent of the United States territory from sea to sea; and, in tracing the probable expansion of the human race from east to west, the mind finds an agreeable resting place on its western limits."

Q. Charlottes I.
LAKE WINIPIC
LAKE MANITOBA
Mandan Villages
M I S S O U R I
Highest Peak
NEW MEXICO
BOLSON DE MAPIMI
Apaches Indians
GULF OF CALIFORNIA
GUADALAXARA
S. LOUIS
GULF
Yellow Stone

MAP
of the
United States
with the contiguous
BRITISH & SPANISH POSSESSIONS
Compiled from the latest & best Authorities
BY
John Melish
ENTERED according to Act of CONGRESS the 6th day of June 1816.
Published by John Melish Philadelphia.
WEST INDIES
CARIBBEAN SEA
LABRADOR
GULF OF ST. LAURENCE
UPPER CANADA
LOWER CANADA
LAKE SUPERIOR
LAKE MICHIGAN
LAKE HURON
LAKE ONTARIO
LAKE ERIE
MICHIGAN TERR.
NEW YORK
ILLINOIS TERRITORY
INDIANA
GEORGIA
ST. GEORGES BANK
SABLE BANK
BERMUDAS ISLANDS
SCALE of MILES
GREAT BAHAMA BANK
GULF OF FLORIDA
Supposed Limits of the Gulph Stream
STATISTICAL TABLE

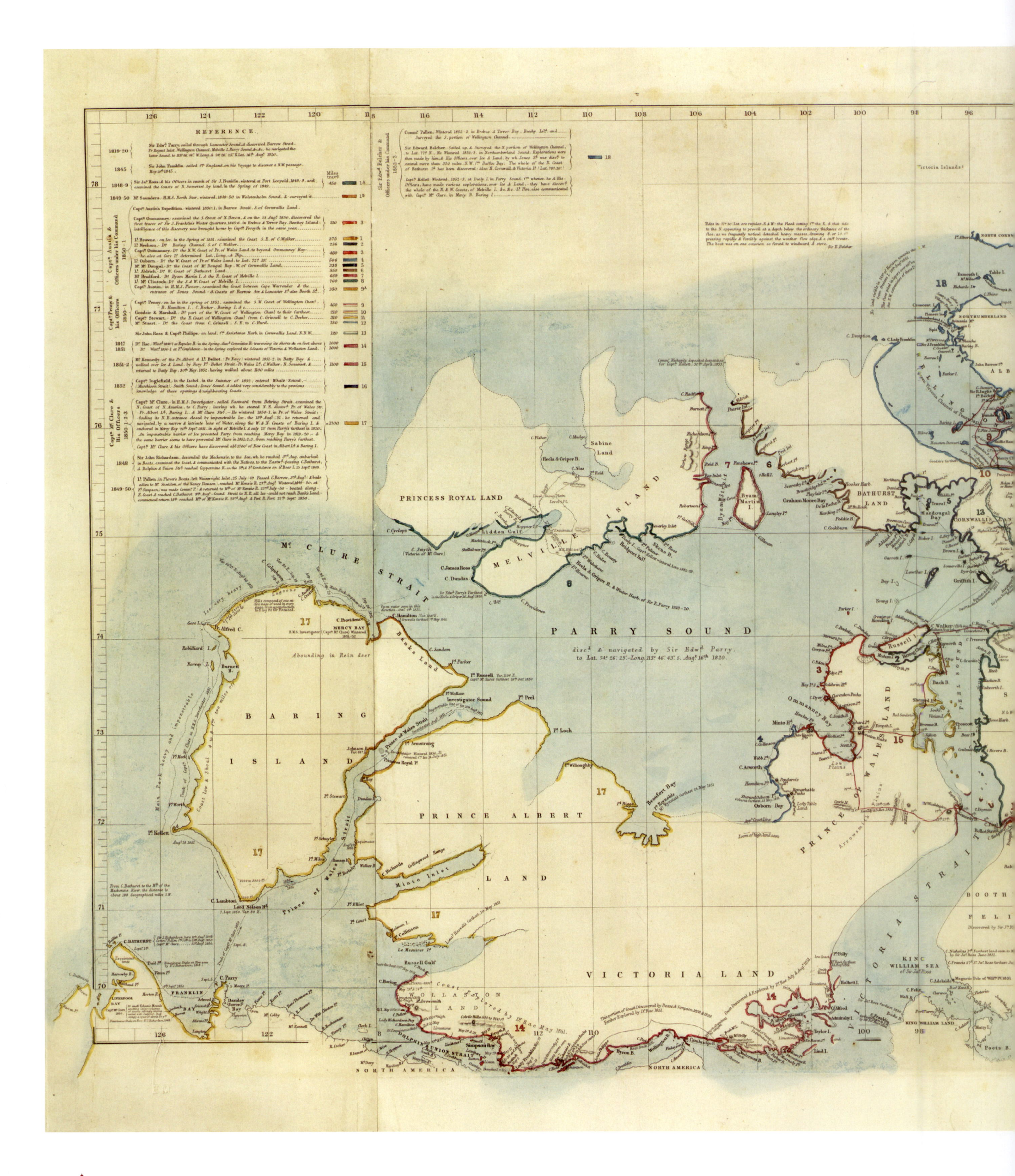
REFERENCE
PRINCESS ROYAL LAND
Sabine Land
MELVILLE ISLAND
Mc CLURE STRAIT
PARRY SOUND
discd & navigated by Sir Edwd Parry,
to Lat. 74° 26' 25" – Long. 113° 46' 43" 5. Aug. 16th 1820.
Abounding in Rein deer
Banks Land
BARING ISLAND
Investigator Sound
Prince of Wales Strait
PRINCE ALBERT LAND
Minto Inlet
VICTORIA LAND
WOLLASTON LAND
Byam Martin I.
BATHURST LAND
CORNWALLIS
Ommanney Bay
PRINCE OF WALES LAND
VICTORIA STRAIT
KING WILLIAM SEA
BOOTH
FELI
Russell Gulf
DOLPHIN & UNION STRAIT
FRANKLIN BAY
LIVERPOOL BAY
C. BATHURST
NORTH AMERICA

As an Epilogue . . .

A synthesis of the principal expeditions in the Arctic Ocean by John Arrowsmith, 1852 [circa 1854]

In 1492, Christopher Columbus was en route for India when a major obstacle appeared. What to do? The Portuguese and Spanish chose to explore a southern route. The French, English, and Dutch, on the other hand, sought a passage across North America. After three centuries of exploration, Mackenzie, Lewis and Clark, Hunt, and Thompson reached the Pacific via routes that were almost impassable. A decision had to be made to trade canoes for wagons. The Platte River, though offering little possibility of navigation, became one of the first routes to the West. At the same time, the English became once again interested in the hypothesis of a northwest passage. On a map dated 1852 but updated around 1854, John Arrowsmith made a synthesis of the main expeditions since those of Edward Parry and John Franklin. It was Sir Robert John Le Mesurier-McLure, in 1851–53, who managed to find the Northwest Passage, during a voyage in search of Sir John Franklin and his crew, who had disappeared in 1847. In fact, Arrowsmith was seeking information on the geography of the Arctic zone. The employees of the Hudson's Bay Company were his main informants, along with the navigators searching for Franklin, whose names appear in a list in the upper left-hand corner of the engraving. Arrowsmith was able to complement, correct, and update his maps with information brought back by Le Mesurier-McLure in 1854. The Northwest Passage was finally crossed in 1903–06 by the Norwegian explorer Roald Amundsen. With global warming, it has become almost navigable a few weeks of the year and thus is of great strategic value today.

The Grammar of Maps

During his first voyage across the Atlantic, as his ship approached the North American coast, baron de Lahontan underwent a rather unpleasant initiation rite: sailors in blackface, dressed in tatters, threw 50 buckets of water into the faces of the voyagers after forcing them to kneel and swear on a collection of nautical charts. A few years later, in the heart of the continent, Lahontan carried in his baggage an astrolabe, a demi-circle, a number of compasses, two large watches, and brushes and drawing paper so that he could make maps of the regions that he visited. When he arrived in the country of the Gnacsitares, an Indian chief offered him a large map drawn on deerskin. Upon his return to Europe, Lahontan published his voyage accounts, in which he included maps that were "very useful and very much in compliance with the tastes of the century," showing "at a glance the true disposition of that country." Along with his keen observations and lucid descriptions of life in North America, he reminded readers, through several examples, of the importance of maps in his times. Used by navigators, army officers, Indians, and the kings of Europe, maps were indispensable artifacts that met a universal need for communication. Their identity was defined essentially by a single function: enabling people to locate themselves in their environment.

Maps take over from words when their descriptive power fails; the advantage of a map is its ability to display instantaneously a synthesis of geographic knowledge. At a glance, a map can show a parcel of land or an entire continent, including its physical and human features, without having to resort to the meandering detours of the written word. A map is a mode of communication that requires a high capacity for abstraction. Making one involves relatively complex perceptual and cognitive operations that seek, by various subterfuges, to imitate geographic reality. With its pictograms, straight and curving lines, forms, texts, and ornaments, cartography has a detailed vocabulary; it also has a grammar – a graphic code that superimposes various layers of information according to known and accepted conventions. Thus, to decode old maps, it may be useful to understand their main components.

Maps are above all scientific objects – mathematical drawings. Thus, it is not unusual to find geographers who are also mathematicians. Maps are a reduced image of the world; the graphic scale informs the viewer of the degree of miniaturization. In general, a professional cartographer will want to shrink the world in regular proportions, according to a uniform, constant scale. Before the metre was devised as a unit of length, this scale was expressed by a variety of measurements. Leagues and miles were the most common, but they had different values in France, England, Spain, and elsewhere in Europe – whence the habit, among some cartographers, of providing several different scales to satisfy all viewers.

Although globes faithfully present Earth's sphere, they are also cumbersome. Maps therefore quickly became the most practical means of representing the world. To portray a curved surface on a flat surface, cartographers invented various projections. The most famous is the Mercator projection, which distorts landmasses distant from the equator but preserves the angles in relation to the longitude and latitude, which is particularly useful for navigation. Once the projection is chosen, the cartographer sets the position of points on a canvas of imaginary lines that structure the graphic space of the map. The most frequently used grid, which serves as the viewer's main point of reference, is composed of parallels and meridians. Parallels are imaginary lines that circle Earth parallel to the equator; meridians are lines that run from pole to pole, crossing the parallels at a perpendicular angle. Once this grid is laid out, the cartographer may fix the locations for which he knows the geographic coordinates in terms of latitude and longitude. Latitude is the measurement of the distance, in degrees, between a given point and the equator. Longitude is the distance, also in degrees, between this same point and the meridian of reference. Today, the Greenwich meridian is universally recognized as the prime meridian. Between the fifteenth and nineteenth centuries, the prime meridian varied from one kingdom to the next, from one cartographer to the next; it might pass through Terceira in the Azores, the Island of Hierro in the Canary Islands, or Paris or London. Two additional parallels are frequently drawn in: the tropics of Cancer and Capricorn, where the Sun passes through its zenith at the solstices.

Sometimes, another, more complex grid was juxtaposed on or replaced the grid of latitudes and longitudes: lines of constant bearing, useful mainly for navigation. Starting from a central point, the cartographer drew a certain number (8, 16, or 32) of straight, equidistant lines called rhumb lines (the rhumb being the space between two lines). The central point, or compass rose, from which all of the lines were drawn resembled a compass card, so that the eye could travel easily back and forth from the map to the compass. For uniform coverage, a

Cartouche from the *Carte du Canada ou de la Nouvelle France*, by Guillaume Delisle, 1703
The engraver of this cartouche not only gives the title of the map, but adds beautifully rendered scenes of sermons, conversions, and scalping.

Engravers' workshop
This vignette from the *Encyclopédie des sciences* portrays the main operations involved in etching and burin engraving: varnishing the plate, blackening the varnish over flame, "biting" the plate in acid, and engraving with an etching tool or burin.

number of compass roses and rhumb lines might be drawn, creating a complex tangles of lines that are quite confusing to the modern eye but made sense to the sailors of the time. The four cardinal points (north, south, east, west) are the references points needed for navigation. To set the orientation, cartographers began to draw a fleur de lys on the north line; only in the late sixteenth century did it start to become customary to place the north at the top of the map and south at the bottom.

More than simple scientific objects, maps were also sometimes magnificent works of art, made for kings, princes, ministers, and other eminent people. Cartographers sought to please those in power in their kingdom and to obtain their patronage through dedications or well-placed coats of arms. Some also adorned their maps with rich illustrations in the centre or on the margin of the map: ships and monsters in the sea; flora, fauna, and indigenous peoples on the landmasses. The title cartouche was invented by the Italian mannerists; the equivalent of the frontispiece of an atlas or book, it became the artist's domain. More or less elaborately ornamented, depending on when the map was produced, this area was the appropriate place to describe the world through interposed symbols emblematic of North America (Indian, missionary, beaver, fish, etc.) and to show the integration of the new lands with European cultural, religious, and scientific values. In addition to the title cartouche, which generally included the title, the maker's name, the publisher's name, the edition, and sometimes who it was made for, the map might also include insets portraying a region, city, or port in greater detail.

From the beginnings of the printing press to the end of the sixteenth century, maps were printed from plates engraved in blocks of wood, then from copper plates using the burin or etching method. The engraver traced the lines using an overlay, marked towns and villages with a punch, and composed the lettering. Once it had been inked and its surface wiped clean, the copper plate was placed on the press and two cylinders, rolled by hand, pressed the plate onto a sheet of paper. When the map was completely dry, an illuminator might colour it with a paintbrush for clients ready to pay for a more expensive product. With the advent of the printing press, cartography became more accessible. It was not unusual to reprint a map with variations and updates. And like books, maps were counterfeited, whence the custom of inscribing "privilège du roi" – at the time, the equivalent of "copyright" – on them. During the Enlightenment, Chevalier Louis de Jaucourt wrote this note in Diderot's Encyclopédie: "Thanks to [engraving and intaglio printing], with a bit of taste, one may without great opulence assemble in a few chosen portfolios more engraved pieces than the wealthiest potentate may have paintings in his galleries." Some Dutch cartographers raised the art of cartographic engraving to new heights. They produced, notably, collections of maps of different parts of the world, a type of work that was called an atlas for the first time by Mercator in 1595. (Contrary to popular belief, the coining of this term was inspired not by the Greek god Atlas, but by an astronomer king of Mauritania of the same name.)

Whether engraved or manuscript, maps were also a form of writing that could serve territorial claims. Although they were intended to be a faithful reflection of geographic reality, they were also used by powers that wished to stamp their domination over a virgin continent with a Christian presence. Iconography and toponymy combined to serve the imperial cause. Certain cartographers displayed flags – Spanish, Portuguese, French, English – testifying to the primacy claimed by their king. Others took advantage of the lack of clarity of treaties to expand or contract the contours of colonial boundaries in the Americas. In times of crisis, they often had no choice but to forsake the objectivity of the Enlightenment, as they were obliged to please the sovereign and his subjects. In short, whatever else it was, the map was a formidable symbolic weapon.

List of Maps and Illustrations

ACJCF	Archives de la Compagnie de Jésus. Province du Canada-français (Saint-Jérôme)	GM	Gilcrease Museum (Tulsa, Oklahoma)
ACPSS	Archives de la Compagnie des Prêtres de Saint-Sulpice (Paris)	HL	Huntington Library (San Marino, Californie)
APO	Archives publiques de l'Ontario	LC	Library of Congress (Washington)
LAC	Library and Archives Canada (Ottawa)	MC	Musée de la civilisation (Quebec)
BAnQ	Bibliothèque et Archives nationales du Québec	MG	McGill University Library, Rare Books and Special Collections Division (Montreal)
BAV	Biblioteca Apostolica Vaticana (Vatican)	MN	Museo naval (Madrid)
BL	British Library (London)	MQB	Musée du quai Branly (Paris)
BnF	Bibliothèque nationale de France	SM	Stewart Museum (Montreal)
BRB	Bibliothèque royale de Belgique (Bruxelles)	NL	Newberry Library (Chicago)
BSHM	Bibliothèque du service historique de la Marine (Vincennes)	NYPL	New York Public Library, Rare Books Division (New York)
CAOM	Centre des archives d'outre-mer (Aix-en-Provence)	SHAT	Service historique de l'armée de terre (Vincennes)
CHAN	Centre historique des Archives nationales (Paris)	UKHO	United Kingdom Hydrographic Office (Taunton)
CL	William L. Clements Library (Ann Arbor, Michigan)	UdeM	Université de Montréal, Livres rares et collections spéciales (Montreal)
ES	Éditions du Septentrion (Quebec)		

p. 149 — *Sauvage du Canada,* drawing by Claude-Louis Desrais, engraving by Jean-Marie Mixelle, in *Costumes civils actuels de tous les peuples connus* by Sylvain Maréchal (Paris: Pavard, 1788), vol. 5
BAnQ, RES BE 272

p. 150 — *Le lac Ontario avec les lieux circonvoisins & particulierement les cinq nations iroquoises* (1688), manuscript map
BSHM, Recueil 67 no 67

p. 151 — *[Plan des forts et des attaques du village des Chicachas],* by Dupin de Belugard (Rochefort, 1736), manuscript map
CAOM, 04DFC 50B

pp. 152–53 — *Carte du fort Rozalie des Natchez françois ou l'on voit la situation des concessions et habitations telles qu'elles etoient avant le massacre arrivé le 29 novembre 1729,* attributed to Dumont de Montigny (ca. 1740), manuscript map
CHAN, Cartes et plans, N III Louisiane 1 pièce 2

p. 154 — *Guerrier Renard,* watercolour drawing (1731)
BnF, Est. Of-4a-Fol.

pp. 156–57 — *A New Map of the English Empire in America,* by John Senex (1719), engraved map
LAC, NMC 175336

p. 158 — *Carte du Canada ou les terres des François sont marquées de bleu et celle des Anglois de jaune,* by Jean-Baptiste de Couagne (1712), manuscript map
BSHM, Recueil 67 no 13

p. 159 — *[Plan of Vicinity of Detroit Fort]* (after 1701), manuscript map
BSHM, Recueil 67 no 74

p. 160 — *A New and Exact Map of the Dominions of the King of Great Britain on ye Continent of North America,* by Herman Moll (1715), engraved map
LAC, NMC 21061

p. 162 — *A New Map of Part of North America,* by Arthur Dobbs and Joseph Lafrance (London: J. Robinson, 1744), engraved map
BAnQ, RES AD 206

p. 163 — *Nouvelle carte des endroits, où l'on a taché de découvrir en 1746 & 1747 un passage par le nord-ouëst,* engraved map, in *Voyage à la baye de Hudson* by Henry Ellis (Paris: Antoine Boudet, 1749)
BAnQ, RES AF 234

p. 164, top — *Esquimaux faisant du feu et allant à la pêche de veaux marins,* engraving, in *Voyage à la baye de Hudson* by Henry Ellis (Paris: Antoine Boudet, 1749), book 2, p. 22
BAnQ, RES AF 234

p. 164, bottom — *[Eskimo woman and baby],* watercolour drawing, attributed to John White (ca. 1577)
British Museum

p. 165 — *Femme acadienne,* drawing by Claude-Louis Desrais, engraving by Jean-Marie Mixelle, in *Costumes civils actuels de tous les peuples connus* by Sylvain Maréchal (Paris: Pavard, 1788), vol. 5
BAnQ, RES BE 272

pp. 166–67 — *Plan du cours de la riviere du dauphin et du fort du Port Royal y scitué avec la banlieuë dud[it] fort à la Cadie en la Nou[ve]lle-France,* by Jean-Baptiste Labat (Acadia, 1710), manuscript map
BnF, Cartes et plans, Ge SH 133-8-6

pp. 168–69 — *Partie orientale de la Nouvelle France ou du Canada,* by Jacques-Nicolas Bellin (Nuremberg: les Héritiers d'Homan, 1755), engraved map
BAnQ, G 3400 1755 B4

p. 170 — *Partie Orientale du Canada,* by Thomas Jefferys, translated by Georges Louis Le Rouge (Paris, 1755), engraved map
BAnQ, G 3415 1755 L47

p. 172 — *Carte des nouvelles découvertes au nord de la mer du sud, tant à l'est de la Siberie et du Kamtchatka qu'à l'ouest de la Nouvelle France,* by Joseph-Nicolas Delisle and Philippe Buache (Paris, 1752), engraved map
LAC, NMC 21056

p. 173 — *Carte d'une partie du lac Supérieur avec la decouverte de la riviere depuis le grand portage A jusqu'au lac Ouinipigon,* by Christophe Dufrost de La Jemerais, copied by Gaspard-Joseph Chaussegros de Léry (Quebec, 1734), manuscript map
CAOM, COL C11E 16/fol. 317

pp. 174–75 — *Carte physique des terreins les plus élevés de la partie occidentale du Canada,* by Philippe Buache (Paris: Académie des sciences, 1754), engraved map
MG, G3310 1754 B8

p. 176 — *Carte contenant les nouvelles découvertes de l'ouest en Canada, mers, rivieres, lacs et nations qui y habittent en l'année 1737,* attributed to Louis-Joseph Gaultier de la Vérendrye (1737), manuscript map
BSHM, Recueil 67 no 42

p. 177 — *Plan des Chapitoulas* (ca. 1726), manuscript map
BSHM, Recueil 68 no 72

p. 178 — *Carte du Mexique et de la Floride,* by Guillaume Delisle (Paris: published by the author, 1703), engraved map
BAnQ, G 1015 A81 1714

p. 179 — *Carte de la Louisiane et du cours du Mississipi,* by Guillaume Delisle (Paris: published by the author, 1718), engraved map
MG, G3700 1718 L57

pp. 180–81 — *Carte particuliere du fleuve St Louis dix lieües au dessus et au dessous de la Nouvelle Orleans ou sont marqué les habitations et les terrains concedés à plusieurs particuliers au Mississipy* (ca. 1723), manuscript map
NL, Ayer MS map 30 sheet 80

p. 182 — *Plan de la seigneurie et etablissement de la mission des Tamarois,* by Jean-Paul Mercier (1735), manuscript map
MC, collection of the Séminaire de Québec archives, SME 12.1/L-43

p. 183 — *Plan of New Orleans the Capital of Louisiana, with the Disposition of Its Quarters and Canals as They Have Been Traced by Mr de la Tour in the Year 1720,* by Louis Pierre Le Blond de La Tour (London: Thomas Jefferys, 1759), engraved map
BAnQ, G 4014 N68 1759 B7

p. 184 — *[Presumed plan of the Ville-Marie fort],* attributed to Jean Bourdon (ca. 1647), manuscript map
MG, Ms 692

p. 185, top — *Plan de la ville, baye, et environs de Baston, dans la Nouvelle Angleterre,* attributed to Jean Baptiste Louis Franquelin (1693), manuscript map
BnF, Cartes et plans, Ge SH 135-6-6

p. 185, bottom — *Plan du port de Louisbourg dan lisle royalle representé de basse mer,* by Étienne Verrier (1735), manuscript map
CAOM, 03DFC 185B

p. 186, top — *Plan de la ville de Québec,* by Gaspard-Joseph Chaussegros de Léry (1727), manuscript map
BAnQ, Centre d'archives de Québec, Collection initiale, D942-Québec-1727

p. 186, bottom — *Plans des forts faicts par le regiment Carignan Salieres sur la riviere de Richelieu dicte autrement des Iroquois en la Nouvelle France* by François-Joseph Le Mercier, engraved map, in *Relation de ce qui s'est passé en la Nouvelle France, és années 1664 & 1665* (Paris: Sebastien Cramoisy & Sebastien Mabre-Cramoisy, 1666)
BAnQ, Gagnon 971.021 R382re 1664-65

p. 187, top — *A View of Quebec from the South East,* engraving, in *Atlantic Neptune* by Joseph Frederick Wallet DesBarres (London 1781), vol. IIA
BAnQ, Centre de Québec, P450, Collection Literary and Historical Society of Quebec

p. 187, bottom — *Plan of the City & Fortifications of Quebec,* by Gother Mann, drawn by William Hall and Jean-Baptiste Duberger (Quebec 1804), manuscript map
LAC, NMC 11082

p. 188, top — *Veue du fort de Chambly en Canada,* 1721, drawing
NL, Ayer MS map 30 sheet 106

p. 188, bottom — *A Plan of the Town and Fort of Carillon at Ticonderoga,* by Thomas Jefferys (London, 1758), engraved map
BAnQ, G 3804 T5 1758 J4 CAR

p. 189, top — *Sketch of the Fort at Michilimackinac,* by Perkins Magra (1765), manuscript map
CL, Maps 6-N-8

p. 189, bottom — *A North West View of Prince of Wales's Fort in Hudson's Bay,* engraving, in *Journey from Prince of Wales's Fort in Hudson's Bay to the Northern Ocean,* by Samuel Hearne (London: A. Strahan and T. Cadell), 1795
BAnQ, RES AC 67

p. 190 — *Amerique septentrionale avec les routes, distances en miles, villages et etablissements françois et anglois,* by John Mitchell, translated by Georges-Louis Le Rouge (Paris, 1756), engraved map
MG, G3300 1756 M57

p. 191 — *Carte de la Louisiane, cours du Mississipi et pais voisins,* by Jacques-Nicolas Bellin, engraved map, in *Histoire et description générale de la Nouvelle France* by Pierre-François-Xavier de Charlevoix (Paris: Didot, 1744), vol. 2
BAnQ, RES AD 46

pp. 192–93 — *Canada, Louisiane et terres angloises,* by Jean-Baptiste Bourguignon d'Anville (Paris: published by the author, 1755), engraved map
MG, G3300 1755 A54

p. 194 — *[Map of the Ohio Valley],* by George Washington, in *Journal to the Ohio,* 1754, manuscript. Facsimile from *Collections of the Massachusetts Historical Society* (vol. 61) from an original conserved at the Public Record Office, London
LC, G3820 1754.W3 1927 TIL

p. 195, top — *Cours de l'Ohio depuis sa source jusqu'à sa jonction avec la riviere d'Ouabache et les païs voisins* (Paris: Au dépôt des cartes et plans de la marine, 1755), manuscript map
BSHM, Recueil 67 no 90

p. 195, bottom — *Journal de Joseph Gaspard Chaussegros de Léry* (1754–55), manuscript
MC, collection du Séminaire de Québec, fonds Viger-Verreau, P32/O-94D/L-71 (Idra Labrie, photographer)

pp. 196–97 — *Carte du grand fleuve St Laurens dressée et dessignée sur les memoires & observations que le sr Jolliet a tres exactement faites en barq: & en canot en 46 voyages pendant plusieurs années,* by Louis Jolliet and Jean Baptiste Louis Franquelin, manuscript map, 1685
BnF, Cartes et plans, Ge SH 126-1-3(1) Rés

p. 197 — *Carte du cours du fleuve de S. Laurent depuis Québec jusqu'au lac Ontario,* attributed to Jean Deshayes (1686?), manuscript map
BSHM, Recueil 67 no 83

pp. 198–99 — *De la grande riviere de Canada appellée par les Europeens de St Laurens,* by Jean Deshayes (Paris: Nicolas de Fer, 1715), engraved map
BAnQ, G 3312 S5 1715 D4

pp. 200–01 — *Nouvelle carte de la riviere de Canada ou St Laurens de l'isle de Anticoste jusqua Quebec,* by Gerard van Keulen (Amsterdam, 1717?), engraved map
BAnQ, G 3312 S5 1717 K4

pp. 202–03 — *[River of St. Lawrence, from Cock Cove near Point au Paire, up to River Chaudière past Quebec],* by Joseph Frederick Wallet Des Barres (London: 1781), engraved map
BAnQ, G 3312 S5 1781 D41

p. 204 — *A New and Accurate Map of North America, Drawn from the Famous Mr. d'Anville with Improvements from the Best English Maps,* by Peter Bell (London: Carington Bowles, 1771), engraved map
LC, G3300 1771 .B4 Vault

p. 205 — *Sauvage de la nation des Shawanoes dessiné d'après nature au pays des Illinois,* drawing from *Carte generale du cours de la riviere de l'Ohio,* by Joseph Warin (after 1796)
BnF, Cartes et plans, Ge.A 664

p. 206 — *Carte de la rivière Longue et de quelques autres qui se dechargent dans le grand fleuve de Missisipi,* engraved map, in *Nouveaux voyages* by Louis Armand de Lom d'Arce, Baron de Lahontan (The Hague: Frères l'Honoré, 1703), vol. 1
BAnQ, RES AF 185

p. 207, top — *[Map of the region west of the Great Lakes, according to the one drawn by the Savage Ochagach],* copy dating from 1750 of an original map drawn ca. 1729, manuscript map
BSHM, Recueil 67 no 16

p. 207, bottom — *Carte dressée sur le rapport de Onouatory, sauvage Oneyoutte, établi à La Présentation,* manuscript map, in *Journal de Joseph Gaspard Chaussegros de Léry* (1756)
MC, collection du Séminaire de Québec, fonds Viger-Verreau, P32/O-94D/L-71 (Idra Labrie, photographer)

p. 208 — *Sketch Given on May 8th 1806 by the Cut Nose and the Brother of the Twisted Hair,* by William Clark (1806), manuscript map. Reproduced in Reuben Gold Thwaites (ed.), *Original Journals of the Lewis and Clark Expedition, 1804–1806* (New York: Antiquarian Press, 1959), vol. 8, map 41
ES

p. 209 — *[Cape made of bison hide (?) painted, representing a ceremony and a battle scene],* by the Quapaws Indians
MQB, 71.1934.33.7

p. 210 — *Il disegno del discoperto della Nova Franza,* attributed to Paolo Forlani (Venice, 1566), engraved map
LAC, NMC 22900

p. 211, top — *[Map of the east coast of North America],* in a manuscript atlas attributed to Nicolas Vallard (Dieppe, 1547)
HL, HM 29, folio 9

p. 211, bottom — *Rivière de St Laurens depuys Montréal jusqu'à Tadoussac,* attributed to Jean Bourdon (ca. 1641), manuscript map
BnF, Carte et plans, Ge.C 5185 Rés

pp. 212–13 — *Carte du domaine du roy en Canada,* by Pierre Laure (1733, manuscript map)
BSHM, Recueil 67 no 9

pp. 214–15 — *A Map of the British Empire in America with the French and Spanish Settlements Adjacent thereto*, by Henry Popple (London: S. Harding and W. H. Toms, 1733), engraved map
BAnQ, G 1105 P66 1733

p. 216 — *Carte de la nouvelle decouverte que les RR. Peres Jesuites ont fait en l'annee 1672, et continuee par le R. Pere Jacques Marquette de la mesme compagnie, accompagné de quelques Francois en l'annee 1673, qu'on pourra nommer la Manitounie* (... after 1673), manuscript map
BnF, Cartes et plans, Ge.C 5014 Rés

p. 217 — *Partie de l'Amerique septentrionalle ou est compris la Nouvelle France* ... by Jean Baptiste Louis Franquelin (1699), manuscript map
BSHM, Recueil 66 no 20-23

pp. 218–19 — *Bowles's New and Accurate Map of North America and the West Indies*, by Carington Bowles (London, 1784), engraved map
LAC, G3300 1783 .B61 Vault

p. 220 — *A Map Exhibiting all the New Discoveries in the Interior Parts of North America*, by Aaron Arrowsmith (London, 1802), engraved map
BAnQ, G 3300 1802 A77 CAR

p. 221 — *Un homme de l'entrée de Nootka*, by John Webber, engraving, in James Cook, *Troisième voyage de Cook ou Journal d'une expédition faite dans la mer Pacifique du sud & du nord en 1776, 1777, 1778, 1779, et 1780* (Paris, Hôtel de Thou, 1785)
BAnQ, RES BE 95

pp. 222–23 — *Chart of the N[orth] W[est] Coast of America and the N[orth] E[ast] Coast of Asia, Explored in the Years 1778 and 1779*, by Henry Roberts and James Cook (London: William Faden, 1784), engraved map
LAC, NMC 14122

p. 224, top — *Naufrage des deux chaloupes au Port des Français*, engraving, in Jean-François de Galaup, Comte de Lapérouse, *Atlas du voyage de La Pérouse* (Paris: Imprimerie de la République, 1797)
BAnQ, RES BA 29

pp. 224–25 — *[Map made in preparation for La Pérouse's voyage]* (Paris, 1785), manuscript map, copy dated 1790
BnF, Cartes et plans, Ge SH 174-1-1

p. 226 — *Côte Nord-Ouest de l'Amérique*, engraved map, in *Voyage de découvertes, à l'océan Pacifique du nord, et autour du monde* by George Vancouver (Paris: Imprimerie de la République, 1800)
BAnQ, RES AA 21

pp. 228–29 — *A plan of Quebec and environs, with its defences and the occasional entrenched camps of the French commanded by Marquis de Montcalm; shewing likewise the principal works and operations of the British forces, under the command of Major General Wolfe, during the siege of that place in 1759*, by Samuel Holland and Joseph Frederik Wallet DesBarres (London, 1780?), engraved map
BAnQ, G 3452 Q4 1759 D4

pp. 230–31 — *Plan of the settled part of the Province of Quebec* (1761), manuscript map, page showing Île Jésus
LAC, NMC 135043

p. 230, bottom — *A New and Accurate Map of the English Empire in North America*, by the Society of Anti-Gallicans (London: Wm Herbert & Robt. Sayer, 1755), engraved map
BAnQ, G 3300 1755 S6

p. 232 — *L'entrée de la riviere de St Laurent et la ville de Quebec dans le Canada* (ca. 1675), drawing
BnF, Cartes et plans, Ge SH 127-6-1

p. 233, top — *Marine, chantier de construction*, by Louis-Jacques Goussier, from drawings by Jacques-Nicolas Bellin, engraving, in *Recueil de planches sur les sciences, les arts libéraux, et les arts méchaniques, avec leur explication*, vol. 7 (Paris: Briasson et Le Breton, 1769)
BAnQ, RES BB 47 Planches

p. 233, bottom — *Carte de l'Amérique septentrionale pour servir à l'histoire de la Nouvelle France*, by Jacques-Nicolas Bellin (1743), engraved map, in *Histoire et description générale de la Nouvelle-France* by Pierre-François-Xavier de Charlevoix (Paris, 1744)
LAC, NMC 98179

pp. 234–35 — *Partie occidentale de la Nouvelle France ou du Canada*, by Jacques-Nicolas Bellin (les Héritiers d'Homan, 1755), engraved map
BAnQ, G 3400 1755 B42

pp. 236–37 — *Carte de l'Amerique septentrionale*, by Jacques-Nicolas Bellin (Paris, 1755), engraved map. Colour facsimile published by the Association of Canadian Map Libraries and Archives, (Ottawa, 1981)
BAnQ, G 1116 S1A8 1980 vol. II, carte no 78

pp. 240–41 — *A Map Exhibiting all the New Discoveries in the Interior Parts of North America*, by Aaron Arrowsmith (London, 1802), engraved map
BAnQ, G 3300 1802 A77 CAR

p. 242 — *A New Map of the Province of Lower Canada Describing all the Seigneuries, Townships, Grants of Land, &c.*, by Samuel Holland (London: William Faden, 1813), engraved map
BAnQ, G 3450 1813 H65 CAR

pp. 244–45 — *Carte du Canada qui comprend la partie septentrionale des Etats Unis d'Amérique*, by Guillaume Delisle (Paris: Jean-André Dezauche, 1783), engraved map
BAnQ, G 3400 1783 L4

p. 246 — *Carte particulière des embouchures de la rivie. S. Louis et de la Mobile*, plate from Guillaume Delisle, *Carte de la Louisiane* (Paris, 1718)
MG, G3700 1718 L57

p. 247 — *Carte particuliere du cours du Mississipi depuis le Missouri et le pays des Illinois jusqu'a l'embouchure de ce fleuve, pour servir à l'intelligence du voyage du général Collot*, attributed to Joseph Warin (1796), manuscript map
BSHM, Recueil 66 no 7

p. 248 — *A Map of the British Colonies in North America*, by John Mitchell (London, ca. 1783), engraved map with handwritten annotations
BL, K.Top118.49.b

p. 249 — *A Map Exhibiting all the New Discoveries in the interior parts of North America*, by Aaron Arrowsmith (London, 1802), engraved map
BAnQ, G 3300 1802 A77 CAR

p. 250—*Plan of part of the province of Lower Canada*, by Samuel Gale and Jean-Baptiste Duberger (Quebec, 1795), facsimile of an original map conserved at McGill University Library
MG

pp. 252–53 — *[Plan of the province of Quebec]*, by Samuel Holland and John Collins, Québec, [between 1765 and 1770]
BAnQ, Centre d'archives de Québec, E21, Ministère des Terres et Forêts, Arpentage, Divers 11

p. 254 — *Carte des possessions angloises & françoises du continent de l'Amérique septentrionale*, by Jean Palairet (London, 1763), engraved map
LC, G3300 1763 .P3 Vault

p. 255 — *The United States of America Laid Down from the Best Authorities, Agreeable to the Peace of 1783*, by John Wallis (London, 1783), engraved map
LC, G3700 1783 .W3 Vault

pp. 256–257 — *A New Topographical Map of the Province of Lower Canada*, by William Vondenvelden and Louis Charland (London, 1803), engraved map
BAnQ, G 3450 1803 V65 CAR

p. 258 — *Traité de paix entre le roi, le roi d'Espagne et le roi de la Grande-Bretagne, conclu à Paris le 10 février 1763* (Paris: Imprimerie Royale, 1763)
BAnQ, RES AD 113

p. 259 — *A Winter View in the Athapuscow Lake*, engraving, in Samuel Hearne, *A Journey from Prince of Wales's Fort in Hudson's Bay to the Northern Ocean* (London: A. Strahan and T. Cadell, 1795)
BAnQ, RES AC 67

p. 260 — *Copy of a map presented to Congress*, by Peter Pond (New York, 1785), manuscript map, copied by St. John de Crèvecœur for La Rochefoucault
BSHM, Recueil 67 no 30

p. 261 — *A Map exhibiting Mr. Hearne's Tracks in his two journies for the discovery of the Copper Mine River, in the years 1770, 1771, and 1772*, engraved map, in Samuel Hearne, *A Journey from Prince of Wales's Fort in Hudson's Bay to the Northern Ocean* (London: A. Strahan and T. Cadell, 1795)
BAnQ, RES AC 67

pp. 262–63 — *A Map of America between latitudes 40 and 70 north, and longitudes 45 and 180 west, exhibiting Mackenzie's track from Montreal to Fort Chipewyan & from thence to the North Sea in 1789, & to the West Pacific Ocean in 1793*, by Alexander Mackenzie (London, 1801), engraved map
BAnQ, G 3400 1801 M3

pp. 264–65 — *Map of the North-West Territory of the Province of Canada*, by David Thompson (1814), manuscript map
APO, F 443, R-C(U), AO 1541

p. 266 — *[Carrying the canoe]*, engraving, in Samuel Hearne, *A Journey from Prince of Wales's Fort in Hudson's Bay to the Northern Ocean* (London: A. Strahan and T. Cadell, 1795)
BAnQ, RES AC 67

p. 267 — *A man and woman of the Naudowessie*, engraving, in Jonathan Carver, *Travels through the interior parts of North-America, in the Years 1766, 1767 and 1768* (Dublin, Printed for S. Price and other, 1779), p. 230
BAnQ, Gagnon 917 C331tr

p. 268 — *A sketch of the north bend of the Missouri*, by David Thompson (1798)
LC, G4127.M5 1798 .B4 TIL Vault

p. 269 — *[Map in preparation for the voyage of Lewis and Clark]*, by Nicholas King (1803?), manuscript map
LC, G4126.S12 1803 .K5 TIL Vault

pp. 270–71 — *A Map of the Discoveries of Capt. Lewis & Clark from the Rockey Mountain and the River Lewis to the Cap of Disappointement or the Coloumbia River at the North Pacific Ocean*, by Robert Frazer, 1807, engraved map
LC, G4126.S12 1807 .F5 Vault

pp. 272–73 — *A Map of Lewis and Clark's Track across the Western Portion of North America*, copied by Samuel Lewis from original drawings by William Clark (London: Longman, Hurst, Rees, Orme & Brown, 1814), engraved map
LC, G4126.S12 1814 .L41 Vault

p. 273 — *[Sacagawea Golden Dollar Coin]*, 2000

p. 274 — *Missouri Territory formerly Louisiana*, by Mathew Carey, 1814, engraved map
LC, G4050 1814 .C3 TIL

p. 275 — *[Map of the Louisiana Purchase]*
ES

pp. 276–77 — *Map of the United States of America with the contiguous British and Spanish possessions*, by John Melish (Philadelphia, 1816), engraved map
BAnQ, G 3700 1816 M45 CAR

p. 279 — *Discoveries in the Arctic Sea between Baffin Bay, Melville Island & Cape Bathurst*, by John Arrowsmith (London, 1852)
LAC, NMC 21054

p. 280 — Title cartouche of *Carte du Canada ou de la Nouvelle France* by Guillaume Delisle (Paris, 1703)
BAnQ, G 1015 A81 1714

p. 281 — *Gravure en taille-douce*, engraving, in *Recueil de planches sur les sciences, les arts libéraux, et les arts méchaniques, avec leur explication*, vol. 5 (Paris: Briasson, David et Le Breton), 1767
BAnQ, RES BB 47 Planches

Bibliography

Monographs

ALLEN, John Logan (ed.). *North American Exploration.* 3 vols. Lincoln: University of Nebraska Press 1997.

BENSON, Guy Meriwether. *Exploring the West from Monticello: A Perspective in Maps from Columbus to Lewis and Clark.* Charlottesville: Department of Special Collections, University of Virginia Library 1995.

BOORSTIN, Daniel Joseph. *The Discoverers.* New York: Random House 1983.

BOUDREAU, Claude. *La cartographie au Québec, 1760-1840.* Sainte-Foy: Presses de l'Université Laval, "Géographie historique" collection 1994.

BUISSERET, David (ed.). *From Sea Charts to Satellite Images: Interpreting North American History through Maps.* Chicago: University of Chicago Press 1990.

—. *Mapping the French Empire in North America.* Chicago: Newberry Library 1991.

BURDEN, Philip D. *The Mapping of North America: A List of Printed Maps, 1511-1670.* Rickmansworth: Herts, Raleigh Publications 1996.

CHARBONNEAU, André, Yvon DESLOGES, and Marc LAFRANCE. *Québec, the Fortified City: From the 17th to the 19th century.* Ottawa: Parks Canada 1982.

CORBIN, Alain, and Hélène RICHARD (eds.). *La mer, terreur et fascination.* Paris: Bibliothèque nationale de France 2004.

CUMMING, William Patterson, Raleigh Ashlin SKELTON, and David Beers QUINN. *The Discovery of North America.* London: Elek 1971.

CUMMING, William Patterson, et al. *The Exploration of North America, 1630-1776.* Toronto: McClelland and Stewart 1974.

DAHL, Edward, and Jean-François GAUVIN. *Sphaerae mundi: Early Globes at the Stewart Museum.* Sillery: Septentrion, and Montreal: McGill-Queens University Press 2000.

DAWSON, Nelson-Martin. *L'atelier Delisle: l'Amérique du Nord sur la table à dessin.* Sillery: Septentrion 2000.

Dictionary of Canadian Biography. Ramsay Cook, general editor. Toronto: University of Toronto Press 2005. [www.biographi.ca].

GOETZMANN, William H., and Glyndwr WILLIAMS. *The Atlas of North American Exploration: From the Norse Voyages to the Race to the Pole.* New York: Prentice Hall General Reference 1992.

GOSS, John. *The Mapping of North America: Three Centuries of Map-Making 1500–1860.* Secaucus: Wellfleet Press 1990.

GOULD, PETER, AND Antoine BAILLY (eds.), *Le pouvoir des cartes: Brian Harley et la cartographie.* Paris: Anthropos 1995.

HARRIS, R. Cole (ed.). *Historical Atlas of Canada. From the Beginning to 1800, Volume 1.* Toronto: University of Toronto Press 1987.

HAVARD, Gilles, and Cécile VIDAL. *Histoire de l'Amérique française.* Paris: Flammarion 2003.

HAYES, Derek. *Historical Atlas of Canada: Canada's History Illustrated with Original Maps.* Vancouver: Douglas & McIntyre, and Seattle: University of Washington Press 2002.

—. *Historical Atlas of the Pacific Northwest, Maps of exploration and Discovery: British Columbia, Washington, Oregon, Alaska, Yukon.* Vancouver: Cavendish Books 1999.

—. *Historical Atlas of the United States, with Original Maps.* Berkeley: University of California Press 2007.

JOHNSON, Adrian Miles. *America Explored: A Cartographical History of the Exploration of North America.* New York: Viking 1974.

Journal de la France et des Français. Chronologie politique, culturelle et religieuse de Clovis à 2000. Vol. 1. Paris: Gallimard 2001.

LACOURSIÈRE, Jacques, Jean PROVENCHER, and Denis VAUGEOIS. *Canada-Québec: synthèse historique, 1534-2000.* Sillery: Septentrion 2001.

LA RONCIÈRE, Monique de, and Michel MOLLAT DU JOURDIN. *Sea Charts of the Early Explorers: Thirteenth to Seventeenth Century.* Translated by L. Le R. Dethan. New York: Thames and Hudson 1984.

LE CARRER, Olivier. *Océans de papier : histoire des cartes marines, des périples antiques au GPS*, Grenoble, Glénat, 2006.

LEVENSON, Jay A. *Circa 1492: Art in the Age of Exploration.* Washington/New Haven: National Gallery of Art/Yale University Press 1991.

LITALIEN, Raymonde. *Les explorateurs de l'Amérique du Nord, 1492-1795.* Sillery/Paris: Septentrion/Klincksieck 1993.

PASTOUREAU, Mireille. *Voies océanes: de l'ancien aux nouveaux mondes.* Paris: Hervas 1990.

PELLETIER, Monique (ed.). *Couleurs de la Terre: des mappemondes médiévales aux images satellitales.* Paris: Seuil/Bibliothèque nationale de France 1998.

PROULX, Gilles. *Between France and New France: Life Aboard the Tall Sailing Ships.* Toronto: Dundurn Press 1984.

THOMSON, Donald Walter. *Men and Meridians: The History of Surveying and Mapping in Canada. Vol. 1, Prior to 1867.* Ottawa: Queen's Printer 1966.

TRUDEL, Marcel. *Atlas de la Nouvelle-France/An Atlas of New France.* Quebec City: Les Presses de l'Université Laval, 1968.

—. *Histoire de la Nouvelle-France.* 5 vols. Montreal: Fides 1963–97.

VACHON, André, Victorin CHABOT, and André DESROSIERS. *Dreams of Empire: Canada before 1700.* Translated by John F. Finn. Ottawa: Public Archives Canada, 1982.

—. *Taking Root: Canada from 1700 to 1760*, Ottawa: Public Archives Canada 1985.

VIDAL, Laurent, and Émilie d'ORGEIX (eds.). *Les villes françaises du Nouveau Monde: des premiers fondateurs aux ingénieurs du roi, XVIe-XVIIIe siècles.* Paris: Somogy 1999.

WILSON, Bruce Gordon. *Colonial Identities: Canada from 1760 to 1815.* Ottawa: National Archives of Canada 1988.

WOODWARD, David, and G. Malcolm LEWIS (ed.). *Cartography in the Traditional African, American, Arctic, Australian, and Pacific Societies.* Chicago: University of Chicago Press 1998.

On-line resources

A number of the maps and illustrations reproduced in this book are available on the Web. Among the larger sites are *Gallica,* by Bibliothèque nationale de France; *American Memory,* by the Library of Congress; the digital collection site of Library and Archives Canada; and the digital collection site of the Bibliothèque et Archives nationales du Québec. Several Web sites also present digitized maps in the form of virtual exhibitions. Among the most remarkable sites for their documentary content are *La France en Amérique,* by the Bibliothèque nationale de France; *Nouvelle-France, horizons nouveaux*, produced jointly by the Direction des Archives de France, Library and Archives Canada, and the Canadian Embassy; and *La Louisiane française: 1682-1803,* by Gilles-Antoine Langlois.

Indexes of the Texts and the Maps

Index of Proper Names in the Texts and Legends

Many of the place names and, especially, names of Indian nations in this index no longer exist today or are completely unknown. To make it easier to interpret these names, we decided to group them together rather than dispersing them alphabetically in the list. Therefore, we listed under the "Indians" entry the many names of Indian nations mentioned in this book; we did the same thing with names of bays, capes, straits, rivers, forts, islands, and lakes. It should be noted that the place names in this index that are from the legends and that are set in small capitals in the legends are names that are in the maps. Readers who wish to have a better idea of the onomastic and toponymic wealth of the maps may be interested in consulting the two indexes following this one. Finally, when a cartographer's name is followed by a page number in boldface type, it means that there is a map by this person on that page. Should also be noted that the names of authors cited in the sources sections are listed in this index.[R.C. and D.V.]

Partial Index of Indian Nations and Tribes in the Maps

The advent of digital technology makes it possible to read maps in greater detail and to complement data collected from the texts of the time. The aim of the index below is simply to draw more attention to the content of the maps. The result is an impressive, even stunning list of names, for which we have supplied some reference points in parentheses, which, in general, contain the most current spellings, such as Dakotas, Kickapoos, and Susquehannocks. In other cases, we list related groups. In most cases, no reference is given, and some might wonder, as we did, if all of these names really correspond to groups of Indians. For these groups, it is useful to turn to the map from which the names come. Take, for example, the Moroa, in Coronelli's 1688 map (p.136). The cartographer has taken the care to note, clearly, "the Moroa or Tamaroa" south of the "Ilinois or Seignelay" River. "Tamaroa" refers us to the Tamarois in the Illinois family. Sometimes, the mystery remains and one must be prudent. Thus, a number of names gathered from the maps, particularly those by Franquelin, have deliberately been omitted from this index. Nevertheless, there is enough to enrich considerably the existing nomenclatures. As the spelling of names of Indian nations is still evolving, we chose to respect those of the maps from which they come. The debate is still open on whether it is Abénaquis or Abenakis, Shoshones or Shoshonis, Ojibwa or Ojibway, Onontagué or Onnontagué, Mandan or Mandane, Atticamegues or Atikamekw. [D.V.]

Partial Index of Place Names in the Maps

To make this index, the authors had to make certain choices that readers should know about in order to facilitate its use. Most of the place names appearing on these old maps are the same today, although in many cases the spelling has changed. Whenever possible, we have kept the old spelling, which is not necessarily what we have chosen to make the entries in this index. For example, the St. Lawrence River was long written "St Laurens." In general, for similar cases, the headword of the entry is the spelling most common today; we then gave the old spellings, which, for many entries, are multiple. Readers may also note that for certain entries, we did not use alphabetical order for the list of spelling variants; rather, we went by resemblance to the name; for instance, for *Labrador,* the name "Terre de Cortereal" appears last, not in alphabetical order, since it is the furthest from the root common to the different spellings.

We also had to simplify certain abbreviations to avoid insoluble classification problems; "S," "S.," and "St." (saint) on the maps were unified as "St." in the index; "B" and "B." for "bay" are found under the entry *Bay;* "I" and "I." (island) are found under *Island;* "R," "R.," and "Riv.," under *River;* and so on. However, we did differentiate "Isle," "Ysle," and "Baye." These spellings are interesting because they indicate the seniority of a name on a map. Readers will also note that we respected the old usage with regard to hyphens and accents, which, in general, were not used.

These "breaches" in the classic rules of dictionaries and indexes are intended to facilitate the transition between what is known by present-day readers and what is found on the maps, whose authors were much less concerned with consistency of spelling than we are today. [R.C.]

MAPPING A CONTINENT WAS TYPESET
IN MINION 11 AND NEUTRA 10.3 AND 8.5
IN A LAYOUT DESIGNED AND PRODUCED BY JOSÉE LALANCETTE
AND PRINTED ON HORIZON SILK 160M PAPER
ON THE PRESSES OF LITHOCHIC IN QUEBEC CITY
IN SEPTEMBER 2007
AT THE PLEASURE OF GILLES HERMAN AND DENIS VAUGEOIS,
PUBLISHERS AT LES ÉDITIONS DU SEPTENTRION,
WHO ARE PARTICULARLY PLEASED TO CELEBRATE IN THIS WAY
THE 400TH ANNIVERSARY OF THE FOUNDATION OF QUEBEC
BY SAMUEL DE CHAMPLAIN
WITH THE CONTRIBUTION OF CHIEFS ANADABIJOU, TESSOUAT, AND CAPITANAL AND A BROTHERLY
WELCOME FROM THE ETCHEMINS, MONTAGNAIS, ALGONQUINS, AND OTHER NATIONS OF THE AMERICAS
WHO, WITH THE PEOPLE FROM NORMANDY, BRITTANY, ST. MALO,
LA ROCHELLE, AND THE BASQUE, QUICKLY JOINED BY THE FRENCH OF PERCHE, SAINTONGE,
AND L'ÎLE-DE-FRANCE, AND MANY OTHERS, WERE TO GIVE BIRTH TO A NEW PEOPLE

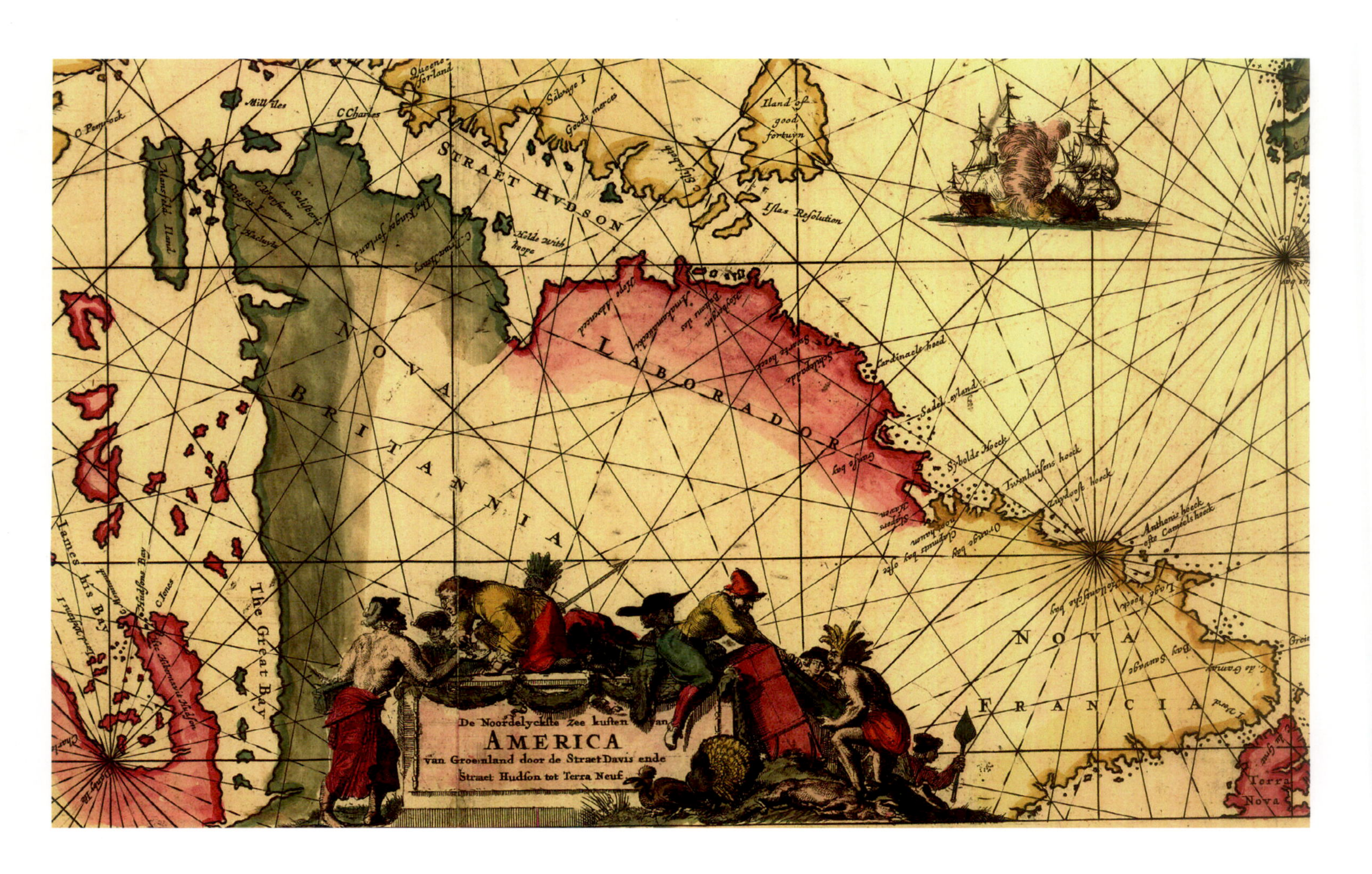

Hoche laga
A M E R I C Æ
Matouowesarin
Charioquet
Rio St. Laurens al. De Groote Rivier van Nieu Nederlandt
N O V A B E L G I C A sive N
Quebecq
T R I O N A L I S
Modus muniendi apud Mahikanenses
Maniere van Woonplaetsen ofte Dorpen der Mahicans ende andre Natien haer geburen
Armeomecks
Canoma-kers
Mackwaas
Mahikans
Sennecaas.
Minnessinck ofte t'Landt van Bacham
N E D E R L A N
t'Hooge Landt van Esopus
Gacheos
Capitanasses
Matanackouses
Konekotays
P A R S
Minquaas.
Minquaas
Sasquesahanough.
Wappinges
Waranawankongs
Waoranecks
Wecke
Pachami
Siwanoys
Sanhicans
t'Lange Eylant
Matovancons
Aquauachuques
Sauwanoos
Ermomex
Naraticons
Tockwoghs
V I R G I N I Æ P A R S
Zuydt Rivier
Cape May
C. Cornelius
C. Hinlopen
Chesepeiack Bay
Smyths Iles
C. Charle
Milliaria Germanica communia
A. Het Fort B. de Kerck C. de Wintmolen D.